Probability Concepts in Engineering

Emphasis on Applications in
Civil & Environmental Engineering

2ND EDITION

Probability Concepts in Engineering*
Emphasis on Applications in Civil & Environmental Engineering

ALFREDO H-S. ANG

Emeritus Professor,
University of Illinois at Urbana-Champaign
* and University of California, Irvine*

WILSON H. TANG

Chair Professor,
Hong Kong University of Science & Technology

WILEY

JOHN WILEY & SONS, INC.

*This title is the 2nd edition of *Probability Concepts in Engineering Planning and Design*, Vol. I: Basic Principles.

ASSOCIATE PUBLISHER	Daniel Sayre
ACQUISITIONS EDITOR	Jennifer Welter
SENIOR PRODUCTION EDITOR	William A. Murray
MARKETING MANAGER	Frank Lyman
COVER DESIGN	Hope Miller
ILLUSTRATION COORDINATOR	Mary Alma
MEDIA EDITOR	Stefanie Liebman

COVER PHOTO

The modern-style Caiyuanba Bridge is a tie-arch bridge located in Chongqing, China over the Yangtze River. It has a main arch span of 420 meters with two decks. The upper deck carries six lanes of traffic and two pedestrian paths; the lower deck carries two monorail tracks. Both the girder and the box-arch ribs are constructed of steel.

The cover image was provided by T.Y. Lin International (San Francisco, California), designer of the main span of the Caiyuanba Bridge. The authors and publisher wish to express their thanks to T.Y. Lin International for the use of the image.

This book was set in Times Roman by TechBooks and printed and bound by Courier Digital Solutions. The cover was printed by Courier Digital Solutions.

This book is printed on acid free paper. ∞

To order books or for customer service please, call 1-800-CALL WILEY (225-5945).

ISBN-13 978-0-471-72064-5
ISBN-10 0-471-72064-X

Printed in the United States of America

14

Dedicated to Myrtle Mae and Bernadette

Preface

OBJECTIVES AND APPROACH

The first edition of this book (originally titled *Probability Concepts in Engineering Planning and Design,* Vol. I, Basic Principles, published in 1975) has served to provide the basics of probability and statistical concepts in terms that are more easily understood by engineers and engineering students. The basic principles are presented and illustrated through problems of relevance to engineering and the physical sciences (particularly to civil and environmental engineering), and the exercise problems in each chapter are designed to further enhance the understanding of the basic concepts and reinforce a working knowledge of the concepts and methods. We firmly believe that the easiest and most effective way for engineers to learn a new set of abstract principles, such as those of probability and statistics, to the extent as to be able to apply them in modeling and formulating engineering problems, is through varied illustrations of applications of such principles. Moreover, the first exposure of engineers to probabilistic concepts and methods should be in physically meaningful terms; this is necessary to properly emphasize and motivate the recognition of the significant roles of the relevant mathematical concepts in engineering.

NEW TO THIS SECOND EDITION

This second edition follows the same general approach expounded in the first edition; however, the text in all the chapters has been improved and has been thoroughly revised, updated, or completely replaced from that of the first edition. In particular, almost all the illustrative examples, as well as the problems, in each chapter of the first edition have been replaced with new ones. Also, where actual data are illustrated, they have been updated; more recent or new data are presented or added in this edition. Several new topics and sections have been added or expanded, including the following:

- **Two Types of Uncertainty** In this edition, we have emphasized (introduced in Chapter 1) the importance of distinguishing the two broad types of uncertainties; namely, the *aleatory* and the *epistemic* uncertainties, and the need to evaluate their respective significances separately in engineering applications. Engineers and engineering students need to be made aware of this approach, especially in practical engineering decision making. Nonetheless, the tools for such evaluations require the same basic principles of probability and statistics as presented here.
- **Extreme Values** Chapter 4 now includes the distributions of *extreme values,* which are of special interest to engineers dealing with natural and extreme hazards.
- **Hypothesis Testing** The topic of *Hypothesis Testing* is added as part of Chapter 6.
- **Anderson–Darling Method** The *Anderson–Darling* method for goodness-of-fit test is now included in Chapter 7 (of relevance when the tails of distributions are important).
- **Confidence Intervals in Regression Analysis** Linear regression in Chapter 8 has been expanded to include the determination of *confidence intervals*.
- **Regression and Correlation Analyses** Chapter 9 on Bayesian probability now includes Bayesian *regression and correlation analyses*.

- **Computer-Based Numerical and Simulation Methods** The new chapter (Chapter 5) on *Computer-Based Numerical and Simulation Methods in Probability* should make this second edition more in tune with modern-day engineering education. The numerical and simulation methods presented in this chapter, particularly with reference to Monte Carlo simulations, extend the practical applicability of probability concepts and methods for formulating and solving engineering problems, beyond those possible with purely analytical tools. These numerical methods are particularly powerful with the present-day availability of personal computers and related commercial software, and should serve to augment the analytical methods, thus extending the general usefulness and utility of probability and statistics in engineering.

- **Quality Assurance** The chapter on *Elements of Quality Assurance and Acceptance Sampling* (Chapter 9 of the first edition) is now designated as Chapter 10 but is available only on the Web. The material in this chapter is beyond the scope of this book, which is devoted to the basic fundamentals of probability and statistics; however, it is useful in the specialized area of quality assurance. For these reasons, the material in this chapter is made available online at the Wiley Web site, www.wiley.com/college/ang.

INTENDED AUDIENCE

The material in the book is intended for a first course on applied probability and statistics for engineering students at the sophomore or junior level, or for self-study, stressing probabilistic modeling and the fundamentals of statistical inferences. The primary aim is to provide an in-depth understanding of the fundamentals for the proper application in engineering. Only knowledge of elementary calculus is required, and thus the material can be taught to undergraduate engineering students at any level. It may be used for a course taught either in the engineering departments or offered for engineers by the departments of mathematics and statistics.

The book is self-contained and thus is also suitable for self-study by practicing engineers who desire a reading and working knowledge of the basic concepts and tools of probability.

SUGGESTED SYLLABUS

One-semester course A suggested outline for a one-semester (or one-quarter) course may be as follows: Chapter 1 (assigned as required reading with guidance from instructors) through Chapter 5 stressing the modeling of probabilistic problems, plus Chapters 6 through 8 stressing the fundamentals of inferential statistics, can be covered in a one-semester course.

One-quarter course For a one-quarter course, the same chapters may be covered with less emphasis on selected sections (e.g., discuss fewer types of useful probability distributions) and limit the number of illustrations in each chapter.

Senior-level course For a course at the senior level, all the chapters, including the first part of Chapter 9, may be covered in one semester.

The extensive variety of problems at the end of each chapter provides wide choices for class assignments and also opportunities for self-measuring a reader's understanding.

INSTRUCTOR RESOURCES

These instructor resources are available on the Instructor section of the Web site at www.wiley.com/college/ang. They are available only to instructors who adopt the text:

- **Solutions Manual:** Solutions to all the exercise problems in the text.

- **Image Gallery:** All figures and tables from the text, appropriate for use in Power-Point presentations.

These resources are password protected. Visit the Instructor section of the book Web site to register for a password to access these materials.

MATHEMATICAL RIGOR

We have not emphasized mathematical rigor throughout the book; such rigor may be supplemented with treatises on the mathematical theory of probability and statistics. We are concerned mainly with the practical applications and relevance of probability concepts to engineering. The necessary mathematical concepts are developed in the context of engineering problems and through illustrations of probabilistic modeling of physical situations and phenomena. In this regard, only the essential principles of mathematical theory are discussed, and these principles are explained in non-abstract terms in order to stress their relevance to engineering. This is necessary and essential to enhance the appreciation and recognition of the practical significance of probability concepts.

MOTIVATION

Uncertainties are unavoidable in the design and planning of engineering systems. Properly, therefore, the tools of engineering analysis should include methods and concepts for evaluating the significance of uncertainty on system performance and design. In this regard, the principles of probability (and its allied fields of statistics and decision theory) offer the mathematical basis for modeling uncertainty and the analysis of its effects on engineering design.

Probability and statistical decision theory have especially significant roles in all aspects of engineering planning and design, including: (1) the modeling of engineering problems and evaluation of systems performance under conditions of uncertainty; (2) systematic development of design criteria, explicitly taking into account the significance of uncertainty; and (3) the logical framework for quantitative risk assessment and risk-benefit tradeoff analysis relative to decision making. Our principal aim is to emphasize these wider roles of probability and statistical decision theory in engineering, with special attention on problems related to construction and industrial management; geotechnical, structural, and mechanical design; hydrologic and water resources planning; energy and environmental problems; ocean engineering; transportation planning; and problems of photogrammetric and geodetic engineering.

The principal motivation for developing this revised edition of the book is our firm belief that the principles of probability and statistics are of fundamental importance to all branches of engineering, although the examples and exercise problems included in this text are mostly from civil and environmental engineering. These principles are essential for the quantitative analysis and modeling of uncertainties in the assessment of risk, which is central in the modern approach to decision making under uncertainty.

The concepts and methods expounded in this book constitute only the basics necessary for the proper treatment of uncertainties. These basic principles may need to be supplemented with more advanced tools for specialized applications. See Volume II of Ang and Tang (1984) for some of these advanced topics.

Over the years, we have received numerous compliments from former students and professional colleagues regarding the way we elucidated the concepts and methods in the first edition, particularly for those wishing to learn and apply the principles of probability and statistics. In this regard, we are encouraged that the first edition of this book has contributed to the education of several generations of engineering students, and of professional colleagues through self-studies. The work for this second edition of the book is also inspired by the hope that this work will continue to contribute to the education of future generations of engineering students in the practical roles and significance of probability and statistics in engineering, enhanced further nowadays by the general availability of personal computers and associated commercial software.

VOLUME II

The first edition of this text was published in two volumes. For the second edition, only Volume I (this text) is being revised. If you would like to obtain a copy of the original Volume II, you may contact Professor Ang directly at ahang2@aol.com.

ACKNOWLEDGEMENTS

Finally, it is our pleasure to acknowledge the many constructive comments and suggestions offered by the prepublication reviewers of our original manuscript, including:

C. H. Aikens, University of Tennessee

B. Bhattacharya, University of Delaware

V. Cariapa, Marquette University

A. Der Kiureghian, University of California, Berkeley

S. Ekwaro-Osire, Texas Tech University

B. Ellingwood, Georgia Institute of Technology

T. S. Hale, Ohio University

P. A. Johnson, Pennsylvania State University

J. Lee, University of Louisiana

M. Maes, University of Calgary

S. Mattingly, University of Texas, Arlington

P. O'Shaughnessy, The University of Iowa

C. Polito, Valparaiso University

J. R. Rowland, University of Kansas

Y. K. Wen, University of Illinois, Urbana-Champaign

as well as a number of other anonymous reviewers. Many of their suggestions have served to improve the final manuscript. We also greatly appreciate the many compliments from several of the reviewers, including the phrase "the authors seem to understand what Socrates knew a long time ago ... 'Analytical tools that are understood have a higher probability of being used' "; relating our work to Socrates certainly represents the height of compliments. Last but not least, our thanks to T. Hu, H. Lam, J. Zhang and L. Zhang for their assistance in the solutions to some of the examples, and in the preparation of the Solutions Manual for the problems in the book.

A. H-S. Ang and W. H. Tang

Contents

*Available online at the Wiley Web site www.wiley.com/college/ang

CHAPTER 1

Roles of Probability and Statistics in Engineering

▶ 1.1 INTRODUCTION

In dealing with real world problems, uncertainties are unavoidable. As engineers, it is important that we recognize the presence of all major sources of uncertainty in engineering. The sources of uncertainty may be classified into two broad types: (1) those that are associated with natural randomness; and (2) those that are associated with inaccuracies in our prediction and estimation of reality. The former may be called the *aleatory* type, whereas the latter the *epistemic* type. Irrespective of the type of uncertainty, probability and statistics provide the proper tool for its modeling and analysis. In the ensuing chapters we will present the fundamental principles of probability and statistics, and illustrate their applications in engineering-type problems. The main aim of this work is to present the concepts and methods of probability and statistics for the modeling and formulation of engineering problems under uncertainty; this is in contrast to books that are devoted to statistical data analysis, although the fundamentals of statistics are also presented here.

The effects of uncertainties on the design and planning of an engineering system are important, to be sure; however, the quantification of such uncertainty and the evaluation of its effects on the performance and design of the system, should properly include the concepts and methods of probability and statistics. Furthermore, under conditions of uncertainty, the design and planning of engineering systems involve risks, which involves probability and associated consequences, and the formulation of related decisions may be based on quantitative risk-benefit trade-offs which are properly also within the province of applied probability and statistics. In this light, and with reference to problems containing randomness and uncertainty, the significance of the concepts of probability and statistics in engineering parallels those of the principles of physics, chemistry, and mechanics in the formulation and solution of engineering problems.

In light of the above, we see that the role of probability and statistics is quite pervasive in engineering; it ranges from the description of basic information to the development and formulation of bases for design and decision making. Specific examples of such imperfect information, and of applications in engineering design and decision-making problems, are described in the following sections.

▶ 1.2   UNCERTAINTY IN ENGINEERING

The presence of uncertainty in engineering, therefore, is clearly unavoidable; the available data are often incomplete or insufficient and invariably contain variability. Moreover, engineering planning and design must rely on predictions or estimations based on idealized models with unknown degrees of imperfections relative to reality, and thus involve additional uncertainty. In practice, we might identify two broad types of uncertainty: namely, (i) uncertainty associated with the randomness of the underlying phenomenon that is exhibited as variability in the observed information, and (ii) uncertainty associated with imperfect models of the real world because of insufficient or imperfect knowledge of reality. As we said earlier, these two types of uncertainty may be called, respectively, the *aleatory uncertainty* and the *epistemic uncertainty*. See Ang (1970, 2004) for a basic framework for defining and treating these two types of uncertainties. The two types of uncertainty may be combined and analyzed as a total uncertainty, or treated separately. In either case, the principles of probability and statistics apply equally.

We might point out that there are good reasons to view the significance of the two types of uncertainty, and their respective effects on engineering, differently. First of all, the aleatory (data-based) uncertainty is associated with the inherent variability of basic information, which is part of the real world (within our ability to observe and describe). Much of the aleatory uncertainty that civil engineers must deal with are inherent in nature and, therefore, may not be reduced or modified. On the other hand, epistemic (or knowledge-based) uncertainty is associated with imperfect knowledge of the real world, and may be reduced through application of better prediction models and/or improved experiments. Second, the respective consequences of these two types of uncertainty may also be different—the effect of the aleatory randomness leads to a calculated probability or risk, whereas the effect of the epistemic type expresses an uncertainty in the estimated probability or risk. In many application areas of engineering and the physical sciences, the uncertainty (or error bounds) of a calculated risk or probability is as important as the risk itself; e.g., the National Research Council (1994) has emphasized the importance of quantifying the uncertainty in the calculated risk, and a number of U.S. government agencies, such as the U.S. Department of Energy (1996), the Environmental Protection Agency (1997), NASA (2002), NIH (1994), as well as in the UK (2000), have applied this approach in the quantitative assessment of risk. In some practical applications, however, the two types of uncertainty are combined and their aggregate effects estimated accordingly. Again, irrespective of whether the two types of uncertainties are combined or treated separately, the concepts and methods covered in the ensuing chapters are equally applicable.

Finally, there should be no problem in delineating between the two types of uncertainty—the aleatory type is essentially data-based, whereas the epistemic type is knowledge based. For practical purposes, the epistemic uncertainty may be limited to the estimation of the mean or median values, even though in theory it includes inaccuracies in the prescribed form of probability distributions and in all the parameters.

1.2.1   Uncertainty Associated with Randomness—the Aleatory Uncertainty

Many phenomena or processes of concern to engineers, or that engineers must contend with, contain randomness; that is, the expected outcomes are unpredictable (to some degree). Such phenomena are characterized by field or experimental data that contain significant *variability* that represents the natural randomness of an underlying phenomenon; i.e., the observed measurements are different from one experiment (or one observation) to another, even if conducted or measured under apparently identical conditions. In other words, there is a range of measured or observed values of the experimental results; moreover, within

TABLE 1.1   Rainfall Intensity Data Recorded over a Period of 29 Years

Year No.	Rainfall Intensity, in.	Year No.	Rainfall Intensity, in.	Year No.	Rainfall Intensity, in.
1	43.30	11	54.49	21	58.71
2	53.02	12	47.38	22	42.96
3	63.52	13	40.78	23	55.77
4	45.93	14	45.05	24	41.31
5	48.26	15	50.37	25	58.83
6	50.51	16	54.91	26	48.21
7	49.57	17	51.28	27	44.67
8	43.93	18	39.91	28	67.72
9	46.77	19	53.29	29	43.11
10	59.12	20	67.59		

this range certain values may occur more frequently than others. The variability inherent in such data or information is statistical in nature, and the realization of a specific value (or range of values) involves probability. The inherent variability in the observed or measured data can be portrayed graphically in the form of a *histogram or frequency diagram*, such as those shown in Figs. 1.1 through 1.23, all of which demonstrate information on physical phenomena of relevance particularly to civil and environmental engineering. Furthermore, if two variables are involved, the joint variability may similarly be portrayed in a *scattergram*.

A histogram simply shows the relative frequencies of the different observed values of a single variable. For example, for a specific set of experimental data, the corresponding histogram may be constructed as follows.

From the range of the observed data set, we may select a range on one axis (for a two-dimensional graph) that is sufficient to cover the largest and smallest values among the set of data, and divide this range in convenient intervals. The other axis can then represent the number of observations within each interval among the total number of observations, or the fraction of the total number. For example, consider the 29 years of annual cumulative rainfall intensity in a watershed area recorded over a period of 29 years as presented in Table 1.1.

An examination of these data will reveal that the observed rainfall intensities range from 39.91 to 67.72 in. Therefore, choosing a uniform interval of 4 in. between 38 and 70 in. the number of observations within each interval and the corresponding fraction of the total observations are calculated as summarized in Table 1.2.

The uniform intervals indicated in Table 1.2 may then be scaled on the abscissa, and the corresponding number of observations (column 2 in Table 1.2) can be shown as a bar

TABLE 1.2   Number and Fraction of Total Observations in Each Interval

Interval	No. of Observations	Fraction of Total Observations
38–42	3	0.1034
42–46	7	0.2415
46–50	5	0.1724
50–54	5	0.1724
54–58	3	0.1034
58–62	3	0.1034
62–66	1	0.0345
66–70	2	0.0690

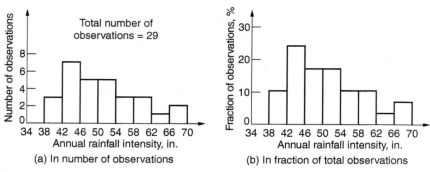

Figure 1.1   Histograms of Annual Rainfall Intensity.

on the vertical axis, as illustrated in the histogram of Fig. 1.1a for the rainfall intensity of the watershed area. Alternatively, the vertical bar may be in terms of the fraction of the total observations (column 3 in Table 1.2) and would appear as shown in Fig. 1.1b. Oftentimes, there may be reasons to compare an empirical frequency diagram, such as a histogram, with a theoretical frequency distribution (such as a *probability density function*, PDF, discussed later in Chapter 3).

For this purpose, the area under the empirical frequency diagram must be equal to unity; we obtain this by dividing each of the ordinates in a histogram by its total area; e.g., we obtain the empirical frequency function of Fig. 1.1a by dividing each of the ordinates by $29 \times 4 = 116$; whereas the corresponding empirical frequency function may also be obtained from Fig. 1.1b by dividing each of the ordinates by $4 \times 1 = 4$. In either case, we would obtain the empirical frequency function of Fig. 1.1c for the rainfall intensity in

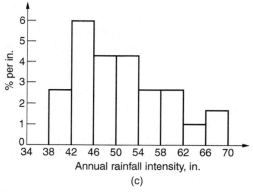

Figure 1.1c   Empirical Frequency Function.

the watershed area. We may then observe that the total area under the empirical frequency function is equal to 1.0, and thus the area over a given range may be used to estimate the probability of rainfall intensity within the given range.

A large number of physical phenomena are represented in Figs. 1.1 through 1.23; these are purposely collected here to demonstrate and emphasize the fact that the state of most engineering information contains significant variability. For examples, the properties of most materials of construction vary widely; in Figs. 1.2 and 1.3 we present the histograms demonstrating the variabilities in the bulk density of soils and the water–cement (w/c) ratio of concrete specimens, respectively, whereas in Figs. 1.4 and 1.5 are shown the yield strength of reinforcing bars and the ultimate shear strength of steel fillet welds.

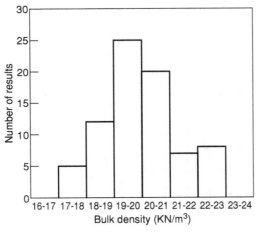

Figure 1.2 Bulk density of residual soils (after Winn et al., 2001).

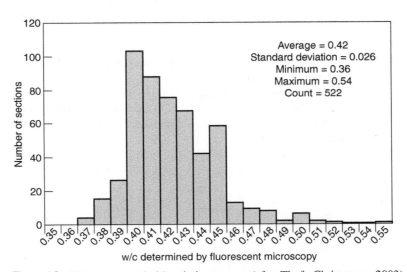

Figure 1.3 Water–cement (w/c) ratio in concrete (after Thoft–Christensen, 2003).

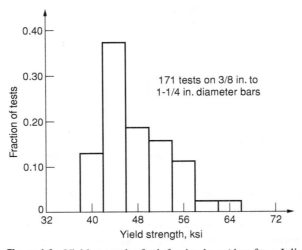

Figure 1.4 Yield strength of reinforcing bars (data from Julian, 1957).

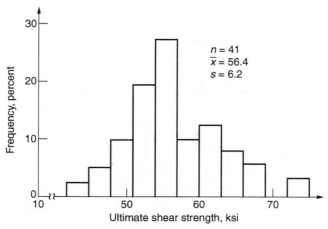

Figure 1.5 Ultimate shear strength of fillet welds (after Kulak, 1972).

Similarly, in the case of construction timber, we see in Fig. 1.6 examples of the histograms of the modulus of elasticity of Southern pine and Douglas fir timber, whereas in Fig. 1.7 we show the histogram of the modulus of elasticity of grouted masonry. As expected, there are wide variabilities in the moduli of elasticity of these two materials; timber is a natural organic material, whereas masonry is a highly heterogeneous mixture of cement and natural sand.

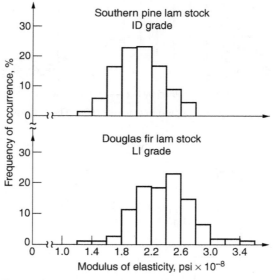

Figure 1.6 Modulus of elasticity of construction lumber (after Galligan and Snodgrass, 1970).

Of great interest in reinforced concrete construction is the possible corrosion of the reinforcing steel; in Fig. 1.8, we show the histogram of the corrosion activity (measured by the electrical current density) of steel in concrete structures. In rock mechanics, homogeneous rock mass contains fissures. We see in Fig. 1.9 that the trace lengths of discontinuities in a rock mass can vary widely.

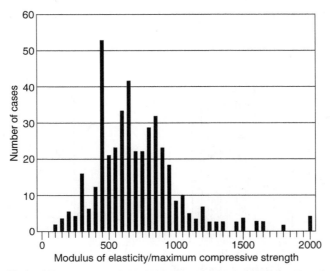

Figure 1.7 Modulus of elasticity of grouted masonry (after Brandow et al., 1997).

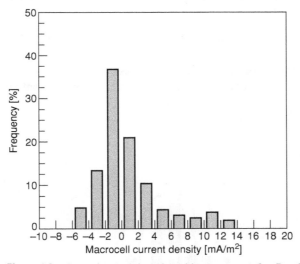

Figure 1.8 Corrosion activity of steel in concrete (after Pruckner & Gjorv, 2002).

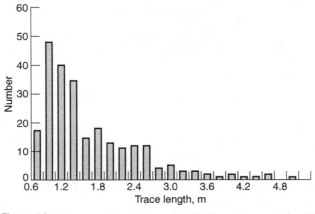

Figure 1.9 Trace lengths of discontinuities in rock mass (after Wu and Wang, 2002).

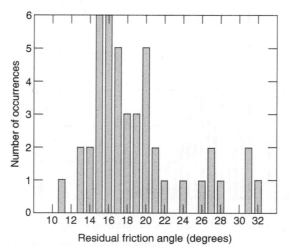

Figure 1.10 Residual friction angle of mudstone (after Becker et al., 1998).

In geotechnical engineering, we can expect significant variability in the information pertaining to soil properties. For example, Fig. 1.10 shows the histogram of the residual angle of friction of soft mudstone, whereas Fig. 1.11 shows the variability of the compressive strength of sandstone underlying the foundation of the Confederation Bridge in Canada.

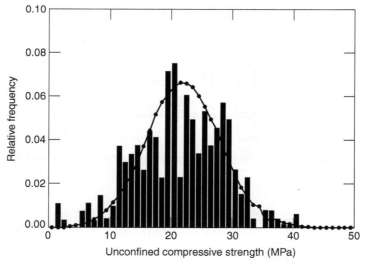

Figure 1.11 Compressive strength of sandstone (after Becker et al., 1998).

Furthermore, the efficiencies of pile groups are also highly variable as shown in Figs. 1.12a and 1.12b, respectively, for pile groups in clay and sand.

Significant variabilities are also present in the loads on structures; these are illustrated in Fig. 1.13 showing the wind-induced pressure fluctuations on tall buildings observed during two typhoons (hurricanes), and in Fig. 1.14 is shown the variability of earthquake-induced shear stresses in soils. In both of these figures, the dispersions are scaled in terms of respective standard deviations.

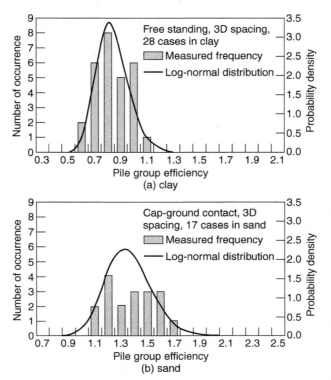

Figure 1.12 Pile group efficiencies in clay and sand (after Zhang et al., 2001).

In environmental engineering, the water quality depends on certain parameters such as the stream temperature and dissolved oxygen (*DO*) deficit. Figure 1.15 illustrates the variability of dissolved oxygen (*DO*) deficit in the Ohio River, whereas in Fig. 1.16 is shown the variability of the weekly maximum temperature of four streams in four different states of the United States of America—Maine, Washington, Minnesota, and Oklahoma.

In the case of transportation engineering, we show in Fig. 1.17 the empirical frequency distribution of *O-D* (origin–destination) trip lengths in Sioux Falls, SD, and in Fig. 1.18 the histogram of the estimated impact speed of passenger car accidents.

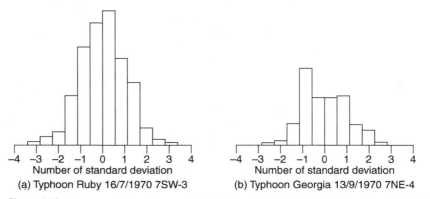

Figure 1.13 Pressure fluctuations on tall buildings during typhoons (after Lam Put, 1971).

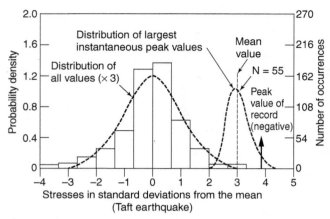

Figure 1.14 Earthquake-induced shear stresses in soils (after Donovan, 1972).

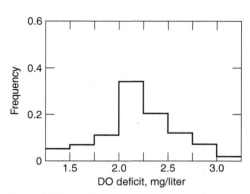

Figure 1.15 Dissolved oxygen (DO) deficit in river (after Kothandaraman & Ewing, 1969).

In Fig. 1.19 we see the histograms of the measured roughness profiles of a rough road and a smooth road, in terms of the respective rms (root-mean-square) values. The cost of injuries from highway work zone accidents in the United States can be expected to vary greatly; this is evidenced in Fig. 1.20.

It is of interest to observe that engineered structures can sometimes fail and cause economic losses as well as loss of human lives. Figure 1.21 shows the statistics of dam failures in the United States as a function of life (in years) after completion. Clearly, we observe that most failures occur during the first year after completion of construction of a dam.

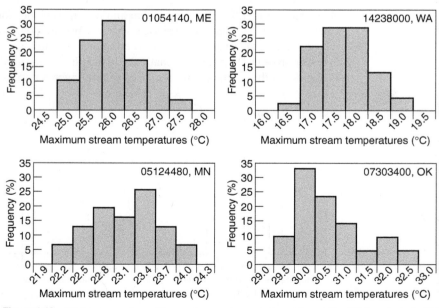

Figure 1.16 Histograms of weekly maximum stream temperature (after Mohseni et al., 2002).

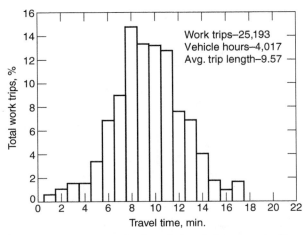

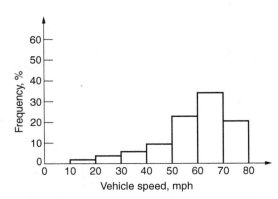

Figure 1.17 Origin–destination trip length frequency in Sioux Falls, SD (U.S. Dept of Comm. 1965).

Figure 1.18 Impact speeds of passenger car accidents (after Viner, 1972).

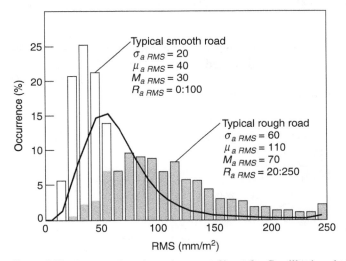

Figure 1.19 Measured road roughness profiles (after Rouillard et al., 2000).

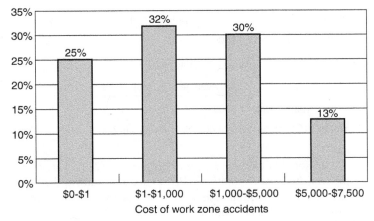

Figure 1.20 Cost of work zone accidents (after Mohan & Gautam, 2002).

11

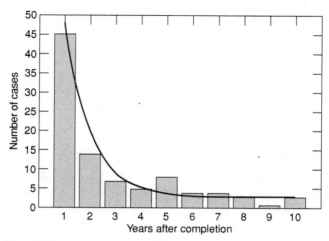

Figure 1.21 Statistics of dam failures in the United States (after van Gelder, 2000).

Finally, the variabilities of information in construction engineering and construction management are illustrated in Figs. 1.22 and 1.23, respectively. Specifically, we show in the figures the completion time of building houses in England and the variability of bid prices in highway construction.

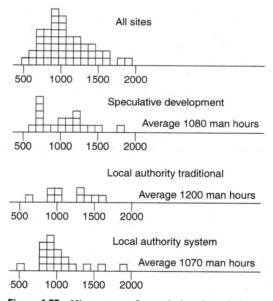

Figure 1.22 Histograms of completion times in house building (after Forbes, 1969).

In some of the figures, e.g., Figs. 1.11, 1.12, 1.14, 1.19, 1.21, and 1.23, the theoretical *probability density functions* (PDFs) are also shown; the significance of these theoretical functions and their relations to the corresponding experimental frequency diagrams will be discussed in greater depth in Chapters 3 and 7.

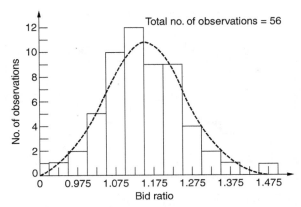

Figure 1.23 Distribution of bid prices in highway construction (after Cox, 1969).

When two (or more) variables are involved, each variable may have its own variability, whereas there may also be joint variability of the two variables. Observed data of pairs of values of the two variables can be portrayed in a two-dimensional graph in the form of a *scattergram* of the observed data points. For example, in Fig. 1.24 is shown the scattergram of the modulus of elasticity versus the corresponding strength of Douglas fir timber, whereas, in Fig. 1.25 we see the scattergram of the tensile strength of concrete versus temperature.

In Fig. 1.26 we observe the scatter of the data points of the mean annual discharge of a stream versus the corresponding drainage area near Honolulu. In Fig. 1.27 is shown the scattergram of the plasticity index versus the liquid limit of soils, which is of fundamental interest to geotechnical engineers.

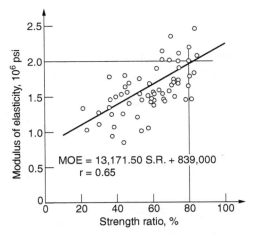

Figure 1.24 Modulus of elasticity vs. strength of timber (after Littleford, 1967).

In Fig. 1.28, we show the scattergram between the concentration of chlorophyll and phosphorous concentration; this information is important to environmental engineers concerned with the productivity of lakes. In Fig. 1.29 we see a typical scattergram of real estate land value plotted against population density.

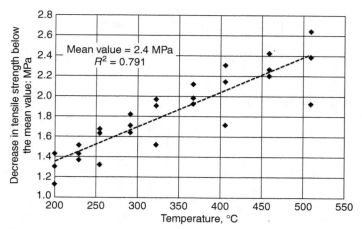

Figure 1.25 Tensile strength of concrete vs. temperature (after dos Santos et al., 2002).

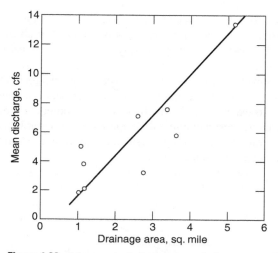

Figure 1.26 Mean annual discharge vs. drainage area of streams (after Todd & Meyer, 1971).

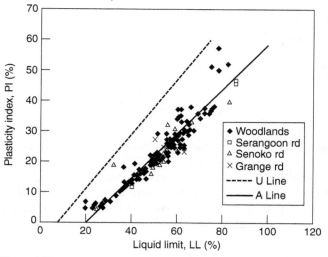

Figure 1.27 Plasticity index vs. liquid limit of solids (after Winn et al., 2001).

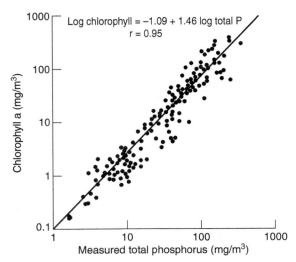

Figure 1.28 Chlorophyll vs. phosphorous concentrations (after Jones & Bachmann, 1976).

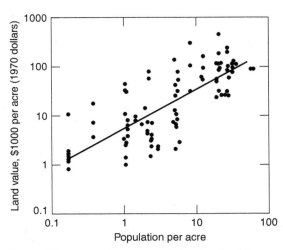

Figure 1.29 Land value vs. population density (after Wynn, 1969).

In estimating the maximum wind speeds, the calculated speed may not be perfect; this is illustrated in Fig. 1.30, which shows the scattergram between the calculated and corresponding observed maximum wind speeds. Also, in traffic engineering, the relationship between the daily conflicts and the total entry volume at traffic intersections is illustrated in Fig. 1.31, which shows the scatter of the data points.

Finally, in assessing the hazards from glacier lake outbursts, which can potentially cause dangerous outburst floods and debris flow in mountainous regions, the area and mean depth of a glacier lake can be used to estimate the volume of a lake. For this purpose, the scattergram of Fig. 1.32 shows such a relationship between the mean depth and the area.

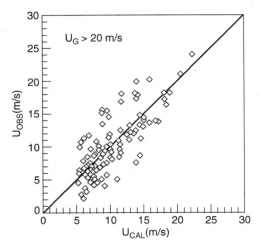

Figure 1.30 Calculated vs. observed wind speeds (after Matsui et al., 2002).

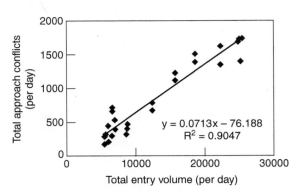

Figure 1.31 Traffic conflicts vs. volume (after Katamine, 2000).

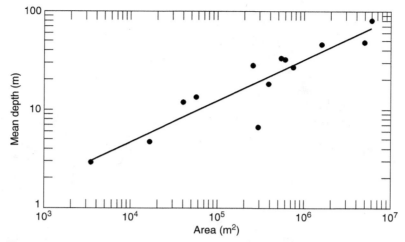

Figure 1.32 Mean depth vs. area of glacier lakes (after Huggel et al., 2002).

A large number of histograms and scattergrams are shown above; the purpose of illustrating such a large number is to demonstrate clearly that variability of engineering information is invariably present and unavoidable in many areas of engineering applications.

We might emphasize that the variability exhibited in any histogram is due to randomness in nature, and thus is an aleatory type of uncertainty. The following example may serve to introduce how such information may be handled in practice.

▶ **EXAMPLE 1.1**

Quite often, when we have a finite set of observational data (called a *sample*), it is of interest to estimate the average of the sample (called the *sample mean*) and a measure of its variability or dispersion (called the *sample variance*); the latter is the aleatory uncertainty corresponding to the data set. Consider, for example, the set of 29 observed annual rainfall intensities in the watershed area tabulated earlier in Table 1.1a. Clearly, we can estimate the sample average of the observations simply as follows:

$$\bar{x} = \frac{1}{29}(43.30 + 53.02 + \cdots + 67.72 + 43.11) = 50.70 \text{ in.}$$

and the corresponding sample variance, which is the average of the squared deviation from the mean (we shall define the sample variance more thoroughly in Chapter 6), is

$$s_X^2 = \frac{1}{29}[(43.30 - 50.70)^2 + (53.02 - 50.70)^2 + \cdots + (67.72 - 50.70)^2 + (43.11 - 50.70)^2]$$

$$= 57.34$$

Finally we obtain the corresponding *sample standard deviation*, $s_X = \sqrt{57.34} = 7.57$ in. This sample standard deviation is, therefore, a measure of the dispersion of the annual rainfall intensity and represents the corresponding randomness or aleatory uncertainty of the rainfall intensity in the watershed area.

As the above average annual rainfall intensity is estimated from the set of 29 observations, there is also epistemic uncertainty underlying the estimated average, called the *sampling error*. In this case, this is the uncertainty underlying the estimated average annual rainfall intensity of 50.70 in. This sampling error (defined in Chapter 6 as a function of the sample size) is equal to $7.57/\sqrt{29} = 1.41$ in.

In this example, we see that the variability (or randomness) in the observed data contains the aleatory uncertainty, whereas the sampling error in the estimated average contributes to the epistemic uncertainty. There are other sources of epistemic uncertainty as illustrated below in Examples 1.2 and 1.3. ◀

1.2.2 Uncertainty Associated with Imperfect Knowledge—the Epistemic Uncertainty

In engineering, we have to rely on idealized models of the real world in our analysis and predictions for the purpose of making decisions or for planning and developing criteria for the design of an engineering system. These idealized models, which may be mathematical or simulation models (e.g., mathematical formulas, equations, numerical algorithms, computer programs) or even laboratory models, are imperfect representations of the real world. Consequently, the results of analysis, estimations, or predictions obtained on the basis of such models are inaccurate (with some unknown degree of error) and thus also contain uncertainty. Such uncertainties are, therefore, knowledge based and are of the epistemic type. Quite often, this epistemic uncertainty may be more significant than the aleatory uncertainty.

In performing a prediction or estimation with an idealized model, the objective is invariably to obtain a specific quantity of interest; this may be the mean-value or median value of a variable. Therefore, in considering the epistemic uncertainty it is reasonable (in practice) to limit our consideration to the inaccuracy in calculating or estimating the *central value*, such as the mean-value or median value. We might emphasize that we can expect errors also in the estimation of the other parameters as well as in the specification of the distribution, and thus there are also epistemic uncertainties in these latter estimations; however, such uncertainties are of second-order importance relative to that in the central value.

For examples, consider the problems in the following illustrations.

▶ **EXAMPLE 1.2**

Consider the calculation of the deflection of a prismatic cantilever beam under the concentrated load P as shown in Fig. E1.2. For engineering purposes, the deflection at the end of the beam B is usually calculated on the basis of the simple beam theory, which gives

$$\Delta_B = \frac{PL^3}{3EI} \tag{1.1}$$

in which

$E =$ the modulus of elasticity of the material, and

$I =$ the moment of inertia of the beam cross section.

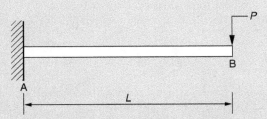

Figure E1.2 A cantilever beam subjected to concentrated load P.

Equation 1.1 is based on several idealized assumptions, which are as follows:

1. The material is linearly elastic.

2. Under the load P, plane sections of the beam remain plane.

3. The support of the beam at A is perfectly rigid.

In reality, each of the above assumptions may not be totally valid. For example, depending on the material of the beam and the magnitude of the load, the behavior may not be linearly elastic. Also, for high loads, plane sections of the beam will not remain plane; finally, it is seldom that the support at A can be perfectly rigid. Therefore, the calculation of the deflection, Δ_B, with Eq. 1.1 will involve some undetermined error and thus some uncertainty. Unless there is reason to believe otherwise, it is reasonable to assume that the calculated deflection is the mean deflection of the cantilever beam, implying that the error of Eq. 1.1 is symmetric about the average deflection. However, if necessary, the result may be corrected for any bias in the calculation.

One way to evaluate or assess the error of Eq. 1.1 is to test a number of similar cantilever beams of the same material under the same conditions, and with precisely measured values of P, L, E, and I. The results of the tests should provide us with a basis for evaluating the error underlying the equation. For illustration, suppose (hypothetically) that 10 wood beams were tested and the results (in terms of the ratios of the measured to the calculated deflections) were as follows:

$$\frac{\text{measured } \Delta_B}{\Delta_B \text{ by Eq. 1.1}} = 1.05; 0.95; 1.10; 0.98; 1.15; 0.97; 1.20; 1.00; 1.08; 1.12$$

These test results would yield the following *sample mean* and *sample standard deviation* (see Chapter 6) for the ratio of the measured to the calculated deflection, Δ_B:

$$\text{Mean (or average) ratio} = 1.06$$
$$\text{Standard deviation of ratio} = 0.25$$

According to the above test results, the deflection of the beam calculated with Eq. 1.1 tends to underestimate the correct deflection and, therefore, needs to be corrected by the bias factor of 1.06, whereas its *coefficient of variation* (defined as the standard deviation divided by the mean) representing the epistemic uncertainty is $0.25/1.06 = 0.24$.

Depending on the material, there may also be significant variability in the EI of the beam. For example, the variability of the modulus of elasticity, E, of construction lumber is illustrated in Fig. 1.6. This variability in EI will also lead to aleatory uncertainty in the calculated end deflection Δ_B. Finally, if the EI of the beam is estimated from sampled observations, there will also be sampling error in the estimated mean EI, and this will add further epistemic uncertainty to the calculated deflection of the beam. ◀

▶ **EXAMPLE 1.3**

In Table E1.3, we show the data of the observed settlements of pile groups and the corresponding calculated settlements of the same pile groups reported by Viggiani (2001). The calculated settlements were obtained with a nonlinear model for predicting settlements of the respective pile groups.

Using the data in Table E1.3, we develop later, in Example 8.3 of Chapter 8, the so-called linear regression equation showing the relationship between the observed and calculated settlements as follows:

$$E(Y|x) = 0.038 + 1.064\,x \text{ mm} \tag{1.2}$$

in which $E(Y|x)$ stands for the expected observed settlement Y of a pile group if the calculated settlement is x. In Example 8.3, we also show that the *conditional standard deviation* about the regression equation (i.e., the average dispersion along the regression line) is $s_{Y|x} = 7.784$ mm.

TABLE E1.3   Data of Observed and Calculated Settlements of Pile Groups (in mm)

Obs. S'mt., y_i	Calc. S'mt., x_i	Obs. S'mt., y_i	Calc. S'mt., x_i	Obs. S'mt., y_i	Calc. S'mt., x_i
0.7	4.8	28.1	26.7	12.7	14.2
64	25.1	0.6	0.6	46	42.1
5	3.8	29.2	31.5	3.8	1.58
25	26.4	32	31	4.9	5
29.5	27.8	5.4	3.7	11.7	11.8
3.8	3.5	3.6	4	1.83	3.39
5.9	6.5	35.9	25.9	9.43	3.24
38.1	2.6	11.6	11.3	6.6	4.3
185	174				

Source: After Viggiani, 2001.

Equation 1.2 has the following significance: The equation can be used to determine the mean-value (average) of the actual settlement, $E(Y|x)$, if the calculated settlement is x. However, the conditional standard deviation of $s_{Y|x} = 7.784$ mm represents the error of the nonlinear model for predicting the actual settlements, and thus is the epistemic uncertainty of the proposed calculational method.

The above Eq. 1.2 may be used to determine the expected settlements of similar pile groups at a site based on the settlements calculated by the same nonlinear model described in Viggiani (2001); however, according to the regression equation, the calculated settlements will tend to be on the low side and must be increased by a bias factor of 1.064. Moreover, there will be an average standard deviation (dispersion) of the true settlement of 7.784 mm for a given value of x. Depending on the value of the calculated settlement x, the corresponding coefficient of variation (c.o.v.) may also be estimated; i.e., the c.o.v. would be $7.784/E(Y|x)$. For example, if the calculated settlement for a particular pile group is $x = 45$ mm, we can expect the actual settlement to be $45 \times 1.064 = 47.88$ mm with a c.o.v. of $7.784/47.88 = 0.16$, which is the epistemic uncertainty underlying the model equation. ◀

Figure 1.30 provides a similar example for assessing the epistemic uncertainty of a predictive model. All uncertainties, whether they are aleatoric or epistemic, can be assessed in statistical terms, and the evaluation of their significance on engineering planning and design can be performed systematically and logically using the concepts and methods that are embodied in the theory of probability.

▷ 1.3   DESIGN AND DECISION MAKING UNDER UNCERTAINTY

As we indicated earlier, engineering information is generally of the type illustrated in Figs. 1.1 through 1.32, in which no single observation or measurement is representative; and any evaluation or prediction must be performed on the basis of imperfect models of the real world. That is, uncertainty (aleatory and/or epistemic) is unavoidable in engineering.

Under the preceding situation, how should engineering designs be formulated or decisions affecting a design and planning be determined? Presumably, we may assume consistently *worst conditions* (e.g., specify the highest possible flood, smallest observed fatigue life of materials) and develop conservative designs on this basis. From the standpoint of system performance and safety, this approach may be suitable, and indeed has been the basis for much of engineering designs and planning in the past and can be expected to continue in the future. However, this approach eschews any information on risk and lacks a systematic basis for evaluating the degree of conservativeness; a resulting design that is overly conservative may be excessively costly, whereas one with insufficient conservatism may be inexpensive but will sacrifice performance or safety. The optimal decision ought to be based on a trade-off between cost and benefit, in order to achieve a balance between cost and system performance. As the available information and evaluative models are invariably imperfect or insufficient, and thus contain uncertainties, the required trade-off analysis ought to be performed within the context of probability and risk.

The situations described above are common to many problems in engineering; in the following we describe several examples illustrating some of these problems. The examples are idealized to simplify the presentations; nevertheless, they serve to illustrate the essence of the decision-making aspects of engineering under conditions of uncertainty.

1.3.1 Planning and Design of Transportation Infrastructures

In planning the transportation system of a city, there are numerous decisions required; for example, in the case of a bridge across a river, more than one type of bridge system may be feasible for the crossing. The cost for the design and construction of each type of system will contain uncertainty; moreover, depending on the political situation, the ability of the city to raise the level of available funding may also be uncertain. Consequently, the selection of the type of bridge may be based on the probability of the cost of a given system relative to the probabilities of realistic levels of funding.

In the design of pavements, such as a roadway pavement or airport pavement, one of the principal decision variables (among several others) is the thickness of a pavement system consisting of several layers of subgrade base material and the finished pavement. In general, the usable life of a pavement will depend on the thickness of the system; i.e., the thicker the pavement system, the longer its useful life. Of course, for the same material and workmanship quality, the cost will also increase with the thickness. A thin system will cost less initially, but the subsequent maintenance and replacement costs will be higher. Therefore, the optimal thickness of the pavement system may be determined on the basis of a trade-off between high initial cost with low maintenance, versus low initial cost but high maintenance and replacement costs over the useful life of the pavement. For the purpose of such a trade-off analysis, the relation between the life of a pavement system and its thickness is required. The pavement life, however, is a function also of other variables, including drainage condition, temperature ranges, density, and degree of compaction of the subgrade. Since all these factors are random and contain variability, as depicted in Fig. 1.33 for the compacted subgrade density, the life of the pavement for a given thickness cannot be predicted with complete certainty. Hence, the total cost (including initial and maintenance costs) associated with a given pavement thickness also cannot be estimated with complete confidence; any meaningful trade-off analysis, therefore, should properly include probability and statistical concepts.

1.3.2 Design of Structures and Machines

The strengths of structural material and components are random and thus contain variabilities as illustrated in Figs. 1.4 and 1.5 for concrete and steel; therefore, the calculation of the capacity of a structure (invariably based on an idealized model) will contain both aleatory and epistemic uncertainties. On the other hand, the applied loads on a structure are also invariably random, as illustrated in Fig. 1.13 for wind pressures on tall buildings. Even for this idealized situation, the design of a structure, i.e., involving the determination of the load carrying capacity of the structure, must consider the question of "how safe is safe

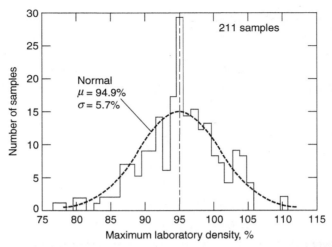

Figure 1.33 Density of compacted vocanic tuff subgrade (after Pettitt, 1967).

enough?"—a question that realistically requires the consideration of risk and the probability of nonperformance or failure.

As a specific example, consider the design of an offshore drilling platform, which is subject to occasional hurricane forces. In such a case, we recognize that aside from the fact that the maximum wind and wave effects during a hurricane are random, as may be inferred from Fig. 1.13, the occurrence frequency of hurricanes in a given region of the ocean is also unpredictable. Hence, in determining the safety level for the design of the platform, the probability of occurrence of strong hurricanes within the specified useful life of the structure must be considered, in addition to the survival probability of the structure under the highest wind and wave forces expected during the life of the structure. Consequently, the level of hurricane force that should be specified in the design, and the required level of protection that would be adequate during a hurricane, are decisions that may require a trade-off between cost and level of protection in terms of risk or failure probability within the lifetime of the platform.

Similarly, in considering the design of structural or machine components that are subjected to repeated or cyclic loads, we recognize that the fatigue life (in number of load cycles until fatigue failure) of a component is also highly variable, even under constant amplitude stress cycles, as illustrated in Fig. 1.34. For this reason, the failure-free life of a component is difficult to predict and may be described only in terms of probability. Therefore, such a component may be designed for a required operational life within a specified *reliability* (probability of no fatigue failure). As expected, the fatigue life is a function of the applied stress amplitudes; in general, the life will increase inversely with the applied stress amplitudes, as we can see in Fig. 1.34. Consequently, if a desired failure-free operational life is specified, a component may be designed to be massive so that the stress amplitudes will be low and thus ensure a high reliability of achieving the desired life; the resulting design will, of course, require more material and higher expense. In contrast, if the component is under-designed, high stresses will be induced resulting in shorter life and more frequent maintenance or replacement in order to maintain the required reliability within the operational life. In this case, the optimal operational life may be determined by minimizing the total expected life-cycle cost of a component, which would include the initial cost of the component, the expected costs of maintenance and replacements (which are functions of

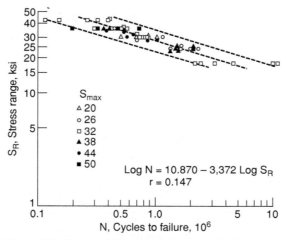

Figure 1.34 Fatigue life of welded beams vs. applied stress (after Fisher et al., 1970).

the specified reliability), and the expected cost associated with the loss of revenue incurred during a repair (which is also a function of reliability). Having decided on the operational design life and a stated reliability, the component may then be proportioned or designed accordingly.

1.3.3 Planning and Design of Hydrosystems

Suppose that the protection of a large agricultural farm from flooding requires the construction of a major culvert at the junction of a roadway and a stream. A decision on the size (i.e., flow capacity) of the culvert is required; clearly, this will depend on the high stream flow, which is a function of the rainfall intensity in the watershed area and the associated runoff. As we have observed in Fig. 1.1, the annual rainfall intensity is highly variable and the estimation of the stream runoff can be expected to be imperfect. Consequently, the prediction of the maximum stream flow over a period of concern (e.g., 10 years) will contain uncertainty, both aleatory and epistemic. Assuming that the flow capacity of a given culvert size can be determined accurately, the size of the culvert would depend on the acceptable probability of flooding within the period of concern.

Clearly, if the culvert is large enough to handle the largest possible stream flow, there would be no danger to flooding; however, the cost of constructing the culvert may be prohibitively high, and even during the heaviest rainfall the culvert may be used only to a fraction of its capacity. That is, over-design would be wasteful and costly. On the other hand, if the culvert is too small, its cost may be low, but the farm is likely to be subject to serious flooding every time there is a heavy rainfall, causing damage to crops and erosion of the upstream soil; i.e., under-design can incur high damage cost.

Therefore, the optimal size of the culvert may be determined so that the total expected life-cycle cost of the project, consisting of the cost of constructing the culvert plus the expected life-cycle loss from flooding and erosion, is minimized. As the expected loss is a function of the probability of flooding, the determination of the total expected life-cycle cost requires probability consideration.

1.3.4 Design of Geotechnical Systems

Properties of soil materials are inherently heterogeneous and highly variable, and natural earth deposits are characterized by irregular layers of various material (e.g., clay, silt, sand, gravel, or combinations thereof) with wide ranges of density, moisture content, and other soil properties that affect the strength and compressibility of the deposit. Similarly, rock formations are often characterized by irregular systems of geologic faults and fissures that can significantly affect the load-bearing capacity of the rock (see Fig. 1.9). In designing foundations on soils or rocks to support structures or facilities, the capacity of the in situ subsoil and/or rock deposit must be determined. Invariably, this determination has to be developed on the basis of available geologic information and data from site exploration with limited soil sampling results or rock core samples.

Because of the natural heterogeneity and irregularity of soil and rock deposits, the capacity of the subsoil could vary widely (see Figs. 1.10 and 1.11) over a foundation site; moreover, because the required load-carrying capacity of a foundation or pile group capacity (as shown in Fig. 1.12) must invariably be estimated on the basis of very limited in-situ data, such estimates are subject to considerable epistemic uncertainty. As a consequence, an estimate may involve some risk of overestimating the actual capacity of the deposit at a site; in this light, the safety of a foundation designed on the basis of such estimates may not be ensured with much confidence unless a sufficient margin of safety is provided in the design. On the other hand, an excessively large safety margin may yield an unnecessarily costly support system. Therefore, the required optimal safety margin for design may be viewed as a problem involving the trade-off between cost and tolerable probability of failure.

Trade-off considerations could be extended to the planning of site exploration and soil testing programs. Obviously, more extensive site exploration and use of sophisticated testing programs will reduce the uncertainties of the site condition and in the estimation of the soil parameters for the site. At a certain point, however, the incremental benefit obtained from further exploration and testing will not yield sufficient increase in the reliability of the performance of the geotechnical system, and hence may not justify its additional cost.

1.3.5 Construction Planning and Management

Many factors in the planning and management of construction projects are subject to high variability and uncertainty, and may not be easily controlled. For example, the required durations of various activities in a construction project will depend on the availability of resources, including labor and equipment and their respective productivities, on weather conditions, and on availability of construction materials. As none of these factors is completely predictable, the durations of the individual activities in *an activity* network as well as the project duration cannot be estimated with much precision or certainty (see, for example, Fig. 1.22) and thus may have to be described as random variables.

Therefore, in preparing a bid for a project, if conservative or pessimistic completion time estimates are assumed, the bid price may be too high, thus reducing the chance of winning the bid. On the other hand, if the bid is prepared on the basis of an optimistic estimate of the project duration, the contractor may lose money or sacrifice profit in a successful bid. What degree of conservativeness should the contractor exercise in order to maximize his or her potential profit? Realistically, because of variabilities and uncertainties in a number of factors, the answer may be based on a consideration of probability; i.e., the bid price may be based on a project duration corresponding to a specified probability of completion (see Fig. 1.23).

1.3.6 Photogrammetric, Geodetic, and Surveying Measurements

All practical engineering measurements are subject to errors, which can be classified as *random* and *systematic* errors. Systematic errors may be eliminated or minimized by evaluating them and applying appropriate corrections. However, the magnitude and propagation of random errors, inherent in making measurements, may be determined and analyzed properly on the basis of statistical methods. Such a probability-based statistical approach is the only reliable means for evaluating the accuracy once measurements are refined beyond instrument capabilities. The statistical basis for these purposes is presented in Chapter 6.

1.3.7 Applications in Quality Control and Assurance

In order to ensure some minimum level of quality, or performance, of engineering products or systems, inspections and quality control are necessary. Clearly, if the standard for acceptance is too stringent, it may unnecessarily increase product cost or problems in compliance, and enforcement may be difficult to achieve; on the other hand, if the standard is too lax, the quality of the finished product may be overly compromised. Also, if the control variables or design variables are random, what constitutes a stringent or nonstringent standard is not immediately clear; in these cases, the standard of acceptance may be developed on the basis of probability considerations.

For example, in constructing an earth embankment, practical standards for acceptability of the compaction need to recognize the variability in the density of compacted soil, as illustrated in Fig. 1.33. Accordingly, an acceptance sampling plan may be developed based on probability considerations and taking into account the inherent variability of compacted earth material.

To control the quality of stream waters, one of the parameters commonly used as a measure of pollution is the concentration of dissolved oxygen (DO) in the stream water, which is also highly variable, as illustrated in Fig. 1.15. Environmental engineers have recognized the need for probabilistic standards for controlling pollution of streams; for example, Loucks and Lynn (1966) proposed the following probabilistic stream standard:

The dissolved oxygen concentration in the stream during any 7 consecutive day period must be such that: (i) the probability of its being less than 4 mg/l for any one day is less than 0.20; and (ii) the probability of its being less than 2 mg/l for any one day is less than 0.1 and for two or more consecutive days is less than 0.05.

To ensure the quality of concrete material in reinforced concrete construction, the building code of the American Concrete Institute (ACI 318-93) requires the following:

The strength level of the concrete will be considered satisfactory if the average of all sets of three consecutive strength test results equal or exceed the required f_c' and no individual strength test result falls below the required f_c' by more than 500 psi. Each strength test result shall be the average of two cylinders from the same sample tested at 28 days or the specified earlier age.

The requirements stated above clearly imply the need for probability and statistics in the quality assurance of concrete material. Similar requirements may be found for the quality assurance of other construction materials.

▶ 1.4   CONCLUDING SUMMARY

In this opening chapter, we emphasized the roles and importance of probability and statistics in decision making for engineering planning and design. It is essential that these roles of probability concepts be recognized in order to appreciate their true significance. In particular, it is important to stress that the description of statistical information and the estimation of statistics, such as the means and variances, are not the only applications of probability theory. Indeed, the much more significant role of probability concepts in engineering lies in providing a unified and logical framework for the quantitative analysis of uncertainty and in the assessment of its significance in system performance, and in the formulation of risk-based trade-off studies relative to decision making, planning, and design.

The examples enumerated and described in Sections 1.2 and 1.3 should serve to illustrate the pervasiveness of probability concepts in engineering planning and design. The many examples presented also serve to illustrate, with real data and realistic engineering problems, that randomness of real-world phenomena and imperfections of engineering models are facts of life. Consequently, uncertainties are unavoidable. The uncertainties associated with randomness are inherent variability, called *aleatory uncertainty*, whereas those underlying imperfect models of reality are knowledge-based, called *epistemic uncertainty*. Even though both types of uncertainty may be represented in statistical terms, their respective significances may be different. Also, as a predictive model is improved the epistemic uncertainty can be reduced; on the other hand, because the inherent variability is part of nature, the aleatory uncertainty may not be reducible.

Finally, it is important to correct any misconception that extensive data are required to apply probability concepts; in fact, the usefulness and relevance of such concepts are equally significant, irrespective of the sufficiency of data or quality of information. In particular, the amount of data affects the definition of the variability (i.e., aleatory uncertainty) of information; however, epistemic uncertainties associated with imperfect models of reality generally require judgmental assessments irrespective of the amount of data. Probability and statistics are the conceptual and theoretical bases for modeling and analyzing uncertainty. The sufficiency of data and quality of information will affect the degree of uncertainty, but should not lessen the usefulness of probability as the proper tool for the analysis of such uncertainty and for the evaluation of its effects on engineering performance and design.

In the ensuing chapters, the essential and fundamental concepts and methods of probability and statistics for the above purposes are systematically developed and expounded.

▶ REFERENCES

Ang, A. H-S., "Extended Reliability Basis of Structural Design under Uncertainties," *Annals of Reliability and Maintainability*, Vol. 9, AIAA, July 1970, pp. 642–649.

Ang, A. H-S., and DeLeon, D., "Modeling and Analysis of Uncertainties for Risk-Informed Decisions in Infrastructure, Engineering," *Structure and Infrastructure Engineering*, Vol. 1, No. 1, Taylor & Francis, March 2005, pp. 19–31.

Becker, D.E., Burwash, W.J., Montgomery, R.A., and Liu, Y., "Foundation Design Aspects of the Confederation Bridge," *Canadian Geotechnical Journal*, Vol. 35, October 1998.

Brandow, G.E., Hart, G., and Virdee, A., "1997 Design of Reinforced Masonry Structures," *Concrete and Masonry Association of California and Nevada*, 1997.

Cox, E.A., "Information Needs for Controlling Equipment Costs," *Highway Research Record*, No. 278, National Research Council, 1969, pp. 35–48.

Donovan, N.C., "A Stochastic Approach to the Seismic Liquefaction Problem," *Proc. 1st Int. Conf. on Application of Statistics and Probability*, Hong Kong University Press, 1972.

dos Santos, J.R., Branco, F.A., and de Brito, J., "Assessment of Concrete Structures Subjected to Fire—The FB Test," *Magazine of Concrete Research*, Vol. 54, Thomas Telford Publishers June 2002.

Environmental Protection Agency, "Policy for Use of Probabilistic Analysis in Risk Assessment: Guiding Principles for Monte Carlo Analysis," EPA/630/R-97/001, May 1997.

Fisher, J.W., Frank, K.H., Hirt, M.A., and McNamee, M., "Effects of Weldments on the Fatigue Strength of Steel Beams," NCHRP Rept. No. 102, National Research Council, 1970.

Forbes, W.S., "A Survey of Progress in House Building," *Building Technology and Management*, Vol. 7(4), April 1969, pp. 88–91.

Galligan, W.L., and Snodgrass, D.V., "Machine Stress Rated Lumber: Challenge to Design," *Journal of Structural Division*, ASCE, Vol. 96, December 1970.

Huggel, C., Kaab, A., Haeberli, W., Teysseire, P., and Paul, F., "Remote Sensing Based Assessment of Hazards from Glacier Lake Outbursts: A Case Study in the Swiss Alps," *Canadian Geotechnical Journal*, Vol. 39, March 2002.

Jones, J.R., and Bachmann, R.W., "Prediction of Phosphorous and Chlorophyll Levels in Lakes," *Journal of the Water Pollution Control Fereration*, Vol. 48, 1976.

Julian, O.G., "Synopsis of First Progress Report of Committee on Factors of Safety," *Journal of Structural Division*, ASCE, Vol. 83, July 1957, p. 1316.

Katamine, N.M., "Various Volume Definitions with Conflicts at Unsignalized Intersections," ASCE *Journal of Transportation Engineering*, Vol. 126, January/February 2000.

Kothandaraman, V., and Ewing, B.B., "A Probabilistic Analysis of Dissolved Oxygen-Biochemical Oxygen Demand Relationship in Streams," *Journal of Water Resources Control Federation*, Part 2, February 1969, pp. 73–90.

Kulak, G.L., "Statistical Aspects of Strength of Connection," *Proc. ASCE Specialty Conf. on Safety and Reliability of Metal Structures*, November 1972, pp. 83–105.

Lam Put, R., "Dynamic Response of a Tall Building to Random Wind Loads," *Proc., 3rd Int. Conf. on Wind Effects on Buildings and Structures*, Tokyo, September 1971.

Littleford, T.W., "A Comparison of Flexural Strength-Stiffness Relationships for Clear Wood and Structural Grades of Lumber," *Information Report VP-X-30*, Forest Products Lab., B.C., Canada, December 1967.

Loucks, D.P., and Lynn, W.R., "Probabilistic Models for Predicting Stream Quality," *Water Resources Research*, Vol. 2, No. 3, September 1966, pp. 593–605.

Matsui, M., Ishihara, I., and Hibi., K., "Directional Characteristics of Probability Distribution of Extreme Wind Speeds by Typhoon Simulation," *Journal of Wind Engineering and Industrial Aerodynamic*, Vol. 90, Elsevier Science, Ltd., 2002.

Mohan, S.B., and Gautam, P., "Cost of Highway Work Zone Injuries," *Practice Periodical on Structural Design and Construction*, Vol. 7, May 2002.

Mohseni, O., Erickson, T.R., and Stefan, H.G., "Upper Bounds for Stream Temperatures in the Contiguous United States," *Journal of Environmental Engineering*, Vol. 128, January 2002.

National Aeronautics and Space Administration, "Probabilistic Risk Assessment Procedures Guide for NASA Managers and Practitioners," August 2002.

National Institutes of Health, "Science and Judgment in Risk Assessment: Needs and Opportunities," Environmental Health Perspectives, Vol. 102, No. 11, November 1994.

National Research Council, "Science and Judgment in Risk Assessment," National Academy Press, Washington, DC, 1994.

Pettitt, J.H.D., "Statistical Analysis of Density Tests," *Journal Highway Div.*, ASCE, Vol. HW2, November 1967.

Pruckner, F., and Gjorv, O.E., "Patch Repair and Macrocell Activity in Concrete Structures," *ACE Materials Journal*, Vol. 99, March–April 2002.

Rouillard, V., "Classification of Road Surface Profiles," *Journal Transportation Engineering*, ASCE, Vol. 126, January/February, 2000, pp. 41–45.

Thoft-Christensen, P., "Stochastic Modeling of the Diffusion Coefficient for Concrete," *Reliability and Optimization of Structural Systems*, Swets & Zeitlinger, Lisse, 2003.

Todd, D.K., and Meyer, C.F., "Hydrology and Geology of the Honolulu Aquifer," *Journal of Hydraulics Div.*, ASCE, Vol. 97, February 1971.

United Kingdom Health and Safety Executive (HSE), "Use of Risk Assessment in Government Departments," U.K. Interdepartmental Liaison Group on Risk Assessment, 2000.

U.S. Department of Energy, "Characterization of Uncertainties in Risk Assessment with Special Reference to Probabilistic Uncertainty Analysis," EH-413-068/0496, April 1996.

van Gelder, P.H.A.J.M., "Statistical Methods for the Risk-Based Design of Civil Structures," *Communications on Hydraulic and Geotechnical Engineering*, Delft University of Technology, 2000.

Viggiani, C., "Analysis and Design of Piled Foundations," *Rivista Italiana di Geotecnica*, Vol. 35, 2001.

Viner, J.G., "Recent Developments in Roadside Crush Cushions," *Journal of Transportation Engineering*, ASCE, Vol. 98, February 1972, pp. 71–87.

Winn, K., Rahardjo, H., and Peng, S.C., "Characterization of Residual Soil in Singapore," *Journal of Southeast Asian Geotechnical Society*, Vol. 32, No. 1, April 2001.

Wu, F-Q., and Wang, S-J., "Statistical Model for Structure of Jointed Rock Mass," *Geotechnique*, Vol. 52, Thomas Telford Publishers 2002.

Wynn, F.H., "Shortcut Modal Split Formula," *Highway Research Record*, National Research Council, 1969.

Zhang, L., Tang, W. H., and Ng, C.W.W., "Reliability of Axially Loaded Driven Pile Groups," *ASCE Journal of Geotechnical and Environmental Engineering*, Vol. 127 (12), December 2001.

CHAPTER 2

Fundamentals of Probability Models

2.1.1 Characteristics of Problems Involving Probabilities

Clearly, on the basis of the discussions in Chapter 1, when we speak of probability, we are referring to the likelihood of occurrence of an event relative to other events; in other words, there is (implicitly at least) more than one possibility, because otherwise the problem would be deterministic. For quantitative purposes, therefore, *probability* can be considered a numerical measure of the likelihood of occurrence of an event within an exhaustive set of all possible alternative events.

Accordingly, the first requirement in the formulation of a probabilistic problem is the identification of the set of all possibilities (i.e., the *possibility space*) and the event of interest. Probabilities are then associated with specific events within a particular possibility space. It is important to emphasize this last point, as probabilities are meaningful only within a given possibility space. To illustrate the various aspects of a probabilistic problem, as characterized above, consider the following engineering-oriented problems.

▶ **EXAMPLE 2.1**

A contractor is planning the acquisition of construction equipment, including bulldozers, needed for a new project in a remote area. Suppose that from his prior experience with similar bulldozers, he estimated that there is a 50% chance that each bulldozer can remain operational for at least 6 months. If he purchased three bulldozers for the new project, what is the probability that there will be only 1 bulldozer left operational after 6 months into the project?

First, we observe that at the end of 6 months, the possible number of operating bulldozers will be 0, 1, 2, or 3; therefore, this set of numbers constitute the possibility space of the number of operational bulldozers after 6 months. However, this possibility space is not pertinent to the question referred to above. For this latter purpose, the possibility space must be derived from the possible status of each bulldozer after 6 months, as follows:

Denoting the condition of each bulldozer after 6 months as O for *operational* and N for *nonoperational*, the possible conditions of the three bulldozers would be:

OOO—all three bulldozers are operational

OON—first and second bulldozers are operational, whereas the third one is nonoperational

ONN

NNN—all three bulldozers are nonoperational

NOO

NNO

ONO

NON

Therefore, the pertinent possibility space consists of the eight possible outcomes as indicated above. We observe also that since the condition of a bulldozer is equally likely to be operational or nonoperational after 6 months, the eight possible outcomes are equally likely to occur. It is also worth noting that among the eight possible outcomes, only one of them can be realized at the end of 6 months; this means that the different possibilities are *mutually exclusive* (we shall say more on this point in Section 2.2.1).

Finally, among the eight possible outcomes, the realization of ONN, NON, or NNO is tantamount to the event of interest, namely, *"only one bulldozer is operational."* Because each possible outcome is equally likely to occur, the probability of the event within the above possibility space is 3/8. ◀

▶ **EXAMPLE 2.2**

In designing a left-turn lane for eastbound traffic at a highway intersection, such as shown in Fig. E2.2, the probability of 5 or more cars waiting for left turns at any given time may be needed to determine the required length of the left-turn lane.

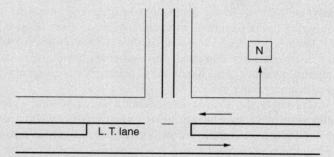

Figure E2.2 Design of a left-turn lane.

For the above purpose, suppose that over a period of 1 week, 60 observations were made at regular time intervals (during periods of heavy traffic) of the number of eastbound motor vehicles waiting for left turns at this intersection, with the following results:

No. of Cars	No. of Observations	Relative Frequency
0	4	4/60
1	16	16/60
2	20	20/60
3	14	14/60
4	3	3/60
5	2	2/60
6	1	1/60
7	0	0
8	0	0

Conceivably, the number of vehicles waiting for left turns, during heavy traffic hours, could be any integer number; however, based on the above observations, it is not likely that there will be seven or more vehicles waiting for left turns at any time.

On the basis of the above observations, the estimated *relative frequency* (in the third column of the above table) may be used approximately as the probability of a particular number of cars waiting for left turns. For example, the probability of the event "*5 or more cars waiting for left turns*" is approximately $2/60 + 1/60 = 3/60$. The estimated probabilities based on relative frequencies are approximate because of "sampling error" which may be significant when the estimate is based on a small number of observations; this is part of the epistemic uncertainty that we discussed earlier in Chapter 1. The accuracy of the estimated probabilities will improve as the total number of observations (*sample* size) increases as we shall discuss further later in Chapter 6. ◄

► **EXAMPLE 2.3**

The simply supported beam *AB* shown in Fig. E2.3 is carrying a load of 100 kg that may be placed anywhere along the span of the beam. The reaction at the support A, R_A, can be any value between 0 and 100 kg depending on the position of the load on the beam; in this case, therefore, any value between 0 and 100 kg is a possible value of R_A, and thus is its possibility space.

An event of interest may be that the reaction is in some specified interval; for example, $(10 \leq R_A \leq 20$ kg) or ($R_A \geq 50$ kg). Therefore, if a particular value of R_A is realized, the event (defined by an interval) containing this value of R_A has occurred, and we can speak of the probability that R_A will, or will not, be in a given interval. For example, if we assume that the 100 kg load is equally likely to be placed anywhere along the beam span, then the probability that the value of R_A will be in a given interval is proportional to the length of the interval; for example,

$$P(10 \leq R_A \leq 20) = 10/100 = 0.10 \quad \text{and} \quad P(R_A \geq 60) = 40/100 = 0.40.$$

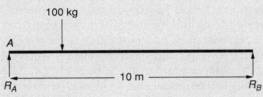

Figure E2.3 A simply supported beam. ◄

► **EXAMPLE 2.4**

Consider the bearing capacity of the footing foundation for a building. Suppose that from prior experience, it is the judgment of the foundation engineer that the bearing capacity of a footing at the building site has a 95% probability of at least 4000 psf (pounds per square foot). If 16 individual footings are required for the building foundation, what is the probability that all the footings will have at least 4000 psf bearing capacity? Conversely, what is the probability that at least one of the 16 footings would have its bearing capacity less than 4000 psf?

In this case, the possibility space consists of $2^{16} = 65,536$ sample points. Suppose that each footing has a probability of 0.95 that its bearing capacity will be at least 4000 psf. Then if the bearing capacities of the different footings are statistically independent of each other, the probability that all the footings will have bearing capacities of at least 4000 psf is then $(0.95)^{16} = 0.440$.

In the second question, "at least one is the *complement* of none." Therefore, the probability of at least one footing with bearing capacity less than 4000 psf is $1 - 0.440 = 0.560$. The concepts of *complement of an event* and *statistical independence* will be discussed later in Sections 2.2 and 2.3, respectively. ◄

From the foregoing examples, we can observe the following special characteristics of probabilistic problems.

1. Every problem is defined with reference to a specific possibility space (containing more than one possible outcome), and each event is composed of one or more outcomes within this possibility space.

2. The probability of an event is a function of the probabilities of the individual outcomes within a given possibility space, and may be derived from the probabilities of these basic outcomes.

In Sections 2.2 and 2.3, we shall present the mathematical tools pertinent to and useful for each of these purposes.

2.1.2 Estimating Probabilities

From the examples illustrated above, it may be observed that in estimating or calculating the probability of an event, a basis for assigning probability measures to the various possible outcomes is necessary. The assignments may be based on prior conditions (such as deduced on the basis of prescribed assumptions), or based on results of empirical observations, or based on subjective judgments.

In Examples 2.1 and 2.3, the probabilities of all the possible outcomes were based on prior assumptions. In the case of Example 2.1, each of the possible operational conditions of the three bulldozers was assumed to be equally probable, each equal to 1/8 (consistent with the prior information that each bulldozer is equally likely to be operational or nonoperational after 6 months), whereas, in Example 2.3 the probability that the reaction R_A will be in a given interval was assumed to be proportional to the interval length (consistent with the assumption that the 100-kg load is equally likely to be placed anywhere along the span of the beam). However, in Example 2.2 the probability of the number of vehicles waiting for left turns is based on the corresponding observed relative frequency, which is determined from available empirical observations. Finally, in Example 2.4, the probability that the bearing capacity of a footing will be greater than 4000 psf is based on the experience and subjective judgment of the foundation engineer.

It should be emphasized that we shall treat probability as a measure necessary and useful for solving engineering problems where more than one possible outcome or event is possible (i.e., problems that are *nondeterministic*). In particular, we shall eschew the philosophical question of the meaning of a probability measure, and limit our concern only to the utilitarian aspects of probability and its mathematical and statistical theories for modeling and analyzing problems under conditions of uncertainty, in the same sense that we use the factor of safety to develop engineering designs without worrying about the real meaning of a safety factor, or employing Newton's second law of motion without being concerned about the meaning of mass and force.

The usefulness of a calculated probability, however, will depend on the appropriateness of the basis for its determination. In this regard, we should emphasize that the validity of the a priori basis for estimating a probability depends on the reasonableness of the underlying assumptions, whereas the validity of the empirical *relative frequency* may depend on the amount of observational data. When data are limited, the relative frequency by itself may give an inaccurate estimate of the true probability, or at best an approximate estimate of the true probability.

A third basis for estimating probability involves the combination of intuitive or subjective assumptions with available empirical observations; the proper vehicle for this combination is the Bayes' theorem (see Section 2.3.5), and the result is known as *Bayesian probability* (presented in Chapter 9).

► 2.2 ELEMENTS OF SET THEORY—TOOLS FOR DEFINING EVENTS

We can recognize that the first requirement in the formulation of a probabilistic problem is the definition of the event of interest within a particular possibility space. The basic mathematical tools for this purpose are contained in the elements of the *theory of sets*. In this section we present the basic elements of set theory followed in Section 2.3 with the fundamentals of the theory of probability as they relate to and are useful for the formulation of probabilistic problems in engineering.

2.2.1 Important Definitions

In the terminology of set theory, the set of all possibilities in a probabilistic problem is collectively a *sample space*, and each of the individual possibilities is a *sample point*. An *event* is then defined as a *subset of the sample space*.

Sample spaces may be *discrete* or *continuous*. In the discrete case, sample points are discrete entities and countable; in the continuous case, the sample space is composed of a continuum of sample points.

A discrete sample space may be *finite* (composed of a finite number of sample points) or *infinite* (i.e., with a countably infinite number of sample points). The possible configurations of the three bulldozers in Example 2.1 are an example of a finite discrete sample space; each of the possible configurations is a sample point, and the eight possible configurations collectively constitute the corresponding sample space. Other examples of finite sample spaces are as follows:

- The potential winner in a competitive bidding for a construction project will be among those firms submitting bids for the project. In this case, the sample space is generally finite and consists of all the firms submitting bids for the project, whereas each of the firms is a sample point.

- The number of days in a year with potentially measurable precipitation in Seattle is finite and conceivably will range from 0 to 365 days. Each day of the year is a sample point, and the number of days in the year plus one constitute the sample space.

- In determining the percentage of flights that arrive more than 15 minutes late at the O'Hare International Airport, the total number of flights landing at O'Hare in a 24-hour day is a finite sample space and each of the flights is a sample point within this sample space.

Examples of discrete sample spaces with countably infinite number of sample points are the following:

- The number of flaws in a given length of welding—there may be none or only a few flaws in the weld, or the number of flaws could be very large. The actual number of flaws in a weld could conceivably be infinite.

- The number of cars crossing a toll bridge until the next accident on the bridge over a period of one year. An accident may possibly occur with the first car crossing the bridge, or there may not be any accident in the year.

In a continuous sample space, the number of sample points is always infinite. For example:

- In considering the location on a toll bridge where a traffic accident may occur, each of the possible locations is a sample point, and the sample space would be the continuum of possible locations on the bridge.

• If the bearing capacity of a clay soil deposit is between 1.5 tsf (tons per square foot) and 4.0 tsf, then any value within the range 1.5 to 4.0 is a sample point, and the entire continuum of values in this range constitutes the sample space.

Irrespective of whether a sample space is discrete or continuous, however, an event is always a subset of the appropriate sample space. Therefore, an event will always contain one or more sample points (unless it is an *impossible* event which is a null set), and *the realization of any of these sample points constitutes the occurrence of the corresponding event*. Finally, again, when we speak of probability, we are always referring to an event within a particular sample space.

The following examples should serve to clarify the preceding definitions and concepts in more definitive and quantitative terms.

▶ **EXAMPLE 2.5**   Consider again a simply supported beam AB as shown in Fig. E2.5a.

(a) If a concentrated load of 100 kg can be placed only at any of the 1-meter interval points on the beam, the sample space of the reaction R_A will be as follows:

$$(0, 10, 20, 30, 40, 50, 60, 70, 80, 90, 100 \text{ kg})$$

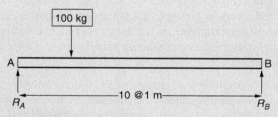

Figure E2.5a   Beam AB.

(b) The sample space of R_A and R_B; that is, all pairs of R_A and R_B such that $R_A + R_B = 100$ belong to the sample space. Graphically, this sample space is shown in Fig. E2.5b.

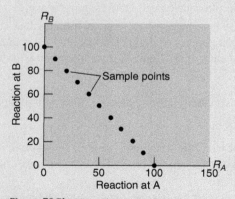

Figure E2.5b   Sample space of R_A and R_B.

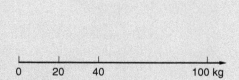

Figure E2.5c   Sample space of R_A.

(c) If the 100-kg load can be placed anywhere along the length of the beam, the sample space of R_A can be represented by the straight line between 0 and 100 (Fig. E2.5c), whereas the corresponding sample space of R_A and R_B is the diagonal line shown in Fig. E2.5d. In Fig. 2.5c, an event may be defined as $(20 < R_A < 40)$, whereas in Fig. 2.5d an event for (R_A, R_B) may be between $(20, 80)$ and $(40, 60)$.

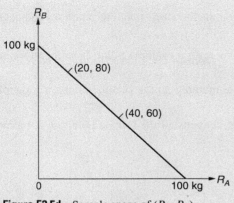

Figure E2.5d Sample space of (R_A, R_B).

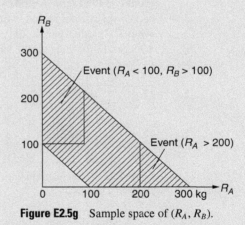

Figure E2.5e Sample space of R_A, or R_B.

(d) Next, consider that the load can be 100 kg, 200 kg, or 300 kg and can be placed anywhere along the beam. In this case, the sample space of R_A, or R_B, contains all the values between 0 and 300 kg as represented by the line shown in Fig. E2.5e, whereas the sample space of (R_A, R_B) is represented by the three lines shown in Fig. E2.5f.

(e) Finally, if the load can be any value between 100 and 300 kg and can be placed anywhere along the beam, the sample space of R_A or R_B is also the straight line of Fig. E2.5e, whereas the sample space of (R_A, R_B) would be the hatched area in Fig. E2.5g.

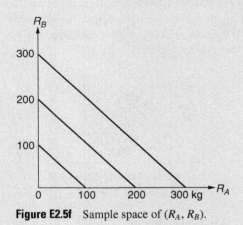

Figure E2.5f Sample space of (R_A, R_B).

Figure E2.5g Sample space of (R_A, R_B).

The event within the sample space of Fig. E2.5g, $(R_A > 200 \text{ kg})$ is the triangular region shown in Fig. E2.5g, whereas the event $(R_A < 100; R_B > 100)$ is the trapezoid in the same figure. ◀

▶ **EXAMPLE 2.6**

From historical data of floods for a river, suppose the annual maximum flood levels above the mean river flow range from 1 m to 5 m. If the annual maximum flow is measured in an increment of 0.1 m, then the sample space of the annual flow would contain the 51 sample points (1.0, 1.1, 1.2, ..., 4.8, 4.9, 5.0 m). The event of annual flood flow exceeding 3.0 m, therefore, would contain the 20 sample points defined by (3.1, 3.2, ..., 4.8, 4.9, 5.0 m).

On the other hand, if the annual maximum flow can be any level from 1 m to 5 m, then the sample space of the annual flow would be the continuum of infinite values between 1 m and 5 m. Similarly, the event of flood flow exceeding 3.0 m will be the continuum of values between 3 m and 5 m. ◀

Special Events

We define the following special events and adopt the corresponding notations indicated below:

- *Impossible event*, denoted ϕ, is the event with no sample point. It is, therefore, an *empty* set in a sample space.
- *Certain event*, denoted S, is the event containing all the sample points in a sample space; i.e., it is the sample space itself.
- *Complementary event* $\overline{E}$, of an event E contains all the sample points in S that are not in E.

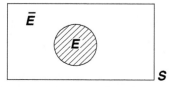

Figure 2.1 A Venn diagram of sample space S.

The Venn Diagram

A sample space and its subsets (or events) can be represented pictorially with a *Venn diagram*. As illustrated in Fig. 2.1, the sample space S is represented by a rectangle; an event E is then represented by the circle (or any closed region) within this rectangle, and the part of the rectangle that is outside this closed region is the corresponding complementary event $\overline{E}$. In other words, the event E contains all the sample points within the closed region, whereas $\overline{E}$ contains all the sample points in S that are outside of E.

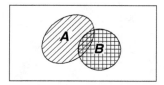

Figure 2.2a Venn diagram with two events A and B.

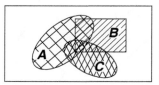

Figure 2.2b Venn diagram with three events A, B, and C.

A Venn diagram with two (or more) events is illustrated in Fig. 2.2.

In many practical problems, the event of interest may be a combination of several other events. For instance, in Example 2.1 the event of *at least two bulldozers in operating condition after 6 months* may be of interest. This event would be the combination of *two or three bulldozers in operating condition*. Such an event involves the *union* of the two individual events.

There are only two ways that events may be combined, or an event may be derived from other events; namely, by the *union* or *intersection*. Consider two events E_1 and E_2. The union of E_1 and E_2, denoted $E_1 \cup E_2$, is the occurrence of E_1 or E_2 or both. (In set theory, *or* is used in an *inclusive* sense, which means *or/and*.) This means that $E_1 \cup E_2$ is another event containing all the sample points belonging to either E_1 or E_2.

The Venn diagram for the union of E_1 and E_2 would be the hatched region shown in Fig. 2.3. It follows, therefore, that the region outside of the hatched region within S is the complementary event $\overline{E_1 \cup E_2}$, i.e., the complement of the event $(E_1 \cup E_2)$.

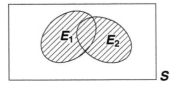

Figure 2.3   Venn diagram for $E_1 \cup E_2$.

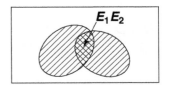

Figure 2.4   Venn diagram of $E_1 E_2$.

The union of three or more events, as shown Fig. 2.2b, means the occurrence of at least one of them, and is the subset of the sample points within the three hatched regions of the individual events A, B, and C.

The *intersection* of two events E_1 and E_2, denoted $E_1 \cap E_2$ or simply $E_1 E_2$, is an event representing the joint occurrence of E_1 and E_2; in other words, $E_1 E_2$ is the subset of sample points belonging to both E_1 and E_2. The Venn diagram of $E_1 E_2$ would be the double hatched region shown in Fig. 2.4.

Finally, the intersection of three or more events is the occurrence of all of them and would be the subset of sample points belonging to all three individual events.

Examples of Union of Events

- In describing the state of supply of construction materials, if E_1 represents the shortage of concrete, and E_2 represents the shortage of steel, then the union $E_1 \cup E_2$ is the shortage of concrete or steel, or both. In this case, the complementary event, $\overline{E_1 \cup E_2}$ means no shortage of construction material, i.e., concrete and steel are both available, whereas $\overline{E_1} \cup \overline{E_2}$ means there is no shortage of concrete or there is no shortage of steel (observe the subtle difference).

- The transportation of cargoes between Chicago and New York may be by air, highway, or railway. If the availability of each of these three modes of transportation is denoted, respectively, as A, H, and R, the available means of transporting cargoes between Chicago and New York is $(A \cup H \cup R)$, i.e., cargoes may be shipped by air or highway or railway.

Examples of Intersection of Events

- Referring to the first example above, the event $E_1 E_2$ would mean the shortage of both concrete and steel, and $\overline{E_1} \overline{E_2}$ means no shortage of both materials.

- Referring to the second example above, the event AHR means that all three modes of transportation between Chicago and New York are available. Observe also that the event $A \overline{H}\, \overline{R}$ means that only air transportation is available.

▶ **EXAMPLE 2.7**

Suppose there are two highway routes from city A to city B as shown in Fig. E2.7a. Let E_1 represent the event that Route 1 is open and E_2 that Route 2 is open to traffic. Then $E_1 \cup E_2$ means that Route 1 is open or Route 2 is open; in other words, at least one of the two routes is open.

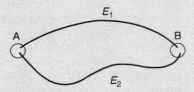

Figure E2.7a   Two routes from A to B.

The intersection, E_1E_2, means that both Routes 1 and 2 are open, whereas $E_1\overline{E}_2$ means that Route 1 is open but Route 2 is closed, and $\overline{E}_1\overline{E}_2$ means that both routes are closed perhaps because of heavy snow.

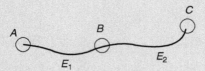

Figure E2.7b Route 1 from A to B; Route 2 from B to C.

Next consider the three cities A, B and C with Route 1 connecting A to B and Route 2 connecting B to C as shown in Fig. E2.7b. If $\overline{E}_1$ and $\overline{E}_2$, respectively, mean that Route 1 is closed and Route 2 is closed, then the union $\overline{E}_1 \cup \overline{E}_2$ means that Route 1 is closed or Route 2 is closed, which also means that it is not possible to go from city A to city C.

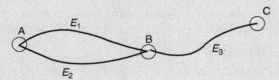

Figure E2.7c Routes 1 and 2 from A to B; Route 3 from B to C.

Finally, assume that there are two alternative routes from city A to city B, and only Route 3 from city B to city C, as shown in Fig. E2.7c. In this case, there are two possible ways to go from city A to city C; namely, $E_1E_3 \cup E_2E_3$: alternatively, this can be expressed as the event $(E_1 \cup E_2)E_3$. Observe that the event $\overline{E}_1\overline{E}_2 \cup \overline{E}_3$ means that there will be no transportation from city A to city C. ◀

▶ **EXAMPLE 2.8**

Consider the last case of Example 2.5, in which the load ranges from 100 kg to 300 kg and the sample space of the two reactions (R_A, R_B) as shown in Fig. E2.5g.

$$\text{If} \quad A = \text{the event that } R_A > 100 \text{ kg; and}$$
$$B = \text{the event that } R_B > 100 \text{ kg}$$

the events A and B would be the respective subsets of the point pairs shown in Figs. E2.8a and E2.8b.

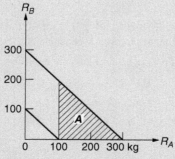

Figure E2.8a Subset of event A.

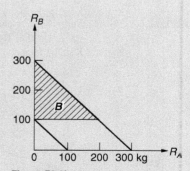

Figure E2.8b Subset of event B.

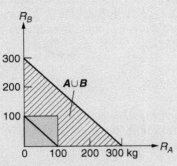

Figure E2.8c The union $A \cup B$.

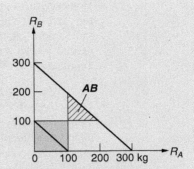

Figure E2.8d The intersection AB.

The union $A \cup B$ is the hatched region in Fig. E2.8c, whereas the intersection AB would be the hatched region in Fig. E2.8d.

Note that in this example, Figs. E2.8a through E2.8d also serve as the corresponding Venn diagrams. ◀

Mutually Exclusive Events

Two events are *mutually exclusive* if the occurrence of one precludes the occurrence of the other event; that is, the occurrence of both events is impossible. The corresponding subsets of sample points, therefore, will have no overlap as shown in the Venn diagram of Fig. 2.5; in other words, the subsets are disjoint.

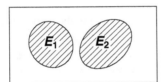

Figure 2.5 Mutually exclusive events E_1 and E_2.

If two events are mutually exclusive, then their intersection is an impossible event; i.e., $E_1 E_2 = \phi$.

Examples of events that are naturally mutually exclusive include the following:

1. A car making a right turn and making a left turn at a street intersection.

2. Occurrence of flood and occurrence of drought of a river at a given time.

3. Collapse and no damage of a building under a strong earthquake.

Similarly, three or more events are mutually exclusive if the occurrence of one event precludes the occurrence of all the other events. Examples are the following:

1. If there are three competing locations for a single airport, then the three choices for the final location of the airport are mutually exclusive.

2. In Example 2.1, the number of bulldozers that will remain operational after 6 months are mutually exclusive.

3. In Example 2.2, the number of vehicles waiting for left turns at the intersection are mutually exclusive.

However, in Example 2.7, the conditions of the different routes are not mutually exclusive as the closing of one route does not necessarily preclude the closing of another route. Likewise, in Example 2.8, the events A and B are not mutually exclusive because both R_A and R_B can exceed 100 kg if the load is sufficiently high, say > 200 kg; also the intersection AB is not a null set.

Collectively Exhaustive Events

Two or more events are *collectively exhaustive* if the union of the events make up the underlying sample space. For example, the event E and its complement $\overline{E}$, as shown in Fig. 2.1, are collectively exhaustive; clearly, $E \cup \overline{E} = S$.

▶ **EXAMPLE 2.9**

Two construction companies *a* and *b* are bidding for projects. Define A as the event that Company *a* wins a bid, and B as the event that Company *b* wins a bid. Let us sketch the Venn diagrams for the sample spaces of the following:

1. Company *a* is submitting a bid for one project, and Company *b* is submitting its own bid for another project. The Venn diagram would be as shown in Fig. E2.9a.

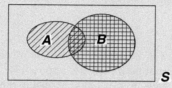

Figure E2.9a Sample space of A and B.

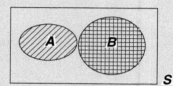

Figure E2.9b Sample space with mutually exclusive A and B.

In this case, it is possible for both companies to win their respective bids as represented by the intersection of A and B.

2. Companies *a* and *b* are submitting bids for the same project, and there are also other bidders for the project. The corresponding Venn diagram would be as shown in Fig. E2.9b.

In this case, Company *a* or *b* may be awarded the project, or one of the other companies may be awarded the project. If Company *a* wins its bid, then no other companies, including Company *b* can also be the winner of the award. That is, the occurrence of A precludes the occurrence of B; thus, the events A and B are mutually exclusive, as shown in Fig. E2.9b where there is no overlapping region between A and B. Moreover, the complement of $A \cup B$, i.e., $\overline{A \cup B}$, means that one of the other companies wins the award.

3. Company *a* and Company *b* are the only companies submitting competing bids for a single project. The Venn diagram for this case would appear as shown in Fig. E2.9c.

In this particular case, since Company *a* and Company *b* are the only bidders for the single project, and only one of them can be the winner, the events A and B are mutually exclusive, and are also collectively exhaustive or $A \cup B = S$. Thus, the sample space contains only the two sets A and B as shown in Fig. E2.9c.

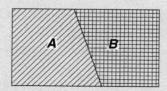

Figure E2.9c Sample space with A and B only. ◀

► **EXAMPLE 2.10**   There are three possible sites, denoted as *Site a*, *Site b* and *Site c* for the construction of a new airport for a major city. Define the following:

$$A = Site\ a\ \text{is selected for the airport;}$$
$$B = Site\ b\ \text{is selected for the airport;}\ \text{and}$$
$$C = Site\ c\ \text{is selected for the airport.}$$

If the above three sites are the only feasible sites for the airport, then the available alternatives for the location of the airport is the union $A \cup B \cup C$. However, if one of the alternative sites is selected, then this precludes the selection of the two other sites; therefore, the events A, B, and C are mutually exclusive. The Venn diagram, therefore, will appear as shown in Fig. E2.10.

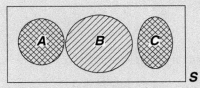

Figure E2.10   Sample space of events A, B, and C.

In this case, we also observe the following:

$A\overline{B}\,\overline{C}$ means that *Site a* is selected and not *Site b* or *Site c*.

$\overline{A \cup B \cup C}$ means that none of the three sites is selected as location for the airport.   ◄

2.2.2   Mathematical Operations of Sets

In Section 2.2.1, we observed that two or more sets, or events, can be combined in only two ways; namely, by taking the *union* or taking the *intersection*. These two operations and the process of taking the *complement* of an event constitute the basic operations involving sets. The notations that we have adopted to designate sets and the associated operations are as follows:

$$\cup = \text{the union}$$
$$\cap = \text{the intersection}$$
$$\supset = \text{contains}$$
$$\subset = \text{belongs to, or is contained in}$$
$$\overline{E} = \text{the complement of } \boldsymbol{E}$$

With these notations, the mathematical rules governing the operations of sets are the following:

Equality of Sets
Two sets are *equal* if and only if both sets contain exactly the same sample points. On this basis, we observe that

$$A \cup \phi = A$$

in which ϕ is the null (or empty) set.

Also,

$$A \cap \phi = \phi$$

Furthermore,

$$A \cup A = A$$

$$A \cap A = A$$

and, for the sample space S,

$$A \cup S = S$$

whereas

$$A \cap S = A$$

On Complementary Sets

With regard to an event E and its complement $\overline{E}$, we observe the following:

$$E \cup \overline{E} = S \qquad (2.1)$$

whereas

$$E \cap \overline{E} = \phi \qquad (2.2)$$

and

$$\overline{\overline{E}} = E$$

that is, the complement of the complementary event yields the original event.

Commutative Rule

The union and intersection of sets are *commutative*; that is, for two sets A and B,

$$A \cup B = B \cup A$$

Also,

$$A \cap B = B \cap A$$

Associative Rule

The union and intersection of sets are also *associative*; that is, for three sets, A, B, and C,

$$(A \cup B) \cup C = A \cup (B \cup C)$$

Also,

$$(AB)C = A(BC)$$

Distributive Rule

Finally, the union and intersection of sets are *distributive*; that is, for three sets A, B, and C,

$$(A \cup B) \cap C = A \cap C \cup B \cap C \quad \text{or} \quad AC \cup BC$$

and also,

$$(AB) \cup C = (A \cup C) \cap (B \cup C)$$

We might observe that the above commutative, associative, and distributive rules for sets are similar to the same algebraic rules for numbers. In particular, the operational rules governing the addition and multiplication of numbers apply (with certain equivalences) to the union and intersection of sets. With the following equivalences—*union for addition and intersection for multiplication* (i.e., $\cup \to +$ and $\cap \to \times$)—the rules of conventional algebra apply to operations of sets or events. Moreover, in accordance with the hierarchy of algebraic operations, intersection takes precedence over union of sets, unless parenthetically indicated otherwise.

It should be emphasized that the above equivalences are only valid in an operational sense; conventional algebraic operations such as addition and multiplication have no meaning for sets or events. Moreover, there are *no equivalent operations for subtraction or division of sets*. On the other hand, there are operations and operational rules that apply to sets that have no counterparts in conventional algebra. For example, for a set A,

$$A \cup A = A \quad \text{and} \quad A \cap A = A$$

Another case in point is the second of the distributive rule described above, which says that

$$(A \cup C)(B \cup C) = AB \cup AC \cup BC \cup CC$$

but,

$$BC \cup CC = C$$

Similarly,

$$AC \cup C = C.$$

Hence, the final result is

$$(A \cup C)(B \cup C) = AB \cup C$$

whereas in conventional algebra we would have

$$(a + c)(b + c) = ab + ac + bc + c^2 \neq ab + c$$

Finally, another important rule that applies to sets but has no counterpart in conventional algebra is the *de Morgan's* rule, as described below.

De Morgan's Rule

This rule relates to sets and their complements. For two sets, or events, E_1 and E_2, the de Morgan's rule says that

$$\overline{E_1 \cup E_2} = \overline{E_1} \cap \overline{E_2}$$

The general validity of this relation can be shown with the Venn diagrams in Fig. 2.6.

The unhatched region in Fig. 2.6a is clearly $\overline{E_1 \cup E_2}$. The two Venn diagrams in Fig. 2.6b show, respectively, the complementary sets $\overline{E_1}$ and $\overline{E_2}$, the intersection of which is the double-hatched region in Fig. 2.6c. From Figs. 2.6a and 2.6c, we see the equality of the two sets $\overline{E_1 \cup E_2} = \overline{E_1} \cap \overline{E_2}$, thus verifying the de Morgan's rule.

In more general terms, the de Morgan's rule is

$$\overline{E_1 \cup E_2 \cup \cdots \cup E_n} = \overline{E_1} \cap \overline{E_2} \cap \cdots \cap \overline{E_n} \tag{2.3a}$$

Applying Eq. 2.3a to the complements of $\overline{E_1}, \overline{E_2}, \ldots, \overline{E_n}$, we have

$$\overline{\overline{E_1} \cup \overline{E_2} \cup \cdots \cup \overline{E_n}} = E_1 E_2 \cdots E_n$$

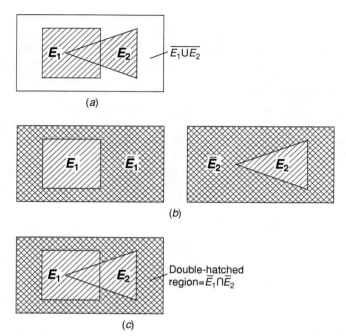

Figure 2.6 Venn diagrams showing de Morgan's rule.

Thence, taking the complements of both sides of the above equation, the de Morgan's rule can be stated also as

$$\overline{E_1 E_2 \cdots E_n} = \overline{E_1} \cup \overline{E_2} \cup \cdots \cup \overline{E_n} \tag{2.3b}$$

In light of Eqs. 2.3a and 2.3b, we can establish the following *duality relation*: The complement of the unions and intersections of events is equal to the intersections and unions of the respective complements of the same events.

The following examples illustrate the above duality relation:

$$\overline{A \cup BC} = \overline{A} \cap \overline{BC} = \overline{A}(\overline{B} \cup \overline{C})$$
$$\overline{(A \cup B)C} = \overline{(A \cup B)} \cup \overline{C} = \overline{A}\,\overline{B} \cup \overline{C}$$
$$\overline{(\overline{AB} \cup C)(\overline{A} \cup \overline{C})} = \overline{AB\overline{C} \cup AC} = \overline{AB\overline{C}} \cap \overline{AC}$$

For further illustrations of the de Morgan's rule, consider the following engineering examples:

▶ **EXAMPLE 2.11**

Consider a simple chain consisting of two links as shown in Fig. E2.11. Clearly, the chain will fail to carry the load *F* if either link breaks; thus, if

$$E_1 = \text{the breakage of link 1}$$
$$E_2 = \text{the breakage of link 2}$$

Figure E2.11 A two-link chain.

Then

$$\text{Failure of chain} = E_1 \cup E_2$$

No failure of the chain, therefore, is the complement $\overline{E_1 \cup E_2}$. However, no failure of the chain also means that both links survive (no breakage); that is,

$$\text{No failure of chain} = \overline{E_1} \cap \overline{E_2}$$

Therefore, we have

$$\overline{E_1 \cup E_2} = \overline{E_1} \cap \overline{E_2}$$

which is a demonstration of the validity of the de Morgan's rule applied to an engineering problem. ◄

► **EXAMPLE 2.12**

The water supply for two cities C and D comes from the two sources A and B as shown in Fig. E2.12. Water is transported by pipelines consisting of branches 1, 2, 3, and 4. Assume that either one of the two sources, by itself, is sufficient to supply the water for both cities.

Denote:
$E_1 = $ failure of branch 1;
$E_2 = $ failure of branch 2;
$E_3 = $ failure of branch 3; and
$E_4 = $ failure of branch 4.

Failure of a pipe branch means there is serious leakage or rupture of the branch.

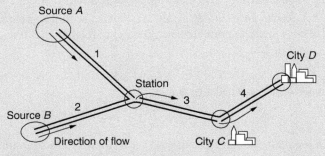

Figure E2.12 A water supply system.

Shortage of water in city C would be represented by $E_1 \cap E_2 \cup E_3$, and its complement $\overline{E_1 E_2 \cup E_3}$ means that there is no shortage of water in city C. Applying the de Morgan's rule, we have

$$\overline{E_1 E_2 \cup E_3} = (\overline{E_1} \cup \overline{E_2})\overline{E_3}$$

The last event above means that there is no failure in branch 1 or branch 2 and also no failure in branch 3. Similarly, shortage of water in city D would be the event $E_1 E_2 \cup E_3 \cup E_4$. Therefore, no shortage of water in city D is

$$\overline{E_1 E_2 \cup E_3 \cup E_4} = (\overline{E_1} \cup \overline{E_2})\overline{E_3}\overline{E_4}$$

which means that there is sufficient supply at the station, i.e., $(\overline{E_1} \cup \overline{E_2})$, and there are no failures in both branches 3 and 4, represented by $\overline{E_3}\overline{E_4}$. ◄

▶ 2.3   MATHEMATICS OF PROBABILITY

In all of our discussions thus far, we have tacitly assumed that a nonnegative measure, called *probability*, is associated with every event in a particular sample space. Implicitly, we have also assumed that such probability measures possess certain properties and follow certain operational rules. Formally, these properties and rules are embodied in the *mathematical theory of probability*. As with other branches of mathematics, the theory of probability is based on certain fundamental assumptions, or *axioms*, that are not subject to proofs; these are as follows (see, e.g., Feller, 1957; Parzen, 1960; Papoulis, 1965):

Axiom 1: For every event E in a sample space S, there is a probability

$$P(E) \geq 0 \tag{2.4}$$

Axiom 2: The probability of the *certain event S* is

$$P(S) = 1.0 \tag{2.5}$$

Axiom 3: Finally, for two events E_1 and E_2 that are *mutually exclusive*,

$$P(E_1 \cup E_2) = P(E_1) + P(E_2) \tag{2.6}$$

Equations 2.4 through 2.6 constitute the basic axioms of probability theory. These are essential assumptions and, therefore, cannot be violated. However, these axioms and the resulting theory must be consistent with and useful for real-world problems. In this latter regard, we may observe the following:

- The probability of an event, $P(E)$, is a relative measure, i.e., relative to other events in the same sample space. For this purpose, it is natural and convenient to assume such a measure to be nonnegative as prescribed in Eq. 2.4.
- Because an event, E, is always defined within a prescribed sample space S, it is convenient to normalize its probability relative to S (the certain event), as specified in Eq. 2.5.

Therefore, on the basis of Eqs. 2.4 and 2.5, it follows that the probability of an event E is bounded between 0 and 1.0; that is,

$$0 \leq P(E) \leq 1.0$$

With regard to the third axiom of Eq. 2.6, we may observe intuitively that from a relative frequency standpoint, if an event E_1 occurs n_1 times among n repetitions of an experiment, and another event E_2 occurs n_2 times in the same n repetitions, in which E_1 and E_2 cannot occur simultaneously (they are mutually exclusive), then E_1 or E_2 will have occurred $(n_1 + n_2)$ times among the n repetitions of the experiment. Thence, on the basis of relative frequency, we have (for large n)

$$P(E_1 \cup E_2) = \frac{n_1 + n_2}{n} = \frac{n_1}{n} + \frac{n_2}{n}$$
$$= P(E_1) + P(E_2)$$

It should be emphasized that the mathematical theory of probability provides the logical bases for developing the relationships among probability measures. As expected, all such relationships and any theoretical results are based on the three basic axioms stated in Eqs. 2.4 through 2.6.

2.3.1 The Addition Rule

As an event E and its complement $\overline{E}$ are mutually exclusive, we obtain on the basis of Eq. 2.6

$$P(E \cup \overline{E}) = P(E) + P(\overline{E})$$

but since $E \cup \overline{E} = S$, we have, on the basis of Eq. 2.5, $P(E \cup \overline{E}) = P(S) = 1.0$. Hence, we obtain the first useful relation

$$P(\overline{E}) = 1 - P(E) \tag{2.7}$$

More generally, if two events E_1 and E_2 are not mutually exclusive, the *addition rule* is

$$P(E_1 \cup E_2) = P(E_1) + P(E_2) - P(E_1 E_2) \tag{2.8}$$

The general addition rule of Eq. 2.8 can be shown to follow from Eq. 2.6.

For this purpose, we first observe from Fig. 2.7 that $E_1 \cup E_2 = E_1 \cup \overline{E_1} E_2$.

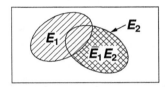

Figure 2.7 Union of E_1 and $\overline{E_1} E_2$.

As can be seen in Fig. 2.7, E_1 and $\overline{E_1} E_2$ are mutually exclusive. Therefore, according to Eq. 2.6,

$$P(E_1 \cup \overline{E_1} E_2) = P(E_1) + P(\overline{E_1} E_2)$$

But $\overline{E_1} E_2 \cup E_1 E_2 = (\overline{E_1} \cup E_1) E_2 = S E_2 = E_2$; and $E_1 E_2$ and $\overline{E_1} E_2$ are clearly mutually exclusive. Hence,

$$P(\overline{E_1} E_2) = P(E_2) - P(E_1 E_2)$$

from which we obtain Eq. 2.8.

▶ **EXAMPLE 2.13**

A contractor is starting two new projects—jobs 1 and 2. There is some uncertainty on the scheduled completion of each job; at the end of 1 year, the condition of completing each of the jobs may be defined as follows:

$$A = \text{definitely completed}$$
$$B = \text{completion questionable}$$
$$C = \text{definitely incomplete}$$

PROBLEMS

1. Describe the sample space for the states of completion of the two jobs; i.e., identify all the possible situations regarding the completion of both jobs 1 and 2 at the end of 1 year in terms of the notations above; e.g., AA means both jobs will be definitely completed in 1 year.

The pertinent Venn diagram is shown in Fig. E2.13 where all the sample points are contained in S. If E_1 is the event that job 1 will definitely be completed in 1 year, then

$$E_1 \supset (AA, AB, AC)$$

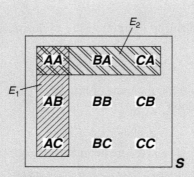

Figure E2.13 Sample space with E_1 and E_2.

Similarly, if **E_2** is the event that job 2 will definitely be completed in 1 year,

$$E_2 \supset (AA, BA, CA)$$

2. Assuming that each state of completion of both jobs is equally likely at the end of 1 year (i.e., each sample point has a probability of 1/9), what is the probability that at least one job will definitely be completed at the end of 1 year?

In this case, the event of interest is the union $E_1 \cup E_2$. Observe first that the intersection $E_1 E_2 \supset (AA)$; therefore, according to Eq. 2.8,

$$P(E_1 \cup E_2) = \frac{3}{9} + \frac{3}{9} - \frac{1}{9} = \frac{5}{9}$$

We can also observe from Fig. E2.13 that $(E_1 \cup E_2) \supset (AA, AB, AC, BA, CA)$; its probability is also 5/9, thus verifying the above result.

3. Only one of the two projects will definitely be completed at the end of one year **E**; this event contains the following sample points,

$$E \supset (AB, AC, BA, CA)$$

Its probability, therefore, is 4/9. ◀

▶ **EXAMPLE 2.14**   For the purpose of designing the left turn (L.T.) lane in Example 2.2, the 60 observations of the number of vehicles waiting for left turns at the intersection yielded the results shown in Table E2.14.
    Define

$$E_1 = \text{more than 2 vehicles waiting for left turns}$$
$$E_2 = \text{at most 4 vehicles waiting for left turns}$$

The different number of vehicles waiting for left turns at the intersection are obviously mutually exclusive events. Then, using the relative frequencies in Table E2.14 to represent the corresponding probabilities, we obtain, approximately,

$$P(E_1) = \frac{14}{60} + \frac{3}{60} + \frac{2}{60} + \frac{1}{60} = \frac{20}{60}$$

whereas

$$P(E_2) = \frac{4}{60} + \frac{16}{60} + \frac{20}{60} + \frac{14}{60} + \frac{3}{60} = \frac{57}{60}$$

TABLE E2.14 Observations of Vehicles Waiting for Left Turns

No. of Vehicles Waiting	No. of Observations	Relative Frequency
0	4	4/60
1	16	16/60
2	20	20/60
3	14	14/60
4	3	3/60
5	2	2/60
6	1	1/60
7	0	0
8	0	0

Also, the intersection

$$E_1 E_2 \supset (3, 4)$$

with corresponding probability

$$P(E_1 E_2) = \frac{14}{60} + \frac{3}{60} = \frac{17}{60}$$

Then, according to Eq. 2.8,

$$P(E_1 \cup E_2) = \frac{20}{60} + \frac{57}{60} - \frac{17}{60} = \frac{60}{60} = 1.0$$

In this case, we observe that the union $E_1 \cup E_2$ contains all the possible number of vehicles waiting for left turns, i.e., the entire sample space. Hence, its probability is 1.0. ◄

► **EXAMPLE 2.15**

In Example 2.8, two events associated with the reactions at A and B are defined as

$$A = (R_A > 100 \, \text{kg})$$
$$B = (R_B > 100 \, \text{kg})$$

These are represented by the respective subsets in the sample space of Fig. E2.15. For this illustration, we may assume that the sample points are all equally likely to occur. This implies that the probability of an event within this sample space is proportional to its "area" relative to the sample space.

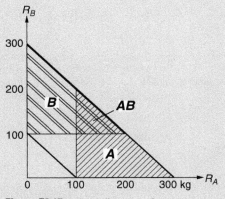

Figure E2.15   Venn diagram of events.

Total area of sample space $= \frac{1}{2}[(300)^2 - (100)^2] = 40,000$

Then, referring to Fig. E2.15, we have

$$P(A) = \frac{\frac{1}{2}(200)^2}{40,000} = \frac{1}{2}$$

Similarly, $P(B) = \frac{1}{2}$

whereas

$$P(AB) = \frac{\frac{1}{2}(100)^2}{40,000} = \frac{1}{8}$$

Therefore, according to Eq. 2.8, we obtain,

$$P(A \cup B) = \frac{1}{2} + \frac{1}{2} - \frac{1}{8} = \frac{7}{8}$$

From Fig. E2.15, we also obtain the area of the union

$$(A \cup B) = 40,000 - \frac{1}{2}(100)^2 = 35,000$$

from which we also obtain $P(A \cup B) = \dfrac{35,000}{40,000} = \dfrac{7}{8}$ ◀

Extending the addition rule, Eq. 2.8, to three events E_1, E_2, E_3 we would have,

$$\begin{aligned}
P(E_1 \cup E_2 \cup E_3) &= P[(E_1 \cup E_2) \cup E_3] \\
&= P(E_1 \cup E_2) + P(E_3) - P[(E_1 \cup E_2)E_3] \\
&= P(E_1) + P(E_2) + P(E_3) - P(E_1 E_2) \\
&\quad - P(E_1 E_3) - P(E_2 E_3) + P(E_1 E_2 E_3)
\end{aligned} \tag{2.9}$$

The above procedure of the addition rule may be extended to the union of any number of events; however, for n events the probability of the union may be obtained more conveniently by applying the de Morgan's rule, as follows:

$$\begin{aligned}
P(E_1 \cup E_2 \cup \cdots \cup E_n) &= 1 - P(\overline{E_1 \cup E_2 \cup \cdots \cup E_n}) \\
&= 1 - P(\overline{E}_1 \overline{E}_2 \cdots \overline{E}_n)
\end{aligned} \tag{2.10}$$

If the n events are all mutually exclusive, however, extension of *Axiom 3*, Eq. 2.6, yields

$$P(E_1 \cup E_2 \cup \cdots \cup E_n) = \sum_{i=1}^{n} P(E_i) \tag{2.6a}$$

▶ **EXAMPLE 2.16**

The major airline industry is subject to labor strikes by the *pilots, mechanics*, and the *flight attendants* or by two or more of these labor groups. Using the following notations,

$$A = \text{strike by the pilots}$$
$$B = \text{strike by the mechanics}$$
$$C = \text{strike by the flight attendants}$$

determine the probability of a labor strike in the major airline industry in the next 3 years. Assume the following respective probabilities of strikes by the three individual groups: $P(A) = 0.03$; $P(B) = 0.05$;

$P(C) = 0.05$, and that strikes by the different labor groups are *statistically independent* (see Sects. 2.3.2 and 2.3.3), which means according to Eq. 2.15,

$$P(AB) = P(A)P(B); \qquad P(AC) = P(A)P(C); \qquad P(BC) = P(B)P(C);$$
$$\text{and } P(ABC) = P(A)P(B)P(C)$$

SOLUTION A strike in the industry will occur in the next 3 years if any one or more of the three labor groups go on strike during this period; therefore, we are interested in the union of A, B, and C whose probability according to Eq. 2.9 is

$$P(A \cup B \cup C) = 0.05 + 0.03 + 0.05 - (0.05 \times 0.03) - (0.05 \times 0.05)$$
$$- (0.03 \times 0.05) + (0.05 \times 0.03 \times 0.05)$$
$$= 0.1246$$

The solution may also be obtained (more conveniently) with Eq. 2.10 as follows:

$$P(A \cup B \cup C) = 1 - P(\overline{A}\,\overline{B}\,\overline{C})$$

Note that $\overline{A}$, $\overline{B}$, and $\overline{C}$ are also statistically independent; i.e., $P(\overline{A}\,\overline{B}\,\overline{C}) = P(\overline{A})P(\overline{B})P(\overline{C})$. Thus, we also obtain

$$P(A \cup B \cup C) = 1 - (0.95 \times 0.97 \times 0.95)$$
$$= 0.1246 \qquad \blacktriangleleft$$

2.3.2 Conditional Probability

There are occasions when the probability of an event may depend on the occurrence (or nonoccurrence) of another event. If this dependence exists, the relevant probability is a *conditional probability*. For this purpose, we shall use the notation

$$P(E_1|E_2) = \text{the probability of } \boldsymbol{E_1} \text{ assuming the occurence of } \boldsymbol{E_2}, \text{ or simply}$$
$$\text{the probability of } \boldsymbol{E_1} \text{ given } \boldsymbol{E_2}.$$

In the Venn diagram of Fig. 2.8, we may observe intuitively the following:

The conditional probability $P(E_1|E_2)$ may be interpreted as the likelihood of realizing a sample point of $\boldsymbol{E_1}$ that is in $\boldsymbol{E_2}$. In other words, we are effectively interested in the event $\boldsymbol{E_1}$ within the "reconstituted sample space" $\boldsymbol{E_2}$. Therefore, the conditional probability pertains to the sample points of $\boldsymbol{E_1}$ relative to those of $\boldsymbol{E_2}$ and thus must be normalized with respect to $\boldsymbol{E_2}$; hence, with the appropriate normalization, we obtain the conditional probability

$$P(E_1|E_2) = \frac{P(E_1 E_2)}{P(E_2)} \tag{2.11}$$

It may be well to emphasize that the conditional probability, as defined in Eq. 2.11, is merely a generalization of the (unconditional) probability of an event. When we speak of

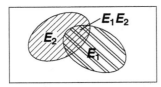

Figure 2.8 Venn diagram for E_1 given E_2.

the probability of an event E, it is implicitly conditioned on the sample space S. To be more explicit, $P(E)$ should be expressed as

$$P(E|S) = \frac{P(ES)}{P(S)}$$

But since $ES = E$ and $P(S) = 1.0$,

$$P(E|S) = P(E)$$

In other words, conditioning on the sample space S is presumed to be understood; however, when the occurrence of an event depends on the occurrence or nonoccurrence of another event, the reconstituted sample space must be explicitly identified.

We also observe the following regarding the conditional probability of a complementary event:

$$P(E_1|E_2) + P(\overline{E_1}|E_2) = \frac{P(E_1 E_2)}{P(E_2)} + \frac{P(\overline{E_1} E_2)}{P(E_2)}$$

$$= \frac{1}{P(E_2)}\{P[(E_1 \cup \overline{E_1})E_2]\}$$

$$= \frac{P(E_2)}{P(E_2)} = 1.0$$

Therefore,

$$P(\overline{E_1}|E_2) = 1 - P(E_1|E_2) \qquad (2.12)$$

which is a generalization of Eq. 2.7. It is important to recognize that in Eq. 2.12 the conditioning event E_2 is the same reconstituted sample space on both sides of the equation. For this reason, one must make sure, when applying Eq. 2.12, that the event (e.g., E_1) and its complement refer to the same conditioning event E_2.

Observe, for example, the following:

$$P(E_1|\overline{E_2}) \neq 1 - P(\overline{E_1}|E_2)$$

$$P(\overline{E_1}|E_2) \neq 1 - P(E_1|\overline{E_2})$$

$$P(\overline{E_1}|\overline{E_2}) \neq 1 - P(E_1|E_2)$$

▶ **EXAMPLE 2.17**

There are two highways from City A to City B as shown in Fig. E2.17.
Route 1 is on flat terrain, whereas *Route 2* is a scenic route that goes through mountainous terrain. During severe winter seasons, one or both routes may be closed to traffic because of heavy snowfalls.

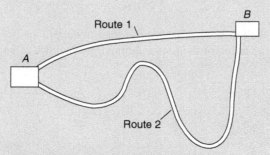

Figure E2.17   Highway routes from A to B.

Denote the following:

$$E_1 = \textit{Route 1} \text{ is open to traffic}$$
$$E_2 = \textit{Route 2} \text{ is open to traffic}$$

Between the two routes, *Route 2* is obviously more likely to be closed than *Route 1* during the winter months. Moreover, the condition of *Route 1* during a severe snow storm may depend on whether or not *Route 2* is open to traffic. Suppose, during a severe snow storm, the probabilities that the routes will be open are, respectively,

$$P(E_1) = 0.75,$$

whereas

$$P(E_2) = 0.50$$

and the probability that both routes will be open is

$$P(E_1 E_2) = 0.40$$

Then, if *Route 2* is open during a snow storm, the probability that *Route 1* is also open is, according to Eq. 2.11,

$$P(E_1 | E_2) = \frac{P(E_1 E_2)}{P(E_2)} = \frac{0.40}{0.50} = 0.80$$

On the other hand, if *Route 2* is closed during a severe snow storm, the probability that *Route 1* is also closed may be determined as follows:

$$P(\overline{E}_1 | \overline{E}_2) = \frac{P(\overline{E}_1 \overline{E}_2)}{P(\overline{E}_2)}$$

in which

$$P(\overline{E}_1 \overline{E}_2) = 1 - P(\overline{\overline{E}_1 \overline{E}_2}) = 1 - P(E_1 \cup E_2)$$
$$= 1 - [P(E_1) + P(E_2) - P(E_1 E_2)]$$
$$= 1 - (0.75 + 0.50 - 0.40)$$
$$= 1 - 0.85 = 0.15$$

Therefore, we obtain

$$P(\overline{E}_1 | \overline{E}_2) = \frac{0.15}{0.50} = 0.30$$

and the probability that *Route 1* is open given that *Route 2* is closed is

$$P(E_1 | \overline{E}_2) = 1 - 0.30 = 0.70$$

◀

▶ **EXAMPLE 2.18**

Suppose that motor vehicles approaching a certain intersection are twice as likely to go *straight ahead* than to make a *right turn*; and *left turns* are only half as likely as *right turns*.

As a vehicle approaches the intersection, the possible directions may be defined as follows:

$$E_1 = \text{straight ahead}$$
$$E_2 = \text{turning right}$$
$$E_3 = \text{turning left}$$

and the respective probabilities will be

$$P(E_1) = \frac{4}{7}; \qquad P(E_2) = \frac{2}{7}; \qquad \text{and} \qquad P(E_3) = \frac{1}{7}$$

At the intersection, if a vehicle is definitely making a turn, the probability that it will be a right turn is (observe that the three alternative directions are mutually exclusive)

$$P(E_2|E_2 \cup E_3) = \frac{P[E_2(E_2 \cup E_3)]}{P(E_2 \cup E_3)} = \frac{P(E_2 \cup E_2 E_3)}{P(E_2 \cup E_3)}$$

$$= \frac{P(E_2)}{P(E_2) + P(E_3)} = \frac{2/7}{3/7} = \frac{2}{3}$$

On the other hand, if the vehicle is definitely making a turn at the intersection, the probability that it will not turn right, according to Eq. 2.12, is

$$P(\overline{E_2}|E_2 \cup E_3) = 1 - P(E_2|E_2 \cup E_3)$$

$$= 1 - \frac{2}{3} = \frac{1}{3}$$

◀

Statistical Independence

If the occurrence, or nonoccurrence, of an event does not affect the probability of occurrence of another event, the two events are *statistically independent*. In other words, the probability of occurrence of one event does not depend on the occurrence or nonoccurrence of another event. Therefore, if two events E_1 and E_2 are statistically independent,

$$P(E_1|E_2) = P(E_1)$$

and

$$P(E_2|E_1) = P(E_2) \tag{2.13}$$

It might be prudent to point out here the difference between events that are *statistically independent* versus those that are *mutually exclusive*. The difference is profound, and there should be no confusion—statistical independence between two events refers to the probability of their joint occurrence, whereas two events are mutually exclusive when their joint occurrence is impossible, as the occurrence of one event precludes the occurrence of the other. In other words, $P(E_2|E_1) = 0$ if E_1 and E_2 are mutually exclusive. Finally, statistical independence of two or more events pertains to the probability of the joint event, whereas mutual exclusiveness refers to the definition of the events.

2.3.3 The Multiplication Rule

From Eq. 2.11, it follows that the probability of the joint event $E_1 E_2$ is

$$P(E_1 E_2) = P(E_1|E_2)P(E_2)$$

or

$$P(E_1 E_2) = P(E_2|E_1)P(E_1) \tag{2.14}$$

Therefore, with Eq. 2.13, if E_1 and E_2 are statistically independent events, the above *multiplication rule* becomes[*]

$$P(E_1 E_2) = P(E_1)P(E_2) \tag{2.15}$$

---

[*]Mathematically, statistical independence is generally defined in the form of Eq. 2.15 or 2.15a.

For three events, the multiplication rule would give,

$$P(E_1 E_2 E_3) = P(E_1 | E_2 E_3) P(E_2 E_3)$$
$$= P(E_1 | E_2 E_3) P(E_2 | E_3) P(E_3) \tag{2.14a}$$

whereas if the three events are statistically independent,

$$P(E_1 E_2 E_3) = P(E_1) P(E_2) P(E_3) \tag{2.15a}$$

We would expect that if two events E_1 and E_2 are statistically independent, their complements $\overline{E_1}$ and $\overline{E_2}$ would also be statistically independent; i.e.,

$$P(\overline{E}_1 \overline{E}_2) = P(\overline{E}_1) P(\overline{E}_2) \tag{2.16}$$

In fact, we can verify this assertion in the case of two events as follows:

$$P(\overline{E}_1 \overline{E}_2) = P(\overline{E_1 \cup E_2}) = 1 - P(E_1 \cup E_2)$$
$$= 1 - [P(E_1) + P(E_2) - P(E_1)P(E_2)]$$
$$= [1 - P(E_1)][1 - P(E_2)]$$
$$= P(\overline{E}_1) P(\overline{E}_2)$$

We might emphasize that all the mathematical rules pertaining to the probabilities of events apply equally to the conditional probabilities of events that are conditioned on the same reconstituted sample space, including the *addition rule* and the *multiplication rule*. In particular, we may observe the following:

$$P(E_1 \cup E_2 | A) = P(E_1 | A) + P(E_2 | A) - P(E_1 E_2 | A) \tag{2.17}$$

$$P(E_1 E_2 | A) = P(E_1 | E_2 A) P(E_2 | A) \tag{2.18}$$

and if E_1 and E_2 are statistically independent events, given event A,

$$P(E_1 E_2 | A) = P(E_1 | A) P(E_2 | A)$$

▶ **EXAMPLE 2.19**

Consider again the chain system of Example 2.11 consisting of two links as shown in Fig. E2.19 subjected to a force $F = 300$ kg.

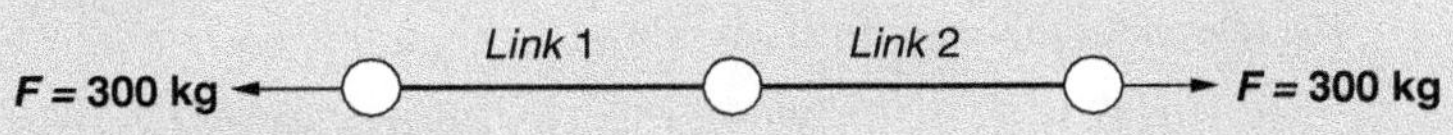

$F = 300$ kg ◀——○——— Link 1 ———○——— Link 2 ———○——▶ $F = 300$ kg

Figure E2.19 A two-link chain.

If the fracture strength of a link is less than 300 kg, it will fail by fracture. Suppose that the probability of this happening to either of the two links is 0.05. Clearly, the chain will fail if one or both of the two links should fail by fracture. To determine the probability of failure of the chain, define

$$E_1 = \text{fracture of } link\ 1$$
$$E_2 = \text{fracture of } link\ 2$$

Then, $P(E_1) = P(E_2) = 0.05$ and the probability of failure of the chain is

$$P(E_1 \cup E_2) = P(E_1) + P(E_2) - P(E_1 E_2)$$
$$= 0.05 + 0.05 - P(E_2 | E_1) P(E_1)$$
$$= 0.10 - 0.05 P(E_2 | E_1)$$

We observe that the solution requires the conditional probability $P(E_2|E_1)$, which is a function of the mutual dependence between E_1 and E_2. If there is no dependence or they are statistically independent, $P(E_2|E_1) = P(E_2) = 0.05$. In this case, the probability of failure of the chain is

$$P(E_1 \cup E_2) = 0.10 - 0.05 \times 0.05 = 0.0975$$

On the other hand, if there is complete or total dependence between E_1 and E_2, which means that if one link fractures the other will also fracture, then $P(E_2|E_1) = 1.0$. In such a case, the probability of failure of the chain becomes

$$P(E_1 \cup E_2) = 0.10 - 0.05 \times 1.0 = 0.05$$

In this latter case, we see that the failure probability of the chain system is the same as the failure probability of a single link.

Therefore, we can state that the probability of failure of the chain system ranges between 0.05 and 0.0975. ◀

▶ **EXAMPLE 2.20**

The foundation of a tall building may fail because of inadequate bearing capacity or by excessive settlement. Let B and S represent the respective modes of foundation failure, and assume $P(B) = 0.001$, $P(S) = 0.008$, and $P(B|S) = 0.10$, which is the conditional probability of bearing capacity failure given that there is excessive settlement. Then, the probability of failure of the foundation is

$$\begin{aligned} P(B \cup S) &= P(B) + P(S) - P(BS) \\ &= P(B) + P(S) - P(B|S)P(S) \\ &= 0.001 + 0.008 - (0.1)(0.008) \\ &= 0.0082 \end{aligned}$$

However, the probability that the building will experience excessive settlement but without bearing capacity failure is

$$\begin{aligned} P(S \cap \overline{B}) &= P(\overline{B}|S)P(S) \\ &= [1 - P(B|S)]P(S) \\ &= (1.0 - 0.1)(0.008) = 0.0072 \end{aligned}$$

In this problem, we might observe that the conditional probability $P(B|S)$ cannot be larger than 1/8. The reason for this assertion is as follows:

$$P(B|S)P(S) = P(S|B)P(B)$$

$$P(B|S) = \frac{0.001}{0.008}P(S|B) = \frac{1}{8}P(S|B)$$

Because $P(S|B) \leq 1.0$, the value of $P(B|S)$ is limited to a maximum of 1/8. ◀

▶ **EXAMPLE 2.21**

Two rivers, a and b, flow through the neighborhood of a paper mill that is allowed to dispose its waste into both rivers. The dissolved oxygen, DO, level in the water downstream is an indication of the degree of pollution of the rivers caused by the disposed waste from the mill. Let

A = the event that *River a* is found to have unacceptable level of pollution

B = the event that *River b* is found to have unacceptable level of pollution

Based on the records of the respective DO levels in the two rivers tested over the past year, it was determined that on a given day, the likelihood of unacceptable water pollution in each of the rivers is as follows:

$$P(A) = 20\% \quad \text{and} \quad P(B) = 33\%$$

whereas the probability that both rivers will have unacceptable levels of pollution on the same day is 0.10; i.e.,

$$P(AB) = 10\%$$

On a given day, the probability that at least one of the rivers will have an unacceptable level of pollution is

$$P(A \cup B) = 0.20 + 0.33 - 0.10 = 0.43$$

If *River a* is tested to have an unacceptable pollution level, the probability that *River b* will also have unacceptable pollution level will be

$$P(B|A) = \frac{P(AB)}{P(A)} = \frac{0.10}{0.20} = 0.50$$

whereas the probability that *River a* will have an unacceptable pollution level assuming that *River b* has been tested to have an unacceptable level is

$$P(A|B) = \frac{P(AB)}{P(B)} = \frac{0.10}{0.33} = 0.30$$

A related question: On any given day, what is the probability that only one of the two rivers will have an unacceptable pollution level?

SOLUTION This means that either one of the two rivers is polluted and the other is not; thus, the probability is

$$P(A\overline{B} \cup \overline{A}B) = P(\overline{B}|A)P(A) + P(\overline{A}|B)P(B)$$
$$= 0.50 \times 0.20 + 0.70 \times 0.33 = 0.33 \quad ◀$$

▶ **EXAMPLE 2.22**

The electrical power for a city is supplied by two generating plants—*Plant a* and *Plant b*. Each of the plants has sufficient capacity to supply the average daily power requirement of the entire city. However, during peak hours of a day, the capacities of both plants are needed; otherwise, there will be *brownouts* in parts of the city. Denote the following events:

$$A = \text{failure of } Plant\ a$$
$$B = \text{failure of } Plant\ b$$

and assume

$$P(A) = 0.05$$
$$P(B) = 0.07$$
$$P(AB) = 0.01$$

If one of the two units should fail on a given day, what is the probability of failure of the other unit on the same day?

SOLUTION The probability of failure of the second unit will depend on which unit fails first. For example, if *Plant a* fails first, we have

$$P(B|A) = \frac{P(AB)}{P(A)} = \frac{0.01}{0.05} = 0.20$$

On the other hand, if *Plant b* should fail first, then

$$P(A|B) = \frac{P(AB)}{P(B)} = \frac{0.01}{0.07} = 0.14$$

Related questions and answers: What is the probability of a *brownout* in the city on a given day? *Brownouts* will occur when one or both of the plants fail; therefore, the pertinent probability is

$$P(A \cup B) = P(A) + P(B) - P(AB) = 0.05 + 0.07 - 0.01$$
$$= 0.11$$

If there is a brownout within the city during the peak hours of a day, the probability that it is caused solely by the failure of *Plant a* is

$$P(A\overline{B}|A \cup B) = \frac{P[A\overline{B}(A \cup B)]}{P(A \cup B)} = \frac{P(A\overline{B}A \cup AB\overline{B})}{P(A \cup B)} = \frac{P(A\overline{B})}{P(A \cup B)}$$
$$= \frac{P(\overline{B}|A)P(A)}{P(A \cup B)} = \frac{(1 - 0.20)0.05}{0.11} = 0.36$$

Similarly, if there is a brownout, the probability that it is caused solely by the failure of *Plant b* is

$$P(B\overline{A}|A \cup B) = \frac{P(\overline{A}B)}{P(A \cup B)} = \frac{P(\overline{A}|B)P(B)}{P(A \cup B)}$$
$$= \frac{(1 - 0.14)0.07}{0.11} = 0.54$$

Finally, the probability that a brownout will be caused by the failures of both plants would be

$$P(AB|A \cup B) = \frac{P[AB(A \cup B)]}{P(A \cup B)} = \frac{P(ABA \cup ABB)}{P(A \cup B)} = \frac{P(AB)}{P(A \cup B)}$$
$$= \frac{0.01}{0.11} = 0.09$$

◀

▶ **EXAMPLE 2.23**

Suppose that before a section (say 1/10 km in length) of a newly constructed highway pavement is accepted by the State Department of Highways, the thickness of a 25-cm pavement is inspected for specification compliance by taking ultrasonic readings at every 1/10-km point of the constructed pavement, as shown in Fig. E2.23. Each 1/10-km section will be accepted if the measured thickness is at least 23 cm; otherwise, the entire section will be rejected or a penalty will be imposed.

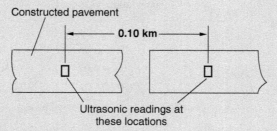

Figure E2.23 Inspection of constructed pavement.

From past experience, 90% of constructed highway works by the same contractor were found to be in compliance with specifications. The ultrasonic thickness determination is only 80% reliable; i.e., there is a 20% chance that a conclusion based on the ultrasonics test may be erroneous.

Use the following notations:

$$G = \text{constructed thickness of pavement is at least 23 cm}$$

$$A = \text{ultrasonic measured thickness is} \geq 23 \text{ cm}$$

The statement that "the ultrasonic test is 80% reliable" means

$$P(G|A) = 0.80$$

and also,

$$P(\overline{G}|\overline{A}) = 0.80$$

This may also be interpreted to mean that an ultrasonic reading could have an error of 20%, which means

$$P(\overline{G}|A) = P(G|\overline{A}) = 0.20$$

Based on the contractor's past construction record, we may assume that 90% of his constructed pavement will have acceptable ultrasonics readings; hence, $P(A) = 0.90$.

The probability that a particular 1/10-km section of the pavement is well constructed *and* will be accepted by the Highway Department is, therefore,

$$P(GA) = P(G|A)P(A)$$
$$= (0.80)(0.90) = 0.72$$

The more pertinent and practical questions are perhaps the following:

1. What is the probability that a well constructed section of the pavement will be accepted by the Highway Department on the basis of the ultrasonics test?

2. Conversely, what is the probability that a poorly constructed section will be rejected?

SOLUTIONS The probability that a well-constructed section of the pavement will be accepted is

$$P(A|G) = \frac{P(G|A)P(A)}{P(G)}$$

Since $G = G(A \cup \overline{A}) = GA \cup G\overline{A}$, where GA and $G\overline{A}$ are mutually exclusive,

$$P(G) = P(GA) + P(G\overline{A}) = P(G|A)P(A) + P(G|\overline{A})P(\overline{A})$$
$$= 0.80 \times 0.90 + 0.20 \times 0.10 = 0.74$$

Thus,

$$P(A|G) = \frac{0.80 \times 0.90}{0.74} = 0.97$$

Conversely, the probability that a poorly constructed section will be rejected is,

$$P(\overline{A}|\overline{G}) = \frac{P(\overline{G}|\overline{A})P(\overline{A})}{P(\overline{G})}$$
$$= \frac{0.80 \times 0.10}{1 - 0.74} = 0.31$$

◀

2.3.4 The Theorem of Total Probability

On occasion, the probability of an event, say A, cannot be determined directly; its occurrence will depend on the occurrence or nonoccurrence of other events such as

$E_i, i = 1, 2, \ldots n$, and the probability of A will depend on which of the E_i's has occurred. On such an occasion, the probability of A would be composed of the conditional probabilities (conditioned on each of the E_i's) and weighted by the respective probabilities of the E_i's. Such problems require the *theorem of total probability*.

Before formally presenting the mathematical theorem, let us examine the following example to illustrate the essential elements of the theorem.

▶ **EXAMPLE 2.24**

The flooding of a river in the spring season will depend on the accumulation of snow in the mountains during the past winter season. The accumulation of snow may be described as *heavy, normal*, and *light*. Clearly, if the snow accumulation in the mountains is *heavy*, the probability of flooding in the following spring will be high, whereas, if the snow accumulation is *light*, this probability will be low. Flooding, of course, may also be caused by rainfalls in the spring. With the following notations,

$$F = \text{occurrence of flooding in the river}$$
$$H = \text{heavy accumulation of snow}$$
$$N = \text{normal accumulation of snow}$$
$$L = \text{light (including no) accumulation of snow}$$

Assume also

$$P(F|H) = 0.90; \qquad P(F|N) = 0.40; \qquad P(F|L) = 0.10$$

whereas in a given winter season,

$$P(H) = 0.20; \qquad P(N) = 0.50; \qquad P(L) = 0.30$$

Then the probability of flooding in the river during the following spring season will be

$$P(F) = P(F|H)P(H) + P(F|N)P(N) + P(F|L)P(L)$$
$$= 0.90 \times 0.20 + 0.40 \times 0.50 + 0.10 \times 0.30$$
$$= 0.41$$

◀

In Example 2.24, we can observe the following:

1. The accumulations of snow in the past winter, namely, *heavy, normal*, and *light*, are mutually exclusive.

2. The probabilities of the three levels of snow accumulation, namely, H, N, and L add up to 1.0.

Therefore, the three events H, N, and L are mutually exclusive and also collectively exhaustive.

Formally, consider n events that are mutually exclusive and collectively exhaustive, namely $E_1, E_2, \ldots, E_n$. That is, $E_1 \cup E_2 \cup \cdots \cup E_n = S$. Then, if A is an event in the same sample space S as shown in Fig. 2.9, we derive the *theorem of total probability* as follows:

$$A = AS$$
$$= A(E_1 \cup E_2 \cup \cdots \cup E_n)$$
$$= AE_1 \cup AE_2 \cup \cdots \cup AE_n$$

where $AE_1, AE_2, \ldots, AE_n$ are also mutually exclusive as can be seen in the Venn diagram of Fig. 2.9.

Then,

$$P(A) = P(AE_1) + P(AE_2) + \cdots + P(AE_n)$$

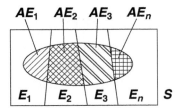

Figure 2.9 Intersection of A and $E_1, E_2, \ldots, E_n$ in sample space S.

and by virtue of the multiplication rule, Eq. 2.14, we obtain the *theorem of total probability* as

$$P(A) = P(A|E_1)P(E_1) + P(A|E_2)P(E_2) + \cdots + P(A|E_n)P(E_n) \qquad (2.19)$$

Again, in applying the above total probability theorem, it is important to observe that the conditioning events $E_1, E_2, \ldots E_n$ must be mutually exclusive and collectively exhaustive.

▶ **EXAMPLE 2.25**

Hurricanes along the Gulf of Mexico and the eastern seaboard of the United States occur every year, mostly in the summer and fall. These hurricanes are classified into five categories from C1 through C5; to be classified as a hurricane, the wind speed must be at least 75 miles per hour (120 km/hr). The frequencies of hurricanes, of course, would decrease with the categories; for example, a Category C5 hurricane, with sustained wind ≥ 150 mph (242 km/hr), would very seldom occur.

We might observe first that the five categories of hurricanes, C1, C2, C3, C4, C5, are mutually exclusive, as it is reasonable to assume that no two categories can occur at the same time and cover all possible hurricanes; thus, these five categories plus nonhurricane winds (denoted C0) are also collectively exhaustive.

Assume that annually there can be at most one hurricane striking a particular area in the southern coast of Louisiana along the Gulf of Mexico, and the annual occurrence probabilities of the different hurricane categories are as follows:

$$P(C1) = 0.35; \qquad P(C2) = 0.25; \qquad P(C3) = 0.14; \qquad P(C4) = 0.05; \qquad P(C5) = 0.01$$

Structural damage to an engineered building in the reference area can be expected to occur depending on the category of hurricane that the building will be subjected to. Suppose the conditional probabilities of damage to the building are as follows:

$$P(D|C1) = 0.05; \qquad P(D|C2) = 0.10; \qquad P(D|C3) = 0.25; \qquad P(D|C4) = 0.60;$$
$$P(D|C5) = 1.00; \qquad \text{and} \qquad P(D|C0) = 0$$

hence the annual probability of wind damage to the building would be

$$P(D) = P(D|C1)P(C1) + P(D|C2)P(C2) + P(D|C3)P(C3) + P(D|C4)P(C4) + P(D|C5)P(C5)$$
$$= 0.05 \times 0.35 + 0.10 \times 0.25 + 0.25 \times 0.14 + 0.60 \times 0.05 + 1.00 \times 0.01 = 0.1175$$

Therefore, annually the probability of hurricane wind damage to the building is about 12%. We might observe from the above calculations that the greatest contributions to the annual damage probability are from hurricanes of Categories 3 and 4, $P(D|C3)P(C3) = 0.035$ and $P(D|C4)P(C4) = 0.030$. Observe also that even though damage to the building will certainly occur under a Category 5 hurricane, $P(D|C5) = 1.00$, the occurrence of such hurricanes is very rare, annually $P(C5) = 0.01$, which means that it might occur (on the average) only about once in every 100 years. ◀

▶ **EXAMPLE 2.26**

Figure E2.26 shows the eastbound directions of two interstate highways I_1 and I_2 merging into another highway I_3. Interstates I_1 and I_2 have the same traffic capacities; however, the rush-hour traffic volume on I_2 is about twice that on I_1 so that during rush hours, the probabilities of traffic congestion, denoted, respectively, as E_1 and E_2, are as follows:

$$P(E_1) = 0.10; \qquad P(E_2) = 0.20$$

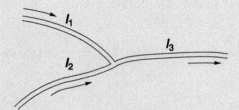

Figure E2.26 Eastbound interstate highways.

Also, when one route has excessive traffic, the chance of excessive traffic on the other route can be expected to increase; assume that these conditional probabilities are:

$P(E_1|E_2) = 0.40$, whereas, from *Bayes' theorem* (see Eq. 2.20), we must have

$P(E_2|E_1) = 0.80$

We might wish also to determine the probability of traffic congestion on the third route I_3, $P(E_3)$.

1. First, let us assume that the capacity of I_3 is the same as that of I_1 or I_2, and that when I_1 and I_2 are both carrying less than their respective traffic capacities, there is a 20% probability that I_3 will experience excessive traffic; i.e., $P(E_3|\overline{E}_1\overline{E}_2) = 0.20$.

We would expect that the relevant probability will depend on the traffic conditions on I_1 and I_2, which may be E_1E_2, $E_1\overline{E}_2$, $\overline{E}_1E_2$, or $\overline{E}_1\overline{E}_2$; observe that these four joint events are mutually exclusive and collectively exhaustive. Their respective probabilities are then

$$P(E_1E_2) = P(E_1|E_2)P(E_2) = 0.40 \times 0.20 = 0.080$$
$$P(E_1\overline{E}_2) = P(\overline{E}_2|E_1)P(E_1) = (1-0.80)0.10 = 0.020$$
$$P(\overline{E}_1E_2) = P(\overline{E}_1|E_2)P(E_2) = (1-0.40)0.20 = 0.120$$
$$P(\overline{E}_1\overline{E}_2) = 1 - 0.08 - 0.02 - 0.12 = 0.780$$

Clearly, the traffic on I_3 will be congested when the traffic on I_1 or I_2 or both are excessive. Then we obtain the total probability of traffic congestion on I_3 to be

$$P(E_3) = P(E_3|E_1 E_2)P(E_1E_2) + P(E_3|E_1 \overline{E}_2)P(E_1\overline{E}_2) + P(E_3|\overline{E}_1 E_2)P(\overline{E}_1E_2)$$
$$+ P(E_3|\overline{E}_1 \overline{E}_2)P(\overline{E}_1\overline{E}_2)$$
$$= 1.00 \times 0.08 + 1.00 \times 0.02 + 1.00 \times 0.12 + 0.20 \times 0.78$$
$$= 0.08 + 0.02 + 0.12 + 0.156$$
$$= 0.376$$

2. Next, assume that the traffic capacity of I_3 is twice that of I_1 or I_2. In this case, if only one of the two routes I_1 or I_2 has excessive traffic, the probability of congestion on I_3 will be 25%, i.e., $P(E_3|E_1\overline{E}_2) = P(E_3|\overline{E}_1E_2) = 0.25$. If both routes have excessive traffic, the probability that the capacity of I_3 will be exceeded is 95%, meaning $P(E_3|E_1E_2) = 0.95$.

In this latter case, the probability of traffic congestion on I_3 is then

$$P(E_3) = 0.95 \times 0.08 + 0.25 \times 0.02 + 0.25 \times 0.12 + 0 \times 0.78$$
$$= 0.111$$

◀

In the next example, we will consider a problem involving double conditional events in association with the application of the total probability theorem.

▶ **EXAMPLE 2.27**

The town of Urbana is in the county of Champaign, Illinois, which is located in the plains area of the midwestern United States. High-wind storms can occur in this area with maximum wind speeds exceeding 60 mph (96 km/hr), and occasionally such storms can spawn tornadoes creating even higher wind speeds.

Suppose we wish to assess the annual probability of structural damage to residential houses in Urbana, and based on historical data were able to determine the following information:

High-wind storms in the county of Champaign occur about once every other year, and during such a storm the probability that it will be accompanied by tornadoes is 0.25. The probability that tornadoes occurring in the county of Champaign will strike the town of Urbana is 0.15.

Based on engineering calculations, we also determined the following:

The probability of severe damage to houses in Urbana during a wind storm (in the absence of tornadoes) is 0.05; however, when tornadoes are spawned in the county but do not strike the town of Urbana, the probability of severe damage to houses is 0.10, whereas if the tornadoes strike the town, the probability of severe damage to one or more houses in Urbana will be a certainty.

With the above information, we define the following:

$$F = \text{severe damage to houses in Urbana}$$
$$S = \text{occurrence of storm in Champaign County}$$
$$T = \text{occurrence of tornadoes in Champaign County during a storm}$$
$$H = \text{a tornado striking the town of Urbana}$$

and also,

$$P(S) = 0.50; \qquad P(T|S) = 0.25; \qquad P(H|ST) = 0.15$$

The Venn diagram would appear as in Fig. E2.27.

In this case, the following events are mutually exclusive and collectively exhaustive:

$$STH, ST\overline{H}, S\overline{T}, \overline{S}\,\overline{T}, \text{ whereas } \overline{S}T = \phi \text{ a null set;}$$

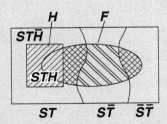

Figure E2.27 Venn diagram of S, T, H, and F.

with corresponding probabilities:

$$P(STH) = P(H|ST)P(ST) = P(H|ST)P(T|S)P(S)$$
$$= 0.15 \times 0.25 \times 0.50 = 0.01875$$
$$P(ST\overline{H}) = P(\overline{H}|ST)P(ST) = P(\overline{H}|ST)P(T|S)P(S)$$
$$= 0.85 \times 0.25 \times 0.50 = 0.10625$$
$$P(S\overline{T}) = P(\overline{T}|S)P(S) = (1 - 0.25)(0.50) = 0.3750$$
$$P(\overline{S}\,\overline{T}) = 1 - 0.01875 - 0.10625 - 0.3750 = 0.5000$$

and the probabilities of damage to houses depending on the above different conditions are

$$P(F|S\overline{T}) = 0.05; \quad P(F|STH) = 1.00; \quad P(F|ST\overline{H}) = 0.10$$

Therefore, the annual probability of wind storm damage to houses in Urbana would be

$$P(F) = P(F|STH)P(STH) + P(F|ST\overline{H})P(ST\overline{H}) + P(F|S\overline{T})P(S\overline{T})$$
$$+ P(F|\overline{S}\,\overline{T})P(\overline{S}\,\overline{T})$$
$$= 1.00 \times 0.01875 + 0.10 \times 0.10625 + 0.05 \times 0.3750 + 0 \times 0.500$$
$$= 0.048$$

◀

▶ **EXAMPLE 2.28**   A tower may be subjected to earthquake loads which could be of high intensity (event H) or of long duration (event L). It is estimated that if the load has long duration, the probability that its intensity is high is 0.7. Also, if the load has high intensity, there is 20% probability that it will be of short duration. Finally, the probability of having a long duration earthquake load is 0.3.

The designer estimated that the probability of failure when the tower is subjected to a short duration–high intensity earthquake is 0.05, whereas, this probability is doubled if the earthquake is of long duration but low intensity. Also, he is certain that the tower will fail if subjected to an earthquake with both high intensity and long duration, and that it will survive with certainty if subjected to an earthquake of low intensity and short duration.

PROBLEMS OF INTEREST ARE:

(a) Are the events H and L mutually exclusive?

(b) Are the events H and L statistically independent?

(c) Are the events H and L collectively exhaustive?

Finally, calculate the probability of failure of this tower when subjected to an earthquake.

SOLUTIONS   From the problem statements, we observe the following probabilities: $P(H|L) = 0.7 \Rightarrow P(\overline{H}|L) = 0.3$; also, $P(\overline{L}|H) = 0.2$ and $P(L|H) = 0.8$. Furthermore, $P(L) = 0.3$ and $P(\overline{L}) = 0.7$. Note that $P(H)$ is not given, but it can be found from

$$P(H) = P(LH)/P(L|H)$$
$$= P(H|L)P(L)/P(L|H)$$
$$= 0.7 \times 0.3/0.8$$
$$= 0.2625$$

Also, the conditional damage probabilities are

$$P(F|H\overline{L}) = 0.05, \ P(F|\overline{H}L) = 0.1, \ P(F|HL) = 1, \ \text{and} \ P(F|\overline{H}\,\overline{L}) = 0$$

(a) Since $P(H|L) = 0.7 \neq 0$, H can happen given that L occurs. So H and L are not mutually exclusive.

(b) $P(L|H) = 1 - P(\overline{L}|H) = 1 - 0.20 = 0.80$, but the unconditional probability $P(L) = 0.3 \neq P(L|H)$; hence, L is more likely to occur given the occurrence of H, so H and L are not statistically independent.

(c)

$$
\begin{aligned}
P(H \cup L) &= P(H) + P(L) - P(HL) \\
&= P(H) + P(L) - P(L|H)P(H) \\
&= 0.2625 + 0.3 - 0.8 \times 0.2625 \\
&= 0.353
\end{aligned}
$$

which is less than 1.0. Hence, H and L are not collectively exhaustive.

(d) Let F denote the failure of the tower. The total probability of F is obtained by summing the contributions from each of the four events, namely HL, $H\overline{L}$, $\overline{H}L$, $\overline{H}\,\overline{L}$, which are collectively exhaustive. Hence, by applying the theorem of total probability,

$$
\begin{aligned}
P(F) &= P(F|HL)P(HL) + P(F|H\overline{L})P(H\overline{L}) + P(F|\overline{H}L)P(\overline{H}L) + P(F|\overline{H}\,\overline{L})P(\overline{H}\,\overline{L}) \\
&= P(F|HL)P(H|L)P(L) + P(F|H\overline{L})P(\overline{L}|H)P(H) + P(F|\overline{H}L)P(\overline{H}|L)P(L) + 0 \times P(\overline{H}\,\overline{L}) \\
&= 1 \times 0.7 \times 0.3 + 0.05 \times (1 - 0.8) \times 0.2625 + 0.1 \times 0.3 \times 0.3 \\
&= 0.222
\end{aligned}
$$

◀

2.3.5 The Bayes' Theorem

In deriving the theorem of total probability, Eq. 2.19, the probability of the event A depends on which of the conditioning events E_i, $i = 1, 2, \ldots, n$, has occurred. On the other hand, we may be interested in the probability of a particular E_i given the occurrence of A. In a sense, this is the "inverse" probability, which is given by the Bayes' theorem, which may be derived as follows:

Applying Eq. 2.14 to the joint event AE_i, we have

$$P(A|E_i)P(E_i) = P(E_i|A)P(A)$$

from which we obtain the "inverse" probability

$$P(E_i|A) = \frac{p(A|E_i)P(E_i)}{P(A)} \tag{2.20}$$

which is the Bayes' theorem. In Eq. 2.20, if $P(A)$ is expanded using the total probability theorem, Eq. 2.20 becomes

$$P(E_i|A) = \frac{P(A|E_i)P(E_i)}{\displaystyle\sum_{j=1}^{n} P(A|E_j)P(E_j)} \tag{2.20a}$$

▶ **EXAMPLE 2.29**

Aggregates for the construction of a reinforced concrete building are supplied by two companies, *Company a* and *Company b*. Orders are for *Company a* to deliver 600 truck loads a day and 400 truck loads a day from *Company b*. From prior experience, it is expected that 3% of the material from *Company a* will be substandard, whereas 1% of the material from *Company b* are substandard.

Define the following events:

$$A = \text{Aggregates supplied by } Company\ a$$
$$B = \text{Aggregates supplied by } Company\ b$$
$$E = \text{Aggregates are substandard}$$

Then,

$$P(A) = \frac{600}{600 + 400} = 0.60; \quad \text{and} \quad P(B) = \frac{400}{600 + 400} = 0.40$$

whereas
$$P(E|A) = 0.03; \quad \text{and} \quad P(E|B) = 0.01$$

and the probability of substandard aggregates is

$$P(E) = P(E|A)P(A) + P(E|B)P(B)$$
$$= 0.03 \times 0.60 + 0.01 \times 0.40 = 0.022$$

However, if a load of aggregates is found to be substandard, the probability that it is from *Company a* is determined by the Bayes's theorem, Eq. 2.20a, as

$$P(A|E) = \frac{P(E|A)P(A)}{P(E|A)P(A) + P(E|B)P(B)}$$
$$= \frac{0.03 \times 0.60}{0.03 \times 0.60 + 0.01 \times 0.40} = 0.82$$

It is well to observe the difference between $P(E|A)$ and $P(A|E)$. The former is the proportion of aggregates from *Company a* that are substandard, whereas the latter is the probability that aggregates found to be substandard are from *Company a*. The difference is clearly significant. ◀

The Bayes' theorem provides a valuable and useful tool for revising or updating a calculated probability as additional data or information becomes available. The following examples will serve to illustrate this concept, including how prior information (which may be based on subjective judgments) can be combined with test results to update a calculated probability.

▶ **EXAMPLE 2.30**

In order to ensure the quality of concrete material used in a reinforced concrete construction, concrete cylinders are collected at random from concrete mixes delivered to the construction site by a mixing plant. Past records of concrete from the same plant show that 80% of concrete mixes are good or of satisfactory quality.

To further ensure that the concrete delivered on site is of good quality, the engineer requires that one cylinder among those collected each day be tested (after 7 days of curing) for minimum compressive strength. The test method is not perfect—its reliability is only 90%, meaning the probability that a good-quality concrete cylinder will pass the test is 0.90, or that a poor-quality cylinder can pass the test is 0.10. Define the following events:

$$G = \text{good quality concrete}$$
$$T = \text{a concrete cylinder passes the test}$$

According to the available information, we have

$$P(G) = 0.80; \quad \text{and} \quad P(T|G) = 0.90; \quad P(T|\overline{G}) = 0.10$$

Then, if a concrete cylinder passes the test, the probability of good-quality concrete delivered on site is updated as follows:

$$P(G|T) = \frac{P(T|G)P(G)}{P(T|G)P(G) + P(T|\overline{G})P(\overline{G})} = \frac{0.90 \times 0.80}{0.90 \times 0.80 + 0.10 \times 0.20} = 0.973$$

Therefore, with a positive test result, the probability of good-quality concrete used in the construction is increased from 80% to 97.3%.

Now, suppose the engineer is not satisfied with just testing one cylinder, and requires that a second cylinder be tested. If the second cylinder tested also gave a positive result, the probability that the concrete is of good quality becomes

$$P(G|T_2) = \frac{P(T_2|G)P(G)}{P(T_2|G)P(G) + P(T_2|\overline{G})P(\overline{G})}$$

$$= \frac{0.90 \times 0.973}{0.90 \times 0.973 + 0.10 \times 0.027} = 0.997$$

The above probability is updated sequentially. The updating may also be performed in a single step by using the two test results together; in this latter case, denoting

$$T_1 = \text{first cylinder is tested positive}$$

$$T_2 = \text{second cylinder is tested positive}$$

we should obtain the same result, as demonstrated with the following calculations:

$$P(G|T_1T_2) = \frac{P(T_1T_2|G)P(G)}{P(T_1T_2|G)P(G) + P(T_1T_2|\overline{G})P(\overline{G})}$$

$$= \frac{0.90 \times 0.90 \times 0.80}{0.90 \times 0.90 \times 0.80 + 0.10 \times 0.10 \times 0.20} = 0.997$$

In this second calculation, we implicitly assumed that the two tests are statistically independent; i.e.,

$$P(T_1T_2|G) = P(T_1|G)P(T_2|G)$$

However, if the test result of either one of the two cylinders tested was negative, the probability of good-quality concrete would be updated as follows:

$$P(G|T_1\overline{T}_2) = \frac{P(T_1\overline{T}_2|G)P(G)}{P(T_1\overline{T}_2|G)P(G) + P(T_1\overline{T}_2|\overline{G})P(\overline{G})}$$

$$= \frac{0.90 \times 0.10 \times 0.80}{0.90 \times 0.10 \times 0.80 + 0.10 \times 0.90 \times 0.20} = 0.80$$
◄

► 2.4 CONCLUDING SUMMARY

This chapter presents the basic fundamentals of the mathematics of probability useful for modeling engineering and physical problems. The principles are all presented with examples illustrating the modeling of physical problems of relevance to engineering.

In particular, we learned that in the formulation and solution of a problem involving probability, two things are paramount: (1) the definition of the possibility space and the identification of the event within this space; and (2) the evaluation of the probability of the event. The relevant mathematical bases and tools useful for these purposes are the elementary theory of sets and the basic theory of probability. The basic elements of both of these theories have been developed in nonabstract terms that should be more transparent and more easily comprehensible to engineers and physical scientists.

Simple elements of set theory are used to define an event and its *complement*, and the operational rules of sets and subsets govern the combination of events, which consist of the *union* and *intersection* of two or more events to form another event. Similarly, based on three simple assumptions (or axioms), the mathematical theory of probability provides the operational rules for developing logical relationships among the probabilities of different events within a given possibility space, which are basically the *addition rule*, the *multiplication rule*, the *total probability theorem*, and the *Bayes' theorem*.

In essence, the concepts and tools developed in this chapter constitute the essential fundamentals necessary for correctly applying probability in engineering. In the ensuing chapters, particularly Chapters 3 and 4, additional analytical tools will be developed based on the fundamental concepts expounded in this chapter.

► PROBLEMS

2.1 Suppose the travel time between two major cities A and B by air is 6 or 7 hr if the flight is nonstop; however, if there is one stop, the travel time would be 9, 10, or 11 hr. A nonstop flight between A and B would cost $1200, whereas with one stop the cost is only $550. Then, between cities B and C, all flights are nonstop requiring 2 or 3 hours at a cost of $300. For a passenger wishing to travel from city A to city C,

(a) What is the possibility space or sample space of his travel times from A to B? From A to C?

(b) What is the sample space of his travel cost from A to C?

(c) If T = travel time from city A to city C, and S = cost of travel from A to C, what is the sample space of T and S?

2.2 The settlement of a bridge pier, say Pier 1, is estimated to be between 2 and 5 cm. Similarly, the settlement of an adjacent pier, Pier 2, is also estimated to be between 4 and 10 cm. There will, therefore, be a possibility of differential settlements between these two adjacent piers.

(a) What would be the sample space of this differential settlement?

(b) If the differential settlements in the above sample space are equally likely, what would be the probability that the differential settlement will be between 3 and 5 cm?

2.3 The direction of the prevailing wind at a particular building site is between due East ($\theta = 0°$) and due North ($\theta = 90°$). The wind speed V can be any value between 0 and ∞.

(a) Sketch the sample space for wind speed and direction.

(b) Denote the events:

$$E_1 = (V > 35 \text{ kph});$$
$$E_2 = (15 \text{ kph} < V \le 45 \text{ kph})$$
$$E_3 = (\theta \le 30°)$$

Identify the events E_1, E_2, E_3, and $\overline{E}_1$ within the sample space of Part (a).

(c) Use new sketches to identify the following events:

$$A = E_1 \cap E_3; \quad B = E_1 \cup E_2; \quad C = E_1 \cap E_2 \cap E_3$$

Are the events A and B mutually exclusive? How about between the events A and C?

2.4 A cylindrical tank is used to store water for a town as shown in the figure below. On any given day, the water supply inflow per day may fill the tank with an additional 6, 7, or 8 ft of water. The daily demand or consumption of the water for the town will draw down the water level in the tank by an amount equivalent to 5, 6, or 7 ft of the water in the tank.

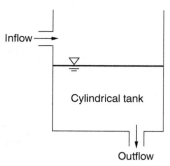

Cylindrical water tank.

(a) What are the possible combinations of inflow and outflow of water in the tank in a given day?

(b) If the water level in the tank is 7 ft from the bottom at the start of a day, what are the possible water levels in the tank at the end of the day?

(c) If the amounts of inflow and outflow of water for the tank are both equally likely, what would be the probability that there would be at least 9 ft of water remaining in the tank at the end of the day?

2.5 A 20-ft cantilever beam is shown in the figure below. Load $W_1 = 200$ lb, or $W_2 = 500$ lb, or both may be applied at the midpoint B or at the end of the beam C. The bending moment induced at the fixed support A, M_A, will depend on the magnitudes of the loads at B and C.

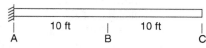

20-ft cantilever beam.

(a) Determine the sample space of M_A.

(b) Define the following events:

$$E_1 = (M_A > 5,000 \text{ ft-lb})$$
$$E_2 = (1,000 \le M_A < 12,000 \text{ ft-lb})$$
$$E_3 = (2,000, 7,000 \text{ ft-lb})$$

Are the events E_1 and E_2 mutually exclusive? Explain why.

(c) Assume the following respective probabilities for the positions of the two loads:

$$P(W_1 \text{ at } B) = 0.25$$
$$P(W_1 \text{ at } C) = 0.60$$
$$P(W_2 \text{ at } B) = 0.30$$
$$P(W_2 \text{ at } C) = 0.50$$

Assuming that the positions of W_1 and W_2 are statistically independent; what are the respective probabilities associated with each of the possible values of M_A?

(d) Determine the probabilities of the following events:

$$E_1, E_2, E_3, E_1 \cap E_2, E_1 \cup E_2, \text{ and } \overline{E}_2$$

2.6 Two cities 1 and 2 are connected by route A, and route B connects cities 2 and 3 as shown in the figure below. Let us denote the eastbound lanes as A_1 and B_1, and the westbound lanes as A_2 and B_2.

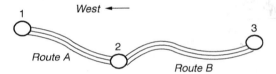

Routes connecting three cities.

Suppose the probability is 95% that one of the two lanes in route A will not require major resurfacing of the pavement for at least 2 years; the corresponding probability for a lane in route B is only 85%.

(a) Determine the probability that route A will require major resurfacing in the next 2 years. Do the same for route B. Assume that if one lane of a route needs major resurfacing, the chance that the other lane of the same route will also need resurfacing is 3 times its original probability.

(b) Assuming that the need for resurfacing in routes A and B are statistically independent, what is the probability that the road between cities 1 and 3 will require major resurfacing in two years?

2.7 From past experience, it is known that, on the average, 10% of welds performed by a particular welder are defective. If this welder is required to do three welds in a day,

(a) What is the probability that none of the welds will be defective?

(b) What is the probability that exactly two of the welds will be defective?

(c) What is the probability that all the welds for a day are defective?

It is assumed that the condition of each weld is independent of the conditions of the other welds.

2.8 On a given day, casting of concrete structural elements at a construction project depends on the availability of material. The required material may be produced at the job site or delivered

from a premixed concrete supplier. However, it is not always certain that these sources of material will be available. Furthermore, whenever it rains at the site, casting cannot be performed. On a given day, define the following events:

E_1 = there will be no rain

E_2 = production of concrete material at the job site is feasible

E_3 = supply of premixed concrete is available

with the following respective probabilities:

$$P(E_1) = 0.8; \qquad P(E_2) = 0.7; \qquad P(E_3) = 0.95;$$
$$\text{and} \qquad P(E_3|\overline{E}_2) = 0.6$$

whereas E_2 and E_3 are statistically independent of E_1.

(a) Identify the following events in terms of E_1, E_2, and E_3:

(i) A = casting of concrete elements can be performed on a given day.

(ii) B = casting of concrete elements cannot be performed on a given day.

(b) Determine the probability of the event B.

(c) If production of concrete material at the job site is not feasible, what is the probability that casting of concrete elements can still be performed on a given day?

2.9 A construction firm purchased 3 tractors from a certain company. At the end of the fifth year, let E_1, E_2, E_3 denote, respectively, the events that tractors no. 1, 2, and 3 are still in good operational condition.

(a) Define the following events at the end of the 5th year, in terms of E_1, E_2, and E_3, and their respective complements:

A = only tractor no. 1 is in good condition.

B = exactly one tractor is in good condition.

C = at least one tractor is in good condition.

(b) Past experience indicates that the chance of a given tractor manufactured by this company having a useful life longer than 5 years (i.e., in good condition at the end of the 5th year) is 60%. If one tractor needs to be replaced (not in good operational condition) at the end of the 5th year, the probability of replacement for one of the other two tractors is 60%; if two tractors need to be replaced, the probability of replacement of the remaining one is 80%.

Evaluate the probabilities of the events A, B, and C.

2.10 A contractor has two subcontractors for his excavation work. Experience shows that in 60% of the time, subcontractor A was available to do a job, whereas subcontractor B was available 80% of the time. Also, the contractor is able to get at least one of these two subcontractors 90% of the time.

(a) What is the probability that both subcontractors will be available to do the next job?

(b) If the contractor learned that subcontractor A is not available for the job, what is the probability that the other subcontractor will be available?

(c) Suppose E_A denotes the event that subcontractor A is available, and E_B denotes that subcontractor B is available.
    (i) Are E_A and E_B statistically independent?
    (ii) Are E_A and E_B mutually exhaustive?
    (iii) Are E_A and E_B collectively exhaustive?

2.11 An underground site is being considered for the storage of hazardous waste. Within the next 100 years, there is a 1% chance that the hazardous material could leak outside of the storage containment. Two adjacent towns, A and B, rely on ground water for their water supply. The water to each town will be contaminated if there is a leakage in the waste storage and if there exists a continuous seam of sand between the storage containment and the given town. Observe that the presence of a continuous seam of sand would allow the contaminant to move freely and quickly access a region.

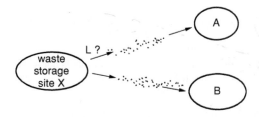

Suppose there is a 2% chance of a continuous seam of sand from the storage site to A, and the probability of a continuous seam of sand to B is slightly higher and equals 3%. However, if a continuous seam of sand exists between X and A, the probability of a continuous seam of sand between X and B is increased to 20%. Assume that the event of leakage from the storage is independent of the presence of seams of sand. Consider the period over the next 100 years.
(a) What is the probability that water in town A will be contaminated?
(b) What is the probability that at least one of the two towns' water will be contaminated?

2.12 Towns A, B, and C lie along a river, as shown in the figure below, which may be subject to overflow (flooding). The annual probabilities of flooding are 0.2, 0.3, and 0.1 for towns A, B, and C, respectively. The events of flooding in each of the towns A, B, and C are not statistically independent. If town C is flooded in a given year, the probability that town B will also be flooded that same year is increased to 0.6; and if both towns B and C are flooded in a given year, the probability that town A will also be flooded that year is increased to 0.8. However, if town C does not experience flooding in a given year, the probability that both towns A and B will also not suffer any flooding in that year is 0.9. In a given year, if all three towns are flooded, it is regarded as a *disaster year* for the year. Suppose the flooding events between any two years are statistically independent. Answer the following:
(a) What is the probability that a given year in the region is a disaster year?

(b) If town B is flooded in a given year, what is the probability that town C is also flooded?
(c) What is the probability that at least one town is flooded in a given year?

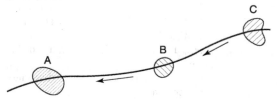

2.13 Successful completion of a construction project depends on the supply of materials and labor as well as the weather condition. Consider a given project that can be successfully completed if either one of the following conditions prevails:
(a) Good weather and at least labor or materials are adequately available.
(b) Bad weather but both labor and materials are adequately available.
    Define:

$$G = \text{Good weather}$$
$$G' = \text{Bad weather}$$
$$L = \text{Adequate labor supply}$$
$$M = \text{Adequate materials supply}$$
$$C = \text{Successful completion}$$

Suppose $P(L) = 0.7$; $P(G) = 0.6$ and L is independent of both M and G. If the weather is good, adequate supply of materials is guaranteed, whereas the probability of adequate supply of material is only 50% if bad weather prevails.
(a) Formulate the event of successful completion in terms of G, L, and M.
(b) Determine the probability of successful completion. (*Ans.* 0.74)
(c) If the project was successfully completed, what is the probability that labor supply had been inadequate? (*Ans.* 0.243)

2.14 The number of accidents at rail-highway grade-crossings reported for a province over the last 10 years are summarized and classified as follows:

		Type of Accident	
		(R) Run into Train	(S) Struck by Train
Time of Occurrence	Day (D)	30	60
	Night (N)	20	20

Suppose there are 1000 rail–highway grade-crossings in province XY.
(a) What is the probability that an accident will occur at a given crossing next year? (*Ans.* 0.013)

(b) If an accident is reported to have occurred in the daytime, what is the probability that it is a "struck by train" accident? (*Ans. 2/3*)

(c) Suppose that 50% of the "run into train" accidents are fatal and 80% of the "struck by train" accidents are fatal, what is the probability that the next accident will be fatal? (*Ans. 0.685*)

(d) Suppose D = event that the next accident occurs in the daytime

R = event that the next accident is a "run into train" accident

  (i) Are D and R mutually exclusive? Justify.

  (ii) Are D and R statistically independent? Justify. (*Ans. no*)

2.15 The promising alternative energy sources currently under development are fuel cell technology and large-scale solar energy power. The probabilities that these two sources will be successfully developed and commercially viable in the next 15 years are, respectively, 0.70 and 0.85. The successful development of these two energy sources are statistically independent. Determine the following:

(a) The probability that there will be energy supplied by these alternative sources in the next 15 years.

(b) The probability that only one of the two alternative energy sources will be commercially viable in the next 15 years.

2.16 An examination of the 10-year record of rainy days for a town reveals the following:

  1. 30% of the days are rainy days.

  2. There is a 50% chance that a rainy day will be followed by another rainy day.

  3. There is a 20% chance that two consecutive rainy days will be followed by a third rainy day.

A house is scheduled for painting starting next Monday for a period of 3 days.

(a) Let

E_1 = Monday is a rainy day

E_2 = Tuesday is a rainy day

E_3 = Wednesday is a rainy day

Express the events corresponding to the three probabilities indicated above; i.e., 1, 2, and 3, in terms of E_1, E_2, E_3.

(b) What is the probability that it will rain on both Monday and Tuesday?

(c) What is the probability that Wednesday will be the only dry day during the painting period?

(d) What is the probability that there will be at least one rainy day during the 3-day painting period?

2.17 Mr. X who works in the office building D is selected for observation in a study of the parking problem on a college campus. Each day, assume that Mr. X will check the parking lots A, B, and C in that sequence, as shown in the figure below, and will park his car as soon as he finds an empty space. Assume also that there are only these three parking lots available and no street parking is allowed, among which lots A and B are free—whereas lot C is metered.

Suppose that from prior statistical observations, the probabilities of getting a space on each weekday morning in lots A, B, and C are 0.20, 0.15, and 0.80, respectively. However, if lot A is full, the probability that Mr. X will find a space in lot B is only 0.05. Also, if both lots A and B are full, Mr. X will only have a probability of 40% of getting a parking space in lot C.

Determine the following:

(a) The probability that Mr. X will not find free parking on a weekday morning.

(b) The probability that Mr. X will be able to park his car on campus on a weekday morning.

(c) If Mr. X successfully parked his car on campus on a weekday morning, what is the probability that his parking is free?

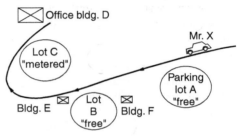

A study of campus parking.

2.18 A building may fail by excessive settlement of the foundation or by collapse of the superstructure. Over the life of the building, the probability of excessive settlement of the foundation is estimated to be 0.10, whereas the probability of collapse of the superstructure is 0.05. Also, if there is excessive settlement of the foundation, the probability of superstructure collapse will be increased to 0.20.

(a) What is the probability that building failure will occur over its life?

(b) If building failure should occur during its life, what is the probability that both failure modes will occur?

(c) What is the probability that only one of the two failure modes will occur over the life of the building?

2.19 The automobile brake system consists of the following components: the master cylinder, the wheel cylinders, and the brake pads. Failure of any one or more of these components will constitute failure of the brake system. Within a period of 4 years or 50,000 miles without maintenance, the failure probabilities of the master cylinder, wheel cylinders, and brake pads are 0.02, 0.05, and 0.50, respectively. The probability that both the master cylinder and the wheel cylinders will fail within the same period or mileage is 0.01. The failure of the brake pads is statistically independent of the failures in the master and wheel cylinders.

(a) What is the probability that only the wheel cylinders will fail within 4 years or 50,000 miles?

(b) What is the probability that the brake system will fail within 4 years or 50,000 miles?

(c) When there is failure in the brake system, what is the probability that only one of the three components failed?

2.20 Let E_1, E_2, and E_3 denote the events of excessive snowfall in the first, second, and third winters, respectively, from this fall. Statistical records of snowfalls indicate that during any winter, the probability of excessive snow is 0.10. However, if excessive snowfall occurred in the previous winter, the probability of excessive snowfall in the following winter is increased to 0.40, whereas if the preceding two winters are both subjected to excessive snowfalls, the probability of excessive snow in the following winter will be 0.20.

(a) From the information given above, determine the following:

$$P(E_1),\ P(E_2),\ P(E_2|E_1),\ P(E_3|E_1\ E_2),\ P(E_3|E_2)$$

(b) What is the probability that excessive snowfall will occur in at least one of the next two winters?

(c) What is the probability that excessive snowfall will occur in each of the next three winters?

(d) If the preceding winter did not experience excessive snowfall, what is the probability that the subsequent winter will not suffer excessive snowfall? In other words, determine $P(\overline{E}_2|\overline{E}_1)$. *Hint*: Start out with the following relationship:

$$P(\overline{E}_1 \cup \overline{E}_2) = 1 - P(\overline{\overline{E}_1 \cup \overline{E}_2}) = 1 - P(E_1 E_2)$$

2.21 Weather records for a certain region indicate that there is a tendency for two hot summers to occur in a row. From the statistics for the region, the following information has been established: (i) the chance of any given summer being hot is 0.20; (ii) if the weather in a given summer is hot, the following summer will also be hot with a probability of 0.40; and (iii) the weather in any given year depends only on the weather in the previous year.

(a) What is the probability that there will be three hot summers in a row in the given region?

(b) If the summer this year is not hot, what is the probability that the summer next year will not be hot?

(c) What is the probability that there will be at least one hot summer in the next 3 years?

2.22 A metropolitan city derives its electrical power from three generating plants A, B, and C. Each plant has a generating capacity of 50 MW (megawatt). The chances that the generating plants will be shut down (for periodic maintenance, due to overload, accidents, etc.) in any given week are, respectively, 5%, 5%, and 10%. The operations of plants A and B are related in such a way that in case of a shutdown at one plant, there is a probability of 50% that the other plant will also shut down due to overload, whereas the operation of plant C is independent of the other two plants.

(a) During a severe storm, lightning knocked out the power lines from plant A, and repair of the lines will take at least 1 week. What is the chance that the city will suffer a complete blackout (no power) in the week caused by the storm?

(b) What is the probability that the city will have no power in any given week?

(c) What is the probability that the available power supply to the city is less than or equal to 100 MW in a given week?

2.23 The probability that a strong earthquake will cause damage to a certain structure has been estimated to be 0.02.

(a) What is the probability that the structure can survive three such strong earthquakes?

(b) What is the probability that the structure will be damaged during the second of two such strong earthquakes?

It may be assumed that damages to the structure caused by successive earthquakes are statistically independent.

2.24 A team of two engineers, A and B, was assigned to check a set of computations. The two work simultaneously but separately and independently. The probability of engineer A spotting a given error is 0.8, whereas that for B is 0.9.

(a) Suppose there is only one error in the computation. What is the probability that this error will be spotted by this team? (*Ans. 0.98*)

(b) If the error in part (a) was identified, what is the probability that it was discovered by A alone? (*Ans. 0.082*)

(c) Suppose there is an alternative team consisting of three engineers C_1, C_2, and C_3, each of whom works separately and independently and has a probability of 0.75 of spotting a given error. Would this team of three engineers be selected instead, if the objective is to maximize the chance of spotting the error? Please justify. (*Ans. yes*)

(d) In part (a), suppose there were two errors in the computations. What is the probability that both errors will be spotted by the team of two engineers? Assume that the events of spotting between any two errors are statistically independent. (*Ans. 0.960*)

2.25 A self-standing antenna disk is supported by a lattice structure that is anchored to the ground at the base. During a wind storm, the disk may be damaged as a result of anchorage failure and/or failure of the lattice structure. Suppose the following information is known:

1. The probability of anchorage failure during a wind storm is 0.006.

2. If the anchorage should fail, the probability of failure of the lattice structure will be 0.40, whereas the probability of failure of the anchorage given that failure has occurred in the lattice structure is 0.30.

Determine the following:

(a) The probability of damage to the antenna disk during a wind storm.

(b) The probability that only one of the two potential failure modes will occur during a wind storm.

(c) If the disk is damaged during a wind storm, what is the probability that it was caused only by anchorage failure?

2.26 The probability of a severe fire (denoted event F) occurring in a new hospital is assumed to be low. The insurance company estimates that the probability of a fire occurring in a year is $P(F) = 0.01$. However, for additional safety, a very sensitive fire alarm system was installed. This system will always sound an alarm (*A*) whenever there is a fire; but because of its high sensitivity, it may also cause false alarms with a probability of $P(A|\overline{F}) = 0.1$. Assume that there is no possibility for more than one fire in a year.

(a) List the set of mutually exclusive and collectively exhaustive events.

(b) Calculate the probabilities for each of the events listed in (a).

(c) What is the probability that the alarm system will be triggered in one year?

(d) What is the probability of a real fire given that the alarm sounded?

2.27 The probability of contracting a certain disease (event *D*) during one's lifetime is very small, with probability $P(D) = 0.001$. However, if left untreated the disease is always fatal. Fortunately, modern medical science has provided a diagnostic test *T* to detect the presence of the disease; however, the test is not always correct. If a person has the disease, there is only 85% probability that the test will be positive, i.e., $P(T|D) = 0.85$. Also, there is a small probability of 2% that the test will be positive even when a patient does not have the disease, i.e., $P(T|\overline{D}) = 0.02$.

(a) Identify the set of mutually exclusive and collectively exhaustive events.

(b) Calculate the probabilities for each of the events identified in (a).

(c) What is the probability of a positive test result?

(d) If a person's test result is positive, what is the probability that he/she has the disease?

2.28 A car is traveling from point M to point Q according to the route indicated in the figure below. The driver has to pass through three intersections, namely at A, B, and C where traffic lights are installed. Information on the traffic lights relative to the driver is as follows:

1. The light at A is equally likely to be either *red* or *green*.

2. The light at B is equally likely to be either *red* or *green*; however, if the driver encounters *green* at A, he will meet a *green* light at B with a probability of 0.80.

3. The driver will encounter a *red* or *green* or *left turn only* at C. The probability of meeting *left turn only* at C is 0.20. The signal lights at C are statistically independent of those at A and B, and left turns at C are permitted only when the *left turn only* signal is on.

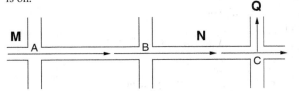

(a) In terms of the following: G_A = light signal at A is *green*; G_B = light signal at B is *green*; and G_{LT} = *left turn only* signal is on at C, define the following events:

E_1 = stopped by *red* lights between M and N.

E_2 = stopped by signal at C from M to Q.

E_3 = stopped by *red* signal once between M and N.

(b) Compute the probability that the driver will be stopped by traffic signals at least once traveling from M to Q.

(c) Compute the probability that the driver will be stopped by traffic signals at most once going from M to N.

2.29 From previous records on winter weather for a town, the probability that it snows on a given day is 0.2. Records also reveal that low temperature (say less than 0°F) occurs on 5% of the days, whereas windy days occur 10% of the time. If temperature is low, the probability of snow on that day is increased to 30%. Let S, C, and W denote the events of snow, low temperature and wind, respectively, on a given day. The event W may be assumed to be statistically independent of S and C.

(a) What is the probability of having a *doomsday* (i.e., snow, low temperature, and wind all occurring on the same day)? (*Ans. 0.0015*)

(b) Suppose a construction project cannot progress if it is windy or cold, but it is not affected by snow. What is the probability that construction work will be stopped on a given day? (*Ans. 0.145*)

(c) On a day without snow, what is the probability that construction work cannot proceed? (*Ans. 0.139*)

(d) What is the probability of a *nice* winter day (i.e., no snow, no gusty wind, and no low temperature)?

(e) Suppose U denotes the event of uncomfortable wind chill condition on a given day. If either C or W (but not both) occurs, the probability of U is 50%, whereas, if both C and W occur, the probability of U is 100%, and U will not occur if both C and W do not occur. What is the probability of the event U on a given winter day?

2.30 Lead and bacteria are the two common sources of contamination in a water distribution system. Suppose 4% of the water distribution systems are contaminated by lead, and only 2% of the water distribution systems are contaminated by bacteria. Assume that the events of lead and bacterial contamination are statistically independent.

(a) Determine the probability that a water distribution system selected at random for inspection is contaminated. (*Ans. 0.0592*)

(b) If a system is indeed contaminated, what is the probability that it is caused by lead only? (*Ans. 0.662*)

2.31 The 15-ft diameter tank shown below rests on a concrete base. When there is water in the tank, it will be at the 10-ft or 20-ft level, and the probability of either level is 0.40. The weight of the tank is 100,000 lb and the weight of the water is 11,000 lb per foot of water level in the tank. The frictional resistance against the horizontal force is equal to the total weight of the tank and contents times the coefficient of friction, or

$$F = CW$$

where

$$C = \text{the coefficient of friction}$$
$$W = \text{total weight of tank and contents}$$

(a) If C is equally likely to be 0.10 or 0.20, identify the sample space of the frictional resistance to horizontal sliding of the tank.

(b) If the total maximum horizontal force during a wind storm is 15,000 lb, what is the probability that the tank will be displaced (slide) horizontally during the storm? The value of C is statistically independent of the total weight.

(c) Suppose that the total maximum horizontal force during a wind storm (that may cause sliding) may be 15,000 lb or 20,000 lb with respective probabilities of 60% and 40%; in such a case, what would be the probability of sliding of the tank? Assume that the maximum wind force is statistically independent of the frictional resistance.

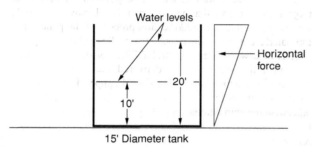

15' Diameter tank

2.32 A landfill containment system is shown in the figure below. A thick layer of clay (which is a highly impermeable material) was placed between the landfill and the surrounding soil stratum to prevent the contaminants leaking from the landfill into the soil stratum resulting from rainfall infiltration. A layer of synthetic material called geomembrane was also placed above the clay material to provide additional protection against leakage of contaminants. Nevertheless, the quality of workmanship during construction may not be completely satisfactory. First, the clay might have been compacted poorly; second, the geomembrane might have holes punctured by sharp stones that were not detected during inspection. Moreover, extremely heavy rainfall could happen during the operation of the landfill, which could induce excessive pore pressure on the geomembrane/clay layers.

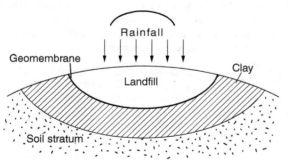

The engineer in this case believes that leakage will happen "during extremely heavy rainfall, and either the clay was not well compacted or there were holes in the geomembrane" (event I). Leakage could also occur "under ordinary rainfalls (i.e., without extremely heavy rainfall), but only when the clay was not well compacted and the geomembrane contained holes" (event II).

Let   W = event of well-compacted clay; and this occurs with 90% probability

H = event of geomembrane containing holes; and this is 30% likely

E = event of extremely heavy rainfall; and the likelihood is 20%

The quality of construction has no effect on the future amount of rainfall. However, if the geomembrane contained holes, the probability of a well-compacted clay is reduced to 60%.

(a) Express event I in terms of the symbols defined above. Repeat for event II.

(b) Determine the probability of event I. Repeat for event II. (*Ans. 0.056, 0.096*)

(c1) Are W and H mutually exclusive? Are they statistically independent? Provide explanations to support your answers. (*Ans. no, no*)

(c2) Are the events I and II mutually exclusive? Are they collectively exhaustive? (*Ans. yes, no*)

(d) Determine the probability of leakage for this landfill containment system. (*Ans. 0.152*)

2.33 A community is concerned about the supply of energy for the coming winter. Suppose there are three major sources of energy for the community, namely electrical power, natural gas, and oil. Let E, G, and O denote, respectively, the shortages of these sources of energy for the next winter. Also, it is estimated that the respective probabilities of these shortages are as follows:

$$P(E) = 0.15; \qquad P(G) = 0.10; \qquad \text{and} \qquad P(O) = 0.20$$

Furthermore, if there is a shortage of oil, the probability that there will be shortage of electrical energy will be doubled. The shortage of gas may be assumed to be statistically independent of shortages in oil and electricity.

(a) What is the probability that there will be shortages in all the three sources of energy next winter?

(b) What is the probability that there will be shortages in at least one of the following sources next winter: gas and electricity?

(c) If there is a shortage of electricity next winter, what is the probability that there will also be shortages in both gas and oil?

(d) What is the probability that at least two of the three sources of energy will be in short supply next winter?

2.34 An aerial reconnaissance system consists of three remote sensing components A, B, and C such that the failure of any one of the components will constitute the failure of the system.

Flying at normal attitudes, the probabilities of failure of the components over a period of 10 years are, respectively, 0.05, 0.03, and 0.02, whereas operating at ultrahigh altitudes, the corresponding failure probabilities would be 0.07, 0.08, and 0.03. In 60% of the time, the reconnaissance system will be operating at normal altitudes, and 40% of the time it will be used at ultrahigh altitudes.

The system is such that if component A should fail, the failure probability of component B will be twice its original probability. On the other hand, the failure of component C is statistically independent of A or B.

If the reconnaissance system fails within 10 years, what is the probability that it was caused by the failure of component B?

2.35 In any given year, the winter in a Midwest city can be cold (C) and wet (W). On the average, 50% of the winters in this city are cold and 30% of the winters are wet. Moreover, 40% of the cold winters are also wet. An unpleasant winter (U) is one when the weather is either cold or wet or both.
(a) Are the events C and W statistically independent? Justify.
(b) What is the probability of an unpleasant winter in a given year?
(c) What is the probability that the winter in any given year will be cold but not wet?
(d) If the winter in a given year is indeed unpleasant, what is the probability that it will be both cold and wet?

2.36 A city commuter can take one of three possible routes A, B, or C to work. On a given weekday morning, during rush hours, the chances that routes A, B, and C will have traffic congestion, H, are, respectively, 60%, 60%, and 40%. Routes A and B are close by each other such that if the traffic is congested in one route the chance of congestion in the other route increases to 85%, whereas the condition in route C is unaffected (i.e., independent) by the traffic conditions in routes A and B. Also, if the traffic in all the three routes are congested, the chance that the commuter will be late, L, for work is 90%, otherwise it will be 30%.
(a) What is the probability that exactly one of the three routes will be congested on a weekday morning?
(b) What is the probability that the commuter will be late for work on a given weekday morning?
(c) If it is known that the traffic in route A is congested, how would this fact change the probability in (b)?

2.37 The damage in a structure after an earthquake can be classified as none (N), light (L) or heavy (H). For a new undamaged structure, the probability that it will suffer light and heavy damages after an earthquake is 0.2 and 0.05, respectively. However, if a structure was already lightly damaged, its probability of getting heavy damage during the next earthquake is increased to 0.5.
(a) For a new structure, what is the probability that it will be heavily damaged after two earthquakes? Assume that no repair was performed after the first earthquake. (*Ans. 0.188*)
(b) If a structure is indeed heavily damaged after two earthquakes, what is the probability that the structure was either undamaged or lightly damaged before the second earthquake? (*Ans. 0.733*)
(c) If the structure is restored to its undamaged condition after each earthquake, what is the probability that the structure will ever experience heavy damage during three earthquakes? (*Ans. 0.143*)

2.38 In a typical month, the demand for cement material in a city may be low (*L*), average (*A*), or high (*H*) with respective probabilities of 0.60, 0.30, and 0.10. The various suppliers of

cement material in the city can definitely handle a low demand (*L*), but if the demand is average (*A*) or high (*H*), the supply may be inadequate with probabilities of 0.10 and 0.50, respectively.
(a) What is the probability of shortage of cement material in a given month?
(b) If a shortage occurred in a month, what is the probability that the demand had been average?
(c) What is the probability of shortage of cement in at least 1 month over a 2-month period? Assume that the demand and supply of cement material are statistically independent between consecutive months.

2.39 Prefabricated wall panels are shipped to a construction project site. Suppose that one shipment is delivered daily to the site. Due to errors in the fabrication process of the panels, it is estimated that the number of defective wall panels in a shipment is 0, 1, or 2 with respective probabilities of 0.2, 0.5, and 0.3.

When the wall panels are delivered, they are inspected by the supervisor at the construction site. The supervisor will accept the entire shipment if at most one panel is found to be defective; otherwise the shipment is rejected.
(a) Determine the probability that a shipment will be accepted on a given day.
(b) What is the probability that exactly one shipment will be rejected in a week consisting of 5 working days?
(c) The inspection procedure, however, is not perfect; it is only 80% likely that a defective panel will be correctly identified. Assuming that the identification of any two defective panels are statistically independent, determine the probability that a shipment will be accepted on a given day.

2.40 A construction project is about 2 months away from the scheduled completion date. Based on the progress of the project to date, the contractor estimated that the project can be completed on schedule without difficulty if good weather continues for the next 2 months; whereas if the weather in the next 2 months is normal, the probability of on-time completion will be 90%. This probability will be reduced to 20% if bad weather prevails, in which case he has the option of launching a crash program to improve the on-time completion probability to 80%. However, because of uncertainty in the labor market, there is only a 50–50 chance that he can successfully launch a crash program when needed. Assume that no crash program is necessary if the weather condition is good or normal. Suppose also that the weather bureau predicts that the relative likelihoods of good, normal, and bad weather conditions in the next 2 months are 1:2:2.
(a) What is the probability that the construction project will be completed on schedule?
(b) If the project was not completed on schedule, what is the probability that the weather condition had been normal?

2.41 A country is examining its energy situation for the next 10 years. Suppose this country depends on oil and natural gas as its principal sources of energy. Energy experts in the country estimated that there is a 40% probability that gas supply will be low in the next 10 years, whereas the probability of low oil supply is 20%. However, if oil supply is low, the probability that gas supply will also be low is increased to 50%.

The demand for energy in the next 10 years is also projected, and the experts estimated that the probabilities of low, normal, and high demand in the next 10 years will be 0.3, 0.6, and 0.1, respectively. Depending on the level of energy demand and the supplies of oil and gas in the next 10 years, an "energy crisis" might occur according to the following table.

Energy Demand	Oil and Gas Supply	Occurrence of Energy Crisis
Low	Both low	Yes
Normal	At least one low	Yes
High	Regardless of supply situation	Yes

(a) If there is normal energy demand in the next 10 years, what is the probability of an "energy crisis"?
(b) What is the probability of an "energy crisis" in the next 10 years?

2.42 In an offshore area in the Gulf of Mexico, the probabilities that the area will be hit by one or two hurricanes in a given year are, respectively, 20% and 5%. The probability of more than two hurricanes in a year is negligible. An oil production platform in this area consists of two substructures, a jacket truss and a deck. If the hurricane occurs only once in a given year, the annual probability that both substructures will not suffer any damage is 99%; this probability drops to 80% if two hurricanes should occur in a year.
(a) What is the probability that the platform (i.e., at least one of the substructures) will suffer some damage if the area is hit by one hurricane in a given year? Assume that damages to the two substructures are statistically independent. What would be this probability if two hurricanes should occur in the given year?
(b) What is the probability that the platform will suffer some hurricane damage next year?
(c) If no damage to the platform was found at the end of a year, what is the probability that no hurricanes occurred that year?

2.43 The common five-axle semitrailer trucks dominate the heavy truck traffic over a bridge that has been selected for structural monitoring and testing. For simplicity, the heavy vehicles are classified as *empty* (E), *half-full* (H), or *fully loaded* (F). Only one vehicle at a time can be observed on the bridge at any instant, but it can be in either of two lanes, lane A or lane B. The likelihood of a truck being in lane A is five times greater than in lane B. The lane in which a truck crosses the bridge is independent of the load it carries.

Of concern are stresses in a girder under lane A. If a fully loaded truck, F, crosses the bridge in lane A, a critical fatigue stress is certain to occur, whereas if the same truck were to cross in lane B, the likelihood of a critical stress is reduced to 60%. The likelihood that a half-loaded truck, H, will produce a critical stress is 40% when in lane A, but only 10% when in lane B. No

other loading conditions will produce critical fatigue stresses in the girder.
(a) Determine the sample space of possible bridge loading conditions relative to the fatigue stresses in the girder, i.e., consisting of load and lane position.
(b) Given that a vehicle is in lane B, what is the probability that it is fully loaded (event F)?
(c) What is the probability of the occurrence of critical fatigue stresses in the girder?
(d) If a critical stress is observed, what is the probability that it was caused by a vehicle in lane A?

2.44 There are two possible modes of failure of a reinforced concrete (R/C) beam; namely, shear failure, which can occur suddenly without warning, and bending failure which is preceded by large deflection. Experience indicates that 5% of all failures of R/C beams are by the shear mode and the rest by bending mode. Laboratory tests show that 80% of all shear failures exhibit diagonal cracks at the end of a beam prior to failure, whereas only 10% of bending failures show such diagonal cracks.
(a) What is the probability that an R/C beam will show diagonal cracks before failure?
(b) After a severe earthquake, upon inspection of an R/C building, an engineer found cracks at the end of one of the beams in the building. Should he recommend immediate repair if the owner's instruction is that immediate repair is necessary only if shear failure is likely to occur or that failure probability exceeds 75%?

2.45 Traffic signals were installed at an intersection involving two one-way streets. Suppose 85% of the vehicles will decelerate when they see the amber light, whereas 10% will accelerate and 5% are "indecisive" and simply continue with the same speed. Five percent of those who accelerated will eventually run a red light, and only 2% of the "indecisive" drivers will be forced to run a red light. All of those who decelerated are able to stop before the red light.
(a) For a vehicle encountering the amber light at this intersection, what is the probability that it will run the red light?
(b) If a vehicle were found to have run a red light, what is the probability that the driver had accelerated?
(c) The likelihood of an accident resulting from a vehicle running the red light (referred to as a problem vehicle) is studied as follows. Suppose in 60% of the time, vehicles are waiting on the other street at the start of their green light cycle, ready to cross the intersection. Most of these drivers, say 80%, are cautious before they entered the intersection, whereas the rest are not cautious. Given the presence of a problem vehicle in the intersection zone, a cautious driver can avoid the problem vehicle 95% of the time, whereas 20% of the noncautious drivers will collide with the problem vehicle. What is the probability that a problem vehicle will lead to an accident?
(d) Suppose the annual traffic flow in one of these one-way streets is 100,000 vehicles and 5% of them would encounter the amber signal light. Estimate the number of accidents per year

in the intersection that would be traced back to vehicles running through a red light on that street.

2.46 A major construction company has three branches A, B, and C operating in different parts of the country. The chances that the branches will be profitable in any given year are 70%, 70%, and 60%, respectively. The operations of branches A and B are related such that if one makes a profit the probability of the other branch also making a profit increases to 90%, whereas branch C is independent of both A and B. At the end of each year, if at least two branches are profitable, the chance that employees will receive a bonus is 80%; otherwise the chance of a bonus will only be 20%.

(a) What is the probability that exactly two branches will make profits in a given year?

(b) Determine the probability that the company employees will receive a bonus this year.

(c) If it is known for sure that branch A will end this year in the red (i.e., not making any profit), how would this fact change the probability of part (b)?

2.47 A contractor finds that some of his building projects involve difficult foundation conditions, and others are considered easy. His current projects are in three counties in Illinois; namely, in Champaign, Ford, and Iroquois. The contractor has statistics from previous jobs in these counties which enable him to make the following observations with confidence:

   1. The probability of any building project picked at random will have a difficult foundation problem is 2/3.
   2. 1/3 of the projects are in Ford County.
   3. 2/5 of his projects are in Iroquois County and involve difficult foundation conditions.
   4. Of all his projects in Ford County, 50% involve difficult foundation conditions.
   5. The events of a project in Champaign County and the foundation condition are statistically independent.

(a) What is the probability of having the next project in Ford County that has easy foundation conditions?

(b) What is the probability of having the next project in Champaign County with easy foundation conditions?

(c) If we know only that his next project will have easy foundation conditions, what is the probability that it is in Iroquois County?

2.48 The three major loadings on a nuclear power plant, as far as safety is concerned, are those due to severe earthquakes (E), loss of coolant accident (L), and thermal transients (T). For a typical plant, the chances of occurrence of E, L, and T in a given year are, respectively, 0.0001, 0.0002, and 0.00015. Also, severe earthquakes sometimes cause L due to pipe breaks; it is estimated that the chance of L increases to 10% if severe earthquakes occur in the same year. T is assumed to be independent of both E and L.

(a) What is the probability that in a given year, all three types of loadings will occur?

(b) If at least one of the major loadings occurs, the plant has to be shut down for a period of time and the utility company will suffer some loss of revenue due to power outage. What is the probability of this loss in a given year?

(c) If the utility company incurs this loss of revenue in a given year, what is the probability that severe earthquakes occurred that year?

2.49 At a construction project, the amount of material used in a day's construction is either 100 units or 200 units, with corresponding probabilities of 0.60 and 0.40. If the amount of material required in a day is 100 units, the probability of shortage of material is 0.10, whereas if the amount of material required is 200 units, the probability of shortage of material is 0.30.

(a) What is the probability of shortage of material in a given day? (*Ans. 0.18*)

(b) If there is a shortage of material in a given day, what is the probability that the amount of material required that day is 100 units? (*Ans. 1/3*)

2.50 A space vehicle is designed to land on Mars. Assume that the ground condition on Mars is either hard or soft. If hard ground is encountered during landing, the vehicle will be successfully landed with probability 0.9, whereas, if soft ground is encountered, the corresponding probability of a successful landing is only 0.5. Based on the available information, it is judged that the chance of hitting hard ground is three times that of hitting soft ground.

(a) What is the probability of a successful landing? (*Ans. 0.8*).

(b) Suppose a stick can be projected to test the ground condition before landing. It will penetrate into soft ground with probability 0.9, and hard ground with probability of only 0.2. If the stick were observed to penetrate into the ground, what is the probability that the ground is hard? (*Ans. 0.4*)

(c) What is the probability of a successful landing if the stick penetrated the ground? (*Ans. 0.66*)

2.51 Cement and reinforcing steel (rebars) are essential materials for constructing a reinforced concrete building. During construction, the probabilities of encountering shortages of these materials (e.g., caused by strikes at the factories) are, respectively, 0.10 and 0.05. However, if cement is available, the probability of shortage of rebars is reduced to one-half of the original probability.

(a) What is the probability that there will be shortage of either (or both) construction material?

(b) What is the probability that only one of the two materials will be available?

(c) If there is shortage of material during construction, what is the probability that it will be shortage of reinforcing steel?

The construction materials referred to above must be transported from the factories to the construction site either by trucks or trains. Past records show that 60% of the materials are transported by trucks and the remaining 40% by trains. Also, the probability of on-time delivery by trucks is 0.75, whereas the corresponding probability by trains is 0.90.

(d) What is the probability that materials to the construction site will be delivered on schedule?
(e) If there is delay in the transportation of construction materials to the site, what is the probability that it will be caused by truck transportation?

2.52 Water for a city is supplied from two sources; namely, Source A and Source B. During the summer season, the probability that the supply from Source A will be below normal is 0.30; the corresponding probability for Source B is 0.15. However, if Source A is below normal, the probability that Source B will also be below normal during the same summer season is increased to 0.30.

   The probability of water shortage in the city will obviously depend on the supplies from the two sources. In particular, if only Source A is below normal supply, the probability of water shortage is 0.20, whereas if only Source B is below normal the corresponding probability of shortage is 0.25. Obviously, if none of the sources are below normal, there would be no chance of shortage, whereas if both sources are below normal during the summer, the probability of water shortage in the city would be 0.80.

   During a summer season, determine the following:
(a) The probability that there will be below normal supply of water from either or both sources.
(b) The probability that only one of the two sources will be below normal supply.
(c) The probability of water shortage in the city during the summer season.
(d) If water shortage should occur in the city, what is the probability that it was caused by below-normal supplies from both sources?
(e) If there is no water shortage in the city during the summer, what is the probability there was normal supply from Source A?

2.53 A consulting engineer must meet a deadline for a project consisting of two independent phases:
(1) *Field work*—If weather conditions are favorable, the probability that the field work will be completed on schedule is 0.90. Otherwise, the probability of on-schedule completion is reduced to 0.50. The probability of unfavorable weather is 0.60.
(2) *Computations*—Two independent computers are available to perform the required calculations. Each computer has a reliability of 70% (i.e., the probability of working is 0.70). If only one of the computers is working, the probability of completing the computations on time is 0.60, whereas, if both computers are working, this probability increases to 0.90. Furthermore, if both of the computers are not working, the engineer must perform his calculations using desk calculators which are 100% reliable, but will decrease the probability of completing the computations to 0.40.
(a) What is the probability that the field work will be completed on schedule?
(b) What is the probability that the computations will be completed on time?

(c) What is the probability that the engineer will meet his deadline for the project?

2.54 At a rock quarry, the time required to load crushed rocks onto a truck is equally likely to be either 2 or 3 minutes. Also, the number of trucks in a queue waiting to be loaded can vary considerably; data from 40 previous observations taken at random show the following:

No. of Trucks in Queue	No. of Observations	Relative Frequency
0	7	0.175
1	5	0.125
2	12	0.3
3	11	0.275
4	4	0.1
5	1	0.025
6	0	0
	Total = 40	

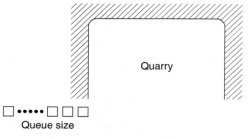

A quarry site.

The time required to load a truck is statistically independent of the queue size.
(a) If there are two trucks in the queue when a truck arrives at the quarry, what is the probability that its "waiting time" will be less than 5 minutes?
(b) Before arriving at the quarry, and thus not knowing the size of the queue upon arrival, what is the probability that the waiting time of the truck will be less than 5 minutes?

2.55 A small old bridge is susceptible to damages from heavy trucks. Suppose the bridge can have room for at most two trucks, one in each lane. The event of possible damage to the bridge when two trucks are present simultaneously is investigated next.

   Suppose 10% of the trucks are overloaded (i.e., above legal load limit) and the event of overloading is statistically independent between trucks. The damage probability of the bridge is 30% when both trucks are overloaded; the probability is 5% if only one is overloaded and 0.1% when both trucks are not overloaded.
(a) What is the probability of damage to the bridge while supporting two trucks?
(b) If the bridge is damaged, what is the probability that it was caused by overloaded truck (or trucks)? [*Hint*: Determine first the probability that damage was not caused by overloaded truck(s).]

(c) Return to part (a). Suppose the county board can allocate a sum of money for strengthening the bridge such that the probability of damage will be half of the existing bridge. Alternatively, that sum of money could be used to increase the inspection frequency of trucks such that the fraction of overloaded trucks entering the bridge is decreased from 10% to 6%. Which alternative is better if the objective is to minimize the probability of damage to the bridge while supporting two trucks?

2.56 A geologic anomaly embedded underneath a site could induce geotechnical failure if the anomaly is sufficiently large and consists of undesirable soil properties. Suppose an engineer estimates that there is a 30% likelihood that anomaly may be present at a given site on the basis of the geology in the region. An exploration program may be performed at the site to verify the presence of an anomaly.

One plan calls for the use of geophysical techniques. If an anomaly is present, such techniques will have 50% probability of detecting the anomaly; otherwise, no signal will be registered.
(a) If the geophysical technique is used but it failed to detect any anomaly, what is the probability that the occurrence of an anomaly is still possible at the site? (*Ans. 0.176*)
(b) At this point, a more discriminating plan is used such that the probability of detecting an anomaly is as high as 80% if an anomaly is present. Suppose this new plan also did not detect any anomaly.

    (i) How confident is the engineer now about his claim that the site is free of any anomaly? (*Ans. 0.959*)
    (ii) A foundation system will be built at the site. The engineer estimates that the foundation should be 99.99% safe if there is no anomaly. However, if an anomaly exists, the reliability of the foundation is reduced to 80%. What is the probability of failure of this foundation system? (*Ans. 0.008*)
    (iii) Suppose failure of the foundation system could bring a loss of one million dollars, whereas survival of the foundation system will not result in any loss. What is the expected loss associated with a failure in the foundation? How much of this expected loss can be saved if the site can be verified to be anomaly free? (*Hint*: Expected loss = probability of failure × failure loss) (*Ans. $8300, $8200*)

2.57 Past records show that a batch of mixed concrete supplied by a certain manufacturer can be of good quality (*G*), average quality (*A*), or bad quality (*B*), with respective probabilities of 0.30, 0.60, and 0.10. Suppose that the probability of failure of a reinforced concrete component would be 0.001, 0.01, or 0.1 depending on whether the quality of concrete is good (*G*), average (*A*), or bad (*B*).
(a) What would be the probability of failure of a reinforced concrete structural component cast with a batch of concrete supplied by the manufacturer?
(b) A test may be performed to give more information on the quality of concrete supplied by the manufacturer. The probabilities of passing the test for good, average, and bad quality concrete are 0.90, 0.70, and 0.20, respectively. If a batch of concrete passed the test,

    (i) What is the probability that it will be of good quality?
    (ii) In this case, i.e., concrete passed the test, what would be the probability of failure of a structural component cast from this batch of concrete?

2.58 The maximum intensity of the next earthquake in a city may be classified (for simplicity) as low (*L*), medium (*M*), or high (*H*) with relative likelihoods of 15:4:1. Suppose also that buildings may be divided into two types; poorly constructed (*P*) and well constructed (*W*). About 20% of all the buildings in the city are known to be poorly constructed for earthquake resistance.

It is estimated that a poorly constructed building will be damaged with a probability of 0.10, 0.50, or 0.90 when subjected to a low-, medium-, or high-intensity earthquake, respectively. However, a well-constructed building will survive a low-intensity earthquake, although it may be damaged when subjected to a medium- or high-intensity earthquake with probability of 0.05 or 0.20, respectively.
(a) What is the probability that a well-constructed building will be damaged during the next earthquake?
(b) What proportion of the buildings in this city will be damaged during the next earthquake?
(c) If a building in the city is damaged after an earthquake, what is the probability that the building was poorly constructed?

2.59 A transit system consists of one-way trains running between four stations as shown in the figure below. The distances between stations are as indicated in the figure. The probabilities concerning origin and destination of passengers are summarized in the following matrix.

Origin	Destination			
	1	2	3	4
1	0	0.1	0.3	0.6
2	0.6	0	0.3	0.1
3	0.5	0.1	0	0.4
4	0.8	0.1	0.1	0

For example, a passenger originating from Station 1 will get off at Station 2, 3, or 4 with probabilities 0.1, 0.3 and 0.6, respectively. Furthermore, the fraction of trips originating from Stations 1, 2, 3, and 4 are 0.25, 0.15, 0.35, and 0.25, respectively.

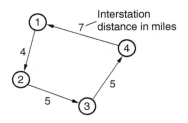

(a) What is the probability that a passenger will leave the train at Station 3? (*Ans. 0.145*)

(b) What is the expected trip length for a passenger boarding at Station 1? (*Note*: "Expected value of X" $= \Sigma_{all} x_i p_i$ where p_i is probability of the outcome x_i). (*Ans. 11.5 miles*)

(c) What proportion of passenger trips will exceed 10 miles? (*Ans. 0.5*).

(d) What fraction of the passengers departing the train at Station 3 originated from Station 1? (*Ans. 0.517*)

2.60 Three machines A, B, and C produce, respectively, 6%, 30%, and 10% of the total number of items of a factory. From past records, the percentages of defective outputs of these machines are, respectively, 2% 3%, and 4%. An item is selected at random and is found to be defective.

(a) What is the probability that the particular defective item was produced by machine A? (*Ans. 0.48*)

(b) What is the probability that the defective item was produced by either machine A or B? (*Ans. 0.84*)

2.61 An engineering Company E is submitting bids for two projects A and B; the probabilities of winning are estimated to be, respectively, 0.50 and 0.30. Also, if the company wins one bid, his chance of winning the other bid is reduced to one-half of the original probability.

(a) What is the probability of Company E winning at least one of its bids?

(b) If Company E wins at least one of its bids, what is the probability that it will be for project A and not project B?

(c) If Company E is awarded only one project, what is the probability that it will be project A?

Furthermore, on the basis of past performance, it is estimated that the probability of Company E completing project A within a target time is 0.75, whereas, if another company is awarded the project, the probability of on-time completion of project A is only 0.50.

(d) What is the probability that project A will be completed on time?

(e) If the project A is completed on time, what is the probability that it was done by Company E?

2.62 Defective concrete on a construction project can be caused by either poor aggregates or poor workmanship (such as selection and grading of material, pouring, curing) or both. The quality of aggregates is also affected by the quality of workmanship, and vice versa.

On a given project, the probability of poor aggregates is 0.20. The probability of poor workmanship if the aggregates are poor is 0.30. The probability of poor aggregates if there is poor workmanship is 0.15.

(a) What is the probability of poor workmanship on the project?

(b) What is the probability of at least one of the causes of defective concrete occurring on the project?

(c) Determine the probability that only one of the two possible causes will occur.

(d) In the above project, if there is only poor aggregates and not poor workmanship, the probability of defective concrete is 0.15. If there is only poor workmanship but good aggregates, the cor-

responding probability would be 0.20. However, if both causes are present, the probability of defective concrete would be 0.80; if none of the causes is present, the corresponding probability is 0.05. Determine the probability of defective concrete on the project.

(e) If there is defective concrete on the project, what is the probability that it is caused by both poor aggregates and poor workmanship?

2.63 The structure shown in the figure below could be subject to settlement problems. The likelihood of having a settlement problem (event A) depends on the subsoil condition, in particular, whether a weak zone exists in the subsoil or not. If there is a small weak zone (event S), the probability of A is 0.2; if the weak zone is large (event L), the probability of A becomes 0.6; last, if no weak zone exists (event N), then the probability of A is only 0.05. Based on their experiences with the geology of the neighborhood and the soil information from the preliminary site exploration program, the engineers in this case believe that there is a 70% chance of no weak zone in the stratum underlying the structure; however, if there is a weak zone, it would be twice as likely to be small than large.

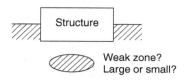

(a) What is the probability that the structure will have a settlement problem? (*Ans. 0.135*)

(b) Suppose an additional boring can be performed at the site to gather more information about the presence of the weak material. The engineers judge that: If a large weak zone exists, it is 80% likely that the boring will encounter it; this encounter probability drops to 30% for a small weak zone. Obviously, the boring will not encounter any weak material if the weak zone does not exist at all. Suppose the additional boring failed to encounter any weak material:

    i. What is the probability of the presence of a large weak zone? (*Ans. 0.023*)

    ii. What is the probability of a small weak zone? (*Ans. 0.163*)

    iii. In this case, what is the probability that the structure will have a settlement problem? (*Ans. 0.087*)

2.64 A dam is proposed to be built in a seismically active area as shown in the following figure:

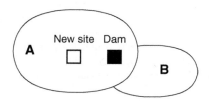

Two regions, A and B, can be identified in the vicinity such that earthquakes in either area could cause damage to the proposed dam. Earthquakes occur independently between regions A and B. Suppose the annual probabilities of earthquake occurrences in regions A and B are 0.01 and 0.02, respectively. Moreover, the chance of two or more earthquakes occurring annually in each region is negligible.

(a) What is the probability of an earthquake occurring in the vicinity of the dam in a given year? (*Ans. 0.03*)

(b) If an earthquake occurred in A (but not in B), the likelihood of damage to the dam is 0.3; however, if an earthquake occurred in B (but not in A), the likelihood of damage is only 0.1. Furthermore, if earthquakes occurred in both regions, the dam would have a 50–50 chance of damage. What is the probability that the dam will be damaged in a given year? (*Ans. 0.00502*)

(c) Suppose the dam can be relocated close to the center of region A, such that earthquakes in region B will not cause any damage to the dam. However, the likelihood of damage due to an earthquake in region A will increase to 0.4. Should the dam be sited in this new location if the objective is to minimize the probability of incurring damage? Please substantiate your answer.

(d) Would the decision in part (c) change if the new site is also susceptible to: (i) a landslide caused by severe rainstorms with an annual probability of 0.002, and (ii) a 0.001 annual probability of subsidence due to poor supporting subsoil structures? Explain. Assume that the dam will be damaged during landslide or subsidence. Also, the events of damage caused by earthquake, landslide, and subsidence are statistically independent.

2.65 Three oil companies, **A**, **B**, and **C**, are exploring for oil in an area. The probabilities that they will discover oil are, respectively, 0.40, 0.60, and 0.20. If **B** discovers oil, the probability that **A** will also discover oil is increased by 20%, whereas this does not affect the chance of **C** discovering oil (i.e., **C** is independent of **B**). Moreover, assume that **C** is also independent of **A**.

(a) What is the probability that oil will be discovered in the area by one or more of the three companies?

(b) If oil is discovered in the area, what is the probability that it will be discovered by Company **C**?

(c) What is the probability that only one of the three companies will discover oil in the area?

2.66 Delays at the airport are common phenomena for air travelers. The likelihood of delay often depends on the weather condition and time of the day. The following information is available at a local airport:

1. In the morning (AM), flights are always on time if good weather prevails; however, during bad weather, half of the flights will be delayed.

2. For the rest of the day (PM), the chances of delay during good and bad weather are 0.3 and 0.9, respectively.

3. 30% of the flights are during AM hours, whereas 70% of the flights are during PM hours.

4. Bad weather is more likely in the morning; in fact, 20% of the mornings are associated with bad weather, but only 10% of PM hours are subject to bad weather.

5. Assume only two kinds of weather, namely, good or bad. Define events as follows:

$$A = \text{AM (morning)}$$
$$P = \text{PM (rest of the day)}$$
$$D = \text{Delay}$$
$$G = \text{Good weather}$$
$$B = \text{Bad weather}$$

Answer the following:

(a) What fraction of the flights at this airport will be delayed? Observe that this is the same as the probability that a given flight will be delayed. (*Ans. 0.282*)

(b) If a flight is delayed, what is the probability that it is caused by bad weather? (*Ans. 0.330*)

(c) What fraction of the morning flights at this airport will be delayed? (*Ans. 10%*)

2.67 Defects in the articles produced by a manufacturing company can be the result of one of the following independent causes:

1. Malfunction of the machinery, which occurs 5% of the production time

2. Carelessness of workers, which occurs 8% of the production time

Defective articles are produced only by these two causes. Also, if only the machinery malfunctions, the probability of a defective article is 0.10, whereas whenever there is only carelessness of workers, the probability of a defective article is 0.20. However, when both causes occur, the probability of a defective article is 0.80.

(a) What is the probability of producing a defective article in the company?

(b) If a defective article is discovered, what is the probability that it was caused by carelessness of workers?

2.68 Leakage of contaminated material is suspected from a given landfill. Monitoring wells are proposed to verify if leakage has occurred. The location of two wells, A and B, are shown in the following figure:

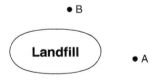

If leakage has occurred, it will be observed by Well A with 80% probability, whereas Well B is 90% likely to detect the leakage. Assume that either well will not register any contaminant if there is no leakage from the landfill. Before the wells are installed, the engineer believes that there is a 70% chance that leakage has occurred.

(a) Suppose Well A has been installed and no contaminants were observed. How likely will the engineer now believe that leakage has occurred? (*Ans. 0.318*)

(b) Suppose both wells have been installed. Assume that the events of detecting leakage between the wells are statistically independent.

> **(i)** What is the probability that contaminants will be observed in at least one of the wells? (*Ans. 0.686*)
>
> **(ii)** If contaminants were not observed by the wells, how confident is the engineer in concluding that no leakage has occurred? (*Ans. 0.955*)

(c) If the cost of installing Well A is the same as Well B, and the budget is sufficient for the installation of only one well, which well should be installed? Please justify. (*Ans. B*)

2.69 Drinking water may be contaminated by two pollutants. In a given community, the probability of its drinking water containing excessive amount of pollutant A is 0.1, whereas that of pollutant B is 0.2. When pollutant A is excessive, it will definitely cause health problems; however, when pollutant B is excessive, it will cause health problems in only 20% of the population who has low natural resistance to that pollutant. Also, data from many similar communities reveal that the presence of these two pollutants in drinking water is not independent; half of those communities whose drinking water contain excessive amounts of pollutant A will also contain excessive amounts of pollutant B.

Suppose a resident is selected at random from this community, what is the probability that he or she will suffer health problems from drinking the water? Assume that a person's resistance to pollutant B is innate, which is independent of the event of having excessive pollutant in the drinking water.

2.70 A contractor submits bids to three highway jobs and two building jobs. The probability of winning each job is 0.6. Assume that winning among the jobs is statistically independent.

(a) What is the probability that the contractor will win at most one job? (*Ans. 0.087*)

(b) What is the probability that the contractor will win at least two jobs? (*Ans. 0.913*)

(c) What is the probability that he or she will win exactly one highway job, but none of the building jobs? (*Ans. 0.046*)

▶ REFERENCES

Feller, W., *An Introduction to Probability Theory and Its Applications*, Vol. 1, 2nd ed., John Wiley and Sons, New York, 1957.

Papoulis, A., *Probability, Random Variables, and Stochastic Processes*, McGraw-Hill Book Co., New York, 1965.

Parzen, E., *Modern Probability Theory and Its Applications*, J. Wiley & Sons, Inc, New York, 1960.

CHAPTER 3

Analytical Models of Random Phenomena

▷ 3.1 RANDOM VARIABLES AND PROBABILITY DISTRIBUTIONS

In Chapter 2, we developed the fundamental concepts and tools for defining the occurrences of random phenomena and the determination of the associated probabilities. In this chapter, we shall introduce the analytics of these concepts, involving *random variables* and *probability distributions*.

3.1.1 Random Events and Random Variables

A random variable is a mathematical vehicle for representing an event in analytical form. In contrast to a deterministic variable that can assume a definite value, the value of a *random variable* may be defined within a range of possible values. If X is defined as a random variable, then $X = x$, or $X < x$, or $X > x$, represents an event, where $(a < x < b)$ is the range of possible values of X. Intuitively, a random variable may be defined mathematically as a mapping function that transforms (or maps) the events in a possibility space into the number system (the real line).

In engineering and the physical sciences, many random phenomena of interest are associated with the numerical outcomes of some physical quantities. In several of the examples we illustrated earlier in Chapter 2, the number of bulldozers that remain operating after 6 months, the time required to complete a project, and the flood of a river above mean flow level are all outcomes in numerical terms. In the other examples, we also illustrated problems in which the possible outcomes are not in numerical terms—for instance, the failure or survival of a chain, the degree of completion of a project, and the opening and closing of highway routes. These latter events may also be identified in numerical terms by artificially assigning numerical values to each of the possible outcomes—for example, assigning a numerical value of 0 to failure and 1 to survival (or nonfailure) of a chain.

Therefore, the possible outcomes of a random phenomenon can be represented by numerical values, either naturally or assigned artificially. In any case, an outcome or event may then be identified through the value or range of values of a function, which is called a *random variable*. We shall denote a random variable with a capital letter, and its possible values with lowercase letters. The value, or range of values, of a random variable then represents a distinct event; for example, if the values of X represent floods above the mean flow level, then $X > 2$ m stands for the occurrence of a flood higher than 2 m, whereas, if X

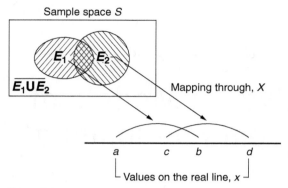

Figure 3.1   Mapping of events in sample space S into real line x.

represents the possible states of a chain (as described in the paragraph above), then $X = 0$ means failure of the chain. In other words, a random variable is a mathematical device for identifying events in numerical terms. Henceforth, in terms of the random variable, we can speak of an event as $(X = a)$, or $(X \le a)$ or $(a < X \le b)$.

More formally, a random variable may be considered as a mathematical function or rule that maps (or transform) events in a sample space into the number system (i.e., the real line). The mapping is unique, and mutually exclusive events are mapped into nonoverlapping intervals on the real line, whereas intersecting events are represented by the respective overlapping intervals on the real line. In Fig. 3.1, the events E_1 and E_2 are mapped into the real line through the random variable X, and thus can be identified, respectively, as indicated below:

$$E_1 = (a < X \le b)$$
$$E_2 = (c < X \le d)$$
$$\overline{E_1 \cup E_2} = (X \le a) + (X > d),$$

and

$$E_1 E_2 = (c < X \le b)$$

Just as a sample space can consist of discrete sample points or a continuum of sample points, a random variable may be *discrete* or *continuous*.

The advantages and purpose of identifying events in numerical terms should be obvious; this will then permit us to conveniently represent events analytically, as well as graphically display the events and their respective probabilities.

3.1.2   Probability Distribution of a Random Variable

As the values or ranges of values of a random variable represent events, the numerical values of the random variable, therefore, are associated with specific probability or probability measures. These probability measures may be assigned according to prescribed rules that are called *probability distributions* or "probability law."

If X is a random variable, its probability distribution can always be described by its *cumulative distribution function* (CDF), which is denoted by

$$F_X \equiv P(X \le x) \qquad \text{for all } x^* \tag{3.1}$$

In Eq. 3.1, X is a *discrete* random variable if only discrete values of x have positive probabilities. Alternatively, X is a *continuous* random variable if probability measures are defined for all values of x. X may also be a *mixed* random variable if its probability distribution is a combination of probabilities at discrete values of x as well as over a range of continuous values of x.

For a discrete random variable X, its probability distribution may also be described by its *probability mass function* (PMF), denoted $p_X(x_i) \equiv P(X = x_i)$, which is simply a function expressing the probability $P(X = x_i)$ for all x_i. In this case, the cumulative distribution function, CDF, is

$$F_X(x) = \sum_{all\ x_i \le x} P(X = x_i) = \sum_{all\ x_i \le x} p_X(x_i) \tag{3.2}$$

However, if X is continuous, probabilities are associated with intervals of x, because events are defined as intervals on the real line x. Therefore, at a specific value of x, such as $X = x$, only the *density of probability* is defined, but there is no probability; i.e., $P(X = x) = 0$. Thus, for a continuous random variable, the probability law is described in terms of the *probability density function* (PDF) denoted as $f_X(x)$ such that the probability of X in the interval $(a, b]$ is

$$P(a < X \le b) = \int_a^b f_X(x)dx \tag{3.3}$$

It follows then that the corresponding distribution function is

$$F_X(x) = P(X \le x) = \int_{-\infty}^x f_X(\tau)\,d\tau \tag{3.4}$$

Accordingly, if $F_X(x)$ has a first derivative, then from Eq. 3.4, its PDF is

$$f_X(x) = \frac{dF_X(x)}{dx} \tag{3.5}$$

We might emphasize again that the PDF, $f_X(x)$, is not a probability but a probability density; this is analogous to a mass density which contains no mass. However,

$$f_X(x)dx = P(x < X \le x + dx)$$

is the probability that X will be in the interval $(x, x + dx]$.

The three types of probability distributions described above are illustrated graphically in Fig. 3.2a–c.

---

*We have adopted the notation of using a capital letter to denote a random variable, and its possible values with the corresponding lower case letter.

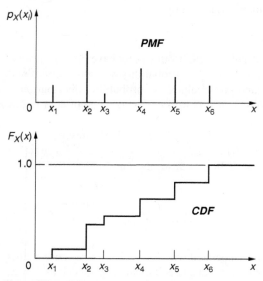

Figure 3.2a Discrete probability distribution.

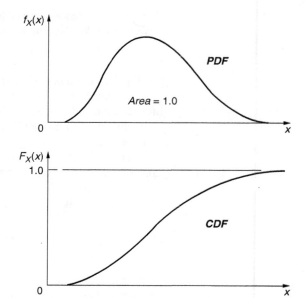

Figure 3.2b Continuous probability distribution.

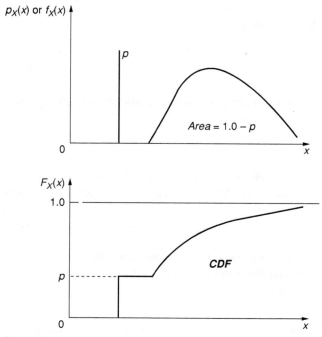

Figure 3.2c Mixed probability distribution.

It may be emphasized that any function used to represent the probability distribution of a random variable must necessarily satisfy the axioms of probability theory, as described earlier in Sect. 2.3.1. Accordingly, the function must be nonnegative and the probabilities associated with all the possible values of the random variable must add up to 1.0. In other words, if $F_X(x)$ is the CDF of X, then it must satisfy the following conditions:

(i) $F_X(-\infty) = 0$;   and $F_X(\infty) = 1.0$

(ii) $F_X(x) \geq 0$, for all values of x, and is nondecreasing with x.

(iii) $F_X(x)$ is continuous to the right with x.

Conversely, any function possessing all the above properties can be a bona fide CDF. By virtue of these properties and Eqs. 3.2 through 3.5, the PMF and PDF are nonnegative functions of x, whereas all the probability masses of a PMF add up to 1.0, and the total area under a PDF is equal to 1.0.

Figure 3.2 presents graphic examples of legitimate probability distributions, and also illustrates the characteristics of the probability distributions of discrete, continuous, and mixed random variables.

We observe that we can write Eq. 3.3 as

$$P(a < X \leq b) = \int_{-\infty}^{b} f_X(x)dx - \int_{-\infty}^{a} f_X(x)dx$$

Similarly, for a discrete X, we have

$$P(a < X \leq b) = \sum_{all\ x_i \leq b} p_X(x_i) - \sum_{all\ x_i \leq a} p_X(x_i)$$

Thus, by virtue of Eqs. 3.2 and 3.4, we have

$$P(a < X \leq b) = F_X(b) - F_X(a) \tag{3.6}$$

▶ **EXAMPLE 3.1**

Let us again consider Example 2.1, which involves a discrete random variable. Using X as the random variable whose values represent the number of operating bulldozers after 6 months, the events of interest are mapped into the real line as shown in Fig. E3.1a. Thus, $(X = 0)$, $(X = 1)$, $(X = 2)$, and $(X = 3)$ now represent the corresponding events of interest.

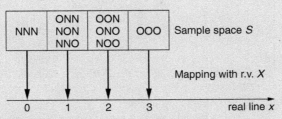

Figure E3.1a Mapping events into real line.

Assuming again that each of the three bulldozers is equally likely to be operating or nonoperating after 6 months, i.e., probability of operating is 0.5, and that the conditions between bulldozers are statistically independent, the PMF of X would be as shown in Fig. E3.1b.

The corresponding CDF would appear as shown in Fig. E3.1c.

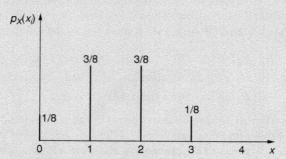

Figure E3.1b PMF of X.

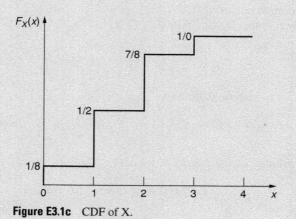

Figure E3.1c CDF of X.

◀

▶ **EXAMPLE 3.2**

For a continuous random variable, consider the 100-kg load in Example 2.5. If the load is equally likely to be placed anywhere along the span of the beam of 10 m, then the PDF of the load position X is uniformly distributed in $0 < x \leq 10$; that is,

$$f_X(x) = c \qquad 0 < x \leq 10$$
$$= 0 \qquad \text{otherwise}$$

where c is a constant. As the area under the PDF must be equal to 1.0, the constant $c = 1/10$. Graphically, the PDF is shown in Fig. E3.2a.

The corresponding CDF of X would be

$$F_X = \int_0^x c\,dx = cx = \frac{x}{10} \qquad 0 < x \leq 10;$$
$$= 1.0 \qquad x > 10;$$
$$= 0 \qquad x < 0$$

Graphically, the above CDF is shown in Fig. E3.2b.

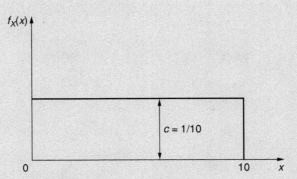

Figure E3.2a PDF of X.

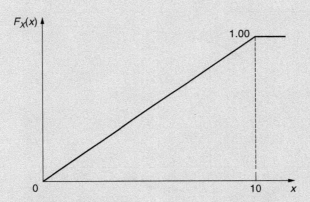

Figure E3.2b CDF of X.

The probability that the position of the load will be between 2 m and 5 m on the beam is

$$P(2 < X \leq 5) = \int_2^5 \frac{1}{10}\, dx = 0.30$$

Alternatively, using Eq. 3.6, we also obtain

$$P(2 < X \leq 5) = F_X(5) - F_X(2) = \frac{5}{10} - \frac{2}{10} = 0.30$$

◄

► **EXAMPLE 3.3**   The useful life, T (in hours) of welding machines is not completely predictable, but may be described by the *exponential* distribution, with the following PDF:

$$\begin{aligned} f_T(t) &= \lambda e^{-\lambda t} && t \geq 0 \\ &= 0 && t < 0 \end{aligned}$$

in which λ is a constant. Graphically, this PDF is shown in Fig. E3.3a.

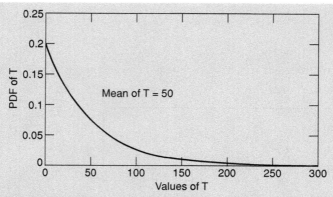

Figure E3.3a Exponential PDF of useful life T.

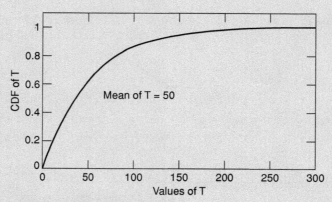

Figure E3.3b CDF of useful life T.

The corresponding CDF would be $F_T(t) = \int_0^t \lambda e^{-\lambda \tau} d\tau = 1 - e^{-\lambda t}$ as shown graphically in Fig. E3.3b. ◀

3.1.3 Main Descriptors of a Random Variable

The probabilistic properties of a random variable would be completely described if the form of the distribution function, in terms of its CDF or PDF (for a continuous random variable) or its PMF (in the case of a discrete random variable) and its associated parameters, are specified. In practice, the exact form of the probability distribution function may not be known; in such cases, approximate description of a random variable in terms of its *main descriptors* can be useful and sometimes may be necessary.

Moreover, even when the distribution function is known, the main descriptors remain useful, because they contain information on the properties of the random variable that are of first importance in many practical applications. Also, the parameters of a known distribution function may be derived as functions of these principal quantities, or in some cases may be the pertinent parameters themselves (see Chapter 6).

Central Values

As there is a range of possible values of a random variable, some *central value* of this range, such as the average value, would naturally be of special interest. In particular, because the different values of the random variable are associated with different probabilities or probability densities, the "weighted average" (i.e., weighted by the respective probability measures) would be of special interest; this weighted average is the *mean value* or the *expected value* of the random variable.

Therefore, if X is a discrete random variable with PMF, $p_X(x_i)$, its mean value, denoted $E(X)$, is

$$E(X) = \sum_{all\ x_i} x_i\, p_X(x_i) \tag{3.7a}$$

whereas if X is a continuous random variable with PDF, $f_X(x)$, its mean value is

$$E(X) = \int_{-\infty}^{\infty} x f_X(x)dx \tag{3.7b}$$

Other quantities that may be used to designate the central value of a random variable include the *mode*, or modal value, and the *median*. The mode, which we shall denote as $\tilde{x}$, is the most probable value of a random variable; i.e., it is the value of the random variable with the largest probability or the highest probability density. The median of a random variable X, which we shall denote as x_m, is the value at which the CDF is 50% and thus larger and smaller values are equally probable; that is,

$$F_X(x_m) = 0.50 \tag{3.8}$$

In general, the mean, mode, and median of a random variable are different, especially if the underlying PDF is skewed (nonsymmetric). However, if the PDF is symmetric and unimodal (single mode), these three quantities will be the same.

Mathematical Expectation

The notion of an expected value, or weighted average, as described in Eq. 3.7, can be generalized for a function of X. Given a function $g(X)$, its expected value $E[g(X)]$, can be obtained as a generalization of Eq. 3.7 as

$$E[g(X)] = \sum_{all\ x_i} g(x_i)p_X(x_i) \tag{3.9a}$$

if X is discrete, whereas if X is continuous,

$$E[g(X)] = \int_{-\infty}^{\infty} g(x)f_X(x)dx \tag{3.9b}$$

In either case, $E[g(X)]$ is known as the *mathematical expectation* of $g(X)$ and is the weighted average of the function $g(X)$.

Measures of Dispersion

Again, as there is a range of possible values of a random variable associated with different probabilities or probability measures, $p_X(x_i)$ or $f_X(x)$, some *measure of dispersion* is needed to indicate how widely or narrowly the values of the random variable are dispersed. Of special interest is a quantity that gives a measure of how closely or widely the values of the variate are clustered around a central value. Intuitively, such a quantity must be a function of the deviations from the central value. However, whether a deviation is above or below the central value should be immaterial; therefore, the function should be an even function of the deviations.

If the deviations are defined relative to the mean value, then a suitable measure of the dispersion is the *variance*. For a discrete random variable X with PMF $p_X(x_i)$, the variance of X is

$$Var(X) = \sum (x_i - \mu_X)^2\, p_X(x_i) \tag{3.10a}$$

in which $\mu_X \equiv E(X)$. We might observe that this is the weighted average of the squared deviations from the mean, or with reference to Eq. 3.9 it is the mathematical expectation of $g(X) = (X - \mu_X)^2$. Therefore, according to Eq. 3.9b, for a continuous X with PDF $f_X(x)$, the variance of X is

$$Var(X) = \int_{-\infty}^{\infty} (x - \mu_X)^2\, f_X(x)dx \tag{3.10b}$$

Expanding the integrand of 3.10b, we have

$$Var(x) = \int_{-\infty}^{\infty} (x^2 - 2\mu_X x + \mu_X^2)\, f_X(x)dx$$
$$= E(X^2) - 2\mu_X E(X) + \mu_X^2 = E(X^2) - 2\mu_X^2 + \mu_X^2$$

The same result would be obtained by expanding Eq. 3.10a. Thus, a useful relation for the variance is

$$Var(X) = E(X^2) - \mu_X^2 \tag{3.11}$$

In Eq. 3.11, the term $E(X^2)$ is known as the *mean square* value of X.

Dimensionally, a more convenient measure of dispersion is the square root of the variance, which is the *standard deviation* σ_X; that is,

$$\sigma_X = \sqrt{Var(X)} \tag{3.12}$$

We might observe that based solely on the value of the variance or standard deviation, it may be difficult to state the degree of dispersion; for this purpose, a measure of the dispersion relative to the central value would be appropriate. In other words, whether the dispersion is large or small is more meaningful if measured relative to the central value. For this reason, for $\mu_X > 0$, the *coefficient of variation* (c.o.v.),

$$\delta_X = \frac{\sigma_X}{\mu_X} \tag{3.13}$$

is often a preferred and convenient nondimensional measure of dispersion or variability.

▶ **EXAMPLE 3.4**

In Example 3.1, the PMF of the number of operating bulldozers after 6 months is shown in Fig. E3.1b. On this basis, we obtain the expected number of operating bulldozers after 6 months as

$$\mu_X = E(X) = 0(1/8) + 1(3/8) + 2(3/8) + 3(1/8)$$
$$= 1.50$$

As the random variable is discrete, the mean value of 1.5 is not necessarily a possible value; in this case, we may only conclude that the mean number of operating bulldozers is between 1 and 2 at the end of 6 months.

The corresponding variance is

$$Var(X) = (0 - 1.5)^2(1/8) + (1 - 1.5)^2(3/8) + (2 - 1.5)^2(3/8) + (3 - 1.5)^2(1/8)$$
$$= 0.75$$

We may also compute the variance using Eq. 3.11, as follows:

$$Var(X) = [0^2(1/8) + 1^2(3/8) + 2^2(3/8) + 3^2(1/8)] - (1.5)^2$$
$$= 0.75$$

The corresponding standard deviation is, therefore,

$$\sigma_X = \sqrt{0.75} = 0.866$$

and the coefficient of variation, c.o.v., is

$$\delta_X = \frac{0.866}{1.50} = 0.577$$

which means that the degree of dispersion is over 50% of the mean value, a relatively large dispersion. ◀

▶ **EXAMPLE 3.5**

In Example 3.3, the useful life, T, of welding machines is a random variable with an exponential probability distribution; the PDF and CDF are, respectively, as follows:

$$f_T(t) = \lambda e^{-\lambda t} \quad \text{and} \quad F_T(t) = 1 - e^{-\lambda t}; \qquad t \geq 0$$

These were also shown graphically in Figs. E3.3a and E3.3b.

The mean life of the welding machines is then

$$\mu_T = E(T) = \int_0^\infty t\lambda e^{-\lambda t} dt$$

Performing the integration by parts, we obtain $\mu_T = 1/\lambda$.

Therefore, the parameter λ of the exponential distribution is the reciprocal of the mean value; i.e., $\lambda = 1/E(T)$.

In this case, the mode is zero, whereas the median life t_m is obtained as follows:

$$\int_0^{t_m} \lambda e^{-\lambda t} dt = 0.50$$

obtaining the median life

$$t_m = \frac{-\ln 0.50}{\lambda} = \frac{0.693}{\lambda}$$

Therefore,

$$t_m = 0.693 \, \mu_T$$

The variance of T is

$$Var(T) = \int_0^\infty (t - 1/\lambda)^2 \lambda e^{-\lambda t} \, dt$$

Integration by parts yields

$$Var(T) = \frac{1}{\lambda^2}$$

From which we obtain the standard deviation of T

$$\sigma_T = \frac{1}{\lambda} = \mu_T$$

This shows that the c.o.v. of the exponential distribution is 100%. ◄

► **EXAMPLE 3.6**

Suppose a construction company has an experience record showing that 60% of its projects were completed on schedule. If this record prevails, the probability of the number of on-schedule completions in the next six projects can be described by the *binomial* distribution (see Sect. 3.2.3) as follows:

If X is the number of projects completed on schedule among 6 future projects, then (see Eq. 3.30),

$$P(X = x) = \binom{6}{x} (0.6)^x (0.4)^{6-x} \qquad x = 0, 1, 2, \ldots, 6$$
$$= 0 \qquad\qquad\qquad \text{otherwise}$$

the factor $\binom{n}{r} = \dfrac{n!}{r!(n-r)!}$ is known as the *binomial coefficient*. In the present problem, $n = 6$ and $r = x$.

The above PMF is shown graphically in Fig. E3.6.

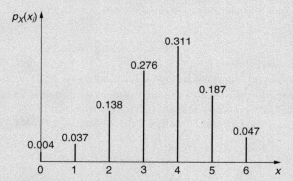

Figure E3.6 PMF of X, no. of projects completed on schedule.

The mean number of projects that will be completed on schedule is

$$E(X) = \sum x \frac{6!}{x!(6-x)!}(0.6)^x(0.4)^{6-x}$$

$$= (1)\frac{6!}{1!(6-1)!}(0.6)(0.4)^5 + (2)\frac{6!}{2!(6-2)!}(0.6)^2(0.4)^4 + (3)\frac{6!}{3!(6-3)!}(0.6)^3(0.4)^3$$

$$+ (4)\frac{6!}{4!(6-4)!}(0.6)^4(0.4)^2 + (5)\frac{6}{5!(6-5)!}(0.6)^5(0.4) + (6)\frac{6!}{6!(6-6)!}(0.6)^6(0.4)^0$$

$$= 0.03686 + 2(0.13830) + 3(0.27640) + 4(0.31110) + 5(0.18660) + 6(0.04666)$$

$$= 3.60$$

Therefore, the expected number of projects among six that can be completed on schedule by the company is between three and four. The corresponding variance is

$$Var(X) = \sum_{x=0}^{6} (x - 3.60)^2 \frac{6!}{x!(6-x)!} (0.6)^x (0.4)^{6-x}$$

$$= (-3.60)^2 \frac{6!}{0!(6-0)!} (0.6)^0 (0.4)^6 + (1 - 3.60)^2 \frac{6!}{1!(6-1)!} (0.6)(0.4)^5$$

$$+ (2 - 3.60)^2 \frac{6!}{2!(6-2)!} (0.6)^2 (0.4)^4 + (3 - 3.60)^2 \frac{6!}{3!(6-3)!} (0.6)^3 (0.4)^3$$

$$+ (4 - 3.60)^2 \frac{6!}{4!(6-4)!} (0.6)^4 (0.4)^2 + (5 - 3.60)^2 \frac{6!}{5!(6-5)!} (0.6)^5 (0.4)$$

$$+ (6 - 3.60)^2 \frac{6!}{6!(6-6)!} (0.6)^6 (0.4)^0$$

$$= 0.0531 + 0.2492 + 0.3539 + 0.0995 + 0.0498 + 0.3658 + 0.2684$$

$$= 1.44$$

The standard deviation, therefore, is

$$\sigma_X = \sqrt{1.44} = 1.20$$

and the coefficient of variation, c.o.v., is

$$\delta_X = \frac{1.20}{3.60} = 0.333$$

In this case, $X = 4$ has the highest probability of 0.311 as shown in Fig. E3.6. Therefore, the most probable number of projects that will be completed on schedule is four. ◀

Measure of Skewness

Another useful and important property of a random variable is the symmetry or asymmetry of its PDF or PMF, and the associated degree and direction of asymmetry. A measure of this asymmetry or *skewness* is the *third central moment*; i.e.,

$$E(X - \mu_X)^3 = \sum_{all \ x_i} (x_i - \mu_X)^3 \ p_X(x_i) \qquad \text{for discrete } X$$

and

$$E(X - \mu_X)^3 = \int_{-\infty}^{\infty} (x - \mu_X)^3 \ f_X(x)dx \qquad \text{for continuous } X$$

The above third moment would be zero if the PDF or PMF of the random variable is symmetric about the mean value μ_X; otherwise, it may be positive or negative. It will be positive if the values of X above μ_X are more widely dispersed than the dispersion of the values below μ_X. Conversely, this third moment will be negative if the relative degree of dispersion is reversed.

Therefore, the skewness of the PDF or PMF of a random variable may be designated as *positive* or *negative* in accordance with the sign of the third central moment $E(X - \mu_X)^3$; the magnitude of this third moment also indicates the corresponding degree of skewness. These properties are illustrated in Fig. 3.3.

A convenient dimensionless measure of skewness is the *skewness coefficient*

$$\theta = \frac{E(X - \mu_X)^3}{\sigma^3} \tag{3.14}$$

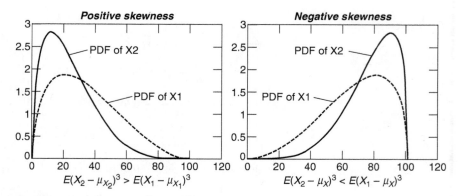

Figure 3.3 Properties of asymmetric PDFs.

The Kurtosis

Finally, another useful property is the fourth central moment of a random variable known as the *kurtosis*. Extending what we have presented above for the second and third central moments, the kurtosis is

$$E(X - \mu_X)^4 = \sum_{all\ x_i} (x_i - \mu_X)^4 p_X(x_i) \qquad \text{for discrete } X$$

and

$$E(X - \mu_X)^4 = \int_{-\infty}^{\infty} (x - \mu_X)^4 f_X(x)dx \qquad \text{for continuous } X$$

Physically, the kurtosis is a measure of the *peakedness* of the underlying PDF of *X*.

Analogies with Geometrical Properties

The mean value and the variance of a random variable are analogous, respectively, to the *centroidal distance* and the *central moment of inertia* of a unit area. To see this, we consider a unit area with an irregular shape defined by the function $y = f(x)$ as shown in Fig. 3.4.

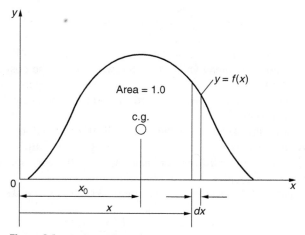

Figure 3.4 An irregular unit area.

The centroidal distance, x_o, of the area is

$$x_o = \frac{\int_{-\infty}^{\infty} x f(x)dx}{area} = \int_{-\infty}^{\infty} x f(x)dx \qquad (3.15)$$

which is also the first moment (about the origin) of the irregular-shaped unit area. The moment of inertia of the area about the vertical axis through the centroid is

$$I_y = \int_{-\infty}^{\infty} (x - x_o)^2 f(x)dx \qquad (3.16)$$

Comparing Eq. 3.7b with Eq. 3.15, and Eq. 3.10b with Eq. 3.16, we see that the expected value is equivalent to the centroidal distance, whereas the variance is equivalent to the moment of inertia of a unit area.

We may, therefore, refer to the mean value as the *first moment* and the variance as the *second (central) moment* of a random variable. Extending this terminology, we may define the *nth moment* of a random variable X as

$$E(X^n) = \int_{-\infty}^{\infty} x^n f_X(x)dx \qquad (3.17)$$

► **EXAMPLE 3.7**

Let us calculate the skewness of the PMF of Example 3.6. As this is a discrete random variable, we calculate the sknewness as follows:

From Example 3.6, we have $\mu_X = 3.60$ and $\sigma_X = 1.11$.

$$E(X - \mu_X)^3 = \sum_{i=0}^{6} (x_i - 3.60)^3 p_X(x_i)$$

$$= (0 - 3.60)^3(0.004) + (1 - 3.60)^3(0.037) + (2 - 3.60)^3(0.138) + (3 - 3.60)^3(0.276)$$

$$+ (4 - 3.60)^3(0.311) + (5 - 3.60)^3(0.187) + (6 - 3.60)^3(0.047)$$

$$= -0.279$$

Therefore, the distribution of X is negatively skewed with a skewness coefficient, according to Eq. 3.14, of

$$\theta = -\frac{0.279}{1.11^3} = -0.204$$

◄

► **EXAMPLE 3.8**

For the exponential distribution of the useful life of welding machines, T, of Example 3.5, the mean life of the machines is μ_T. Then, the third central moment of the PDF is (using μ for μ_T),

$$E(T - \mu)^3 = \int_0^{\infty} (t - \mu)^3 \left(\frac{1}{\mu} e^{-t/\mu}\right) dt$$

$$= \frac{1}{\mu} \int_0^{\infty} (t^3 - 3t^2\mu + 3t\mu^2 - \mu^3) e^{-t/\mu} dt$$

$$= \mu^3 e^{-t/\mu} \left\{ -\left[\left(\frac{t}{\mu}\right)^3 + 3\left(\frac{t}{\mu}\right)^2 + 6\left(\frac{t}{\mu}\right) + 6\right] + 3\left[\left(\frac{t}{\mu}\right)^2 + 2\left(\frac{t}{\mu}\right) + 2\right] \right.$$

$$\left. - 3\left[\left(\frac{t}{\mu}\right) + 1\right] + 1 \right\}_0^{\infty}$$

$$= -\mu^3 e^{-t/\mu} \left[\left(\frac{t}{\mu}\right)^3 + \left(\frac{t}{\mu}\right) + 2\right]_0^{\infty} = 2\mu^3$$

We recall from Example 3.5 that the standard deviation of the exponential distribution is $\sigma_T = \mu_T$. Therefore, the skewness coefficient of this distribution, according to Eq. 3.14, is $\theta = 2.0$. ◀

▶ 3.2 USEFUL PROBABILITY DISTRIBUTIONS

Theoretically, any function possessing all the properties described earlier in Sect. 3.1.1 can be used to represent the probability distribution of a random variable. For practical purposes, however, there are a number of both discrete and continuous distribution functions that are especially useful because of one or more of the following reasons:

1. The function is the result of an underlying physical process and can be derived on the basis of certain physically reasonable assumptions.
2. The function is the result of some limiting process.
3. It is widely known, and the necessary probability and statistical information (including probability tables) are widely available.

Several of these probability distribution functions are presented, and their special properties are described and illustrated in this section.

3.2.1 The Gaussian (or Normal) Distribution

The best known and most widely used probability distribution is undoubtedly the *Gaussian distribution*, which is also known as the *normal distribution*. Its PDF for a continuous random variable X, is given by

$$f_X(x) = \frac{1}{\sigma\sqrt{2\pi}}\exp\left[-\frac{1}{2}\left(\frac{x-\mu}{\sigma}\right)^2\right] \qquad -\infty < x < \infty \tag{3.18}$$

where, μ and σ are the parameters of the distribution. In this case, for the Gaussian distribution, these parameters are also the mean and standard deviation, respectively, of the random variable X. A popular and convenient short notation for this distribution is $N(\mu, \sigma)$, which we shall also adopt in specifying the normal PDF.

The significance of the two parameters, μ and σ, may be observed graphically from Fig. 3.5. In Fig. 3.5a, μ is held constant at 60 and σ varies with 10, 20, and 30, whereas in Fig. 3.5b, μ varies with values of 30, 45, and 60 for a constant $\sigma = 10$.

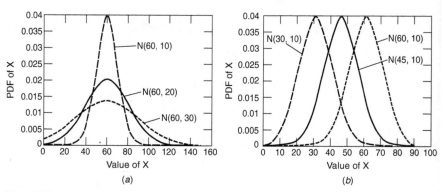

Figure 3.5 Gaussian PDFs (a) with varying σ; (b) with varying μ.

The Standard Normal Distribution

A Gaussian probability distribution with the parameters $\mu = 0$ and $\sigma = 1.0$ is known as the *standard normal* distribution and is denoted appropriately as $N(0, 1)$. Its PDF is accordingly,

$$f_X(x) = \frac{1}{\sqrt{2\pi}} e^{-(1/2)x^2} \qquad -\infty < x < \infty \tag{3.18a}$$

Because of its wide usage, a special notation $\Phi(s)$ is commonly used to designate the cumulative distribution function (CDF) of the *standard normal variate* S; i.e.,

$$\Phi(s) = F_S(s)$$

Referring to Fig. 3.6, we will denote the shaded area p under the PDF as $\Phi(s_p)$.

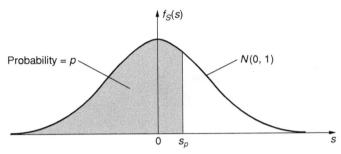

Figure 3.6 CDF of $N(0,1)$.

Conversely, the value of the standard normal variate at a cumulative probability p would be denoted as

$$s_p = \Phi^{-1}(p)$$

These notations will be adopted throughout this book.

In Fig. 3.7, we can see that the areas (or probabilities) covered within ± 1, ± 2, and ± 3 σ's, i.e., $\pm$ number of standard deviations about the mean $\mu = 0$ of the standard normal distribution are, respectively, equal to 68.3%, 95.4%, and 99.7%. The same probabilities are also valid for any general normal distributions, i.e., $N(\mu, \sigma)$, with the same $\pm$ number of σ's about the mean μ.

The CDF of $N(0, 1)$, i.e., $\Phi(s)$, is tabulated widely as tables of normal probabilities, such as Table A.1 of Appendix A. It is important to observe that the probabilities in Table A.1 start at $x = 0$, and are given only for positive values of the random variable. By virtue of symmetry of the standard normal PDF about zero, the probabilities for negative values of the variate can be obtained as

$$\Phi(-s) = 1 - \Phi(s) \tag{3.19}$$

Furthermore, values of s corresponding to probabilities $p < 0.5$ are negative and may be obtained using Table A.1 as

$$s = \Phi^{-1}(p) = -\Phi^{-1}(1 - p) \tag{3.20}$$

The real utility of $\Phi(s)$ as given in Appendix A.1 is to calculate the probabilities of any Gaussian distribution. That is, the probabilities of any normal distribution can be evaluated readily using $\Phi(s)$ as follows. Suppose a normal variate X with distribution $N(\mu, \sigma)$; the probability of $(a < X \leq b)$ is

$$P(a < X \leq b) = \frac{1}{\sigma\sqrt{2\pi}} \int_a^b \exp\left[-\frac{1}{2}\left(\frac{x-\mu}{\sigma}\right)^2\right] dx$$

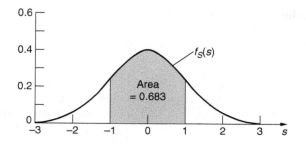

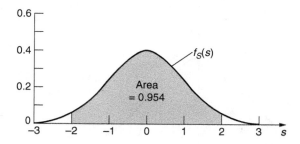

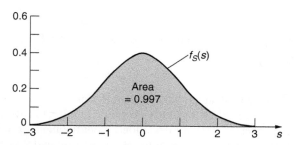

Figure 3.7 PDFs of $N(0, 1)$ with areas covering 1, 2, 3 σ's.

Clearly, this is the area under the general normal PDF between a and b as shown in Fig. 3.8. Theoretically, the above probability can be obtained by performing the integral directly; however, this can be evaluated more readily using the tabulated values of $\Phi(s)$ given in Table A.1. For this purpose, we make the following change of variables:

$$s = \frac{x - \mu}{\sigma} \qquad \text{and} \qquad dx = \sigma ds$$

Then, the above integral becomes

$$P(a < X \le b) = \frac{1}{\sigma\sqrt{2\pi}} \int_{(a-\mu)/\sigma}^{(b-\mu)/\sigma} e^{-(1/2)s^2} \sigma \, ds$$

$$= \frac{1}{\sqrt{2\pi}} \int_{(a-\mu)/\sigma}^{(b-\mu)/\sigma} e^{-(1/2)s^2} \, ds$$

which we may recognize is the area under the PDF of $N(0, 1)$ between $(a - \mu)/\sigma$ and $(b - \mu)/\sigma$. Therefore, according to Eq. 3.6, the above probability may be evaluated as

$$P(a < X \le b) = \Phi\left(\frac{b - \mu}{\sigma}\right) - \Phi\left(\frac{a - \mu}{\sigma}\right) \qquad (3.21)$$

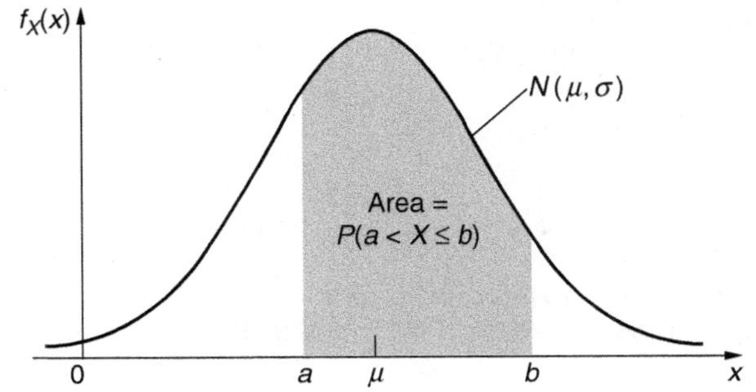

Figure 3.8 Area of $N(\mu, \sigma)$ between a and b.

▶ **EXAMPLE 3.9**

The drainage from a community during a storm is a normal random variable estimated to have a mean of 1.2 million gallons per day (mgd) and a standard deviation of 0.4 mgd; i.e., $N(1.2, 0.4)$ mgd. If the storm drain system is designed with a maximum drainage capacity of 1.5 mgd, what is the underlying probability of flooding during a storm that is assumed in the design of the drainage system?

Flooding in the community will occur when the drainage load exceeds the capacity of the drainage system; therefore, the probability of flooding is

$$P(X > 1.5) = 1 - P(X \leq 1.5) = 1 - \Phi\left(\frac{1.5 - 1.2}{0.4}\right) = 1 - \Phi(0.75)$$

$$= 1 - 0.7734 = 0.227$$

In the above, we obtained $\Phi(0.75) = 0.7734$ from Table A.1 in Appendix A.

Also of related interest are the following: (i) The probability that the drainage during a storm will be between 1.0 mgd and 1.6 mgd, which is

$$P(1.0 < X \leq 1.6) = \Phi\left(\frac{1.6 - 1.2}{0.4}\right) - \Phi\left(\frac{1.0 - 1.2}{0.4}\right) = \Phi(1.0) - \Phi(-0.5)$$

$$= 0.8413 - [1 - \Phi(0.5)] = 0.8413 - (1 - 0.6915)$$

$$= 0.533$$

(ii) The 90-percentile drainage load from the community during a storm. This is the value of the random variable at which the cumulative probability is less than 0.90, which we would obtain as

$$P(X \leq x_{0.90}) = \Phi\left(\frac{x_{0.90} - 1.2}{0.40}\right) = 0.90$$

Therefore,

$$\frac{x_{0.90} - 1.2}{0.40} = \Phi^{-1}(0.90)$$

from Table A.1, we obtain $\Phi^{-1}(0.90) = 1.28$; thus,

$$x_{0.90} = 1.28(0.40) + 1.2 = 1.71 \text{ mgd}$$

◀

▶ **EXAMPLE 3.10**

Statistical data of vehicular accidents show that the *annual vehicle miles* (i.e., miles per vehicle per year) driven between traffic accidents (of all severities) can be represented by a normal random variable with a mean of 15,000 miles per year and a c.o.v. of 25%.

The standard deviation is $\sigma = 0.25 \, (15,000) = 3750$ miles per year. Therefore, the distribution of miles driven between accidents is $N(15,000; 3750)$ miles per year.

Then, for a typical driver who drives 10,000 miles per year, the probability of him/her having an accident in a year is

$$P(X < 10,000) = \Phi\left(\frac{10,000 - 15,000}{3,750}\right) = \Phi(-1.33) = 1 - \Phi(1.33)$$
$$= 1 - 0.9082$$
$$= 0.092$$

which means that the probability of accident of an individual driver is around 9% annually, or the probability of no accident is 91%. If the driver has driven 8000 miles in a given year without encountering any accident, what is the probability of his/her having an accident for the remainder of that year? In this case, we have a conditional probability as follows:

$$P(X < 10,000 \mid X > 8,000) = \frac{P(8,000 < X < 10,000)}{P(X > 8,000)} = \frac{0.092 - \Phi\left(\dfrac{8,000 - 15,000}{3,750}\right)}{1 - \Phi\left(\dfrac{8,000 - 15,000}{3,759}\right)}$$

$$= \frac{0.092 - \Phi(-1.87)}{[1 - \Phi(-1.87)]} = \frac{0.092 - (1 - 0.9692)}{1 - (1 - 0.9692)} = \frac{0.061}{0.9692}$$
$$= 0.063$$

Therefore, the probability of accident-free driving for the remainder of the year would be around 94%. ◀

▶ **EXAMPLE 3.11**

In the fabrication of steel beams and columns by a manufacturer, there are unavoidable variabilities in the dimensions (for example, length) of the steel members. Suppose in a building construction, the erection of the beams and columns for the building frame would require that the actual lengths be within a tolerance of ±5 mm of the specified dimensions with a probability or reliability of 99.7%. What is the required precision, in terms of an allowable σ, of the production process?

The variability in the fabrication of steel beams may be assumed to be normally distributed with zero mean (denoting no systematic bias in the fabrication process) and a standard deviation σ, representing the precision of the process. In this case, the reliability of 99.7% is equivalent to ±3σ's, as shown in Fig. 3.7 for the normal PDF. Therefore, to satisfy the required reliability, the tolerance E must be,

$$P(-5 \leq E \leq 5) = \Phi\left(\frac{5 - 0}{\sigma}\right) - \Phi\left(\frac{-5 - 0}{\sigma}\right) = 2\Phi\left(\frac{5}{\sigma}\right) - 1 = 0.997$$

or

$$\frac{5}{\sigma} = \Phi^{-1}(0.9985) = 2.97$$

From which we obtain $\sigma = 1.68$ mm. Hence, the required precision, σ, of the fabrication process is determined to be 1.68 mm. This is an illustration of the 6-σ rule in quality control. ◀

3.2.2 The Lognormal Distribution

The *logarithmic normal* or simply *lognormal* distribution is also a popular probability distribution. If a random variable X has a lognormal distribution, its PDF is

$$f_X(x) = \frac{1}{\sqrt{2\pi}(\zeta x)} \exp\left[-\frac{1}{2}\left(\frac{\ln x - \lambda}{\zeta}\right)^2\right] \qquad x \geq 0 \qquad (3.22)$$

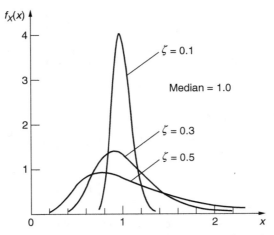

Figure 3.9 The lognormal PDFs for various values of ζ.

where $\lambda = E(\ln X)$ and $\zeta = \sqrt{Var(\ln X)}$, are the parameters of the distribution, which means that these parameters are, respectively, also the mean and standard deviation of $\ln X$. The above PDF is illustrated graphically in Fig. 3.9 for various values of the parameter ζ. Observe that this distribution is strictly for positive values of the variate X.

We may observe from Fig. 3.9 that as the parameter ζ increases, the positive skewness of the PDF also increases.

In Example 4.2 of Chapter 4, we shall show that if X is lognormal with parameters λ and ζ, then $\ln X$ is normal with mean λ and standard deviation ζ; i.e., $N(\lambda, \zeta)$. Because of this logarithmic relationship with the normal distribution, probabilities associated with the lognormal distribution can be determined conveniently also from the table of standard normal probabilities, such as Table A.1. We show this with the following: On the basis of Eqs. 3.3 and 3.22, the probability that X will assume values in an interval $(a, b]$ is

$$P(a < X \le b) = \int_a^b \frac{1}{\sqrt{2\pi}\,\zeta x}\exp\left[-\frac{1}{2}\left(\frac{\ln x - \lambda}{\zeta}\right)^2\right]dx$$

Let $s = \dfrac{\ln x - \lambda}{\zeta}$, such that $dx = \zeta x\,ds$.

Then, the above integral yields

$$P(a < X \le b) = \frac{1}{\sqrt{2\pi}}\int_{(\ln a - \lambda)/\zeta}^{(\ln b - \lambda)/\zeta} e^{-(1/2)s^2}\,ds$$

$$= \Phi\left(\frac{\ln b - \lambda}{\zeta}\right) - \Phi\left(\frac{\ln a - \lambda}{\zeta}\right)$$

(3.23)

which is analogous to Eq. 3.21, and the calculations involve the same standard normal probabilities that can be obtained from Table A.1.

In view of the convenient facility for calculating the probabilities of a lognormal random variable, as indicated in Eq. 3.23, and also because the values of the random variable must be positive (defined only for $x \ge 0$), the lognormal distribution is especially useful in those applications where the values of the variate are known, from physical consideration, to be strictly positive; this includes, for example, the strength and fatigue life of materials, the intensity of rainfall, the volume of air traffic, etc.

The parameters λ and ζ of the lognormal distribution are related to the mean and standard deviation of the random variable X as follows:

$$\mu_X = E(X) = \frac{1}{\sqrt{2\pi}\,\zeta} \int_0^\infty \exp\left[-\frac{1}{2}\left(\frac{\ln x - \lambda}{\zeta}\right)^2\right] dx$$

Now let $y = \ln x$, and $dx = x\,dy$, then

$$\mu_X = \frac{1}{\sqrt{2\pi}\,\zeta} \int_{-\infty}^\infty e^y \exp\left[-\frac{1}{2}\left(\frac{y - \lambda}{\zeta}\right)^2\right] dy$$

$$= \frac{1}{\sqrt{2\pi}\,\zeta} \int_{-\infty}^\infty \exp\left[y - \frac{1}{2}\left(\frac{y - \lambda}{\zeta}\right)^2\right] dy$$

Expanding the square term in the exponential, we obtain

$$\mu_X = \left\{\frac{1}{\sqrt{2\pi}\,\zeta} \int_{-\infty}^\infty \exp\left[-\frac{1}{2}\left(\frac{y - (\lambda + \zeta^2)}{\zeta}\right)^2\right] dy\right\} \exp\left(\lambda + \frac{1}{2}\zeta^2\right)$$

We can recognize that the integral within the braces is the total area under the Gaussian PDF of $N(\lambda + \zeta^2, \zeta)$, which is equal to 1.0. Hence,

$$\mu_X = \exp\left(\lambda + \frac{1}{2}\zeta^2\right) \tag{3.24a}$$

from which we obtain the parameter

$$\lambda = \ln \mu_X - \frac{1}{2}\zeta^2 \tag{3.24b}$$

Similarly, we derive the variance of X as follows:

$$E(X^2) = \frac{1}{\sqrt{2\pi}\,\zeta} \int_{-\infty}^\infty e^{2y} \exp\left[-\frac{1}{2}\left(\frac{y - \lambda}{\zeta}\right)^2\right] dy$$

$$= \frac{1}{\sqrt{2\pi}\,\zeta} \int_{-\infty}^\infty \exp\left\{-\frac{1}{2\zeta^2}[y^2 - 2(\lambda + 2\zeta^2)y + \lambda^2]\right\} dy$$

By completing the square term in the exponential, the above integral yields

$$E(X^2) = \left\{\frac{1}{\sqrt{2\pi}\,\zeta} \int_{-\infty}^\infty \exp\left[-\frac{1}{2}\left(\frac{y - (\lambda + 2\zeta^2)}{\zeta}\right)^2\right] dy\right\} \exp[2(\lambda + \zeta^2)]$$

Again, we can recognize that the quantity inside the braces is the total area under the PDF of $N(\lambda + 2\zeta^2, \zeta)$, which is equal to 1.0. Thus, we have

$$E(X^2) = \exp[2(\lambda + \zeta^2)] = \exp\left[2\left(\lambda + \frac{1}{2}\zeta^2\right) + \zeta^2\right] = \mu_X^2 \cdot e^{\zeta^2}$$

Thence, according to Eq. 3.11 and using Eq. 3.24a, we obtain the variance of X as

$$Var(X) = E(X^2) - \mu_X^2 = \mu_X^2(e^{\zeta^2} - 1)$$

from which we obtain the second parameter

$$\zeta^2 = \ln\left[1 + \left(\frac{\sigma_X}{\mu_X}\right)^2\right] = \ln\left(1 + \delta_X{}^2\right) \qquad (3.25)$$

We should note that if the c.o.v. of X, δ_X, is not large, say ≤ 0.30, $\ln(1 + \delta_X{}^2) \simeq \delta_X{}^2$. In these cases,

$$\zeta \simeq \delta_X \qquad (3.26)$$

The median, instead of the mean, is often used to designate the central value of a lognormal random variable. By the definition of Eq. 3.8, the median x_m is $P(X \leq x_m) = 0.50$.

For the lognormal distribution, this means

$$\Phi\left(\frac{\ln x_m - \lambda}{\zeta}\right) = 0.50$$

Thus,

$$\frac{\ln x_m - \lambda}{\zeta} = \Phi^{-1}(0.50) = 0$$

Hence, in terms of the median of X, the parameter λ is

$$\lambda = \ln x_m \qquad (3.27)$$

Conversely,

$$x_m = e^\lambda \qquad (3.28)$$

Equating the right sides of Eqs. 3.24b and 3.27, and using Eq. 3.25, we obtain the relation between the mean and the median of a lognormal random variable as

$$\mu_X = x_m\sqrt{1 + \delta_X{}^2} \qquad (3.29)$$

which means that the mean value of a lognormal variate is invariably larger than the corresponding median; i.e., $\mu_X > x_m$.

▶ **EXAMPLE 3.12**

In Example 3.9, if the distribution of storm drainage from the community is a lognormal random variable instead of normal, with the same mean and standard deviation, the probability of flooding during a storm would be evaluated as follows.

First, we obtain the parameters λ and ζ of the lognormal distribution as follows: From Eq. 3.25,

$$\zeta^2 = \ln\left[1 + \left(\frac{0.4}{1.2}\right)^2\right] = \ln(1.111) = 0.105$$

Thus,

$$\zeta = 0.324$$

And with Eq. 3.24b,

$$\lambda = \ln 1.20 - \frac{1}{2}(0.324)^2 = 0.130$$

Therefore, the probability of flooding becomes

$$P(X > 1.50) = 1 - P(X \leq 1.50) = 1 - \Phi\left(\frac{\ln 1.5 - 0.130}{0.324}\right) = 1 - \Phi(0.85) = 1 - 0.8023$$

$$= 0.198$$

which may be compared with the probability of 0.227 from Example 3.9, illustrating the fact that the result depends on the underlying distribution of the random variable.

Also, with the lognormal distribution, we obtain the probability that the drainage will be between 1.0 mgd and 1.6 mgd:

$$P(1.0 < X \le 1.6) = \Phi\left(\frac{\ln 1.6 - 0.130}{0.324}\right) - \Phi\left(\frac{\ln 1.0 - 0.130}{0.324}\right) = \Phi(1.049) - \Phi(-0.401)$$

$$= 0.8531 - [1 - \Phi(0.401)] = 0.8531 - (1 - 0.6554)$$

$$= 0.509$$

This may also be compared with the result of Example 3.9 which was 0.533. Finally, the 90% value of the drainage load from the community, with the lognormal distribution, would be

$$P(X \le x_{0.90}) = \Phi\left(\frac{\ln x_{0.90} - 0.130}{0.324}\right) = 0.90$$

from which

$$\frac{\ln x_{0.90} - 0.130}{0.324} = \Phi^{-1}(0.90) = 1.28$$

Therefore, the 90% drainage is

$$x_{0.90} = e^{0.545} = 1.72 \text{ mgd}.$$

With the normal distribution of Example 3.9, the 90% drainage is 1.71 mgd. ◀

▶ **EXAMPLE 3.13**

The time T between breakdowns of a major equipment in an oil platform is defined by a lognormal distribution with a median of 6 months and a c.o.v. of 0.30. In order to ensure a 95% probability that the equipment will be operational at any time, the interval between inspections and repairs can be determined as follows.

In this case, the parameters of the lognormal distribution are: $\lambda = \ln 6 = 1.792$; and $\zeta \simeq 0.30$; then assuming that $t_o =$ the time between inspection and repair, we would require

$$P(T > t_o) = 1 - P(T \le t_o) = 0.95$$

or

$$\Phi\left(\frac{\ln t_o - 1.792}{0.30}\right) = 0.05$$

from which, we have

$$\ln t_o - 1.792 = 0.30\Phi^{-1}(0.05) = 0.30\left[-\Phi^{-1}(0.95)\right] = 0.30(-1.65) = -0.495$$

Hence, the required inspection interval is

$$t_o = e^{1.297} = 3.66 \text{ months}$$

If the equipment is in operational condition at the time it is scheduled for regular maintenance and inspection, the probability that it will remain operational without breakdowns for another 2 months would be

$$P(T > 5.66 | T > 3.66) = \frac{P(T > 5.66) \cap P(T > 3.66)}{P(T > 3.66)} = \frac{P(T > 5.66)}{P(T > 3.66)}$$

$$= \frac{1 - \Phi\left(\dfrac{\ln 5.66 - 1.792}{0.30}\right)}{0.95} = \frac{1 - \Phi(-0.195)}{0.95} = \frac{1 - (1 - 0.577)}{0.95}$$

$$= 0.61$$

Therefore, the probability is 61% (better than a 50% chance) that the equipment will remain operational without breaking down for 2 months beyond its scheduled regular maintenance. It may be observed that the probability of the equipment operating for 5.66 months from the beginning is $P(T > 5.66) = 0.577$. ◄

3.2.3 The Bernoulli Sequence and the Binomial Distribution

In many engineering applications, there are often problems involving the occurrence or recurrence of an event, which is unpredictable, in a sequence of discrete "trials." For example, in allocating a fleet of construction equipment for a project, the anticipated conditions of every piece of equipment in the fleet over the duration of the project would have some bearing on the determination of the required fleet size, whereas in planning the flood control system for a river basin, the annual maximum flow of the river over a sequence of years would be important in determining the design flood level. In these cases, the operational conditions of every piece of equipment, and the annual maximum flow of the river relative to a specified flood level constitute the respective *trials*. We should point out that in these problems, there are only two possible outcomes in each trial; namely, the *occurrence* and *nonoccurrence* of an event—e.g., each piece of equipment *may* or *may not* malfunction over the duration of the project; in each year, the maximum flow of the river *may* or *may not* exceed some specified flood level.

Problems of the type that we just described above may be modeled by a *Bernoulli sequence*, which is based on the following assumptions:

1. In each trial, there are only two possibilities—the *occurrence* and *nonoccurrence* of an event.
2. The probability of occurrence of the event in each trial is constant.
3. The trials are statistically independent.

In the two examples introduced above, we may model each of the problems as a Bernoulli sequence as follows:

- Over the duration of the project, the operational conditions between equipment are statistically independent, and the probability of malfunction for every piece of equipment is the same; then, the conditions of the entire fleet of equipments constitute a Bernoulli sequence.
- If the annual maximum floods between any 2 years are statistically independent and in each year the probability of the flood's exceeding some specified level is constant, then the annual maximum floods over a series of years can be modeled as a Bernoulli sequence.

The Binomial Distribution

In a Bernoulli sequence, if X is the random number of occurrences of an event among n trials, in which the probability of occurrence of the event in each trial is p and the corresponding probability of nonoccurrence is $(1 - p)$, then the probability of exactly x occurrences among the n trials is governed by the *binomial* PMF as follows:

$$P(X = x) = \binom{n}{x} p^x (1 - p)^{n-x} \qquad x = 0, 1, 2, \ldots, n \qquad (3.30)$$

where n and p are the parameters, and

$$\binom{n}{x} = \frac{n!}{x!(n-x)!}$$

is known as the *binomial coefficient*.

The corresponding CDF is

$$F_X(x) = \sum_{k=0}^{x} \binom{n}{k} p^k (1-p)^{n-k} \qquad (3.30a)$$

Values of the CDFs of Eq. 3.30a are tabulated in Table A.2 for specified integer values of n and given probability p.

We can get a better understanding of the basis of Eq. 3.30 by observing the following: By virtue of statistical independence, the probability of realizing a particular sequence of exactly x occurrences and $(n-x)$ nonoccurrences of an event among n trials is $p^x(1-p)^{n-x}$. However, the x occurrences of the event can be permuted among the n trials, so that the number of sequences with x occurrences is $\binom{n}{x}$; for example, if there are x malfunctions among a fleet of n pieces of equipment, the x malfunctions may occur in $\binom{n}{x}$ different sequences among the n machines. Thus, we obtain Eq. 3.30.

▶ **EXAMPLE 3.14**

Five road graders are used in the construction of a highway project. The operational life T of each grader is a lognormal random variable with a mean life of 1500 hr and a c.o.v. of 30% (see Fig. E3.14). Assuming statistical independence among the conditions of the machines, the probability that two of the five machines will malfunction in less than 900 hr of operation can be evaluated as follows.

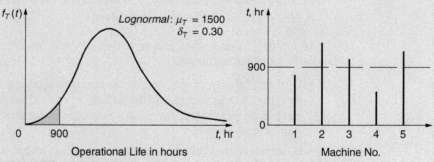

Figure E3.14 Operational life of road graders.

The parameters of the lognormal distribution are: $\zeta \simeq 0.30$; and $\lambda = \ln 1500 - \frac{1}{2}(0.3)^2 = 7.27$. Then, the probability that a machine will malfunction within 900 hr (see Fig. E3.14) is

$$p = P(T < 900) = \Phi\left(\frac{\ln 900 - 7.27}{0.30}\right) = \Phi(-1.56) = 0.0594$$

For the five machines taken collectively, the actual operational lives of the different machines may conceivably be as shown in Fig. E3.14; i.e., as illustrated in the figure, machines No. 1 and 4 have operational lives less than 900 hr, whereas machines No. 2, 3, and 5 have operational lives longer than 900 hr. The corresponding probability of this exact sequence is $p^2(1-p)^5$. But the two malfunctioning machines may happen to any two of the five machines; therefore, the number of sequences with two malfunctioning machines among the five is $5!/2!3! = 10$. Consequently, if X is the number of road graders malfunctioning in 900 hr,

$$P(X = 2) = 10(0.0594)^2(1 - 0.0594)^3 = 0.0294$$

Also, the probability of malfunction among the five graders (i.e., there will be malfunctions in one or more machines) would be

$$P(X \geq 1) = 1 - P(X = 0) = 1 - (0.9406)^5 = 0.264$$

whereas the probability that there will be no more than two machines malfunctioning within 900 hr would be

$$P(X \leq 2) = \sum_{k=0}^{2} \binom{5}{k} 0.0594^k (0.9406)^{5-k} = (0.9406)^5 + 5(0.0594)(0.9406)^4 + 10(0.0594)^2(0.9406)^3$$
$$= 0.7362 + 0.2325 + 0.0294 = 0.9981$$

This last result involves the CDF of the binomial distribution, which is tabulated in Table A.2 for limited values of the parameters. Using Table A.2 with $n = 5$, $x = 2$, and $p = 0.05$, we obtain a value of 0.9988 from this table.* ◀

In spite of its simplicity, the Bernoulli model is quite useful in many engineering applications. There are numerous problems in engineering involving situations with only two alternative possibilities. Aside from those we have described and illustrated above, other problems that can be modeled as respective Bernoulli sequences include the following:

- In a series of piles driven into a soil stratum, each pile *may* or *may not* encounter boulders or hard rock.
- In monitoring the daily water quality of a river on the downstream side of an industrial plant, the water tested daily *may* or *may not* meet the pollution control standards.
- The individual items produced on an assembly line *may* or *may not* pass the inspection to ensure product quality.
- In a seismically active region, a building *may* or *may not* be damaged annually.

In each of these cases, if the situation is repeated, the resulting series may be modeled as a Bernoulli sequence.

We might emphasize that in modeling problems with the Bernoulli sequence, the individual trials must be discrete and statistically independent. In spite of this requirement, however, certain continuous problems may be modeled (approximately at least) with the Bernoulli sequence. For example, time and space problems, which are generally continuous, may be modeled with the Bernoulli sequence by discretizing time (or space) into appropriate intervals and admitting only two possibilities within each interval; what happens in each time (or space) interval then constitutes a trial, and the series of finite number of intervals is then a Bernoulli sequence. We illustrate this with the following example.

▶ **EXAMPLE 3.15**

The annual rainfall (accumulated generally during the winter and spring) of each year in Orange County, California, is a Gaussian random variable with a mean of 15 in. and a standard deviation of 4 in.; i.e., $N(15, 4)$. Suppose the current water policy of the county is such that if the annual rainfall is less than 7 in. for a given year, water rationing will be required during the summer and fall of that year.

---

*Table A.2 is limited to specific values of n and p; for more general values of these parameters, it is more convenient to use computers to evaluate the required probability (see Examples 5.2 and 5.11 of Chapter 5).

Assuming X is the annual rainfall, the probability of water rationing in Orange County in any given year is then

$$P(X < 7) = \Phi\left(\frac{7-15}{4}\right) = \Phi(-2.0) = 1 - \Phi(2.0) = 1 - 0.9772 = 0.0228$$

However, if the county wishes to reduce the probability of water rationing to half that of the current policy, the annual rainfall below which rationing has to be imposed would be determined as follows:

$$P(X < x_r) = \Phi\left(\frac{x_r - 15}{4}\right) = \frac{1}{2}(0.0228) = 0.0114$$

Thus,

$$\frac{x_r - 15}{4} = \Phi^{-1}(0.0114) = -\Phi^{-1}(0.9885) = -2.28$$

Hence, with the new policy, the annual rainfall below which rationing must be imposed is

$$x_r = 15 - (4)(2.28) = 5.88 \text{ in.}$$

Under the current water policy, and assuming that the annual rainfalls between years are statistically independent, the probability that in the next 5 years there will be at least 1 year in which water rationing will be necessary would be determined as follows.

Denoting N as the number of years when rationing would be imposed, the probability would be

$$P(N \geq 1) = 1 - P(N = 0) = 1 - \binom{5}{0}(0.0228)^0 (0.9772)^5 = 0.109$$

Finally, whenever the annual rainfall is less than 7 in. in any given year, the probability of damage to the agricultural crops in the county is 30%. Assuming that crop damages between dry years (i.e., with rainfall less than 7 in.) are statistically independent, the probability of crop damage (denoted D) in the next 3 years may be of interest. In this case, the probability of crop damage would depend on the number of years (between 0 and 3) that the annual rainfall will be less than 7 in.; therefore, the solution requires the theorem of total probability, as follows:

$$P(D) = 1 - P(\overline{D}) = 1 - \left[1.00(0.9772)^3 + (0.70)\binom{3}{1}(0.0228)(0.9772)^2 \right.$$
$$\left. + (0.70)^2 \binom{3}{2}(0.0228)^2 (0.9772) + (0.70)^3 \binom{3}{3}(0.0228)^3\right]$$
$$= 1 - (0.9331 + 0.0457 + 0.0007 + 0) = 1 - (0.9795)$$
$$= 0.020$$

Therefore, the probability of crop damage in the next 3 years is only 2%. ◀

3.2.4 The Geometric Distribution

In a Bernoulli sequence, the number of trials until a specified event occurs for the first time is governed by the *geometric distribution*. We might observe that if the event occurs for the first time on the nth trial, there must be no occurrence of this event in any of the prior $(n - 1)$ trials. Therefore, if N is the random variable representing the number of trials until the occurrence of the event, then

$$P(N = n) = pq^{n-1} \qquad n = 1, 2, \ldots \qquad (3.31)$$

where $q = (1 - p)$. Equation 3.31 is known as the *geometric distribution*.

Recurrence Time and Return Period

In a time (or space) problem that is appropriately discretized into time (or space) intervals, $T = N$, and can be modeled as a Bernoulli sequence, the number of time intervals until the first occurrence of an event is called the *first occurrence time*.

We observe that if the discretized time intervals in the sequence are statistically independent, the time interval till the first occurrence of an event must be the same as that of the time between any two consecutive occurrences of the same event; thus the probability distribution of the *recurrence time* is equal to that of the first occurrence time. Therefore, the recurrence time in a Bernoulli sequence is also governed by the geometric distribution, Eq. 3.31. The *mean recurrence time*, which is popularly known in engineering as the (average) *return period*, therefore, is

$$\overline{T} = E(T) = \sum_{t=1}^{\infty} t \cdot pq^{t-1} = p(1 + 2q + 3q^2 + \cdots)$$

Generally, as $q = (1 - p) < 1.0$, the infinite series within the above parentheses yields $1/(1 - q)^2 = 1/p^2$. Hence, we obtain the *return period*

$$\overline{T} = \frac{1}{p} \tag{3.32}$$

which means that the *mean recurrence time* between two consecutive occurrences of an event is equal to the reciprocal of the probability of the event within one time interval.

It is well to emphasize that the return period is only an average duration between consecutive occurrences of an event and should not be construed as the actual time between the occurrences; the actual time is T, which is a random variable.

▶ **EXAMPLE 3.16**

Suppose that the building code for the design of buildings in a coastal region specifies the 50-yr wind as the "design wind." That is, a wind velocity with a return period of 50 years; or on the average, the design wind may be expected to occur once every 50 yr.

In this case, the probability of encountering the 50-yr wind velocity in any 1 yr is $p = 1/50 = 0.02$. Then, the probability that a newly completed building in the region will be subjected to the design wind velocity for the first time on the fifth year after its completion is

$$P(T = 5) = (0.02)(0.98)^4 = 0.018$$

whereas the probability that the first such wind velocity will occur within 5 yr after completion of the building would be

$$P(T \le 5) = \sum_{t=1}^{5} (0.02)(0.98)^{t-1}$$

$$= 0.02 + 0.0196 + 0.0192 + 0.0188 + 0.0184$$

$$= 0.096$$

We might point out that this latter event (the first occurrence of the wind velocity within 5 yr) is the same as the event of at least one 50-yr wind in 5 yr, which is also the complement of no 50-yr wind in 5 years; thus, the desired probability may also be calculated as $1 - (0.98)^5 = 0.096$. However, the above is quite different from the event of experiencing exactly one 50-yr wind in 5 yr; the probability in this case is given by the binomial probability which would be $\binom{5}{1}(0.02)(0.98)^4 = 0.092$. ◀

▶ **EXAMPLE 3.17**   A fixed offshore platform, shown in Fig. E3.17, is designed for a wave height of 8 m above the mean sea level. This wave height corresponds to a 5% probability of being exceeded per year. The return period of the design wave height is therefore,

$$\overline{T} = \frac{1}{0.05} = 20\,\text{yr}$$

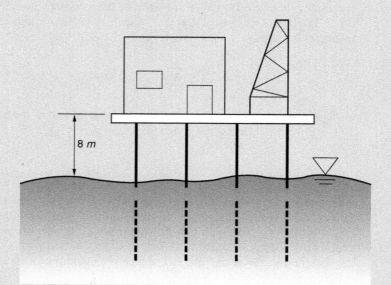

Figure E3.17   An offshore platform.

The probability that the platform will be subjected to the design wave height within the return period is therefore,

$$P(H > 8, \text{ in } 20\,\text{yr}) = 1 - P(H \leq 8, \text{ in } 20\,\text{yr}) = 1 - (0.95)^{20}$$
$$= 0.3585$$

The probability that the first exceedance of the design wave height will occur after the third year is, by the geometric distribution,

$$P(T > 3) = 1 - P(T \leq 3) = 1 - \left[0.05(0.95)^{1-1} + 0.05(0.95)^{2-1} + 0.05(0.95)^{3-1}\right]$$
$$= 1 - (0.05 + 0.0475 + 0.0451) = 1 - 0.1426$$
$$= 0.8574$$

If the first exceedance of the design wave height should occur after the third year as stipulated above, the probability that such a first exceedance will occur in the fifth year is then

$$P(T = 5 \mid T > 3) = \frac{P(T = 5 \cap T > 3)}{P(T > 3)} = \frac{P(T = 5)}{P(T > 3)} = \frac{0.05(0.95)^4}{0.8574} = 0.048 \qquad \blacktriangleleft$$

Observe next that the probability of no event occurring within its return period $\overline{T}$ is

$$P(\text{no occurrence in } \overline{T}) = (1 - p)^{\overline{T}}$$

where $p = 1/\overline{T}$. Expanding the above with the binomial theorem, we have

$$(1 - p)^{\overline{T}} = 1 - \overline{T}p + \frac{\overline{T}(\overline{T} - 1)}{2!} p^2 - \frac{\overline{T}(\overline{T} - 1)(\overline{T} - 2)}{3!} p^3 + \cdots .$$

Furthermore, for large $\overline{T}$, or small p, we may recognize that the series on the right side is approximately equal to $e^{-\overline{T}p}$. Therefore, for large $\overline{T}$,

$$P(\text{no occurrence in } \overline{T}) = e^{-\overline{T}p} = e^{-1} = 0.3679$$

and

$$P(\text{occurrence in } \overline{T}) = 1 - 0.3679 = 0.6321$$

In other words, for a rare event that is defined as one with a long return period, $\overline{T}$, the probability of the event's occurring within its return period is always 0.632. This result is a useful approximation even for return periods that are not very long; for instance, for $\overline{T} = 20$ time intervals, such as in Example 3.17, the probability is

$$P\left(\text{no occurrence in } \overline{T}\right) = \left(1 - \frac{1}{20}\right)^{20} = 0.359$$

which shows that the error in the above exponential approximation is less than 1.5%.

3.2.5 The Negative Binomial Distribution

The geometric PMF is the probability law governing the number of trials, or discrete time units, until the first occurrence of an event in a Bernoulli sequence. The number of time units (or trials) until a subsequent occurrence of the same event is governed by the *negative binomial* distribution. That is, if T_k is the number of time units until the kth occurrence of the event in a series of Bernoulli trials, then

$$P(T_k = n) = \binom{n - 1}{k - 1} p^k q^{n-k} \qquad \text{for } n = k, k + 1, \ldots \ldots$$
$$= 0 \qquad \qquad \text{for } n < k \qquad\qquad (3.33)$$

The basis of Eq. 3.33 is as follows: If the kth occurrence of an event is realized at the nth trial, there must be exactly $(k - 1)$ occurrences of the event in the prior $(n - 1)$ trials and at the nth trial the event also occurs. Thus, from the binomial law, we obtain the probability

$$P(T_k = n) = \binom{n - 1}{k - 1} p^{k-1} q^{n-k} \cdot p$$

thus obtaining Eq. 3.33.

▶ **EXAMPLE 3.18**

In Example 3.16, the probability that the building in the region will be subjected to the design wind for the third time on the tenth year is, according to Eq. 3.33,

$$P(T_3 = 10) = \binom{10 - 1}{3 - 1} (0.02)^3 (0.98)^{10-3} = \binom{9}{2} (0.02)^3 (0.98)^7$$
$$= 36(8 \times 10^{-6})(0.8681) = 0.00025$$

whereas the probability that the third design wind will occur within 5 years would be

$$P(T_3 \leq 5) = \sum_{n=3}^{5} \binom{n-1}{3-1} (0.02)^3 (0.98)^{n-3}$$

$$= \binom{2}{2} (0.02)^3 (0.98)^0 + \binom{3}{2} (0.02)^3 (0.98) + \binom{4}{2} (0.02)^3 (0.98)^2$$

$$= (0.000008) + 3(0.000008)(0.98) + 12(0.000008)(0.96040) = 0.00012 \quad ◀$$

▶ **EXAMPLE 3.19**

A steel cable is built up of a number of independent wires as shown in Fig. E3.19. Occasionally, the cable is subjected to high overloads; on such occasions the probability of fracture of one of the wires is 0.05, and the failure of two or more wires during a single overload is unlikely.

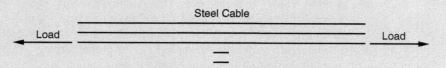

Figure E3.19   A steel cable.

If the cable must be replaced when the third wire fails, the probability that the cable can withstand at least five overloads can be determined as follows.

First, we observe that the third wire failure must occur at or after the sixth overloading. Hence, using Eq. 3.33, the required probability is

$$P(T_3 \geq 6) = 1 - P(T_3 < 6) = 1 - \sum_{n=3}^{5} P(T_3 = n)$$

$$= 1 - \binom{2}{2} (0.05)^3 (0.95)^0 - \binom{3}{2} (0.05)^3 (0.95) - \binom{4}{2} (0.05)^3 (0.95)^2$$

$$= 1 - 0.00116 = 0.9988 \quad ◀$$

3.2.6   The Poisson Process and the Poisson Distribution

Many physical problems of interest to engineers and scientists involve the possible occurrences of events at any point in time and/or space. For example, earthquakes could strike at any time and anywhere over a seismically active region in the world; fatigue cracks may occur anywhere along a continuous weld; and traffic accidents could happen at any time on a given highway. Conceivably, such space–time problems may be modeled also with the Bernoulli sequence, by dividing the time or space into appropriate small intervals, and assuming that an event will either occur or not occur (only two possibilities) within each interval, thus constituting a Bernoulli trial. However, if the event can randomly occur at any instant of time (or at any point in space), it may occur more than once in any given time or space interval. In such cases, the occurrences of the event may be more appropriately modeled with a *Poisson process* or *Poisson sequence*.

Formally, the Poisson process is based on the following assumptions:

1. An event can occur at random and at any instant of time or any point in space.

2. The occurrence(s) of an event in a given time (or space) interval is statistically independent of that in any other nonoverlapping interval.

3. The probability of occurrence of an event in a small interval Δt is proportional to Δt, and can be given by $v\Delta t$, where v is the mean occurrence rate of the event (assumed to be constant).

4. The probability of two or more occurrences in Δt is negligible (of higher orders of Δt).

On the basis of these assumptions, the number of statistically independent occurrences of an event in t (time or space) is governed by the *Poisson* PMF; that is, if X_t is the number of occurrences in a time (or space) interval $(0, t)$, then

$$P(X_t = x) = \frac{(vt)^x}{x!} e^{-vt} \qquad x = 0,1,2,\ldots \tag{3.34}$$

where v is the *mean occurrence rate*, i.e., the average number of occurrences of the event per unit time (or space) interval. It follows then that the mean number of occurrences in t is $E(X_t) = vt$; furthermore, it can be shown that the variance of X_t is also vt. A mathematical derivation of Eq. 3.34 is given in Appendix C. It may be emphasized that the process underlying Eq. 3.34 is a counting process with a constant parameter v.

There are similarities and differences between the Bernoulli sequence and the Poisson process. First of all, if a problem involves distinct entities, such as the number of operational road graders in Example 3.13, the proper model is the Bernoulli sequence. However, in space–time problems, either model may be applicable; in fact, we can show, as illustrated below, that the Bernoulli sequence approaches the Poisson process as the time (or space) interval is decreased. For this purpose, consider the following.

From previous statistical data of traffic counts, an average of 60 cars per hour was observed to make left turns at a given intersection. Then, suppose we are interested in the probability of exactly 10 cars making left turns at the intersection in a 10-min interval.

As an approximation, we may first divide the 1-hr duration into 120 intervals of 30 sec each, such that the probability of a left turn (*L.T.*) in any 30-sec interval would be $p = 60/120 = 0.5$. Then, allowing no more than one *L.T.* in any 30-sec interval, the problem is reduced to the binomial probability of the occurrence of 10 *L.T.* among the maximum possible of 20 *L.T.* in the 10-min interval, in which the probability of a *L.T.* in each 30-sec interval is 0.5. Thus,

$$P(10\ L.T.\ \text{in } 10\ \text{min}) = \binom{20}{10}(0.5)^{10}(0.5)^{20-10} = 0.1762$$

The above solution is grossly approximate because it assumes that no more than one car will be making *L.T.* in a 30-sec interval; obviously, two or more *L.T.s* are possible.

The solution would be improved if we selected a shorter time interval, say, a 10-sec interval. Then, the probability of an *L.T.* in each interval is $p = 60/360 = 0.1667$, and

$$P(10\ L.T.\ \text{in } 10\ \text{min}) = \binom{60}{10}(0.1667)^{10}(0.8333)^{60-10} = 0.1370$$

Further improvements can be made by subdividing time into shorter intervals. If the time t is subdivided into n equal intervals, then the binomial PMF would give

$$P(x\ \text{occurrences in } t) = \binom{n}{x}\left(\frac{\lambda}{n}\right)^x \left(1 - \frac{\lambda}{n}\right)^{n-x}$$

where λ is the *average* number of occurrences of the event in time t. If the event can occur at any time (as in the case of left-turn traffic), the time t would need to be subdivided into a large number of intervals, i.e., $n \to \infty$; then,

$$P(x \text{ occurrences in } t) = \lim_{n \to \infty} \binom{n}{x} \left(\frac{\lambda}{n}\right)^x \left(1 - \frac{\lambda}{n}\right)^{n-x}$$

$$= \lim_{n \to \infty} \frac{n!}{x!(n-x)!} \left(\frac{\lambda}{n}\right)^x \left(1 - \frac{\lambda}{n}\right)^{n-x}$$

$$= \lim_{n \to \infty} \frac{n}{n} \frac{(n-1)}{n} \cdots \frac{(n-x+1)}{n} \cdot \frac{\lambda^x}{x!} \left(1 - \frac{\lambda}{n}\right)^n \left(1 - \frac{\lambda}{n}\right)^{-x}$$

We can observe that

$$\lim_{n \to \infty} \left(1 - \frac{\lambda}{n}\right)^n = 1 - \lambda + \frac{\lambda^2}{2!} - \frac{\lambda^3}{3!} + \cdots = e^{-\lambda}$$

and all the other terms approach 1.0 as $n \to \infty$, except the term $\lambda^x/x!$. Therefore, in the limit, as $n \to \infty$, we have

$$P(x \text{ occurrences in } t) = \frac{\lambda^x}{x!} e^{-\lambda}$$

which is the Poisson distribution of Eq. 3.34, with $\lambda = vt$. On this basis, with $v = 1$ _L.T._ per minute, the probability of 10 _L.T._ in 10 min is then

$$P(X_{10} = 10) = \frac{(1 \times 10)^{10}}{10!} e^{-1 \times 10} = 0.125$$

▶ **EXAMPLE 3.20**

Historical records of severe rainstorms in a town over the last 20 years indicated that there had been an average number of four rainstorms per year. Assuming that the occurrences of rainstorms may be modeled with the Poisson process, the probability that there would not be any rainstorms next year is

$$P(X_t = 0) = \frac{(4 \times 1)^0}{0!} e^{-4} = 0.018$$

whereas the probability of four rainstorms next year would be

$$P(X_t = 4) = \frac{(4 \times 1)^4}{4!} e^{-4} = 0.195$$

We note from this last result that although the average yearly occurrences of rainstorms is four, the probability of actually experiencing four rainstorms in a year is less than 20%. The probability of two or more rainstorms in the next year is

$$(X_1 \geq 2) = \sum_{x=2}^{\infty} \frac{(4 \times 1)^x}{x!} e^{-4 \cdot 1} = 1 - \sum_{x=0}^{1} \frac{4^x}{x!} e^{-4}$$

$$= 1 - 0.018 - 0.074 = 0.908$$

The different probabilities of the number of rainstorms in a year are tabulated below:

x No. of Rainstorms	Probability of x	x No. of Rainstorms	Probability of x
0	0.018	7	0.060
1	0.074	8	0.030
2	0.146	9	0.013
3	0.195	10	0.005
4	0.195	11	0.002
5	0.156	12	0.001
6	0.104	13	0.000

The corresponding PMF is shown graphically in Fig. E3.20.

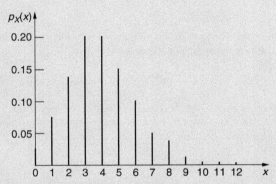

Figure E3.20   PMF of no. of rainstorms in a year.

◄

► **EXAMPLE 3.21**

In designing the left-turn bay at a state highway intersection, the vehicles making left turns at the intersection may be modeled as a Poisson process. If the cycle time of the traffic light for left turns is 1 min, and the design criterion requires a left-turn lane that will be sufficient 96% of the time (which may be the criterion in some states in the United States), the lane distance, in terms of car lengths, to allow for an average left turns of 100 per hour, may be determined as follows.

As stated above, the mean rate of left turns at the intersection is $v = 100/60$ per minute. Let us suppose the design length of the left-turn lane is k car lengths. Then, during a 1-min cycle of the traffic light, the design criterion requires that the probability of no more than k cars waiting for left turns must be at least 96%; therefore, we must have

$$P(X_{t=1} \le k) = \sum_{x=0}^{k} \frac{1}{x!} \left(\frac{100}{60} \times 1 \right)^x e^{-100/60} = 0.96$$

By trial-and-error, we would obtain the following results:

$$\text{If } k = 3, P(X_t \le 3) = \sum_{x=0}^{3} \frac{1}{x!} \left(\frac{100}{60} \right)^x e^{-100/60} = 0.91$$

whereas

$$\text{if } k = 4, P(X_t \le 4) = 0.968$$

Therefore, a left-turn bay of four car lengths at the intersection is sufficient to satisfy the design requirement.   ◄

▶ **EXAMPLE 3.22**

A structure is located in a region where tornado wind force must be considered in its design. Suppose that from the records of tornadoes for the past 20 years, the mean occurrence rate of tornadoes in the region is once every 10 years. Assume that the occurrence of tornadoes can be modeled as a Poisson process.

If the structure is designed to withstand a tornado force with an allowable probability of damage of 20%, the probability that the structure will be damaged in the next 50 years is

$$P(D) = 1 - P(\overline{D}) = 1 - \left[\sum_{n=0}^{\infty} (1 - 0.20)^n \frac{(0.1 \times 50)^n}{n!} e^{-(0.1 \times 50)} \right]$$

$$= 1 - \left[e^{-5.0} + (0.80) \frac{5.0}{1!} e^{-5.0} + (0.80)^2 \frac{(5.0)^2}{2!} e^{-5.0} + (0.80)^3 \frac{(5.0)^3}{3!} e^{-5.0} + \cdots \right]$$

$$= 1 - e^{-5.0} \left[1 + (0.80) \frac{5.0}{1!} + (0.80)^2 \frac{(5.0)^2}{2!} + (0.80)^3 \frac{(5.0)^3}{3!} \cdots \right]$$

we may recognize that the limit of the infinite series inside the brackets is the exponential $e^{0.80 \times 5.0}$. Hence, the probability of damage in 50 years is

$$P(D) = 1 - e^{-5.0} \cdot e^{0.80 \times 5.0} = 1 - e^{-5.0 \times 0.20} = 1 - 0.368 = 0.632$$

Obviously, the above probability of damage is much too high. If the structure were to be upgraded in order to reduce the 50-year probability of damage to 5%, what should be the allowable damage probability against a tornado wind? In this case, we would have

$$P(D) = 0.05$$

which means

$$1 - \sum_{n=0}^{\infty} (1 - p)^n \frac{(0.1 \times 50)^n}{n!} e^{-(0.1 \times 50)} = 1 - e^{-5.0} e^{5.0(1-p)} = 0.05.$$

Thus,

$$1 - e^{-5p} = 0.05$$

or

$$p = 0.010$$

which means that the original structure should be upgraded to reduce the damage probability against a tornado wind of 0.010 or 1%.

Now suppose that the regional government plans to upgrade 10 similar structures to the above standard, i.e., that each structure should have no more than 0.05 probability of damage against tornado wind in 50 years. Then the probability that at most one of the ten upgraded structures will be damaged in the next 50 years would be evaluated as follows.

Assuming that the damages between structures are statistically independent, the solution involves the binomial probability in which the 50-year damage probability of each structure is 0.05. Denoting X as the number of damaged structures in 50 years,

$$P(X \le 1) = \binom{10}{0} (0.05)^0 (0.95)^{10} + \binom{10}{1} (0.05)^1 (0.95)^9$$

$$= 0.914$$

On the other hand, the probability that at least one of the ten structures will be damaged by tornadoes in the next 50 years is

$$P(X \geq 1) = 1 - P(X = 0) = 1 - \binom{10}{0}(0.05)^0(0.95)^{10} = 0.401$$

◀

In the next example, we illustrate a space problem that may be modeled with the Poisson process.

▶ **EXAMPLE 3.23**

A major steel pipeline is used to transport crude oil from an oil production platform to a refinery over a distance of 100 km. Even though the entire pipeline is inspected once a year and repaired as necessary, the steel material is subject to damaging corrosion. Assume that from past inspection records, the average distance between locations of such corrosions is determined to be 0.15 km. In this case, if the occurrence of corrosions along the pipeline is modeled as a Poisson process with a mean occurrence rate of $v = 0.15$/km, the probability that there will be 10 locations of damaging corrosion between inspections is

$$P(X_{100} = 10) = \frac{(0.15 \times 100)^{10}}{10!}e^{-0.15 \times 100} = 0.049$$

whereas the probability of at least five corrosion sites between inspections would be

$$P(X_{100} \geq 5) = 1 - P(X_{100} < 5) = 1 - \sum_{n=0}^{4}\frac{(0.15 \times 100)^n}{n!}e^{-0.15 \times 100}$$

$$= 1 - \left[\frac{(15)^0}{0!}e^{-15} + \frac{(15)^1}{1!}e^{-15} + \frac{(15)^2}{2!}e^{-15} + \frac{(15)^3}{3!}e^{-15} + \frac{(15)^4}{4!}e^{-15}\right]$$

$$= 1 - e^{-15}(1 + 15 + 112.5 + 562.5 + 2109.4) = 1 - 0.0009 = 0.9991$$

In any one of the corrosion sites, there may be one or more cracks that could initiate fracture failure. If the probability of this event occurring at a corrosion site is 0.001, the probability of fracture failure along the entire 100-km pipeline between inspection and repair would be (denote F for fracture failure),

$$P(F) = 1 - P(\overline{F}) = 1 - P(\overline{F} \cap X_{100} \geq 0) = 1 - \sum_{n=0}^{\infty} P(\overline{F}|X_{100} = n)P(X_{100} = n)$$

$$= 1 - \sum_{n=0}^{\infty}(1 - 0.001)^n\frac{(0.15 \times 100)^n}{n!}e^{-15} = 1 - e^{-15}\left[1 + (0.999)\frac{15}{1!} + (0.999)^2\frac{(15)^2}{2!}\right.$$

$$\left. + (0.999)^3\frac{(15)^3}{3!} + \cdots\right]$$

$$= 1 - e^{-15}e^{0.999 \times 15} = 1 - e^{-0.001 \times 15} = 1 - 0.985 = 0.015$$

◀

▶ **EXAMPLE 3.24**

In the last 50 years, suppose that there were two large earthquakes (with magnitudes $M \geq 6$) in Southern California. If we model the occurrences of such large earthquakes as a Bernoulli sequence, the probability of such large earthquakes in Southern California in the next 15 years would be evaluated as follows.

First, we observe that the annual probability of occurrence of large earthquakes is $p = 2/50 = 0.04$. Then,

$$P(X \geq 1) = 1 - P(X = 0) = 1 - \binom{15}{0}(0.04)^0(0.96)^{15} = 0.458$$

Alternatively, if the occurrences of large earthquakes in Southern California were modeled as a Poisson process, we would first determine the mean occurrence rate as $v = 2/50 = 0.04$ per year, and the probability of such large earthquakes in the next 15 years then becomes

$$P(X_{15} \geq 1) = 1 - P(X_{15} = 0) = 1 - \frac{(0.04 \times 15)^0}{0!} e^{-0.04 \times 15} = 0.451$$

The maximum intensity of seismic ground shaking at a given site can be measured in terms of g (gravitational acceleration $= 980$ cm/sec^2). Suppose that during an earthquake of $M \geq 6$, the ground shaking intensity Y at a particular building site has a lognormal distribution with a median of $0.20g$ and a c.o.v. of 0.25. If the seismic capacity of a building is $0.30g$, the probability that the building will suffer damage during an earthquake of magnitude $M \geq 6$ would be

$$P(D|M \geq 6) = P(Y > 0.30g) = 1 - P(Y \leq 0.30g)$$

$$= 1 - \Phi\left(\frac{\ln 0.3g - \ln 0.2g}{0.25}\right) = 1 - \Phi\left(\frac{\ln \frac{0.3g}{0.2g}}{0.25}\right)$$

$$= 1 - \Phi\left(\frac{\ln 1.5}{0.25}\right) = 1 - 0.947 = 0.053$$

Then, in the next 20 years, the probability that the building will not suffer damage from large earthquakes (assuming the Poisson process for occurrences of large earthquakes) would be

$$P(\overline{D} \text{ in 20 years}) = \sum_{n=0}^{\infty} (0.947)^n \frac{(0.04 \times 20)^n}{n!} e^{-0.04 \times 20}$$

$$= e^{-0.80}\left[1 + (0.947)\frac{0.80}{1!} + (0.947)^2\frac{(0.80)^2}{2!} + (0.947)^3\frac{(0.80)^3}{3!} + \cdots\right]$$

$$= e^{-0.80}e^{0.947 \times 0.80} = e^{-0.0424} = 0.958 \qquad \blacktriangleleft$$

We should emphasize that in both the Bernoulli sequence and the Poisson process, the occurrences of an event between trials (in the case of the Bernoulli model) and between intervals (in the Poisson model) are statistically independent. More generally, the occurrence of a given event in one trial (or interval) may affect the occurrence or nonoccurrence of the same event in subsequent trials (or intervals). In other words, the probability of occurrence of an event in a given trial may depend on earlier trials, and thus could involve conditional probabilities. If this conditional probability depends on the immediately preceding trial (or interval), the resulting model is a *Markov chain* (or *Markov process*); the essential principles of the Markov chain are described in Ang and Tang, Vol. 2 (1984).*

3.2.7 The Exponential Distribution

We saw earlier that in the case of a Bernoulli sequence, the probability of the recurrence time between events is described by the geometric distribution (see Sect. 3.2.4). On the other hand, if the occurrences of an event constitute a Poisson process, the recurrence time would be described by the *exponential distribution*.

In the case of a Poisson process, we observe that if T_1 is the time till the first occurrence of an event, then $(T_1 > t)$ means that there is no occurrence of the event in $(0, t)$; therefore,

---

*This reference is out of print, but is available by direct order from <u>ahang2@aol.com</u>.

according to Eq. 3.34,

$$P(T_1 > t) = P(X_t = 0) = e^{-vt}$$

Because the occurrences of an event in nonoverlapping intervals are statistically independent, T_1 is also the recurrence time between two consecutive occurrences of the same event. The CDF of T_1, therefore, is the exponential distribution,

$$F_{T_1}(t) = P(T_1 \le t) = 1 - e^{-vt} \tag{3.35}$$

and its PDF is

$$f_{T_1}(t) = \frac{dF}{dt} = ve^{-vt} \tag{3.36}$$

If the mean occurrence rate, v, is constant, the mean recurrence time, $E(T_1)$, or return period for a Poisson process can be shown to be

$$E(T_1) = 1/v \tag{3.37}$$

This may be compared with the corresponding return period of $1/p$ for the Bernoulli sequence, Eq. 3.32. However, for events with small occurrence rate v, $1/v \simeq 1/p$. We can show this by observing that in a Poisson process with occurrence rate v, the probability of an event occurring in a unit time interval (i.e., $t = 1$) is

$$p = P(X_1 = 1) = ve^{-v} = v(1 - v + \frac{1}{2}v^2 + \cdots)$$

Hence, for small v, $p \simeq v$.

Therefore, for rare events, i.e., events with small mean occurrence rates or long return periods, the Bernoulli and the Poisson models should give approximately the same results.

▶ **EXAMPLE 3.25**

According to Benjamin (1968), the historical record of earthquakes in San Francisco from 1836 to 1961 shows that there were 16 earthquakes with ground motion intensity in MM-scale of VI or higher. If the occurrence of such high-intensity earthquakes in the San Francisco-Bay Area can be assumed to constitute a Poisson process, the probability that the next high-intensity earthquake will occur within the next 2 years would be evaluated as follows.

The mean occurrence rate of high-intensity earthquakes in the region is

$$v = \frac{16}{125} = 0.128 \text{ quake per year}$$

Then, with Eq. 3.35,

$$P(T_1 \le 2) = 1 - e^{-0.128 \times 2} = 0.226$$

The above is equivalent to the probability of the occurrence of such high-intensity earthquakes (one or more) in the next two years. With the Poisson model, this latter probability would be

$$P(X_2 \ge 1) = 1 - P(X_2 < 1) = 1 - P(X_2 = 0)$$

$$= 1 - \frac{(0.128 \times 2)^0}{0!} e^{-0.128 \times 2} = 1 - e^{-0.128 \times 2} = 0.226$$

The probability that no earthquakes of this high intensity will occur in the next 10 years is

$$P(T_1 > 10) = e^{-0.128 \times 10} = 0.278$$

Again, this probability may also be evaluated with the Poisson distribution as

$$P(X_{10} = 0) = \frac{(0.128 \times 10)^0}{0!} e^{-0.128 \times 10} = 0.278$$

The return period of an intensity VI earthquake in San Francisco, according to Eq. 3.37, is therefore,

$$\overline{T_1} = \frac{1}{0.128} = 7.8 \text{ yr}$$

In general, the probability of occurrence of large earthquakes within a given time t is given by the CDF of T_1; in the present case, this is

$$P(T_1 \leq t) = 1 - e^{-0.128t}$$

Graphically, the above CDF is shown in Fig. E3.25.

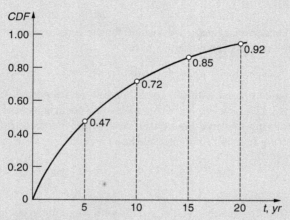

Figure E3.25 CDF of large earthquakes in San Francisco.

In particular, the probability of high-intensity earthquakes occurring within the return period of 7.8 years in the San Francisco area would be

$$P(T_1 \leq 7.8) = 1 - e^{-0.128 \times 7.8} = 1 - e^{-1.0} = 0.632 \qquad \blacktriangleleft$$

In fact, as illustrated above, for a Poisson process the probability of an event occurring (once or more) within its return period is always equal to $1 - e^{-1.0} = 0.632$. This may be compared with the probability of events with long return periods of the Bernoulli model, as discussed earlier in Sect. 3.2.3.

Of course, the exponential distribution is also useful as a general-purpose probability function. We saw this as illustrated earlier in Example 3.3 to describe the useful life of welding machines. In general, the PDF of the exponential distribution is

$$f_X(x) = \lambda e^{-\lambda x} \qquad \text{for } x \geq 0$$
$$= 0 \qquad \text{for } x < 0 \qquad (3.38)$$

where λ is a constant parameter. The corresponding CDF would be

$$F_X(x) = 1 - e^{-\lambda x} \qquad \text{for } x \geq 0$$
$$= 0 \qquad \text{for } x < 0 \qquad (3.39)$$

The mean and variance of X are, respectively,

$$\mu_X = 1/\lambda \qquad \text{and} \qquad \sigma_X^2 = 1/\lambda^2$$

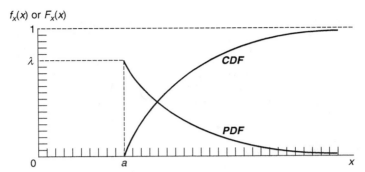

$f_x(x)$ or $F_x(x)$

Figure 3.10 PDF and CDF of the Shifted Exponential Distribution

The Shifted Exponential Distribution

In Eqs. 3.38 and 3.39, the PDF and CDF of the exponential distributions start at $x = 0$. In general, the distribution can start at any positive value of x; the resulting distribution may be called the *shifted exponential distribution*. The corresponding PDF and CDF starting at a would be as follows:

$$f_X(x) = \lambda e^{-\lambda(x-a)} \qquad x \geq a$$
$$= 0 \qquad x < a \tag{3.40}$$

and

$$F_X(x) = 1 - e^{-\lambda(x-a)} \qquad x \geq a$$
$$= 0 \qquad x < a \tag{3.41}$$

Graphically, the above PDF and CDF are illustrated in Fig. 3.10.

The exponential distribution is appropriate for modeling the distribution of the operational life, or time-to-failure of systems under "chance" or constant failure rate condition. In this regard, the parameter λ is related to the *mean life* or *mean time-to-failure E(T)* as

$$\lambda = \frac{1}{E(T)}$$

whereas, for a random variable X with the shifted exponential distribution starting at $x = a$, the mean value of X would be

$$E(X) = a + \frac{1}{\lambda} \qquad \text{or} \qquad E(X - a) = \frac{1}{\lambda}$$

and its standard deviation is $\sigma_x = 1/\lambda$.

► **EXAMPLE 3.26**

Suppose that four identical diesel engines are used to generate backup electrical power for the emergency control system of a nuclear power plant. Assume that at least two of the diesel-powered units are required to supply the needed emergency power; in other words, at least two of the four engines must start automatically during sudden loss of outside electrical power. The operational life T of each diesel engine may be modeled with the shifted exponential distribution, with a rated mean operational life of 15 years and a guaranteed minimum life of 2 years.

In this case, the reliability of the emergency backup system would clearly be of interest. For example, the probability that at least two of the four diesel engines will start automatically during an emergency within the first 4 years of the life of the system can be determined as follows.

First, the probability that any of the engines will start without any problem within 4 years is

$$P(T > 4) = e^{\frac{-1}{(15-2)}(4-2)} = 0.8574$$

Then, denoting N as the number of engines starting during an emergency, the reliability of the backup system within 4 years is

$$P(N \geq 2) = \sum_{n=2}^{4} \binom{4}{n}(0.8574)^n(0.1426)^{4-n} = 1 - \sum_{n=0}^{1} \binom{4}{n}(0.8574)^n(0.1426)^{4-n}$$

$$= 1 - \binom{4}{0}(0.1426)^4 - \binom{4}{1}(0.8574)(0.1426)^3 = 1 - 0.0004 - 0.0099$$

$$= 0.990$$

Therefore, the reliability of the backup system within 4 years is 99%, even though the reliability of each engine is only about 86%. ◀

3.2.8 The Gamma Distribution

In general, the *gamma distribution* for a random variable X has the following PDF,

$$f_X(x) = \frac{v(vx)^{k-1}}{\Gamma(k)}e^{-vx} \qquad x \geq 0$$
$$= 0 \qquad x < 0 \tag{3.42}$$

where v and k are the parameters of the distribution, and $\Gamma(k)$ is the gamma function

$$\Gamma(k) = \int_0^\infty x^{k-1}e^{-x}dx \qquad \text{where } k > 1.0 \tag{3.43}$$

We may recall that the gamma function is the generalization of the factorial to noninteger numbers. In particular, we observe that integration-by-parts will yield, for $k > 1.0$,

$$\Gamma(k) = (k-1)\Gamma(k-1) = (k-1)(k-2)\cdots(k-i)\Gamma(k-i)$$

The mean and variance of the gamma distribution, Eq. 3.42, are respectively,

$$\mu_X = k/v \qquad \text{and} \qquad \sigma_X^2 = k/v^2$$

Calculation of the probability involving the gamma distribution can be performed using tables of the *incomplete gamma function*, which are usually given for the ratio (e.g., Harter, 1963):

$$I(u,k) = \frac{\displaystyle\int_0^u y^{k-1}e^{-y}dy}{\Gamma(k)}$$

Then, the probability of $(a < X \leq b)$ can be obtained as

$$P(a < X \leq b) = \frac{v^k}{\Gamma(k)}\int_a^b x^{k-1}e^{-vx}dx$$

If we let $y = vx$, the above integral becomes

$$P(a < X \leq b) = \frac{1}{\Gamma(k)}\left[\int_0^{vb} y^{k-1}e^{-y}dy - \int_0^{va} y^{k-1}e^{-y}dy\right] \tag{3.43a}$$
$$= I(vb,k) - I(va,k)$$

Therefore, in effect, the *incomplete gamma function ratio* is also the CDF of the gamma distribution.

► **EXAMPLE 3.27**

The gamma distribution may be used to represent the distribution of the *equivalent uniformly distributed load (EUDL)* on buildings. For a particular building, if the mean *EUDL* is 15 psf (pounds per square foot) and the c.o.v. is 25%, the parameters of the appropriate gamma distribution are,

$$\delta = \frac{\sigma}{\mu} = \frac{\sqrt{k}/v}{k/v} = \frac{1}{\sqrt{k}}; \quad \text{thus,} \quad k = \frac{1}{\delta^2} = \frac{1}{(0.25)^2} = 16$$

and

$$v = \frac{k}{\mu} = \frac{16}{15} = 1.067$$

The design live load is generally specified (conservatively) to be on the high side. For instance, if the design *EUDL* is specified to be 25 psf, the probability that this design load will be exceeded according to Eq. 3.43a, is

$$P(L > 25) = 1 - P(L \le 25) = 1 - I(25 \times 1.067, 16) = 1 - I(26.67, 16)$$

$$= 1 - 0.671 = 0.329$$

◄

The Gamma Distribution and the Poisson Process

We should point out that the gamma distribution is related to the Poisson process. If the occurrences of an event constitute a Poisson process in time, then the time till the *k*th occurrence of the event is governed by the gamma distribution. Earlier, in Sect. 3.2.7, we saw that the time until the first occurrence of the event is governed by the exponential distribution.

Let T_k denote the time until the *k*th occurrence of an event; then $(T_k \le t)$ means that there were *k* or more occurrences of the event in time *t*. Hence, on the basis of Eq. 3.34, we obtain the CDF of T_k as

$$F_{T_k}(t) = \sum_{x=k}^{\infty} P(X_t = x) = 1 - \sum_{x=0}^{k-1} \frac{(vt)^x}{x!} e^{-vt}$$

$$= 1 - \left[1 + \frac{(vt)}{1!} + \frac{(vt)^2}{2!} + \cdots + \frac{(vt)^{k-2}}{(k-2)!} + \frac{(vt)^{k-1}}{(k-1)!} \right] e^{-vt}$$

Taking the derivative of the above CDF, it can be shown that the PDF of T_k is as follows:

$$f_{T_k}(t) = \frac{v(vt)^{k-1}}{(k-1)!} e^{-vt} \qquad \text{for } t \ge 0 \tag{3.44}$$

The above gamma distribution with integer *k* is known also as the *Erlang distribution*. In this case, the mean time until the *k*th occurrence of an event is

$$E(T_k) = k/v$$

and its variance is

$$Var(T_k) = k/v^2$$

We can see that for $k = 1$, i.e., for the time until the first occurrence of an event, Eq. 3.44 is reduced to the exponential distribution of Eq. 3.36.

► **EXAMPLE 3.28**

Suppose that fatal accidents on a particular highway occur on the average about once every 6 months. If we can assume that the occurrences of accidents on this highway constitute a Poisson process, with mean occurrence rate of $v = 1/6$ per month, the time until the occurrence of the first accident (or between two consecutive accidents) would be described by the exponential distribution, specifically with the following PDF:

$$f_{T_1}(t) = \frac{1}{6}e^{-t/6}$$

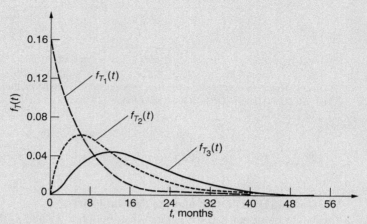

Figure E3.28 PDFs of times until the occurrences of first, second, and third accidents on a highway.

The time until the occurrence of the second accident (or the time between every other accidents) on the same highway is described by the gamma distribution, with the PDF

$$f_{T_2}(t) = \frac{1}{6}(t/6)e^{-t/6}$$

whereas the time until the occurrence of the third accident would also be gamma distributed, with the PDF

$$f_{T_3}(t) = \frac{1}{6}\frac{(t/6)^2}{2!}e^{-t/6}$$

The above PDFs are illustrated graphically in Fig. E3.28, and the corresponding mean occurrence times of T_1, T_2, and T_3 are, respectively, 6, 12, and 18 months. ◄

We might recognize that the exponential and gamma distributions are the continuous analogues, respectively, of the geometric and negative binomial distributions; that is, the geometric and negative binomial distributions govern the first and kth occurrence times of a Bernoulli sequence, whereas the exponential and gamma distributions govern the corresponding occurrence times of a Poisson process.

The Shifted Gamma Distribution

We might point out that most probability distributions are described with two parameters, or even with one parameter, such as the exponential distribution. The *shifted gamma distribution* is one of the few exceptions with three parameters. A three-parameter distribution may be useful for fitting statistical data in which the skewness (involving the third moment) in the data is significant; in particular, the third parameter would be necessary in order to explicitly include the skewness in the observed data.

As an extension of Eq. 3.42, the PDF of the three-parameter shifted gamma distribution, for a random variable X, may be expressed as

$$f_X(x) = \frac{v[v(x-\gamma)^{k-1}]}{\Gamma(k)}\exp[-v(x-\gamma)]; \qquad x \geq \gamma \qquad (3.45)$$

in which $v, k \geq 1.0$, and γ are the three parameters of the distribution. We observe that Eq. 3.45 is reduced to Eq. 3.42 if $\gamma = 0$.

The mean and variance of X are, respectively,

$$\mu_X = \frac{k}{v} + \gamma \quad \text{and} \quad \sigma_X{}^2 = \frac{k}{v^2}$$

▶ **EXAMPLE 3.29**

The three-parameter gamma distribution can be shown to give better fit with statistical data when there is significant skewness in the observed data. For instance, shown below in Fig. E3.29 is the histogram of measured residual stresses in the flanges of steel H-sections. The mean, standard deviation, and skewness coefficient of the measured ratios of residual stress/yield stress are, respectively, 0.3561, 0.1927, and 0.8230.

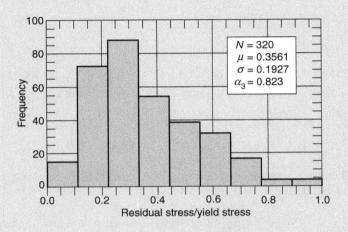

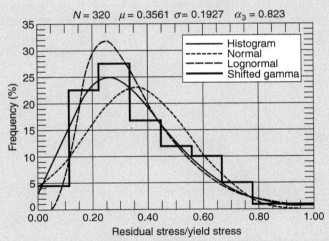

Figure E3.29 Fitting distributions to histogram of residual stresses (after Zhao & Ang, 2003).

Clearly, because the data show significant skewness, a three-parameter distribution is necessary in order to include the skewness for adequately fitting the histogram of the measured residual stresses. As shown in Fig. E3.29, the three-parameter gamma PDF (solid curve) that includes the skewness of 0.8230 has a much closer fit to the histogram than the normal or lognormal distributions which are, of course, two-parameter distributions. This is further verified later in Example 7.10 with the *K-S* goodness-of-fit test. ◀

3.2.9 The Hypergeometric Distribution

The *hypergeometric distribution* arises when samples from a finite population, consisting of two types of elements (e.g., "good" and "bad"), are being examined. It is the basic distribution underlying many sampling plans used in connection with acceptance sampling and quality control.

Consider a lot of N items, among which m are defective and the remaining $(N - m)$ items are good. If a sample of n items is taken at random from this lot, the probability that there will be x defective items in the sample is given by the hypergeometric distribution as follows:

$$P(X = x) = \frac{\binom{m}{x}\binom{N-m}{n-x}}{\binom{N}{n}} \qquad x = 1, 2, \ldots, m \qquad (3.46)$$

The above distribution is based on the following: In the lot of N items, the number of samples of size n is $\binom{N}{n}$; among these, the number of samples with x defectives is

$$\binom{m}{x}\binom{N-m}{n-x}$$

Therefore, assuming that the samples are equally likely to be selected, we obtain the hypergeometric distribution of Eq. 3.46.

▶ **EXAMPLE 3.30**

In a box of 100 strain gages, suppose we suspect that there may be four gages that are defective. If six of the gages from the box were used in an experiment, the probability that one defective gage was used in the experiment is evaluated as follows (in this case, we have $N = 100$, $m = 4$, and $n = 6$); thus,

$$P(X = 1) = \frac{\binom{4}{1}\binom{100-4}{6-1}}{\binom{100}{6}} = 0.205$$

whereas the probability that none of the defective gages in the box was used in the experiment is

$$P(X = 0) = \frac{\binom{4}{0}\binom{100-4}{6}}{\binom{100}{6}} = 0.778$$

and at least one defective gage was used in the experiment would be

$$P(X \geq 1) = 1 - P(X = 0) = 1 - 0.778 = 0.222$$

◀

▶ **EXAMPLE 3.31**

In a large reinforced concrete construction project, 100 concrete cylinders are to be collected from the daily concrete mixes delivered to the construction site. Furthermore, to ensure material quality, the acceptance/rejection criterion requires that ten of these cylinders (selected at random) must be tested for crushing strength after curing for 1 week, and nine of the ten cylinders tested must have a required minimum strength. Is the acceptance/rejection criterion stringent enough?

Whether the acceptance/rejection criterion is too stringent, or not stringent enough, depends on whether it is difficult or easy for poor-quality concrete mixes to go undetected. For example, if there is d percent of defective concrete, then on the basis of the specified acceptance/rejection criterion, the probability of rejection of the daily concrete mixes would be (denoting X as the number of defective cylinders in the test)

$$P(X > 1) = 1 - P(X \le 1)$$

$$= 1 - \left[\frac{\binom{100d}{0}\binom{100(1-d)}{10}}{\binom{100}{10}} + \frac{\binom{100d}{1}\binom{100(1-d)}{9}}{\binom{100}{10}} \right]$$

For example, if there is 10% defectives in the daily concrete mixes, or $d = 10\%$,

$$P(\text{rejection}) = 1 - \left[\frac{\binom{90}{10}}{\binom{100}{10}} + \frac{\binom{10}{1}\binom{90}{9}}{\binom{100}{10}} \right] = 1 - (0.3305 + 0.4080) = 0.2615$$

whereas if $d = 2\%$,

$$P(\text{rejection}) = 1 - \left[\frac{\binom{98}{10}}{\binom{100}{10}} + \frac{\binom{2}{1}\binom{98}{9}}{\binom{100}{10}} \right] = 1 - (0.8091 + 0.1818) = 0.009$$

Therefore, if 10% of the concrete mixes were defective, it is likely (with 26% probability) that the defective material will be discovered with the proposed acceptance/rejection criterion, whereas if 2% of the concrete mixes were defective, the likelihood of the daily mixes being rejected is very low (with 0.009 probability).

Hence, if the contract requires concrete with less than 2% defectives, then the proposed acceptance/rejection criterion is not stringent enough; on the other hand, if material with 10% defectives is acceptable, then the proposed criterion may be satisfactory. ◀

3.2.10 The Beta Distribution

Most probability distributions are for random variables whose range of values are unlimited in either one or both directions. Indeed, thus far all the continuous probability distributions we have seen (except for the uniform PDF) have this characteristic. In some engineering applications, there may be problems in which there are finite lower and upper bound values of the random variables; in these cases, probability distributions with finite lower and upper limits would be appropriate.

The *beta distribution* is one of the few distributions appropriate for a random variable whose range of possible values are bounded, say between a and b. Its PDF is given by

$$f_X(x) = \frac{1}{B(q, r)} \frac{(x - a)^{q-1}(b - x)^{r-1}}{(b - a)^{q+r-1}} \qquad a \le x \le b \tag{3.47}$$

$$= 0 \qquad \text{otherwise}$$

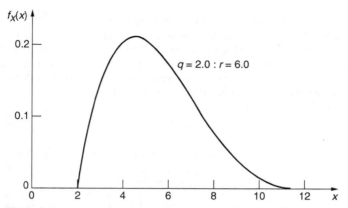

Figure 3.11 The beta distribution with $q = 2.0$ and $r = 6.0$ for $(2 < x < 12)$.

in which q and r are the parameters of the distribution, and $B(q, r)$ is the *beta function*

$$B(q, r) = \int_0^1 x^{q-1}(1 - x)^{r-1}dx \qquad (3.48)$$

which is related to the gamma function of Eq. 3.43 as follows:

$$B(q, r) = \frac{\Gamma(q)\Gamma(r)}{\Gamma(q + r)} \qquad (3.49)$$

Depending on the two parameters q and r, the PDF of the beta distribution will have a different shape. In Fig. 3.11 we show the PDF of a beta distribution for $2 \le x \le 12$ with $q = 2.0$ and $r = 6.0$.

If the lower and upper values of the variate are bounded between 0 and 1.0, i.e., $a = 0$ and $b = 1.0$, Eq. 3.47 becomes

$$\begin{aligned} f_X(x) &= \frac{1}{B(q, r)}x^{q-1}(1 - x)^{r-1} & 0 \le x \le 1.0 \\ &= 0 & \text{otherwise} \end{aligned} \qquad (3.47a)$$

which may be called the *standard beta distribution*.

Figure 3.12 shows the standard beta PDFs with different values of q and r. From this figure, we can observe that the beta PDF assumes different shapes depending on the values

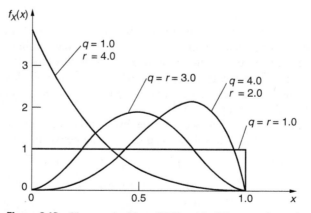

Figure 3.12 The standard beta PDFs with different values of q and r.

of the two parameters q and r. In particular, we also observe that for $q < r$, the PDF is positively skewed, whereas for $q > r$ it is negatively skewed, and for $q = r$ the PDF is symmetric. With these characteristics, the beta distribution is quite versatile and can be used to fit a wide range of histograms of observed data.

The probability associated with a beta distribution can be evaluated in terms of the *incomplete beta function*, which is defined as

$$B_x(q,r) = \int_0^x y^{q-1}(1-y)^{r-1}dy \qquad 0 < x < 1.0 \tag{3.50}$$

and values of the *incomplete beta function ratio* $B_x(q,r)/B(q,r)$ have been tabulated, for example, by Pearson (1934) and Pearson and Johnson (1968). In particular, the CDF of the standard beta distribution is

$$F_X(x) = \frac{1}{B(q,r)} \int_0^x x^{q-1}(1-x)^{r-1}dx = \frac{B_x(q,r)}{B(q,r)}$$

Therefore, the incomplete beta function ratio is the CDF of the standard beta distribution, which we shall denote as

$$\beta(x|q,r) = \frac{B_x(q,r)}{B(q,r)} \tag{3.51}$$

We should point out that the incomplete beta function ratio is generally given for $q \geq r$. For $q < r$, the ratio can be obtained from

$$\beta(x|q,r) = 1 - \beta(x|r,q)$$

For a general beta distribution of Eq. 3.47, the probability between $x = x_1$ and $x = x_2$ may be evaluated as follows:

$$P(x_1 < X \leq x_2) = \frac{1}{B(q,r)} \int_{x_1}^{x_2} \frac{(x-a)^{q-1}(b-x)^{r-1}}{(b-a)^{q+r-1}}dx$$

If we substitute

$$y = \frac{x-a}{b-a}$$

we also have

$$1 - y = \frac{b-x}{b-a} \qquad \text{and} \qquad dy = \frac{dx}{b-a}$$

Hence, the above integral becomes

$$P(x_1 < X \leq x_2) = \frac{1}{B(q,r)}\left[\int_0^{(x_2-a)/(b-a)} y^{q-1}(1-y)^{r-1}dy - \int_0^{(x_1-a)/(b-a)} y^{q-1}(1-y)^{r-1}dy\right]$$

Therefore, denoting $u = (x_2 - a)/(b - a)$ and $v = (x_1 - a)/(b - a)$, the above probability can be evaluated in terms of the CDF of the standard beta distribution, Eq. 3.51, as

$$P(x_1 < X \leq x_2) = \beta(u|q,r) - \beta(v|q,r) \tag{3.52}$$

The mean and variance of X with a beta distribution between a and b are

$$\mu_X = a + \frac{q}{q+r}(b-a) \tag{3.53}$$

$$\sigma_X^2 = \frac{qr}{(q+r)^2(q+r+1)}(b-a)^2 \tag{3.53a}$$

and its coefficient of skewness is

$$\theta_X = \frac{2(r-q)}{(q+r)(q+r+2)} \frac{(b-a)}{\sigma_X} \tag{3.54}$$

Also, the mode of X is

$$\tilde{x} = a + \frac{1-q}{2-q-r}(b-a) \tag{3.55}$$

▶ **EXAMPLE 3.32**

The duration required to complete an activity in a construction project has been estimated by the subcontractor to be as follows:

Minimum duration = 5 days

Maximum duration = 10 days

Expected duration = 7 days

The coefficient of variation of the required duration is estimated to be 10%.

In this case, the beta distribution may be appropriate with $a = 5$ days and $b = 10$ days. The parameters of the distribution would be determined as follows: With Eq. 3.53, we have

$$5 + \frac{q}{q+r}(10-5) = 7$$

giving

$$q = \frac{2}{3}r$$

Then, substituting this into the expression for the variance, Eq. 3.53a, we have

$$\frac{qr}{(q+r)^2(q+r+1)}(10-5)^2 = (0.1 \times 7)^2$$

yielding

$$q = 3.26 \quad \text{and} \quad r = 4.89$$

The probability that the activity will be completed within 9 days is then given by

$$P(T \le 9) = \beta_u(3.26, 4.89)$$

in which $u = (9-5)/(10-5) = 0.8$. From tables of the incomplete beta function ratios (e.g., Pearson and Johnson, 1968) we obtain after suitable interpolation

$$P(T \le 9) = \beta_{0.8}(3.26, 4.89) = 0.993$$

◀

▶ **EXAMPLE 3.33**

In evaluating the reliability of steel bridge components against fatigue damage, the stress range (i.e., the maximum minus the minimum applied stresses) in each loading cycle is the principal load parameter. For this purpose, it is reasonable to assume that the stress range has finite lower and upper bound values.

As a specific example, consider the strain-ranges measured at a particular interstate highway bridge in Illinois under heavy traffic as shown in Fig. E3.33. The corresponding stress ranges (in psi) can be obtained by multiplying the strain range, in micro in./in. of Fig. E3.33 by 30,000 psi/1000 = 30 psi. On this basis, we may model the applied stress ranges, S, on the particular highway bridge with a beta PDF with parameters $q = 2.83$ and $r = 4.39$, and a lower bound value of 0 and an upper bound value of 10,000 psi.

Suppose a steel-wide flange beam of the bridge is subjected to the beta-distributed stress ranges of Fig. E3.33. The fatigue life of metal structures may be described with the so-called SN relation

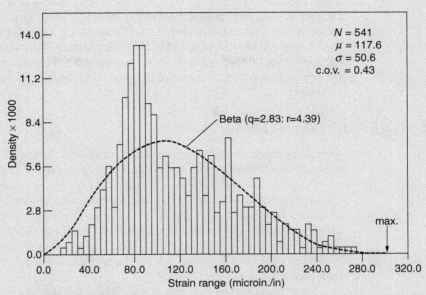

Figure E3.33 Histogram of measured strain range of a beam, Shaffer Creek Bridge, IL under heavy traffic (after Ruhl and Walker, 1975).

specified by two parameters c and m. In the present case of the steel-wide flange beam, these SN parameters are

$$c = 3.98 \times 10^8 \text{ cycles} \quad \text{and} \quad m = 2.75$$

The mean fatigue life when subjected to random stress ranges is given by

$$\bar{n} = \frac{c}{E(S^m)}$$

in which $E(S^m)$ is the mth moment of S; for a beta-distributed S with maximum s_o, this is (see Ang, 1977)

$$E(S^m) = s_o^m \frac{\Gamma(m+q)\Gamma(q+r)}{\Gamma(q)\Gamma(m+q+r)}$$

In the case of the above wide flange beam,

$$E(S^m) = 10^{2.75} \left[\frac{\Gamma(5.58)\Gamma(7.22)}{\Gamma(2.83)\Gamma(9.97)} \right] = 89.486$$

Thus, the mean fatigue life of the beam is

$$\bar{n} = \frac{3.98 \times 10^8}{89.486} = 4.45 \times 10^6 \text{ cycles} \qquad ◄$$

3.2.11 Other Useful Distributions

The probability distributions that we have described thus far are among the most useful and important for practical engineering applications. However, these are far from inclusive; for specific applications other distributions may be more appropriate and useful, including, e.g., the uniform and triangular PDFs. In particular, the class of distributions relating to the extreme values (the maximum and minimum values) are of special significance to many problems in engineering; this class of extreme value distributions is presented in Chapter 4. Also, there are special distributions that are of significance in statistical analysis, including the Student t-distribution, the chi-square (χ^2) distribution, the F-distribution, and the Pearson system (Elderton, 1953). For example, the t-distribution is useful for determining the confidence interval of the population mean when the variance is unknown, whereas the chi-square distribution is useful in the interval estimation of the population variance (see Chapter 6).

▶ 3.3 MULTIPLE RANDOM VARIABLES

The essential concepts of a random variable and its probability distribution can be extended to two or more random variables and their *joint* probability distribution. In order to identify events that are the results of two or more physical processes in numerical terms, the events in a sample space may be mapped into two (or more) dimensions of the real space; implicitly, we can recognize that this requires two or more random variables. Consider, for example, the rainfall intensity at a gage station and the resulting runoff of a river; we may use a random variable X whose values x denote the possible rainfall intensities, and another random variable Y whose values y are the possible runoffs of the river. Accordingly, we shall denote $(X = x, Y = y)$ and $(X \leq x, Y \leq y)$ as the joint events $(X = x \cap Y = y)$ and $(X \leq x \cap Y \leq y)$ defined by the random variables in the xy space. Obviously, this concept can also be extended to multiple random variables.

3.3.1 Joint and Conditional Probability Distributions

As any pair of values of the random variables X and Y represents events, there are probabilities associated with given values of x and y; accordingly, the probabilities for all possible pairs of x and y may be described with the *joint distribution function,* CDF, of the random variables X and Y; namely,

$$F_{X,Y}(x, y) = P(X \leq x, Y \leq y) \tag{3.56}$$

which is the CDF of the joint occurrences of the events identified by $X \leq x$ and $Y \leq y$. In order to comply with the fundamental axioms of probability, the above CDF must satisfy the following:

1. $F_{X,Y}(-\infty, -\infty) = 0$ $\qquad$ $F_{X,Y}(\infty, \infty) = 1.0$

2. $F_{X,Y}(-\infty, y) = 0$ $\qquad$ $F_{X,Y}(\infty, y) = F_Y(y)$
    $F_{X,Y}(x, -\infty) = 0$ $\qquad$ $F_{X,Y}(x, \infty) = F_X(x)$

3. $F_{X,Y}(x, y)$ is nonnegative and is a nondecreasing function of x and y.

For discrete X and Y, the probability distribution may also be described with the joint PMF, which is simply

$$p_{X,Y}(x_i, y_j) = P(X = x_i, Y = y_j) \tag{3.57}$$

and the CDF becomes

$$F_{X,Y}(x, y) = \sum_{\{x_i \leq x, y_j \leq y\}} p_{X,Y}(x_i, y_j) \tag{3.58}$$

whereas, if the random variables X and Y are continuous, the joint probability distribution may also be described with the *joint* PDF, $f_{X,Y}(x,y)$, defined as follows:

$$f_{X,Y}(x, y)dxdy = P(x < X \leq x + dx, y < Y \leq y + dy)$$

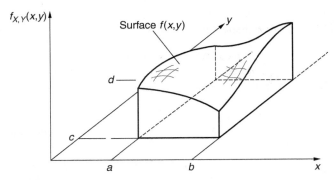

Figure 3.13 Volume under the PDF $f_{X,Y}(x,y)$.

Then,

$$F_{X,Y}(x, y) = \int_{-\infty}^{x} \int_{-\infty}^{y} f_{X,Y}(u, v)dvdu \qquad (3.59)$$

Conversely, if the partial derivatives exist,

$$f_{X,Y}(x, y) = \frac{\partial^2 F_{X,Y}(x, y)}{\partial x \partial y} \qquad (3.60)$$

Also, we observe the following probability:

$$P(a < X \le b, c < Y \le d) = \int_{a}^{b} \int_{c}^{d} f_{X,Y}(u, v)dvdu$$

which is the volume under the surface $f(x,y)$ as shown in Fig. 3.13.

The Conditional and Marginal Distributions

For discrete X and Y, the probability of $(X = x_i)$ may depend on the values of Y, or vice versa; accordingly, by virtue of Eq. 2.11, we have the *conditional PMF*,

$$p_{X|Y}(x_i|y_j) \equiv P(X = x_i|Y = y_j) = \frac{p_{X,Y}(x_i, y_j)}{p_Y(y_j)} \qquad (3.61)$$

if $p_Y(y_j) \ne 0$. Similarly,

$$p_{Y|X}(y_j|x_i) = \frac{p_{X,Y}(x_i, y_j)}{p_X(x_i)} \qquad (3.61a)$$

if $p_X(x_i) \ne 0$.

The PMF of the individual random variables may be obtained also from the joint PMF. Applying the theorem of total probability, Eq. 2.19, we obtain the *marginal* PMF of X as

$$p_X(x_i) = \sum_{\text{all } y_j} P(X = x_i|Y = y_j)P(Y = y_j)$$

$$= \sum_{\text{all } y_j} P(X = x_i, Y = y_j) = \sum_{\text{all } y_j} p_{X,Y}(x_i, y_j) \qquad (3.62)$$

By the same token, the marginal PMF of Y is

$$p_Y(y_j) = \sum_{\text{all } x_i} p_{X,Y}(x_i, y_j) \qquad (3.62a)$$

If the random variables X and Y are statistically independent, meaning the events $X = x_i$ and $Y = y_j$ are statistically independent, then

$$p_{X|Y}(x_i|y_j) = p_X(x_i) \quad \text{and} \quad p_{X|Y}(y_j|x_i) = p_Y(y_j)$$

and Eq. 3.57 becomes

$$p_{X,Y}(x_i, y_j) = p_X(x_i)p_Y(y_j) \tag{3.63}$$

If the random variables X and Y are continuous, the *conditional* PDF of X given Y is

$$f_{X|Y}(x|y) = \frac{f_{X,Y}(x, y)}{f_Y(y)} \tag{3.64}$$

from which we also have

$$f_{X,Y}(x, y) = f_{X|Y}(x|y)f_Y(y) \tag{3.65}$$

or

$$f_{X,Y}(x, y) = f_{Y|X}(y|x)f_X(x)$$

However, if X and Y are statistically independent, i.e., $f_{X|Y}(x|y) = f_X(x)$ and $f_{Y|X}(y|x) = f_Y(y)$, then the joint PDF becomes

$$f_{X,Y}(x, y) = f_X(x)f_Y(y) \tag{3.66}$$

Finally, through the theorem of total probability, we obtain the *marginal* PDFs,

$$f_X(x) = \int_{-\infty}^{\infty} f_{X|Y}(x|y)f_Y(y)dy = \int_{-\infty}^{\infty} f_{X,Y}(x, y)dy \tag{3.67}$$

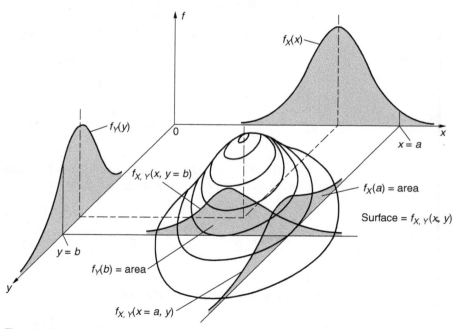

Figure 3.14   Joint and marginal PDFs of two continuous random variables.

Similarly,

$$f_Y(y) = \int_{-\infty}^{\infty} f_{X,Y}(x, y)dx \qquad (3.68)$$

The characteristics of a joint PDF for two random variables, X and Y, and the associated marginal PDFs, are portrayed graphically in Fig. 3.14.

▶ **EXAMPLE 3.34**

Based on a survey of construction labor and its productivity, the work duration (in number of hours per day) and the average productivity (in terms of percent efficiency) were recorded as tabulated below. For simplicity, the data for work duration are recorded as 6, 8, 10, and 12 hr, whereas the average productivity is recorded as 50%, 70%, and 90%. Denoting $X =$ duration, and $Y =$ productivity, the recorded data are as follows:

Duration and Productivity (x, y)	No. of Observations	Relative Frequency
6, 50	2	0.014
6, 70	5	0.036
6, 90	10	0.072
8, 50	5	0.036
8, 70	30	0.216
8, 90	25	0.180
10, 50	8	0.058
10, 70	25	0.180
10, 90	11	0.079
12, 50	10	0.072
12, 70	6	0.043
12, 90	2	0.014
	Total = 139	

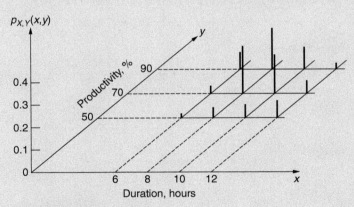

Figure E3.34a Joint PMF $p_{X,Y}(x,y)$.

The above data can be portrayed graphically as the joint PMF of X and Y, as shown in Fig. E3.34a.

The marginal PMF of X, the work duration, is

$$p_X(x) = \sum_{(y_j=50,70,90)} p_{X,Y}(x, y)$$

and would appear as shown in Fig. E3.34b.

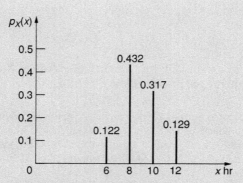

Figure E3.34b   Marginal PMF $p_X(x)$.

For example, the ordinate at $X = 8$ is obtained as

$$p_X(8) = 0.036 + 0.216 + 0.180 = 0.432$$

Similarly, the marginal PMF of Y, for productivity, would be as shown in Fig. E3.34c.

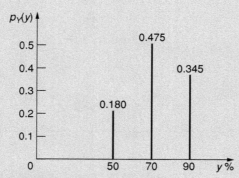

Figure E3.34c   Marginal PMF $p_Y(y)$.

Finally, if the work duration is 8 hr/day, the probability that the average productivity will be 90% is given by the conditional probability of Eq. 3.61a as

$$p_{Y|X}(90\% | \ 8 \ \mathrm{hr}) = \frac{p_{X,Y}(8, 90)}{p_X(8)} = \frac{0.180}{0.432} = 0.417$$

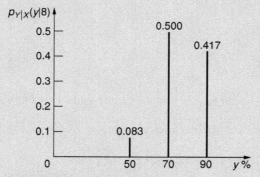

Figure E3.34d   Conditional PMF $p_{Y|X}(y|8)$.

Therefore, the probability of achieving 90% efficiency for an 8 hr/day work duration is less than 42%. The probabilities of other productivity levels for an 8-hr work day are shown in the conditional PMF of Fig. E3.34d. ◄

► **EXAMPLE 3.35**

An example of a joint PDF of two continuous random variables X and Y is the *bivariate normal* density function given by

$$f_{X,Y}(x, y) = \frac{1}{2\pi \sigma_X \sigma_Y \sqrt{1 - \rho^2}} \exp\left[\frac{-1}{2(1 - \rho^2)}\left\{\left(\frac{x - \mu_X}{\sigma_X}\right)^2\right.\right.$$
$$\left.\left. -2\rho\left(\frac{x - \mu_X}{\sigma_X}\right)\left(\frac{y - \mu_Y}{\sigma_Y}\right) + \left(\frac{y - \mu_Y}{\sigma_Y}\right)^2\right\}\right]$$
$$-\infty < x < \infty; \quad -\infty < y < \infty$$

in which ρ is the *correlation coefficient* between X and Y (as defined in Sect. 3.3.2). We can show that the above joint PDF may be written also as

$$f_{X,Y}(x, y) = \frac{1}{\sqrt{2\pi}\sigma_X} \exp\left[-\frac{1}{2}\left(\frac{x - \mu_X}{\sigma_X}\right)^2\right] \frac{1}{\sqrt{2\pi}\sigma_Y\sqrt{1 - \rho^2}}$$
$$\exp\left[-\frac{1}{2}\left(\frac{y - \mu_y - \rho(\sigma_Y/\sigma_X)(x - \mu_X)}{\sigma_Y\sqrt{1 - \rho^2}}\right)^2\right]$$

Then, in light of Eq. 3.64, we see that the conditional PDF of Y given $X = x$ is

$$f_{Y|X}(y|x) = \frac{1}{\sqrt{2\pi}\sigma_Y\sqrt{1 - \rho^2}} \exp\left[-\frac{1}{2}\left(\frac{y - \mu_Y - \rho(\sigma_Y/\sigma_X)(x - \mu_X)}{\sigma_Y\sqrt{1 - \rho^2}}\right)^2\right]$$

whereas the marginal PDF of X is

$$f_X(x) = \frac{1}{\sqrt{2\pi}\sigma_X} \exp\left[-\frac{1}{2}\left(\frac{x - \mu_X}{\sigma_X}\right)^2\right]$$

both of which are Gaussian. In particular, we can observe that the conditional PDF is normal with a mean value of

$$E(Y|X = x) = \mu_Y - \rho(\sigma_Y/\sigma_X)(x - \mu_X)$$

and variance of

$$Var(Y|X = x) = \sigma_Y^2(1 - \rho^2)$$

Similarly, we can show that the conditional PDF of X given $Y = y$ is

$$f_{X|Y}(x|y) = \frac{1}{\sqrt{2\pi}\sigma_X\sqrt{1 - \rho^2}} \exp\left[-\frac{1}{2}\left(\frac{x - \mu_X - \rho(\sigma_X/\sigma_Y)(y - \mu_Y)}{\sigma_X\sqrt{1 - \rho^2}}\right)^2\right]$$

and the marginal PDF of Y is

$$f_Y(y) = \frac{1}{\sqrt{2\pi}\sigma_Y} \exp\left[-\frac{1}{2}\left(\frac{y - \mu_Y}{\sigma_Y}\right)^2\right]$$

◄

3.3.2 Covariance and Correlation

When there are two random variables X and Y, there may be a relationship between the variables. In particular, the presence or absence of a linear statistical relationship is determined as follows: First, we observe the joint second moment of X and Y as

$$E(XY) = \int_{-\infty}^{\infty} \int_{-\infty}^{\infty} xy f_{X,Y}(x, y) dx dy \qquad (3.69)$$

whereas if X and Y are statistically independent, Eq. 3.69 becomes

$$E(XY) = \int_{-\infty}^{\infty} \int_{-\infty}^{\infty} xy f_X(x) f_Y(y) dx dy = \int_{-\infty}^{\infty} x f_X(x) dx \int_{-\infty}^{\infty} y f_Y(y) dy = E(X)E(Y) \qquad (3.70)$$

The joint second central moment is the *covariance* of X and Y; i.e.,

$$\text{Cov}(X, Y) = E[(X - \mu_X)(Y - \mu_Y) = E(XY) - E(X)E(Y)] \qquad (3.71)$$

Therefore, in light of Eq. 3.70, $\text{Cov}(X,Y) = 0$ if X and Y are statistically independent.

The physical significance of the covariance may be inferred from Eq. 3.71. If the Cov (X, Y) is *large and positive*, the values of X and Y tend to be both large or both small relative to their respective means, whereas if the Cov (X,Y) is *large and negative*, the values of X tend to be large when the values of Y are small, and vice versa, relative to their respective means. But if the Cov (X,Y) is small or zero, there is weak or no (linear) relationship between the values of X and Y; or the relationship may be nonlinear. The Cov (X,Y), therefore, is a measure of the degree of linear relationship between two random *variables X and Y*. For practical purposes, however, it is preferable to use the normalized covariance, or *correlation coefficient*, which is defined as

$$\rho = \frac{\text{Cov}(X, Y)}{\sigma_X \sigma_Y} \qquad (3.72)$$

Observe that ρ is dimensionless; its values range between -1.0 and 1.0; i.e.,

$$-1 \leq \rho \leq +1 \qquad (3.73)$$

which we can verify as follows: Schwarz's inequality (Kaplan, 1953; Hardy, Littlewood, and Polya, 1959) says

$$\left[\int_{-\infty}^{\infty} \int_{-\infty}^{\infty} (x - \mu_X)(y - \mu_Y) f_{X,Y}(x, y) dx dy \right]^2$$
$$\leq \int_{-\infty}^{\infty} \int_{-\infty}^{\infty} (x - \mu_X)^2 f_{X,Y}(x, y) dx dy \int_{-\infty}^{\infty} \int_{-\infty}^{\infty} (y - \mu_Y)^2 f_{X,Y}(x, y) dx dy$$

We can recognize that the left-hand side is the $[\text{Cov}(X, Y)]^2$, whereas each of the two double integrals on the right-hand side are, respectively,

$$\int_{-\infty}^{\infty} (x - \mu_X)^2 f_X(x) dx = \sigma_X^2$$

and

$$\int_{-\infty}^{\infty} (y - \mu_Y)^2 f_Y(y) dy = \sigma_Y^2$$

Hence, we have

$$[\text{Cov}(X, Y)]^2 \leq \sigma_X^2 \sigma_Y^2$$

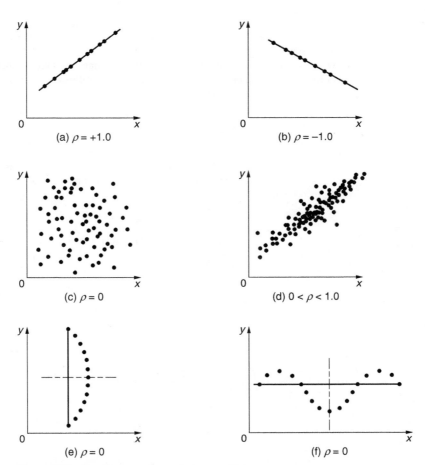

Figure 3.15 Significance of correlation coefficient ρ.

or

$$\rho^2 \le 1.0$$

thus verifying Eq. 3.73.

The significance of the correlation coefficient, ρ, is illustrated graphically in Fig. 3.15. In particular, we observe that when $\rho = \pm 1.0$, the random variables X and Y are linearly related as shown in Fig. 3.15(a) and 3.15(b), respectively, whereas when $\rho = 0$, values of (X, Y) pairs will appear as in Fig. 3.15(c). For intermediate values of ρ, values of (X, Y) pairs will appear as shown in Fig. 3.15(d); the "scatter" in the data points will decrease as ρ increases. Finally, we should also observe from Figs. 3.15(e) and 3.15(f) that when the relation between X and Y is nonlinear, $\rho = 0$ even though there is a perfect functional relationship between the variables.

Therefore, the magnitude of the correlation coefficient, ρ (between 0 and 1), is a statistical measure of the degree of linear interrelationship between two random variables.

Note: It is also well to point out that although ρ is a measure of the degree of linear relationship between two variables, this does not necessarily imply a causal effect between the variables. Two random variables X and Y may both depend on another variable (or variables), such that the values of X and Y may be highly correlated, but the values of one variable may have no direct effect on the values of the other variable. For instance, the flood flow of a river and the productivity of a construction crew may be correlated because both depend on the weather condition; however, the flood flow may have no direct influence on the productivity of the construction crew, or vice versa. For another illustration, consider the next problem in Example 3.36 from the field of mechanics.

► **EXAMPLE 3.36**

A cantilever beam, shown in Fig. E3.36 below, is subjected to two random loads, S_1 and S_2, that are statistically independent with respective means and standard deviations of μ_1, σ_1 and μ_2, σ_2. The shear force Q and bending moment M at the fixed support of the beam are both functions of the two loads, respectively, as follows:

$$Q = S_1 + S_2$$

and
$$M = aS_1 + 2aS_2$$

which are also random variables with respective means and variances as follows (see Sect. 4.3.2):

$$\mu_Q = \mu_1 + \mu_2 \qquad \sigma_Q^2 = \sigma_1^2 + \sigma_2^2$$
$$\mu_M = a\mu_1 + 2a\mu_2 \qquad \sigma_M^2 = a^2\sigma_1^2 + 4a^2\sigma_2^2$$

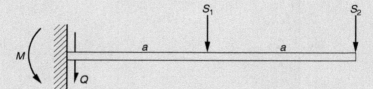

Figure E3.36   A cantilever beam subjected to two loads.

Although the two loads, S_1 and S_2, are statistically independent, Q and M will be correlated; this correlation can be evaluated as follows:

$$E(QM) = E[(S_1 + S_2)(aS_1 + 2aS_2)] = aE(S_1^2) + 3aE(S_1 S_2) + 2aE(S_2^2)$$

But $E(S_1 S_2) = E(S_1)E(S_2)$, and $E(S_1^2) = \sigma_1^2 + \mu_1^2$; $E(S_2^2) = \sigma_2^2 + \mu_2^2$. Thus,

$$E(QM) = a(\sigma_1^2 + \mu_1^2) + 2a(\sigma_2^2 + \mu_2^2) + 3a\mu_1\mu_2$$
$$= a(\sigma_1^2 + 2\sigma_2^2) + \mu_Q\mu_M$$

Therefore,

$$\text{Cov}(Q, M) = E(QM) - \mu_Q\mu_M = a(\sigma_1^2 + 2\sigma_2^2)$$

from which we obtain the corresponding correlation coefficient,

$$\rho_{Q,M} = \frac{\text{Cov}(Q, M)}{\sigma_Q\sigma_M} = \frac{\sigma_1^2 + 2\sigma_2^2}{\sqrt{(\sigma_1^2 + \sigma_2^2)(\sigma_1^2 + 4\sigma_2^2)}}$$

and if $\sigma_1 = \sigma_2$,

$$\rho_{Q,M} = \frac{3}{\sqrt{10}} = 0.948$$

indicating a strong correlation between the shear force Q and bending moment M at the support. This correlation arises because both Q and M are functions of the same loads S_1 and S_2; however, there is no causal relation between Q and M.   ◄

► 3.4 CONCLUDING SUMMARY

The principal concepts introduced in this chapter include the notions of a random variable and its associated probability distributions. A number of the more useful probability distribution functions and their properties are described and illustrated. Foremost among these are the normal (or Gaussian) and lognormal distributions which are widely used in practical applications. Because of their special significance in several areas of engineering, particularly to applications in natural hazards, the fundamental elements of the probability distributions of extreme values, including the respective asymptotic distributions, will be presented subsequently in Chapter 4.

A random variable would be completely described by specifying its probability distribution, either with its PDF or its CDF, including its parameters. However, a random variable may also be described approximately through its moments, such as the mean-value and variance (or standard deviation), which may be called the *main descriptors* of a random variable. For two (or more) random variables, the main descriptors must also include the covariance or correlation coefficient between the variables.

Thus far, and this will continue through Chapter 5, we have been dealing with idealized theoretical models. In particular, we have assumed, tacitly at least, that the probability distribution of a random variable and its parameters, or its main descriptors, are known. In a real problem, of course, these assumptions need to be verified and the parameters must be estimated on the basis of real-world data and information. The principal concepts and methods for these latter purposes are the subjects of Chapters 6 through 8.

► PROBLEMS

3.1 The duration in days of two activities A and B in a construction project are denoted as T_A and T_B, whose PMFs are given graphically below as follows.

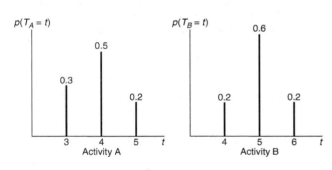

$p(T_A = t)$ $p(T_B = t)$

Activity A   Activity B

Assume that T_A and T_B are statistically independent, and activity B will begin as soon as activity A has been completed. Determine and plot the PMF of the total time T required to complete both activities.

3.2 The profit (in thousand dollars) of a construction project is described by the following PDF:

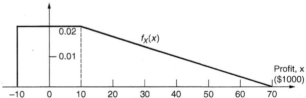

(a) What is the probability that the contractor will lose money on this job? (*Ans. 0.2*).

(b) Suppose the contractor declares that he has made money on this project. What is the probability that his profit was more than forty thousand dollars? (*Ans. 0.1875*)

3.3 The storm runoff X (in cubic meters per second, cms) from a subdivision can be modeled by a random variable with the following probability density function:

$$f_X(x) = c\left(x - \frac{x^2}{6}\right); \qquad \text{for } 0 \leq x \leq 6$$
$$= 0; \qquad\qquad \text{otherwise}$$

(a) Determine the constant c and sketch the PDF. (*Ans. 1/6*).

(b) The runoff is carried by a pipe with a capacity of 4 cms. Overflow will occur when the runoff exceeds the pipe capacity.

If overflow occurs after a storm, what is the probability that the runoff in this storm is less than 5 cms? (*Ans. 0.714*)
(c) An engineer considers replacing the current pipe by a larger pipe having a capacity of 5 cms. Suppose there is a probability of 60% that the replacement would be completed prior to the next storm. What is the probability of overflow in the next storm? (*Ans. 0.1*)

3.4 Severe snow storm is defined as a storm whose snowfall exceeds 10 inches. Let X be the amount of snowfall in a severe snow storm. The cumulative distribution function (CDF) of X in a given town is

$$F_X(x) = 1 - \left(\frac{10}{x}\right)^4 \qquad \text{for } x \geq 10$$
$$= 0; \qquad \text{for } x < 10$$

(a) Determine the median of X.
(b) What is the *expected* amount of snowfall in the town in a severe snow storm?
(c) Suppose a *disastrous snow storm* is defined as a storm with over 15 inches of snowfall. What percentage of the severe snow storms are disastrous?
(d) Suppose the probability that the town will experience 0, 1, and 2 severe snow storms in a year is 0.5, 0.4, and 0.1, respectively. Determine the probability that the town will not experience a disastrous snow storm in a given year. Assume the amounts of snowfall between storms are statistically independent.

3.5 Suppose X is a random variable defined as

$$X = \frac{\text{final cost of project}}{\text{estimated cost of project}}$$

which has a PDF as follows:

$$f_X(x) = \begin{cases} 0 & x < 1 \\ 3/x^2 & 1 \leq x \leq a \\ 0 & x > a \end{cases}$$

(a) Determine the value of a (*Ans. 1.5*)
(b) What is the probability that the final cost of a project will exceed its estimated cost by 25%? (*Ans. 0.4*)
(c) Determine the mean value and standard deviation of X. (*Ans. 1.216, 0.143*)

3.6 The duration of a rainstorm at a given location is described by the following PDF:

$$f_X(x) = \begin{cases} x/8 & 0 < x \leq 2 \\ 2/x^2 & 2 < x < 8 \\ 0 & \text{elsewhere} \end{cases}$$

(a) Determine the mean duration of a rainstorm. (*Ans. 3.106*)
(b) If the current rainstorm started two hours ago, what is the probability that it will be over within another hour? (*Ans. 4/9*)

3.7 The annual maximum snow load X (in lb/ft^2) on buildings with a flat roof in a northern U.S. location can be modeled by a random variable with the following CDF:

$$F_X(x) = 0 \qquad \text{for } x \leq 10$$
$$= 1 - \left(\frac{10}{x}\right)^4 \qquad \text{for } x > 10$$

(a) An engineer recommends a design snow load of 30 lb/ft^2 for a building. What is the probability of roof failure (design load being exceeded) in a given year? Probability that failure will occur for the first time in the fifth year?
(b) If roof failure should occur during 2 or more years over the next 10 yr, the design engineer will face a penalty. What is his chance of going through the next 10 yr without a penalty?

3.8 The CDF of the daily progress in a tunnel excavation project is described graphically below

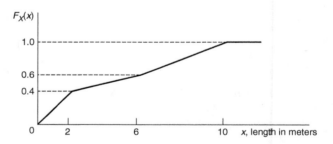

(a) What is the probability that the progress in tunnel excavation will be between 2 and 8 m on a given day?
(b) Determine the median length of tunnel excavated in 1 day.
(c) Plot the probability density function (PDF) of the daily progress in tunnel excavation.
(d) Determine the mean length of tunnel excavated in 1 day.

3.9 The maximum load S (in tons) on a structure is modeled by a continuous random variable S whose CDF is given as follows:

$$F_S(s) = \begin{cases} 0 & \text{for } s \leq 0 \\ -\dfrac{s^3}{864} + \dfrac{s^2}{48} & \text{for } 0 < s \leq 12 \\ 1 & \text{for } s > 12 \end{cases}$$

(a) Determine the *mode* and *mean value* of S.
(b) The strength R of the structure can be modeled by a discrete random variable with the following probability mass function.
Determine the probability of failure, i.e., the probability that loading S is greater than the strength R.

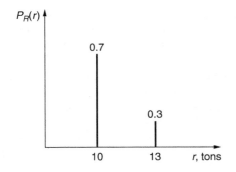

3.10 The annual maximum flood level of a river at a given measuring location has the following CDF:

$$F_X(x) = \begin{cases} 0 & x \le 0 \\ -0.0025x^2 + 0.1x & 0 < x \le 20 \\ 1 & x > 20 \end{cases}$$

(a) Determine and sketch the PDF $f_X(x)$.
(b) What is the median annual maximum flood level?

3.11 The size (in millimeter) of a crack in a structural weld is described by a random variable X with the following PDF:

$$f_X(x) = \begin{cases} x/8 & 0 < x \le 2 \\ 1/4 & 2 < x \le 5 \\ 0 & \text{elsewhere} \end{cases}$$

(a) Sketch the PDF and CDF on a piece of graph paper.
(b) Determine the mean crack size.
(c) What is the probability that a crack will be smaller than 4 mm?
(d) Determine the median crack size.
(e) Suppose there are four cracks in the weld. What is the probability that only one of these four cracks is larger than 4 mm?

3.12 The total load, X (in tons), on the roof of a building has the following PDF:

$$f_X(x) = \begin{cases} 24/x^3 & 3 \le x \le 6 \\ 0 & \text{elsewhere} \end{cases}$$

(a) Determine and plot the CDF of X, i.e., $F_X(x)$.
(b) What is the expected total load?
(c) Determine the coefficient of variation of the total load.
(d) Suppose the roof can carry only 5.5 tons before collapse. What is the probability that the roof will collapse?

3.13 Suppose an offshore platform is designed against the 200-yr wave (i.e., a wave height corresponding to a return period of 200 yr), but it is intended to operate for 30 yr only.
(a) What is the probability that it will be subjected to waves exceeding its design value during the first year of operation? (*Ans. 0.005*)

(b) What is the probability that the platform will not be subjected to waves exceeding its design value during its lifetime of 30 yr? (*Ans. 0.860*)

3.14 A structure with a design life of 50 yr is planned for a site where high-intensity earthquakes may occur with a return period of 100 yr. The structure is designed to have a 0.99 probability of not suffering damage within its design life. Damage effects between earthquakes are statistically independent.
(a) If the occurrence of high-intensity earthquakes at the site is modeled by a Bernoulli sequence, what is the probability of damage to the structure under a single earthquake?
(b) Using the damage probability to a single earthquake from Part (a), what would be the probability of damage to the structure in the next 20 yr, assuming that the occurrences of earthquakes constitute a Poisson process?

3.15 Mechanical failures of a certain type of aircraft constitute a Poisson process. On the average, based on past records, an aircraft will have a failure once every 5000 hr of flight time.
(a) If such aircraft are scheduled for inspection and maintenance after every 2500 flight hours, what is the probability of mechanical failure of an aircraft between inspections?
(b) In a fleet of ten aircraft of the same type, what is the probability that not more than two will have mechanical failures within the above scheduled inspection/maintenance interval? Assume that failures between aircraft are statistically independent.
(c) As the engineer in charge of aircraft safety, you wish to ensure an acceptable probability of failure of no more than 5% for each aircraft. How should the inspection/maintenance schedule in (a) be revised? That is, what should be the revised inspection/maintenance interval (in flight hours)?

3.16 In the fabrication of steel beams, two types of flaws may occur: (1) the inclusion of a small quantity of foreign matter ("slag"); and (2) the existence of microscopic cracks. It has been found by careful laboratory investigation that for a certain size I-beam from a given fabricator, the mean distance between microscopic cracks is 40 ft along the beam, whereas the slag inclusions exist with an average rate of 4 per 100 ft of beam. Each of these types of flaw follows a Poisson process.
(a) For a 20-ft I-beam of this size from this fabricator, what is the chance of finding exactly two microscopic cracks in the beam? (*Ans. 0.076*)
(b) For the same 20-ft beam, what is the chance of finding one or more slag inclusions? (*Ans. 0.551*)
(c) If a 20-ft beam contained more than two flaws, it would be rejected. What is the probability that a 20-ft beam will be rejected? (*Ans. 0.143*)
(d) Four 20-ft I-beams were supplied to a contractor by this fabricator last year. Assume that the flaw conditions among the four beams are statistically independent. What is the probability that only one of the beams had been rejected? (*Ans. 0.360*)

3.17 The air quality in an industrial city may become substandard (poor) at times depending on the weather condition and the amount of factory production. Suppose the event of poor air quality occurs as a Poisson process with a mean rate of once per month. During each time period when the air quality becomes substandard, its pollutant concentration may reach a hazardous level with a 10% probability. The pollutant concentrations between any two periods of poor air quality may be assumed to be statistically independent.

(a) What is the probability of at most two periods of poor air quality during the next 4 and 1/2 months? (*Ans. 0.174*)

(b) What is the probability that the air quality would reach a hazardous level during the next 3 months? (*Ans. 0.259*)

3.18 A country is subject to natural hazards such as floods, earthquakes, and tornadoes. Suppose earthquakes occur according to a Poisson process with a mean rate of 1 in 10 yr; tornado occurrences are also Poisson-distributed with a mean rate of 0.3 per year. There can be either one or no flood each year; hence the occurrence of a flood each year follows a Bernoulli sequence, and the mean return period of floods is 5 yr. Assume floods, earthquakes, and tornadoes can occur independently.

(a) If no hazards occur during a given year, it is referred to as a "good" year. What is the probability of a "good" year?

(b) What is the probability that 2 of the next 5 yr will be good years? (*Ans. 0.287*)

(c) What is the probability of only one incidence of natural hazard in a given year? (*Ans. 0.349*)

3.19 The occurrence of traffic accidents at an intersection may be modeled as a Poisson process, and based on historical records the average rate of accidents is once every 3 yr.

(a) What is the probability that there will be no accident at the intersection for a period of 5 yr?

(b) Suppose that in every accident at the intersection, there is a 5% probability of fatality. Based on the above Poison model, what is the probability of traffic fatality at this intersection over a period of 3 yr?

3.20 A highway traffic condition during a blizzard is hazardous. Suppose one traffic accident is expected to occur in each 50 miles of highway on a blizzard day. Assume that the occurrence of accidents along the highway is modeled by a Poisson process. For a stretch of highway that is 20 miles long, consider the following:

(a) What is the probability that at least one accident will occur on a given blizzard day? (*Ans. 0.33*)

(b) Suppose there are five blizzard days this winter. What is the probability that two out of these five blizzard days are accident free? Assume that accident occurrences between blizzard days are statistically independent. (*Ans. 0.16*)

3.21 The occurrence of accidents at a busy intersection may be described by a Poisson process with an average rate of three accidents per year.

(a) Determine the probability of exactly one accident over a 2-month period. Would this be the same as the probability of exactly two accidents in a 4-month period? Explain.

(b) If fatalities are involved in 20% of the accidents, what is the probability of fatalities occurring at this intersection over a period of 2 months? Assume that events of fatalities between accidents are statistically independent.

3.22 A town is bordered by two rivers as shown in the following figure. Levees A and B were constructed to protect the town from high water in the rivers. The levees were both designed for floods with return periods of 5 and 10 yr, respectively. Assume that the events of flooding from the two rivers are statistically independent.

(a) Determine the probability that the town will encounter flooding in a given year. (*Ans. 0.28*)

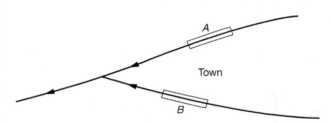

(b) What is the probability that the town will be flooded in at least 2 of the next 5 yr? (*Ans. 0.43*)

(c) Suppose the townspeople desired to reduce the annual probability of flooding to at most 15%. Levee A may be improved to be capable of floods with return periods of 10 or 20 yr with an investment of 5 or 20 million dollars, whereas levee B may be improved for floods with return periods of 20 or 30 yr with corresponding investment of 10 or 20 million dollars. What is the optimal course of action? (*Ans. A to 10 years and B to 20 years*)

3.23 The exterior of a building consists of one hundred 3 m × 5 m glass panels. Past records indicate that on the average one flaw is found in every 50 m^2 of this type of glass panels; also a panel containing two or more flaws will eventually cause breakage problems and have to be replaced. The occurrence of flaws may be assumed to be a Poisson process.

(a) What is the probability that a given panel will be replaced? (*Ans. 0.037*)

(b) Replacement of glass panel is usually expensive. If each replacement costs $5,000, what is the expected cost for replacements of the glass panels in the building? (*Ans. $18,500*)

(c) A higher-grade glass panel, which costs $100 more each, has on the average one flaw in every 80 m^2. Should you recommend using the higher-grade panel, if the objective is to minimize the total expected cost of the glass panels (initial cost and replacement cost)?

3.24 The truck traffic on a certain highway can be described as a Poisson process with a mean arrival rate of 1 truck per minute. The weight of each truck is random, and the probability that a truck is overloaded is 10%.

(a) What is the probability that there will be at least two trucks passing a weigh station on this highway in a 5-min period? (*Ans. 0.96*)

(b) What is the probability that at most one of the next five trucks stopping at the weigh station will be overloaded? (*Ans. 0.92*)

(c) Suppose the weigh station will close for 30 min during lunch hour. What is the probability of overloaded trucks passing the station during the lunch break? (*Ans. 0.95*)

3.25 The occurrence of tornadoes in a county can be modeled as a Poisson process. Twenty tornadoes have touched down in a county within the last twenty years. If there is at least one occurrence of tonadoes in a year, that year is classified as a "tornado year."

(a) What is the probability that next year will be a "tornado year"?

(b) What is the probability that there will be 2 "tornado years" within the next 3 yr?

(c) On the average, over a 10-yr period,
  (i) How many tornadoes are expected to occur?
  (ii) How many "tornado years" are expected to occur?

3.26 Strong earthquakes occur according to a Poisson process in a metropolitan area with a mean rate of once in 50 yr. There are three bridges in the metropolitan area. When a strong earthquake occurs, there is a probability of 0.3 that a given bridge will collapse. Assume the events of collapse between bridges during a strong earthquake are statistically independent; also, the events of bridge collapse between earthquakes are also statistically independent.

(a) What is the probability of at most one strong earthquake occurring in this metropolitan area within the next 20 yr? (*Ans. 0.938*)

(b) During a strong earthquake, what is the probability that exactly one of the three bridges will collapse? (*Ans. 0.441*)

(c) What is the probability of "no bridge collapse from strong earthquakes" during the next 20 yr? (*Ans. 0.769*)

3.27 One of the hazards to an existing underground pipeline is due to improperly conducted excavations. Consider a system consisting of 100 miles of pipeline. Suppose the number of excavations along this pipeline over the next year follows a Poisson process with a mean rate of 1 per 50 miles. Forty percent of the excavations are expected to result in damage to the pipeline. Assume the events of damage between excavations are statistically independent.

(a) What is the probability that there will be at least two excavations along the pipeline next year?

(b) Suppose that two excavations will be performed. What is the probability that the pipeline will be damaged?

(c) What is the probability that the pipeline will not be damaged from excavations next year?

3.28 Flaws in welding may be assumed to occur according to a Poisson process with a mean rate of 0.1 per foot of weld.

(a) Suppose a typical structural connection requires 30 inches of weld and acceptance of such a connection requires the weld to be flawless. What is the probability that a welded connection will be acceptable? (*Ans. 0.779*)

(b) For a welding job consisting of three similar structural connections, what is the probability that at least two connections will be acceptable? (*Ans. 0.875*)

(c) What is the probability that there is altogether only one flaw in the three structural connections? (*Ans. 0.354*)

3.29 The following is the 10-yr record of floods between 1994 and 2003 in Town A.

Year	Number of Floods	Year	Number of Floods
1994	1	1999	0
1995	0	2000	2
1996	1	2001	0
1997	1	2002	0
1998	0	2003	1

The occurrences of floods in the town may be modeled as a Poisson process.

(a) On the basis of the above historical flood data, determine the probability that there will be between one and three floods in Town A over the next 3 yr.

(b) A sewage treatment plant is located on a high ground in the town. The probability that it will be inundated during a flood is 0.02. What is the probability that the treatment plant will not be inundated for a period of 5 yr?

3.30 Highway traffic accidents can be classified into either injury (I) or noninjury (N) accidents. In a given year, the occurrence rates of these two types of accidents along a stretch of highway are 0.01 and 0.05 per mile, respectively. Assume that the occurrence of each type of accidents along the highway follows a Poisson process. Consider a highway that runs between two cities that are 50 miles apart.

(a) Determine the probability that there will be exactly two noninjury accidents in a given year.

(b) Determine the probability that there will be at least three accidents in a given year.

(c) Suppose that exactly two accidents occurred last year. What is the probability that both of them involved injuries?

3.31 The occurrence of thunderstorms in Peoria, Illinois, may be assumed to follow a Poisson process during each of the two seasons, namely,

I. Winter (October to March)
II. Summer (April to September)

A 21-yr record reveals that a total of 173 thunderstorms have taken place during the winter seasons, whereas 840 thunderstorms have occurred during the summer seasons.

(a) Estimate the mean rate of occurrence of thunderstorms per month for the

(i) winter season

(ii) summer season (*Ans. (i) 1.37, (ii) 6.67*)

(b) What is the probability that there will be a total of four thunderstorms during the 2 months of March and April next year? (*Ans. 0.056*)

(c) What is the probability that there will be no December thunderstorms during 2 out of the next 5 yr? (*Ans. 0.267*)

3.32 Geomembrane is often used to provide an effective impervious barrier in a waste containment lining system. The geomembrane has to be sewn together to cover the entire site; defects can thus occur along the seams. Consider a landfill construction project that requires 3000 m of seams and the quality of the seaming operation is such that defects will occur along the seams at a mean rate of one per 200 m. The geomembrane layer is inspected after the installation, and those defects that are detected will be repaired. However, some of the defects will not be detected during the inspection; they will remain and can cause unsatisfactory performance of the lining system. Suppose the current inspection procedure fails to detect 20% of the defects.

(a) What is the mean rate of defects along the seams that remain in the system after an inspection?

(b) Assume that the defects that remained undetected occur according to a Poisson process. What is the probability that there will be more than two defects remaining in the lining system?

(c) Consider a similar but smaller project involving only 1000 m of seams. However, defects in the geomembrane seams are very undesirable for this project. It is required to achieve a 95% probability that the geomembrane lining system will be free of defects after the inspection. Assume that the quality of seaming operation is the same as the uninspected seam (i.e., the same mean rate of defects before inspection), but the inspection effort can be improved to reduce the percent of undetected defects. What is the allowable fraction of undetected defects for this improved inspection procedure?

3.33 The delay time for a given flight is exponentially distributed with a mean of 0.5 hr. Ten passengers on this flight need to take a subsequent connecting flight. The scheduled connection time is either 1 or 2 hr depending on the final destination. Suppose three and seven passengers are associated with these connection times, respectively.

(a) Suppose John is one of the ten passengers needing a connection. What is the probability that he will miss his connection? (*Ans. 0.053*)

(b) Suppose he met Mike on the plane, and Mike also needs to make a connection. However, Mike is going to another destination and thus has a different connection time from John's. What is the probability that both John and Mike will miss their connections? (*Ans. 0.018*)

(c) A friend of John's, named Mary, happens to live close to the airport where John makes his connection. She would like to take this opportunity to meet John at the airport. Suppose she has already waited for 30 min beyond John's scheduled arrival time. What is the probability that John will miss his connection so

that they could have a leisurely dinner together? Assume John's scheduled connection time is 1 hr in part (c). (*Ans. 0.368*)

3.34 Suppose rebars from a supplier are suspected to contain 2% that are below specification. 1000 rebars were delivered by the supplier for the construction of a reinforced concrete structure. To ensure the quality of the rebars, the construction company randomly selected 20 rebars and tested them for specification compliance. Based on the suspected 2% defective rebars, answer the following:

(a) What is the probability that all the 20 rebars tested will pass the test?

(b) What is the probability that at least two of the rebars tested will fail the test?

(c) How many of the rebars delivered need to be tested without a single failure if the construction company wish to have a 90% probability of assurance of the quality of the rebars from the supplier?

3.35 The traffic on the one-way main street shown below may be satisfactorily described by a Poisson process with an average rate of arrival $v = 10$ cars per minute. A driver (indicated by the box) on the side street is waiting to cross the main street. He will cross as soon as he finds a gap of 15 sec.

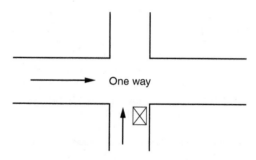

(a) Determine the probability, p, that a gap will be longer than 15 sec.

(b) What is the probability that the driver will cross at the fourth gap?

(c) Determine the mean number of gaps he has to wait until crossing the main street.

(d) What is the probability that he will cross within the first four gaps?

3.36 Suppose cracks exist in a certain material, and the crack length [in micrometers (μm)] of a randomly selected crack is normally distributed with an average length of 71 and a variance of 6.25.

(a) What percentage of cracks are over 74 μm long?

(b) What percentage of cracks exceeding 72 μm are over 77 μm?

3.37 A project is described by the simple activity network shown below, consisting of directed branches representing activities, and nodes representing the beginning/termination of activities.

Activities A and B are independent activities and are scheduled to start simultaneously at day 0. Node "a" represents completion of both activities A and B, and node "b" represents completion of the project. Activity C can start only when both A and B are completed. All durations (in days) required for the respective activities are normal random variables as defined in each of the branches in the figure below.

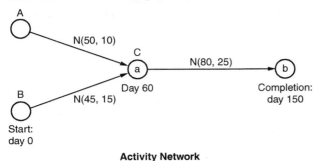

Activity Network

At node "a," activity C is scheduled to start at day 60; however, if there is delay, it cannot start until 15 days beyond its scheduled starting date (e.g., because of reallocation of resources to other projects).

(a) What is the probability that activity C will start on schedule, i.e., 60 days after the starting date of the project?

(b) What is the probability that the project will be completed on target, i.e., 150 days after the starting date?

3.38 The current traffic volume at an airport (number of take-offs and landings) during the peak hour of each day is a normal variate with a mean of 200 planes and a standard deviation of 60 planes.

(a) If the present runway capacity (for landings and take-offs) is 350 planes per hour, what is the current probability of traffic congestion at this airport? Assume that there is one peak hour per day. (*Ans. 0.0062*)

(b) If the mean traffic volume is increasing linearly at the annual rate of 10% of the current volume, with the c.o.v. remaining constant, what would be the probability of congestion at the airport 10 yr hence?

(c) If the projected growth of traffic volume is correct, what airport capacity will be required 10 yr from now in order to maintain the present service condition, i.e., to maintain the current probability of congestion?

3.39 An airport has two runways, one is North–South (N–S), and the other is East–West (E–W). Because of the prevailing winds, the N–S runway is used 80% of the days, and the E–W one the remainder of the time. The selection of the runway for a given day is based on the wind at the beginning of the day, and once selected will not be changed for the entire day. The airport has a single peak period each day; this is the hour between 4:00 and 5:00 PM. During this peak hour, the volume of air traffic varies day to day and may be described with a normal variate N(100, 10). The N–S runway is considered overcrowded if more than

120 planes use it during this period, whereas the E–W runway is considered overcrowded if more than 115 planes use it during the peak period.

(a) What is the probability that the N–S runway is going to be overcrowded on a given day, if at the beginning of the day this runway is selected to be used?

(b) What would be the probability of overcrowding, if at the beginning of a given day the E–W runway were selected to be used?

(c) What is the probability of overcrowding at the airport, if there is no advance information as to which runway will be used on a given day?

3.40 A foundation engineer estimates that the settlement of a proposed structure will not exceed 2 in. with 95% probability. From a record of performance of many similar structures built on similar soil conditions, he finds that the coefficient of variation of the settlement is about 20%. If a normal distribution is assumed for the settlement of the proposed structure, what is the probability that the proposed structure will settle more than 2.5 in.? (*Ans. 0.00047*)

3.41 A student has submitted a concrete cylinder to the "strength contest" in Engineering Open House. Suppose the strength of her concrete cylinder is normally distributed as N(80, 20) in kips. She was scheduled to be the last contestant for load testing. Immediately prior to her cylinder being tested, the two highest strengths in the contest thus far are 100 and 70 kips.

(a) What is the probability that she will be the second-place winner?

(b) Suppose her cylinder is being tested, and it has not shown any sign of distress at a load of 90 kips. What is the probability that she will win first place?

(c) Suppose she submitted a similar cylinder to another contest. Her boyfriend used an alternative procedure to make his concrete cylinder, such that the strength is expected to be only 1% higher than hers, but the c.o.v. is 50% higher. Who is more likely to score a higher strength in that contest? Justify.

3.42 An offshore platform is built to withstand ocean wave forces.

(a) The annual maximum wave height of the ocean waves (above mean sea level) is a random variable with a Gaussian distribution having a mean height of 4.0 m and a c.o.v. of 0.80. What is the probability that the wave height will exceed 6 m in a given year?

(b) If the platform were to be designed for a wave height (above mean sea level) such that it will not be exceeded by ocean waves over a period of 3 yr with a probability of 80%, what should be the platform height above the mean sea level? Wave height exceedances between years are statistically independent.

(c) Suppose that ocean waves exceeding 6 meters will occur according to a Poisson process and that each of these waves could potentially cause damage to the platform with a probability of 0.40. What is the probability that there will be no damage to the platform in 3 yr? Damages to the platform between waves are statistically independent.

3.43 The daily SO_2 concentration in the air for a given city is normally distributed with a mean of 0.03 ppm and a c.o.v. of 40%. Assume statistical independence between the SO_2 concentration for any 2 days. Suppose the criteria for clean air standard require that:
1. The weekly average SO_2 concentration should not exceed 0.04 ppm.
2. The SO_2 concentration should not exceed 0.075 ppm on more than 1 day during a given week.
Determine which one of these two criteria is more likely to be violated in this city. Substantiate your answer with calculated probabilities.

3.44 The daily flow rate of contaminant from an industrial plant is modeled by a normal random variable with a mean value of 10 units and a c.o.v. of 20%. When the contaminant flow rate exceeds 14 units on a given day, it is considered excessive. Assume that the contaminant flow rate between any 2 days is statistically independent.
(a) What is the probability of having excessive contaminant flow rate on a given day? (*Ans. 0.02275*)
(b) Regulation requires the measurement of contaminant flow rate for 3 days. The plant will be charged with a violation if excessive contaminant flow rate is observed during the 3-day period. What is the probability that the plant will not be charged with violation? (*Ans. 0.933*)
(c) Suppose there is a proposal to change the regulation such that contaminant flow rate will be measured for 5 days and the plant will be charged with a violation if excessive contaminant flow rate is observed in more than one of the 5 days. Will the plant be better off with the proposed change? Justify your answer. (*Ans. yes*)
(d) Return to Part (b). Although the plant cannot reduce the standard deviation of the daily contaminant flow rate, it can reduce the mean daily contaminant flow rate by improving the chemical process. Suppose the plant decides to limit the probability of violation to 1%. What should be the daily mean contaminant overflow rate? (*Ans. 8.58*)

3.45 The time between severe earthquakes at a given region follows a lognormal distribution with a coefficient of variation of 40%. The expected time between severe earthquakes is 80 yr.
(a) Determine the parameters of this lognormally distributed recurrence time T. (*Ans. 4.308, 0.385*)
(b) Determine the probability that a severe earthquake will occur within 20 yr from the previous one.
(c) Suppose the last severe earthquake in the region took place 100 yr ago. What is the probability that a severe earthquake will occur over the next year?

3.46 The daily average concentration of pollutants in a stream follows a lognormal distribution with a mean of 60 mg/l and a c.o.v. of 20%.
(a) What is the probability that the average concentration of pollutant in the stream will exceed 100 mg/l (a critical level) on a given day?

(b) Suppose the pollutant concentration between days are statistically independent. What is the probability that the critical level of pollutant concentration will not be reached during a given week?

3.47 Because of spatial irregularities, the depth, H, from the ground surface to the rock stratum may be modeled as a lognormal random variable with a median depth of 20 m and a c.o.v. of 30%. In order to provide satisfactory support, a steel pile must be embedded 0.5 m into the rock as indicated in the figure below.

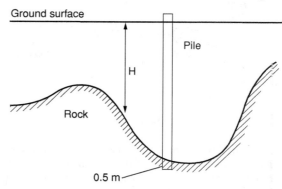

(a) What is the probability that a 25-m-long pile will not anchor satisfactorily in the rock stratum?
(b) If a 25-m pile has been driven 24 m and has not encountered rock, what is the probability that an additional 2 m of pile welded to the original length will be sufficient to anchor this pile satisfactorily in the rock stratum?

3.48 The capacity of a pile supporting a transmission tower is modeled with a lognormal random variable with a mean of 100 tons and a c.o.v. of 20%.
(a) What is the probability that the pile will survive a load of 100 tons?
(b) During the installation of the pile, a pile load test indicated that the pile can support at least a load of 75 tons. What is now the probability that the pile will survive a 100-ton load?
(c) If the pile has survived a recent hurricane during which the load transmitted to the pile is estimated to be 90 tons, what is the probability that the pile can carry a 100-ton load?

3.49 The maximum wind velocity in a tornado at a given city follows a lognormal distribution with a mean of 90 mph and a c.o.v. of 20%.
(a) What is the probability that the maximum wind velocity will exceed 120 mph during the next tornado?
(b) Determine the design tornado wind velocity whose return period is 100 yr. Assume one tornado will strike this city each year.

3.50 Statistical data of breakdowns of computer XXX show that the duration for trouble-free operation of the machine can be described as a gamma distribution with a mean of 40 days and a standard deviation of 10 days. The computer is occasionally taken out for maintenance in order to insure operational condition at any time with a 95% probability.

(a) How often should the computer be scheduled for maintenance? (*Hint: Should it be shorter or longer than the mean of 40 days?*)

(b) If the computer is in good operational condition at the time it is scheduled for regular maintenance, but no maintenance was performed, what is the probability that it will break down within another week beyond its regular maintenance schedule?

(c) Three XXX computers were acquired at the same time by an engineering consulting firm. The computers are operating under the same environment, workload, and regular maintenance schedule. The breakdown times between the computers, however, may be assumed to be statistically independent. What is the probability that at least one of the three machines will break down within the first scheduled maintenance time?

3.51 The capacity of a building to withstand earthquake forces without damage has a gamma distribution with a mean of 2500 tons and a c.o.v. of 35%.

(a) If the building has survived a previous earthquake with a force of 1500 tons without damage, what is the probability that it can withstand a future earthquake with a force of 3000 tons?

(b) The occurrences of earthquakes with a force of 2000 tons constitute a Poisson process with an expected occurrence rate of once every 20 yr. What is the probability that there will be no damage to the building over a life of 50 yr?

(c) In a complex of five similar buildings, each with the same earthquake resistance capacity as described above, what is the probability that at least four of them will not be damaged under an earthquake force of 2000 tons? Assume that the occurrences of damage to the different buildings are statistically independent.

3.52 In a harbor, a merchant ship may need to wait for docking at a quay for loading and unloading operations. The loading/unloading operation of each ship can be either 2 or 3 days with relative likelihoods of 1 to 3, and the operation times between ships are statistically independent. There is only one loading crane and three mooring berths in the quay. Assume the following probabilities for the queue size:

Queue Size (no. of ships)	Probability
0	0.1
1	0.3
2	0.4
3	0.2

(a) What is the probability that the total waiting time of a merchant ship upon arriving at the harbor has to wait longer than 5 days before loading and unloading can commence at the quay, if there are two ships in the queue and the loading/unloading operations are just starting for the first ship in the queue?

(b) What is the probability that the total waiting time of the ship in (a) will be longer than 5 days without knowing the queue size when it arrives at the harbor?

(c) Finally, suppose that the loading/unloading operation time for a ship is a beta-distributed random variable with minimum and maximum operation times of 1.5 days and 4 days, respec-

tively, and shape parameters of 3.00 and 4.50. The operation times between the ships are statistically independent. In this case, what would be the answer to Part (b)? For this part, consider the computer-based methods of Chapter 5.

3.53 A bridge connects two cities A and B. The bridge has a capacity to handle 1000 vehicles per hour. Normally, on a given day, the peak volume of traffic using the bridge can be modeled by a beta distribution with lower and upper bound traffic of 600 and 1100 vehicles per hour and a mean hourly traffic of 750 vehicles and a c.o.v. of 0.20. A jamming condition will develop on the bridge if its capacity is exceeded by the traffic volume, or if an accident should occur on the bridge. The probability of an accident on the bridge is estimated to be 0.02 during any period of peak traffic. It may be assumed that peak traffic volume and the occurrence of an accident are statistically independent.

(a) What is the probability that the bridge will be jammed on a given day?

(b) If the bridge is jammed, what is the probability that it was caused by an accident on the bridge?

3.54 Statistics show that 20% of freshmen students at an engineering school fail after 1 yr. In a class of 30 students, what is the probability that among eight students selected at random, two of them will fail after 1 yr?

3.55 A highway contractor orders the delivery of ten road graders from an equipment rental company. Record shows that the trouble-free operational time of a road grader has a mean time of 35 days with a c.o.v. of 0.25. If the distribution of the trouble-free operational time can be represented by a gamma distribution,

(a) What is the probability that a road grader will be operational without problem for 40 days?

(b) If the rental company has an inventory of 50 road graders, and assuming that 10% of them will have trouble-free operational life of less than 40 days, what is the probability that two among the ten graders will have operational lives less than 40 days?

3.56 An office building is planned and designed with a lateral load-resisting structural system for earthquake resistance in a seismic zone. The seismic capacity (in terms of force factor) of the proposed system is assumed to have a lognormal distribution with a median of 6.5 and a standard deviation of 1.5. The ground motion expected to be generated by the maximum possible earthquake at the building site will have an equivalent force factor of 5.5.

(a) What is the estimated probability of damage to the office building when subjected to the maximum possible earthquake?

(b) If the building should survive (without any damage) a previous moderate earthquake with a force factor of 4.0, what would be its future failure probability under the maximum possible earthquake?

(c) The future occurrences of the maximum possible earthquakes may be modeled by a Poisson process with a return period of 500 yr. If the damage effects between earthquakes are statistically independent, what is the probability of the proposed building surviving a life of 100 yr without damage?

(d) Suppose that the office building will be part of a complex of five identical structures designed with the same earthquake resistance. What is the probability that at least four of the five buildings will survive a life of 100 yr without damage? Survivals among the buildings may be assumed to be statistically independent.

3.57 A water distribution network is shown in the figure below. For a given rate of flow through the network, the performance at a given node is measured by the head pressure at the node. Satisfactory performance at a node requires that its pressure be within a normal range, between 6 and 14 units. Suppose the pressure at node A is a lognormal random variable with a mean value of 10 units and a c.o.v. of 20%.

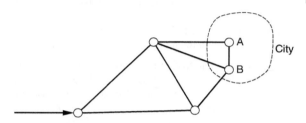

(a) What is the probability of satisfactory performance at node A? (*Ans. 0.955*)

(b) Suppose a city is served by nodes A and B. Ordinarily, the pressure at B has a 90% probability of falling within the normal range. However, the pressure at B is dependent on the pressure at A. In the event that the pressure at A falls outside the normal range, the pressure at B will fall outside of the normal range with twice its ordinary probability. Satisfactory water service to the city requires that at least one of the nodes A and B has normal water pressure. What is the probability of unsatisfactory water service to the city? (*Ans. 0.0089*)

(c) Suppose resources are available to improve the water-distribution systems by either one of the following options:

(I) Reduce the c.o.v. of pressure at A to 15%.

(II) Increase the original probability of satisfactory pressure at B (i.e., within the normal range) to 95%.

Which option is better in terms of minimizing the probability of unsatisfactory water service to the city? (*Ans. I*) Explain.

3.58 The daily water levels (normalized to the respective full condition) of two reservoirs A and B are denoted by two random variables X and Y having the following joint PDF:

$$f(x, y) = (6/5)(x + y^2), \qquad 0 < x < 1; 0 < y < 1$$

(a) Determine the marginal density function of the daily water level for reservoir A.

(b) If reservoir A is half full on a given day, what is the probability that the water level will be more than half full?

(c) Is there any statistical correlation between the water levels in the two reservoirs?

3.59 The joint PMF of precipitation, X (in.) and runoff, Y (cfs) (discretized here for simplicity) due to storms at a given location is as follows:

	$X = 1$	$X = 2$	$X = 3$
$Y = 10$	0.05	0.15	0.0
$Y = 20$	0.10	0.25	0.25
$Y = 30$	0.0	0.10	0.10

(a) What is the probability that the next storm will bring a precipitation of 2 or more in. and a runoff of more than 20 cfs? (*Ans. 0.2*)

(b) After a storm, the rain gauge indicates a precipitation of 2 in. What is the probability that the runoff in this storm is 20 cfs or more? (*Ans. 0.7*)

(c) Are X and Y statistically independent? Substantiate your answer. (*Ans. no*)

(d) Determine and plot the marginal PMF of runoff.

(e) Determine and plot the PMF of runoff for a storm whose precipitation is 2 in.

(f) Determine the correlation coefficient between precipitation and runoff. (*Ans. 0.35*)

▶ REFERENCES

Ang, A. H.-S., "Bases for Reliability Approach to Structural Fatigue," Proc. 2nd. Inf. Conf. on Structural Safety and Reliability, ICOSSA'R '77, Munich, Germany, Werner-Verlag, Dusseldorf, Sept. 1977.

Ang, A.H.-S., and Tang, W.H, "Probability Concepts in Engineering Planning and Design," Vol. 2, *Decision, Risk and Reliability*, John Wiley & Sons, New York, 1984.

Elderton, W.P, *Frequency Curves and Correlation*, 4th ed., Cambridge University Press, Cambridge, England, 1953.

Hardy, G.H, Littlewood, J.E, and Polya, G., *Inequalities*, Cambridge University Press, Cambridge, England, 1959.

Harter, H.L, *New Tables of the Incomplete Gamma Function Ratio and of Percentage Points of the Chi-square and Beta Distributions*,

Aerospace Research Laboratories, U.S. Air Force, U.S. Government Printing Office, Washington, D.C., 1963.

Kaplan, W., *Advanced Calculus*, Addison-Wesley Publishing Co., Cambridge, MA, 1953.

Pearson, E.S, and Johnson, N.L, *Tables of the Incomplete Beta Function*, 2nd ed., Cambridge University Press, Cambridge, England, 1968.

Ruhl, J.A., and Walker, W.H., "Stress Histories for Highway Bridges Subjected to Traffic Loading," *Civil Engineering Studies, Structural Research Series No. 416*, University of Illinois at Urbana-Champaign, March 1975.

Zhao, Y.G, and Ang, A.H.-S., "Three-Parameter Gamma Distribution and Its Significance in Structural Reliability," *Computational Structural Engineering, An International Journal*, Vol. 2, Seoul, Korea, 2002.

CHAPTER 4

Functions of Random Variables

▶ 4.1 INTRODUCTION

In this chapter, we introduce the function of one or more random variables. Engineering problems often involve the determination of functional relations between a dependent variable and one or more basic or independent variables. If any one of the independent variables are random, the dependent variable will likewise be random; its probability distribution, as well as its moments, will be functionally related to and may be derived from those of the basic random variables. As a simple example, the deflection, D, of a cantilever beam of length L subjected to a concentrated load, P, applied at the end of the cantilever (as discussed earlier in Example 1.2) is functionally related to the load P and the modulus of elasticity E of the beam material as follows:

$$D = \frac{PL^3}{3EI}$$

in which I is the moment of inertia of the beam cross section. Clearly, we can expect that if P and E (such as construction timber) are both random variables, with respective PDFs, $f_P(p)$ and $f_E(e)$, the deflection D will also be a random variable with PDF, $f_D(d)$, that can be derived from the PDFs of P and E. Moreover, the moments (such as the mean and variance) of D can also be derived as a function of the respective moments of P and E. In this chapter, we shall develop and illustrate the relevant concepts and procedures for these purposes.

▶ 4.2 DERIVED PROBABILITY DISTRIBUTIONS

4.2.1 Function of a Single Random Variable

Let us first consider the function of a single random variable,

$$Y = g(X) \tag{4.1}$$

In this case, when $Y = y$, $X = g^{-1}(y)$, in which g^{-1} is the inverse function of g. If the inverse function $g^{-1}(y)$ has a single root, and therefore is single valued, then

$$P(Y = y) = P[X = g^{-1}(y)]$$

That is, the PMF of Y is

$$p_Y(y) = p_X[g^{-1}(y)] \tag{4.2}$$

151

We can demonstrate Eq. 4.2 graphically for the variable Y as a function of the discrete random variable X with the following function:

$$Y = X^2 \qquad \text{for } x \geq 0$$

and the PMF of X is as shown in the figure below on the left.

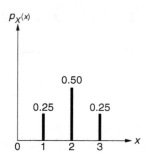

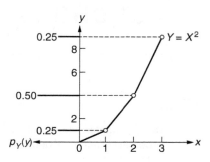

In this case, we can see that according to Eq. 4.2, when $y = 1$, $x = 1$ and $p_Y(1) = p_X(1) = 0.25$, whereas when $Y = 4$, $p_Y(4) = p_X(2) = 0.50$, and when $y = 9$, $p_Y(9) = p_X(3) = 0.25$. At all other values of Y the PMF is zero; e.g., $p_Y(5) = p_X(\sqrt{5}) = p_X(2.24) = 0$. Graphically, the PMF of Y is shown in the figure above on the right.

Also, it follows that

$$P(Y \leq y) = P[X \leq g^{-1}(y)] \qquad \text{if } g(x) \text{ is an increasing function of } x$$

whereas if $g(x)$ is a decreasing function of x, $P(Y \leq y) = P[X \geq g^{-1}(y)]$.
Thus, when y increases with x, the CDF of Y is

$$F_Y(y) = F_X[g^{-1}(y)] \tag{4.3}$$

Therefore for discrete X, we have

$$F_Y(y) = \sum_{\text{all } x_i \leq g^{-1}(y)} p_X(x_i) \tag{4.4}$$

whereas for a continuous X with PDF $f_X(x)$, Eq. 4.3 would be

$$F_Y(y) = \int_{\{x \leq g^{-1}(y)\}} f_X(x)\, dx = \int_{-\infty}^{g^{-1}(y)} f_X(x)\, dx \tag{4.5}$$

In this latter case, for a continuous X, we recall from calculus that by making a change in the variable of integration, Eq. 4.5 becomes

$$F_Y(y) = \int_{-\infty}^{g^{-1}(y)} f_X(x)\, dx = \int_{-\infty}^{y} f_X(g^{-1})\frac{dg^{-1}}{dy}\, dy$$

in which $g^{-1} = g^{-1}(y)$. From which we obtain the PDF of Y as

$$f_Y(y) = \frac{dF_Y(y)}{dy} = f_X(g^{-1})\frac{dg^{-1}}{dy}$$

In the above, when y increases with x, dg^{-1}/dy is positive; however, when y decreases with x, $F_Y(y) = 1 - F_X(g^{-1})$, and then,

$$f_Y(y) = -f_X(g^{-1})\frac{dg^{-1}}{dy}$$

but $\dfrac{dg^{-1}}{dy}$ is negative. Therefore, properly the derived PDF of Y is

$$f_Y(y) = f_X(g^{-1})\left|\frac{dg^{-1}}{dy}\right| \tag{4.6}$$

▶ **EXAMPLE 4.1**

Let us first illustrate the transformation of a discrete distribution as indicated in Eq. 4.2. Consider a cantilever beam with a length L that is subjected to a concentrated load F applied at the free end of the beam as shown in the figure below.

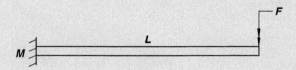

Suppose that the load F consists of a number of boxes each of the same weight of one unit; the number of boxes loaded on the beam, x, varies from 0 to n and is distributed as a binomial distribution in which the probability of a box being loaded is p; this means that the load F has a binomial PMF as follows:

$$p_F(x) = \binom{n}{x} p^x (1-p)^{n-x}; \qquad x = 0, 1, \ldots, n$$

in which x is the number of boxes loaded on the beam. Under a load of x boxes, the bending moment at the fixed support of the cantilever beam is

$$m = x \cdot L$$

and the inverse function is $g^{-1} = x = m/L$. Then, according to Eq. 4.2, the distribution of the bending moment is

$$p_M(m) = p_F[g^{-1}(m)] = P\left(F = \frac{m}{L}\right)$$

$$= \binom{n}{m/L} p^{m/L} (1-p)^{n-m/L}$$

Since $m = xL$, we can easily see that the distribution of the bending moment, M, at the fixed end of the beam has the same binomial PMF as the applied load F. ◀

▶ **EXAMPLE 4.2**

Consider a normal variate X with parameters μ and σ; i.e., $N(\mu, \sigma)$ with PDF

$$f_X(x) = \frac{1}{\sqrt{2\pi}\sigma} \exp\left[-\frac{1}{2}\left(\frac{x-\mu}{\sigma}\right)^2\right]$$

Let $Y = \dfrac{X - \mu}{\sigma}$. Using Eq. 4.6, we determine the PDF of Y as follows.

First, we observe that the inverse function is $g^{-1}(y) = \sigma y + \mu$, and $\dfrac{dg^{-1}}{dy} = \sigma$. Then, according to Eq. 4.6, the PDF of Y is

$$f_Y(y) = \frac{1}{\sqrt{2\pi}\sigma}\exp\left[\frac{-\frac{1}{2}(\sigma y + \mu - \mu)^2}{\sigma^2}\right]|\sigma| = \frac{1}{\sqrt{2\pi}}e^{-\frac{1}{2}y^2}$$

which is the PDF of the standard normal distribution, $N(0, 1)$. ◀

▶ **EXAMPLE 4.3**

Suppose a random variable X has a lognormal distribution with parameters λ and ζ. According to Eq. 4.6, we derive the PDF of the function $Y = \ln X$ as follows.

In this case, the PDF of X is

$$f_X(x) = \frac{1}{\sqrt{2\pi}\zeta}\frac{1}{x}\exp\left[-\frac{1}{2}\left(\frac{\ln x - \lambda}{\zeta}\right)^2\right]$$

and the inverse function is

$$g^{-1}(y) = e^y$$

and

$$\frac{dg^{-1}}{dy} = e^y$$

Therefore, according to Eq. 4.6,

$$f_Y(y) = \frac{1}{\sqrt{2\pi}\zeta}\frac{1}{e^y}\exp\left[-\frac{1}{2}\left(\frac{y - \lambda}{\zeta}\right)^2\right]|e^y| = \frac{1}{\sqrt{2\pi}\zeta}\exp\left[-\frac{1}{2}\left(\frac{y - \lambda}{\zeta}\right)^2\right]$$

which means that the distribution of $Y = \ln X$ is normal with a mean of λ and a standard deviation of ζ; i.e., $N(\lambda, \zeta)$. This result also shows that

$$E(\ln X) = \lambda$$

and

$$Var(\ln X) = \zeta^2.$$ ◀

The inverse function $g^{-1}(y)$ may not be single-valued; i.e., for a given value y there may be multiple values of $g^{-1}(y)$. For instance, if $g^{-1}(y) = x_1, x_2, \ldots, x_k$, then we would have

$$(Y = y) = \bigcup_{i=1}^{k}(X = x_i)$$

And if X is discrete, the PMF of Y is

$$p_Y(y) = \sum_{i=1}^{k}p_X(x_i) \tag{4.7}$$

whereas if X is continuous, the PDF of Y would be

$$f_Y(y) = \sum_{i=1}^{k}f_X\left(g_i^{-1}\right)\left|\frac{dg^{-1}}{dy}\right| \tag{4.8}$$

in which $g_i^{-1} = x_i$ is the ith root of $g^{-1}(y)$.

▶ **EXAMPLE 4.4**   The strain energy in a linearly elastic bar subjected to an axial force S is given by the equation

$$U = \frac{L}{2AE} S^2$$

where:

$L =$ length of the bar
$A =$ cross-sectional area of the bar
$E =$ modulus of elasticity of the material

Using $c = L/2AE$, we can rewrite

$$U = cS^2$$

In this case, the inverse function has two roots which are

$$s = \pm\sqrt{\frac{u}{c}}$$

with derivatives

$$\frac{ds}{du} = \pm\frac{1}{2\sqrt{cu}} \quad \text{and} \quad \left|\frac{ds}{du}\right| = \frac{1}{2\sqrt{cu}}$$

Now, if S is a lognormal variate with parameters λ and ζ, the PDF of U, according to Eq. 4.8, is

$$f_U(u) = \left[f_S\left(\sqrt{\frac{u}{c}}\right) + f_S\left(-\sqrt{\frac{u}{c}}\right) \right] \left|\frac{1}{2\sqrt{cu}}\right|$$

But $f_S(s) = 0$ for $s < 0$; therefore,

$$f_U(u) = f_S\left(\sqrt{u/c}\right)\frac{1}{2\sqrt{cu}} = \frac{1}{\sqrt{2\pi}\,\zeta}\frac{1}{\sqrt{u/c}}\exp\left[-\frac{1}{2}\left(\frac{\ln\sqrt{u/c} - \lambda}{\zeta}\right)^2\right]\frac{1}{2\sqrt{cu}}$$

From which we obtain

$$f_U(u) = \frac{1}{\sqrt{2\pi}(2\zeta)}\frac{1}{u}\exp\left[-\frac{1}{2}\left(\frac{\ln u - \ln c - 2\lambda}{2\zeta}\right)^2\right]$$

which means that the PDF of U is also lognormal with parameters

$$\lambda_U = \ln c + 2\lambda \quad \text{and} \quad \zeta_U = 2\zeta \qquad \blacktriangleleft$$

▶ **EXAMPLE 4.5**   In Example 4.4, if the applied force S is a standard normal variate with the PDF $N(0, 1)$, the PDF of the strain energy U, according to Eq. 4.8, would become as follows:

$$f_U(u) = \left[f_S\left(\sqrt{\frac{u}{c}}\right) + f_S\left(-\sqrt{\frac{u}{c}}\right) \right]\frac{1}{2\sqrt{cu}}$$

In this case, because of the symmetry of the PDF of S about 0, $f_S(-s) = f_S(s)$. Hence,

$$f_U(u) = 2 f_S\left(\sqrt{\frac{u}{c}}\right)\left(\frac{1}{2\sqrt{cu}}\right) = \frac{1}{\sqrt{2\pi\,cu}}\exp\left(-\frac{u}{2c}\right) \qquad u \geq 0$$

which is a *chi-square* distribution with one *degree-of-freedom* (see Chapter 6). Graphically, this PDF would be as shown in Fig. E4.5.

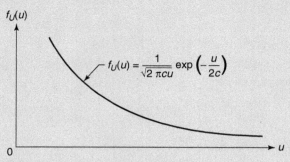

Figure E4.5 PDF of U under S with $N(0, 1)$. ◄

► **EXAMPLE 4.6**

The overall height of earth dams must have sufficient freeboard above the maximum reservoir level in order to prevent waves from washing over the top of the dam. The determination of the overall height must include the wind tide and wave height.

In particular, the wind tide, in feet above the still-water level is given by

$$Z = \frac{F}{1400d}V^2$$

where

$V =$ wind speed in miles per hour

$F =$ fetch, or length of water surface over which the wind blows, in feet

$d =$ average depth of reservoir along the fetch, in feet.

Suppose the wind speed, V, is exponentially distributed with a mean wind speed of v_c; i.e., its PDF is

$$f_V(v) = \frac{1}{v_o}e^{-v/v_o} \qquad v \geq 0$$
$$= 0 \qquad v < 0$$

Then, according to Eq. 4.8, the distribution of the wind tide, Z, is determined as follows:

Denoting $a = F/1400d$, we have $Z = aV^2$
and the inverse functions, $v = \pm\sqrt{\dfrac{z}{a}}$
and

$$\left|\frac{dv}{dz}\right| = \frac{1}{2\sqrt{az}}$$

Then, Eq. 4.8 yields

$$f_Z(z) = \left[f_V\left(\sqrt{\frac{z}{a}}\right) + f_V\left(-\sqrt{\frac{z}{a}}\right) \right]\left(\frac{1}{2\sqrt{az}}\right)$$

However, since $f_V(v) = 0$ for $v < 0$, we have

$$f_Z(z) = f_V\left(\sqrt{\frac{z}{a}}\right) \cdot \frac{1}{2\sqrt{az}} = \frac{1}{2v_o\sqrt{az}}\exp\left(-\frac{\sqrt{z/a}}{v_o}\right) \qquad z \geq 0$$

◄

4.2.2 Function of Multiple Random Variables

A dependent variable may be a function of two or more random variables, in which case it is also a random variable and its probability distribution will depend on those of the independent random variables through the specified functional relationship.

Consider first the case of a function of two random variables X and Y,

$$Z = g(X, Y) \tag{4.9}$$

If X and Y are both discrete random variables, we would have

$$(Z = z) = [g(X, Y) = z] = \bigcup_{g(x_i, y_j)=z} (X = x_i, Y = y_j)$$

from which the PMF of Z is

$$p_Z(z) = \sum_{g(x_i, y_j)=z} p_{X,Y}(x_i, y_j) \tag{4.10}$$

The corresponding CDF of Z would be

$$F_Z(z) = \sum_{g(x_i, y_j)\leq z} p_{X,Y}(x_i, y_j) \tag{4.10a}$$

Sum of Discrete Variates—Consider first the sum of two discrete random variables

$$Z = X + Y$$

in which case, Eq. 4.10 becomes

$$p_Z(z) = \sum_{x_i + y_j = z} p_{X,Y}(x_i, y_j) = \sum_{\text{all } x_i} p_{X,Y}(x_i, z - x_i) \tag{4.11}$$

Sum of Independent Poisson Processes—As an example of the sum of two discrete random variables, consider the sum of two statistically independent variates X and Y that are Poisson-distributed with respective parameters, v and μ, as follows:

$$p_X(x) = \frac{(vt)^x}{x!} e^{-vt}$$

and

$$p_Y(y) = \frac{(\mu t)^y}{y!} e^{-\mu t}$$

According to Eq. 4.11, and since X and Y are statistically independent, we have

$$p_Z(z) = \sum_{\text{all } x} p_X(x) p_Y(z - x) = \sum_{\text{all } x} \frac{(vt)^x (\mu t)^{z-x}}{x!(z-x)!} e^{-(v+\mu)t}$$

$$= e^{-(v+\mu)t} \cdot t^z \sum_{\text{all } x} \frac{v^x \mu^{z-x}}{x!(z-x)!}$$

But the above sum for all x is the binomial expansion of $(v + \mu)^z/z!$. Hence, we obtain the PMF of Z,

$$p_Z(z) = \frac{[(v + \mu)t]^z}{z!} e^{-(v+\mu)t}$$

which means that Z also has a Poisson distribution with parameter $(v + \mu)$. Generalizing this result, we can infer that the sum of n independent Poisson processes is also a Poisson process; specifically, if

$$Z = \sum_{i=1}^{n} X_i$$

where each X_i has a Poisson PMF with parameter v_i, the PMF of Z is also Poisson with parameter

$$v_Z = \sum_{i=1}^{n} v_i$$

It is important to point out, however, that the difference of two independent Poisson processes is not a Poisson process. That is, if X and Y are individually Poisson distributed, the PMF of $Z = X - Y$ does not yield a Poisson distribution.

▶ **EXAMPLE 4.7**

A toll bridge serves three suburban residential districts A, B, and C as shown in Fig. E4.7. During peak hours of each day, the estimated average volumes of vehicular traffic from the three districts are, respectively, 2, 3, and 4 vehicles per minute. If the peak hour traffic from each district can be assumed to be a Poisson process, with respective parameters of $v_A = 2$, $v_B = 3$, $v_C = 4$, the total peak hour traffic at the toll bridge would also be a Poisson process with parameter

$$v = 2 + 3 + 4 = 9 \text{ vehicles/min}$$

The probability of more than nine vehicles crossing the bridge in 1 minute would be

$$P(X_1 > 9) = 1 - \sum_{n=0}^{9} \frac{(9 \times 1)^n}{n!} e^{-9} = 1 - 0.5874 = 0.413$$

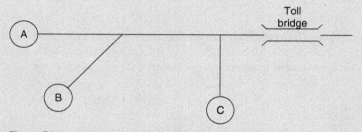

Figure E4.7 Toll bridge serving traffic from three districts. ◀

Now, if the basic random variables X and Y are continuous, the CDF of Z would be

$$F_Z(z) = \iint_{g(x,y) \le z} f_{X,Y}(x, y) \, dx \, dy = \int_{-\infty}^{\infty} \int_{-\infty}^{g^{-1}} f_{X,Y}(x, y) \, dx \, dy$$

where $g^{-1} = g^{-1}(z, y)$. Changing the variable of integration from x to z, the above integral becomes

$$F_Z(z) = \int_{-\infty}^{\infty} \int_{-\infty}^{z} f_{X,Y}(g^{-1}, y) \left| \frac{\partial g^{-1}}{\partial z} \right| dz \, dy \qquad (4.12)$$

Taking the derivative of Eq. 4.12 with respect to z, we obtain the PDF of Z,

$$f_Z(z) = \int_{-\infty}^{\infty} f_{X,Y}(g^{-1}, y) \left| \frac{\partial g^{-1}}{\partial z} \right| dy \qquad (4.13)$$

Alternatively, taking the inverse for y, i.e., $g^{-1} = g^{-1}(x, z)$, we also have for the PDF of Z,

$$f_Z(z) = \int_{-\infty}^{\infty} f_{X,Y}(x, g^{-1}) \left| \frac{\partial g^{-1}}{\partial z} \right| dx \tag{4.13a}$$

Sum of Continuous Variates—Consider the sum of two continuous random variables X and Y,

$$Z = aX + bY$$

in which a and b are constants. Then we have

$$x = \frac{z - by}{a} \quad \text{and} \quad \frac{\partial g^{-1}}{\partial z} = \frac{\partial x}{\partial z} = \frac{1}{a}$$

and Eq. 4.13 would be

$$f_Z(z) = \int_{-\infty}^{\infty} \frac{1}{a} f_{X,Y}\left(\frac{z - by}{a}, y\right) dy \tag{4.14}$$

or with Eq. 4.13a,

$$f_Z(z) = \int_{-\infty}^{\infty} \frac{1}{b} f_{X,Y}\left(x, \frac{z - ax}{b}\right) dx \tag{4.14a}$$

If X and Y are statistically independent variates, the above Eq. 4.14 becomes

$$f_Z(z) = \frac{1}{a} \int_{-\infty}^{\infty} f_X\left(\frac{z - by}{a}\right) f_Y(y) \, dy \tag{4.15}$$

or

$$f_Z(z) = \frac{1}{b} \int_{-\infty}^{\infty} f_X(x) f_Y\left(\frac{z - ax}{b}\right) dx \tag{4.15a}$$

► **EXAMPLE 4.8**

Figure E4.8 shows a frame building subjected to earthquake forces. The building mass m is assumed to be concentrated at the roof level. When subjected to ground shaking during an earthquake, the building will vibrate about its original (at rest) position, inducing velocity components X and Y of the mass, with a resultant velocity $Z = \sqrt{X^2 + Y^2}$.

Assuming that X and Y are statistically independent and are, respectively, standard normal variates with distribution $N(0, 1)$, the probability distribution of the resultant kinetic energy of the building mass during an earthquake would be determined as follows:

The resultant kinetic energy is

$$W = mZ^2 = m(X^2 + Y^2)$$

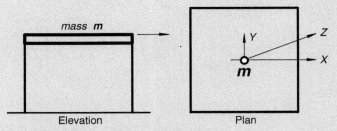

Figure E4.8 Kinetic energy of building under earthquake load.

Denoting $U = mX^2$ and $V = mY^2$, the resultant kinetic energy is then

$$W = U + V$$

From the results of Example 4.5, the PDF of U and V are, respectively,

$$f_U(u) = \frac{1}{\sqrt{2\pi mu}} e^{-u/2m} \qquad u \geq 0$$

$$f_V(v) = \frac{1}{\sqrt{2\pi mv}} e^{-v/2m} \qquad v \geq 0$$

Then, according to Eq. 4.15a, and assuming U and V are statistically independent (based on independence between the velocity components X and Y), we obtain the PDF of W as follows:

With $v = w - u$,

$$f_W(w) = \frac{1}{2\pi m} \int_0^w \frac{1}{\sqrt{u}} e^{-u/2m} \frac{1}{\sqrt{w-u}} e^{-(w-u)/2m} du$$

$$= \frac{1}{2\pi m} e^{-w/2m} \int_0^w u^{-1/2}(w-u)^{-1/2} du$$

Now, let $r = u/w$, and $du = wdr$. Then,

$$f_W(w) = \frac{1}{2\pi m} e^{-w/2m} \int_0^1 r^{-1/2}(1-r)^{-1/2} dr$$

We may observe that the above integral is the beta function $B\left(\frac{1}{2}, \frac{1}{2}\right)$, which is

$$B\left(\frac{1}{2}, \frac{1}{2}\right) = \frac{\Gamma\left(\frac{1}{2}\right)\Gamma\left(\frac{1}{2}\right)}{\Gamma(1)} = \pi$$

Hence,

$$f_W(w) = \frac{1}{2m} e^{-w/2m}$$

which is a chi-square distribution with two *degrees-of-freedom* (see Chapter 6). ◀

Sum (and Difference) of Independent Normal Variates—For two statistically independent normal variates, X and Y, with respective means and standard deviations μ_X, μ_Y and σ_X, σ_Y, the distribution of the sum $Z = X + Y$ would be, according to Eq. 4.15,

$$f_Z(z) = \frac{1}{2\pi \sigma_X \sigma_Y} \int_{-\infty}^{\infty} \exp\left[-\frac{1}{2}\left(\frac{z - y - \mu_X}{\sigma_X}\right)^2 - \frac{1}{2}\left(\frac{y - \mu_Y}{\sigma_Y}\right)^2\right] dy$$

$$= \frac{1}{2\pi \sigma_X \sigma_Y} \exp\left[-\frac{1}{2}\left\{\left(\frac{\mu_Y}{\sigma_Y}\right)^2 + \left(\frac{z - \mu_X}{\sigma_X}\right)^2\right\}\right] \int_{-\infty}^{\infty} \exp\left[-\frac{1}{2}(uy^2 - 2vy)\right] dy$$

where

$$u = \frac{1}{\sigma_X^2} + \frac{1}{\sigma_Y^2} \qquad \text{and} \qquad v = \frac{\mu_Y}{\sigma_Y^2} + \frac{z - \mu_X}{\sigma_X^2}$$

Then substituting,

$$w = y - \frac{v}{u}$$

the integral above becomes

$$\int_{-\infty}^{\infty} \exp\left[-\frac{1}{2}(uy^2 - 2vy)\right] dy = e^{v^2/2u} \int_{-\infty}^{\infty} \exp\left(-\frac{1}{2}uw^2\right) dw$$

$$= \sqrt{\frac{2\pi}{u}} \exp\left(\frac{v^2}{2u}\right)$$

After appropriate algebraic reduction, we obtain the PDF of Z as

$$f_Z(z) = \frac{1}{\sqrt{2\pi\left(\sigma_X^2 + \sigma_Y^2\right)}} \exp\left[-\frac{1}{2}\left(\frac{z - (\mu_X + \mu_Y)}{\sqrt{\sigma_X^2 + \sigma_Y^2}}\right)^2\right]$$

which we can recognize is also a normal PDF with a mean of

$$\mu_Z = \mu_X + \mu_Y$$

and variance,

$$\sigma_Z^2 = \sigma_X^2 + \sigma_Y^2$$

With the same procedure, we can show that $Z = X - Y$ is also Gaussian with a mean of $\mu_X - \mu_Y$ and the same variance $\sigma_Z^2 = \sigma_X^2 + \sigma_Y^2$.

We might point out that the above results remain valid even if the variates are correlated, except that the variance must include the *covariance* between X and Y.

On the basis of the above results, we can infer inductively that if

$$Z = \sum_{i=1}^{n} a_i X_i$$

where a_i are constants, and X_i are statistically independent Gaussian variates $N(\mu_{X_i}, \sigma_{X_i})$, then Z is also Gaussian with mean,

$$\mu_Z = \sum_{i=1}^{n} a_i \mu_{X_i} \qquad (4.16)$$

and variance,

$$\sigma_Z^2 = \sum_{i=1}^{n} a_i^2 \sigma_{X_i}^2 \qquad (4.17)$$

The last result above shows that any linear function of Gaussian variates is also a Gaussian variate. The relationships of Eqs. 4.16 and 4.17, for the mean and variance, however, are not limited to Gaussian variates. We shall observe later in Sect. 4.3.1 that these equations are, in fact, valid for linear functions of any statistically independent random variables irrespective of their distributions.

▶ **EXAMPLE 4.9**

The storm drain of a city is a normal variate with a mean capacity of 1.5 million gallons per day (mgd) and a standard deviation of 0.3 mgd; i.e., the capacity is $N(1.5, 0.30)$. The storm drain serves two independent drainage sources within the city that are also normally distributed as follows: drainage source $A = N(0.70, 0.20)$ mgd, and drainage source $B = N(0.50, 0.15)$ mgd.

The probability that the storm drain capacity D will be exceeded during a storm would be determined as follows:

$$P(D < A + B) = P[D - (A + B) < 0]$$

Since D, A, and B are independent normal variates, the sum $S = D - (A + B)$ is also normal with mean and variance, according to Eqs. 4.16 and 4.17,

$$\mu_S = \mu_D - (\mu_A + \mu_B) = 1.5 - (0.70 + 0.50) = 0.3$$

and

$$\sigma_S = \sqrt{\sigma_D^2 + \sigma_A^2 + \sigma_B^2} = \sqrt{(0.3)^2 + (0.20)^2 + (0.15)^2} = 0.39$$

Therefore, the probability that the drain capacity will be exceeded is

$$P(S < 0) = \Phi\left(\frac{0 - 0.3}{0.39}\right) = \Phi(-0.769) = 1 - \Phi(0.769) = 1 - 0.779 = 0.221$$

Suppose from prior storms, it has been shown that the existing storm drain can carry at least 1.2 mgd. Given that this information is reliable, what is the probability that the storm drain will be able to carry a total drainage of 1.9 mgd during a severe storm?

Clearly, this probability is conditioned on the information that the capacity of the storm drain is >1.2 mgd. Hence, we have

$$P(D > 1.9 \mid D > 1.2) = \frac{P(D > 1.9)}{P(D > 1.2)} = \frac{1 - P(D \le 1.9)}{1 - P(D \le 1.2)} = \frac{1 - \Phi\left(\dfrac{1.9 - 1.5}{0.3}\right)}{1 - \Phi\left(\dfrac{1.2 - 1.5}{0.3}\right)}$$

$$= \frac{1 - \Phi(1.33)}{1 - \Phi(-1.00)} = \frac{1 - 0.9088}{1 - (1 - 0.8413)} = 0.108$$

Therefore, there is little likelihood that the storm drain can carry a drainage of 1.9 mgd.

Now, if another drainage source $C = N(0.8, 0.2)$ mgd is to be added to the existing storm drain, how much must the mean capacity of the current storm drain be increased in order to maintain the same probability of exceedance (i.e., of 0.221) of the existing system? Assume that the standard deviation of the new storm drain will remain at 0.3 mgd.

Denoting the new capacity as D', we must have

$$P(S < 0) = P[D' - (A + B + C) < 0] = 0.221$$

and

$$\mu_S = \mu_{D'} - 0.70 - 0.50 - 0.80 = \mu_{D'} - 2.0$$

$$\sigma_S = \sqrt{(0.30)^2 + (0.20)^2 + (0.15)^2 + (0.20)^2} = 0.44$$

Therefore,

$$P(S < 0) = \Phi\left(\frac{0 - (\mu_{D'} - 2.0)}{0.44}\right) = 0.221$$

or

$$\frac{2.0 - \mu_{D'}}{0.44} = \Phi^{-1}(0.221) = -0.77$$

from which we obtain

$$\mu_{D'} = 2.0 + 0.77 \times 0.44 = 2.34 \text{ mgd}$$

Hence, the mean capacity of the existing storm drain must be increased by $2.34 - 1.5 = 0.84$ mgd. ◄

► **EXAMPLE 4.10**

Products can be transported by rail or trucks between New York and Los Angeles; both modes of transportation go through the city of Chicago. The mean travel times between the major cities for each mode of transportation are indicated in Fig. E4.10.

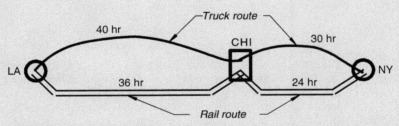

Figure E4.10 Rail and truck routes between New York and Los Angeles

The c.o.v.s of the travel times for the two modes of transportation are 15% and 20%, respectively, for rail and truck. Assume that the travel times between any two cities are statistically independent normal variates. The means and standard deviations of the travel times for the two modes of transportation are, respectively,

$$\mu_T = 40 + 30 = 70 \text{ hr}; \qquad \sigma_T = \sqrt{(0.2 \times 40)^2 + (0.2 \times 30)^2} = 10 \text{ hr}$$

and

$$\mu_R = 36 + 24 = 60 \text{ hr}; \qquad \sigma_R = \sqrt{(0.15 \times 36)^2 + (0.15 \times 24)^2} = 6.49 \text{ hr}$$

If the loading and unloading times in Chicago are 10 hr for trucks and 15 hr for rail, the probabilities that the actual travel times will exceed 85 hr between New York and Los Angeles are as follows.

The loading/unloading times in Chicago are assumed to be deterministic (c.o.v. $= 0$); and must be added to the respective mean travel times. Therefore,

for trucks, $P(T_T > 85) = 1 - P(T_T \leq 85) = 1 - \Phi\left(\dfrac{85 - 80}{10}\right) = 1 - \Phi(0.5) = 1 - 0.691 = 0.309;$

whereas for rail, $\quad P(T_R > 85) = 1 - P(T_R \leq 85) = 1 - \Phi\left(\dfrac{85 - 75}{6.49}\right) = 1 - \Phi(1.54)$

$$= 1 - 0.938 = 0.062.$$

Therefore, the probability that merchandise shipped by rail from New York to Los Angeles will arrive within 85 hours is 0.938, whereas by trucks the corresponding probability is 0.691. ◄

► **EXAMPLE 4.11**

The integrity of the columns is essential to the safety of a high-rise building. The total load acting on the columns may include the effects of the dead load D (primarily the weight of the structure), the live load L (that includes human occupancy, furniture, movable equipment, etc.), and the wind load W.

These individual load effects on the building columns may be assumed to be statistically independent Gaussian variates with the following respective means and standard deviations:

$$\mu_D = 4.2 \text{ tons}; \qquad \sigma_D = 0.3 \text{ ton}$$
$$\mu_L = 6.5 \text{ tons}; \qquad \sigma_L = 0.8 \text{ ton}$$
$$\mu_W = 3.4 \text{ tons}; \qquad \sigma_W = 0.7 \text{ ton}$$

The total combined load S on each of the columns is

$$S = D + L + W$$

which is also Gaussian with mean and standard deviation

$$\mu_S = \mu_D + \mu_L + \mu_W = 4.2 + 6.5 + 3.4 = 14.1 \text{ tons}$$

and

$$\sigma_S = \sqrt{\sigma_D^2 + \sigma_L^2 + \sigma_W^2} = \sqrt{(0.3)^2 + (0.8)^2 + (0.7)^2} = 1.1 \text{ tons}$$

The individual columns were designed with a mean strength that is equal to 1.5 times the total mean load that it carries, and may be assumed to be also Gaussian with a c.o.v. of 15%. The strength of each column R is clearly independent of the applied load. A column will be over-stressed when the applied load S exceeds the strength R; the probability of this event $(R < S)$ occurring is, therefore,

$$P(R < S) = P(R - S < 0) = \Phi\left(\frac{0 - [(1.5 \times 14.1) - 14.1]}{\sqrt{(0.15 \times 1.5 \times 14.1)^2 + (1.1)^2}} \right)$$

$$= \Phi\left(\frac{-7.05}{3.36} \right) = \Phi(-2.10) = 1 - 0.982 = 0.018$$

If we wish to decrease the probability of the event $(R < S)$, and thus increase the safety of the column, we need to increase the strength of the column. ◀

▶ **EXAMPLE 4.12**

The framing of a house may be done by subassembling the components in a plant and then delivering them to the site for framing. Simultaneously, while this subassembly of the components is being fabricated, the preparation of the site, which includes the excavation of the foundation through the construction of the foundation walls, can proceed at the same time. The different activities and their sequence may be represented with the activity network shown in Fig. E4.12; the required work durations for the respective activities are also shown in the table below.

		Completion Time (days)	
Activity	Description	Mean	Std. Dev.
1–2	Excavation of foundation	2	1
2–3	Construction of footings	1	½
3–5	Construction of foundation walls	3	1
1–4	Fabrication and assembly of components	5	1
4–5	Delivery of assembled components to site	2	½

Assume that the required completion times for the different activities are statistically independent Gaussian variates, with the respective means and standard deviations shown in the above table. Clearly, to start framing the house, the foundation walls must be completed and the assembled components must be delivered to the site. The probability that framing of the house can start within 8 days after work started on the job can be determined as follows.

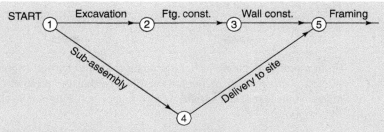

Figure E4.12 House framing activity network.

Denote the durations of the activities listed in the above table as X_1, X_2, X_3, X_4 and X_5, respectively, and let

$$T_1 = X_1 + X_2 + X_3$$

and

$$T_2 = X_4 + X_5$$

In which T_1 and T_2 are also statistically independent. Then, the required probability is

$$P(F) = P[(T_1 \le 8) \cap (T_2 \le 8)] = P(T_1 \le 8)P(T_2 \le 8)$$

in which

$$P(T_1 \le 8) = \Phi\left(\frac{8 - (2 + 1 + 3)}{\sqrt{(1.0)^2 + (0.5)^2 + (1.0)^2}}\right) = \Phi\left(\frac{2}{1.5}\right) = \Phi(1.33) = 0.907$$

and

$$P(T_2 \le 8) = \Phi\left(\frac{8 - (5 + 2)}{\sqrt{(1.0)^2 + (0.5)^2}}\right) = \Phi\left(\frac{1.0}{1.12}\right) = \Phi(0.89) = 0.813$$

Hence, the probability that framing can start within 8 days is

$$P(F) = 0.907 \times 0.813 = 0.74$$

◄

Products and Quotients of Random Variables—If the function is the product of two random variables, say

$$Z = XY$$

then

$$X = Z/Y \qquad \text{and} \qquad \frac{\partial x}{\partial z} = \frac{1}{y}$$

and Eq. 4.13 yields the PDF of Z as

$$f_Z(z) = \int_{-\infty}^{\infty} \left|\frac{1}{y}\right| f_{X,Y}\left(\frac{z}{y}, y\right) dy \qquad (4.18)$$

Similarly, if the function is a quotient of two random variables, say

$$Z = X/Y$$

The PDF of Z would be

$$f_Z(z) = \int_{-\infty}^{\infty} |y| f_{X,Y}(zy, y)\, dy \tag{4.19}$$

From a practical standpoint, the product (and quotient) of lognormal random variables is of special interest. In particular, we observe that the product or quotient of statistically independent lognormal variates is also lognormal. This can be shown as follows.
Suppose

$$Z = \prod_{i=1}^{n} X_i$$

where the X_i's are statistically independent lognormal variates, with respective parameters λ_{X_i} and ζ_{X_i}; then

$$\ln Z = \sum_{i=1}^{n} \ln X_i$$

But each of the $\ln X_i$ is normal (as shown in Example 4.3) with mean λ_i and variance ζ_i^2; hence, $\ln Z$ is the sum of normals and, therefore, is also normal with mean and variance as follows:

$$\lambda_Z = E(\ln Z) = \sum_{i=1}^{n} \lambda_{X_i} \tag{4.20}$$

We might also recall from Eq. 3.27 that $\lambda_{X_i} = \ln x_{m,i}$, in which $x_{m,i}$ is the median of X_i; therefore, λ_Z is also

$$\lambda_Z = \sum_{i=1}^{n} \ln x_{m,i} \tag{4.20a}$$

and, the variance

$$\zeta_Z^2 = \text{Var}(\ln Z) = \sum_{i=1}^{n} \zeta_{X_i}^2 \tag{4.21}$$

Hence, Z is lognormal with the above parameters λ_Z and ζ_Z.

▶ **EXAMPLE 4.13**   The annual operational cost for a waste treatment plant is a function of the weight of solid waste, W, the unit cost factor, F, and an efficiency coefficient, E, as follows:

$$C = \frac{WF}{\sqrt{E}}$$

where W, F, and E are statistically independent lognormal variates with the following respective medians and coefficients of variation (c.o.v.):

Variable	Median	c.o.v.
W	2000 tons/yr	20%
F	$20 per ton	15%
E	1.6	12.5%

As C is a function of the product and quotient of lognormal variates, its probability distribution is also lognormal, which we can show as follows:

$$\ln C = \ln W + \ln F - \frac{1}{2}\ln E$$

According to Eqs. 4.20a and 4.21, $\ln C$ is normal with mean $\lambda_C = \lambda_W + \lambda_F - \frac{1}{2}\lambda_E$ and variance $\zeta_C^2 = \zeta_W^2 + \zeta_F^2 + \left(\frac{1}{2}\zeta_E\right)^2$. Therefore, C is lognormal with

$$\lambda_C = \ln 2000 + \ln 20 - \ln 1.6 = 10.13$$

and

$$\zeta_C = \sqrt{(0.20)^2 + (0.15)^2 + (1/2 \times 0.125)^2} = 0.26$$

On the basis of the above, the probability that the annual cost of operating the waste treatment plant will exceed $35,000 is

$$P(C > 35{,}000) = 1 - P(C \le 35{,}000) = 1 - \Phi\left(\frac{\ln 35{,}000 - 10.13}{0.26}\right)$$

$$= 1 - \Phi(1.28) = 1 - 0.900 = 0.100 \qquad \blacktriangleleft$$

▶ **EXAMPLE 4.14**

The structure and foundation of the high-rise building shown in Fig. E4.14 are both designed to withstand wind-induced pressures with the respective pressure capacities as follows:

Superstructure, $R_S = 40$ psf(lb/ft^2)

Foundation, $R_F = 30$ psf

The peak wind pressure P_W on the building during a wind storm is given by

$$P_W = 1.165 \times 10^{-3} C V^2, \quad \text{in psf}$$

where:

$V = $ maximum wind speed, in fps

$C = $ drag coefficient

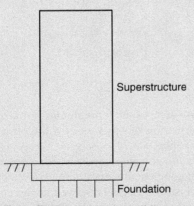

Figure E4.14   A high-rise building.

During a 50-year wind storm, the maximum wind speed V may be assumed to be a lognormal variate with mean speed of $\mu_V = 100$ fps and a c.o.v. $\delta_V = 0.25$. The drag coefficient C is also lognormal with a mean of $\mu_C = 1.80$ and $\delta_C = 0.30$.

Clearly, the distribution of the wind-induced pressure, P_W, is also lognormal with the following parameters for the 50-year wind:

$$\lambda_{P_W} = \ln 1.165 \times 10^{-3} + \lambda_C + 2\lambda_V = -6.755 + (\ln 1.80 - 0.30^2) + 2(\ln 100 - 0.25^2)$$
$$= -6.755 + 0.498 + 2(4.543) = 2.829$$

and

$$\zeta_{P_W} = \sqrt{(2 \times 0.25)^2 + (0.30)^2} = 0.583$$

Whenever the maximum wind pressure exceeds the design pressure capacity of the superstructure or the foundation, there may be damage to the corresponding substructure. These probabilities are, respectively, as follows.

For the superstructure,

$$P(P_W > 40) = 1 - P(P_W \le 40) = 1 - \Phi\left(\frac{\ln 40 - 2.829}{0.583}\right) = 1 - \Phi(1.47) = 1 - 0.929 = 0.071$$

whereas for the foundation,

$$P(P_W > 30) = 1 - P(P_W \le 30) = 1 - \Phi\left(\frac{\ln 30 - 2.829}{0.583}\right) = 1 - \Phi(0.98) = 1 - 0.836 = 0.164$$

In this case, because the wind resistances of the superstructure and foundation are deterministic, the probability of damage to the high-rise system when subjected to the 50-year wind would be

$$P(\text{damage}) = P(S \cup F) = P(P_W > 30) = 0.164$$

Finally, if the occurrences of 50-year wind storms constitute a Poisson process, and damages to the building between storms are statistically independent, the probability of wind damage, D, to the high-rise building for a period of 20 years would be

$$P(D \text{ in } 20 \text{ years}) = 1 - \left[\sum_{n=0}^{\infty} [P(\overline{D})]^n \frac{(0.02 \times 20)^n}{n!} e^{-0.02 \times 20}\right]$$

$$= 1 - e^{-0.02 \times 20}\left[1 + 0.836 \times 0.02 \times 20 + \frac{1}{2}(0.836)^2(0.02 \times 20)^2 + \cdots\right]$$

$$= 1 - e^{-0.02 \times 20} \cdot e^{0.836 \times 0.02 \times 20} = 1 - e^{-0.164 \times 0.02 \times 20}$$

$$= 1 - e^{-0.0656} = 1 - 0.937 = 0.063 \qquad \blacktriangleleft$$

The Central Limit Theorem—One of the most significant theorems in probability theory pertains to the limiting distribution of the sum of a large number of random variables known as the *central limit theorem*. Stated loosely, the theorem says that the sum of a large number of individual random components, none of which is dominant, tends to the Gaussian distribution as the number of component variables (regardless of their initial distributions) increases. Therefore, if a physical process is derived as the combined totality of a large number of individual effects, then according to the central limit theorem, the process would tend to be Gaussian.

The rigorous proof of the theorem is beyond our scope of interest; however, the essence of the theorem may be demonstrated with the following example. Take, for example, the following sum

$$S = \frac{1}{\sqrt{n}} \sum_{i=1}^{n} X_i$$

where the X_i's are statistically independent and identically distributed random variables with PMF

$$P(X_i = 1) = \frac{1}{2}$$

$$P(X_i = -1) = \frac{1}{2}$$

and

$$P(X_i = x) = 0, \text{ otherwise}$$

According to the central limit theorem, the sum S will approach the Gaussian distribution $N(0, 1)$ as $n \to \infty$. This is demonstrated in Fig. 4.1 for the above initial PMF of the X_i's as n increases from 2 to 20.

By virtue of the above central limit theorem, we can infer that the product (or quotient) of independent factors, none of which is dominating, will tend to approach the lognormal distribution. That is, regardless of the distributions of the X_i's, the distribution of the product

$$P = c \prod_{i=1}^{n} X_i$$

will approach the lognormal distribution as $n \to \infty$.

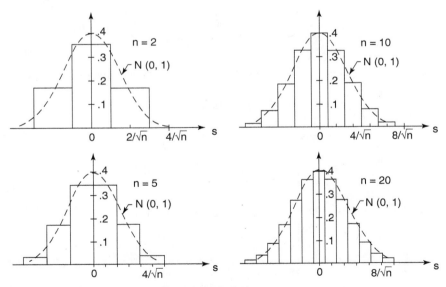

Figure 4.1 A demonstration of the central limit theorem.

▶ **EXAMPLE 4.15**   Let us consider the product of a number of n random variables

$$P = \prod_{i=1}^{n} X_i$$

in which each X_i has the following exponential distribution:

$$f_X(x) = \frac{1}{\lambda}e^{-x/\lambda}; \qquad x \geq 0$$

Using $Y_i = \ln X_i$, we have

$$\ln P = \sum_{i=1}^{n} \ln X_i = \sum_{i=1}^{n} Y_i$$

Therefore, as $n \to \infty$, the distribution of $\sum_{i=1}^{n} \ln X_i$ will approach the Gaussian distribution according to the central limit theorem. We show this numerically for increasing n (through Monte Carlo simulation as presented in Chapter 5) for the exponential distribution with parameter $\lambda = 2$ as follows: Let

$$Y_1 = \ln X_1$$

With a sample of 10,000, we generate the histogram of Y_1 and the corresponding statistics as shown below.

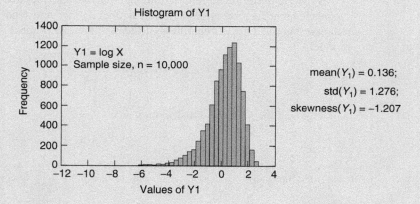

mean(Y_1) = 0.136;
std(Y_1) = 1.276;
skewness(Y_1) = −1.207

For $n = 5$, the sum of $A = \sum_{i=1}^{5} Y_i$; the histogram and corresponding statistics are as follows:

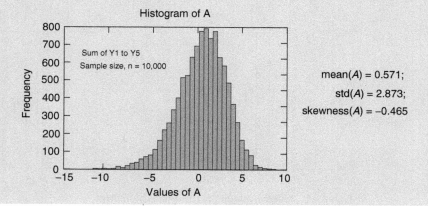

mean(A) = 0.571;
std(A) = 2.873;
skewness(A) = −0.465

and for $n = 100$, the sum of $B = \sum_{i=1}^{100} Y_i$ we obtain the histogram and corresponding statistics as follows:

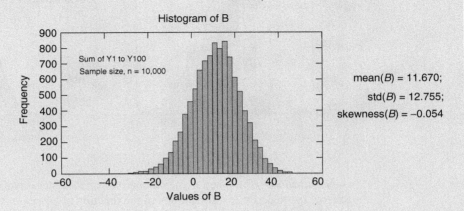

mean(B) = 11.670;
std(B) = 12.755;
skewness(B) = −0.054

From the above, we see that as n increases from 1 to 5 to 100, the sum $\sum_{i=1}^{n} Y_i$ becomes more symmetric as indicated by the skewness coefficient, and approaches zero for $n = 100$. We can also show that the kurtosis approaches zero for increasing n, indicating therefore that the sum $\sum_{i=1}^{n} Y_i$ approaches the Gaussian distribution. Therefore, the product $P = \prod_{i=1}^{n} X_i$ approaches the lognormal distribution. ◀

▶ **EXAMPLE 4.16**

Finally, consider the sum of n uniformly distributed random variables X_i each ranging between 0 and 2; i.e.

$$S = \sum_{i=1}^{n} X_i$$

For $n = 2$, 5, 10, and 100, the respective histograms of S (all normalized with a mean of 100), and the corresponding superimposed PDFs, with the same means and standard deviations, are shown below:

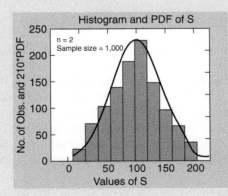 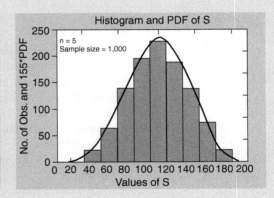

All the histograms shown above are generated with 1000 or 10,000 repetitions (sample size). The convergence to the normal PDF for increasing n is clearly demonstrated.

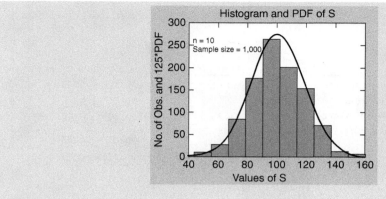 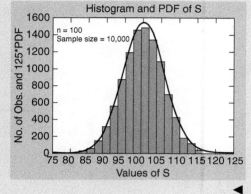

Generalization—The function of two random variables described above in Eq. 4.13 can be generalized to derive the distribution of the function of n random variables. In particular, if

$$Z = g(X_1, X_2, \ldots, X_n) \tag{4.22}$$

then, generalizing Eq. 4.13, we have

$$
\begin{aligned}
F_Z(z) &= \int\limits_{g(x_1, x_2, \ldots x_n) \leq z} \int \cdots \int f_{X_1, X_2, \ldots X_n}(x_1, x_2, \ldots, x_n) dx_1 \ldots dx_n \\
&= \int_{-\infty}^{\infty} \cdots \int_{-\infty}^{g^{-1}} f_{X_1, \ldots, X_n}(x_1, \ldots x_n) dx_1 \ldots dx_n
\end{aligned}
\tag{4.23}
$$

in which $g^{-1} = g^{-1}(z, x_2, \ldots, x_n)$. Changing the variable of integration from x_1 to z, we have

$$F_Z(z) = \int_{-\infty}^{\infty} \cdots \int_{-\infty}^{\infty} \int_{-\infty}^{z} f_{X_1, \ldots, X_n}(g^{-1}, x_2, \ldots, x_n) \left| \frac{\partial g^{-1}}{\partial z} \right| dz \, dx_2 \cdots dx_r$$

from which we obtain

$$f_Z(z) = \int_{-\infty}^{\infty} \cdots \int_{-\infty}^{\infty} f_{X_1, \ldots X_n}(g^{-1}, x_2, \ldots, x_n) \left| \frac{\partial g^{-1}}{\partial z} \right| dx_2 \ldots dx_n \tag{4.24}$$

4.2.3 Extreme Value Distributions

The extremes (i.e., the maximum and minimum values) of natural phenomena are often of special interest and importance in engineering. The statistics of such extremes are of special significance to many problems involving natural hazards, such as the maximum flood of a river in the next 100 years, the maximum earthquake intensity expected at a building site in the next 50 years, and the minimum flow (drought) of a river in the next 25 years. For this purpose, we will summarize here the basic fundamentals of the topic (more complete discussion of the topic can be found in Ang and Tang, Vol. 2, 1984).[*] Moreover, we shall emphasize here only those elements that are particularly of practical importance.

When we speak of extreme values, we are considering the largest and smallest values from a sample of size n within a known population. In this regard, each of the largest or

---

[*]This book is out of print but is available for direct order from ahang2@aol.com.

smallest value will have its own probability distribution, which may be an exact distribution or an asymptotic distribution.

Exact Distributions—Consider a random variable X (or population) with known probability distribution $f_X(x)$ or $F_X(x)$. A sample of size n from this population will have a largest and a smallest value; these extreme values will also have their respective distributions which are related to the distribution of the initial variate X.

A sample of size n is a set of observations $(x_1, x_2, \ldots, x_n)$ representing, respectively, the first, second, $\ldots$, and nth observed values. Because the observed values are unpredictable, they are a specific realization of an underlying set of random variables $(X_1, X_2, \ldots, X_n)$. Therefore, we are interested in the maximum and minimum of $(X_1, X_2, \ldots, X_n)$; i.e., the random variables

$$Y_n = \max(X_1, X_2, \ldots, X_n)$$

and

$$Y_1 = \min(X_1, X_2, \ldots, X_n)$$

Under certain assumptions, the exact probability distributions of Y_1 and Y_n can be derived as follows.

We may observe first that if Y_n is less than some specified value y, then all the other sample random variables $X_1, X_2, \ldots, X_n$ must individually be less than y. Then with the assumption that $X_1, X_2, \ldots, X_n$ are statistically independent and identically distributed as the initial variate X, i.e.,

$$F_{X_1}(x) = F_{X_2}(x) = \cdots = F_{X_n}(x) = F_X(x)$$

The CDF of Y_n, therefore, is

$$\begin{aligned} F_{Y_n}(y) &= P(X_1 \le y, X_2 \le y, \ldots, X_n \le y) \\ &= [F_X(y)]^n \end{aligned} \tag{4.25}$$

and the corresponding PDF is

$$f_{Y_n}(y) = \frac{dF_{Y_n}(y)}{dy} = n[F_X(y)]^{n-1} f_X(y) \tag{4.26}$$

where $F_X(y)$ and $f_X(y)$ are, respectively, the CDF and PDF of the initial variate X.

Similarly, the exact distribution of Y_1, can be derived as follows. In this case, we observe that if Y_1, the smallest among $X_1, X_2, \ldots, X_n$, is larger than y, then all the sample random variables must be individually larger than y. Hence, the *survival function* which is the complement of the CDF, is

$$\begin{aligned} 1 - F_{Y_1}(y) &= P(X_1 > y, X_2 > y, \ldots, X_n > y) \\ &= [1 - F_X(y)]^n \end{aligned}$$

Thus, the CDF of Y_1 is

$$F_{Y_1}(y) = 1 - [1 - F_X(y)]^n \tag{4.27}$$

and the corresponding PDF is

$$f_{Y_1}(y) = n[1 - F_X(y)]^{n-1} f_X(y) \tag{4.28}$$

Clearly, from Eqs. 4.25 through 4.28, we see that the exact distribution of the largest and smallest values from samples of size n is a function of the distribution of the initial variate.

▶ **EXAMPLE 4.17**   Consider the initial variate X with the exponential PDF as follows:

$$f_X(x) = \lambda e^{-\lambda x} \qquad x \geq 0$$

The corresponding CDF of X is

$$F_X(x) = 1 - e^{-\lambda x}$$

Therefore, the CDF of the largest value from samples of size n, according to Eq. 4.25, is

$$F_{Y_n}(y) = (1 - e^{-\lambda y})^n$$

and the corresponding PDF is

$$f_{Y_n}(y) = \lambda n (1 - e^{-\lambda y})^{n-1} e^{-\lambda y}$$

Graphically, the above PDF and CDF of Y_n are shown (for $\lambda = 1.0$) in Fig. E4.17 for different sample sizes n from 1 to 100.

From Fig. E4.17, we can see that the PDF as well as the CDF shift to the right with increasing n. Also, as expected, the mode of the largest value increases with increasing n.

Expanding the above $F_{Y_n}(y)$ by the binomial series expansion, we obtain the series

$$(1 - e^{-\lambda y})^n = 1 - ne^{-\lambda y} + \frac{n(n-1)}{2!} e^{-2\lambda y} - \cdots$$

For large n the above series approaches the double exponential $\exp(-ne^{-\lambda y})$. Hence, for large n the CDF of the largest value from an exponential population approaches the double exponential

$$F_{Y_n}(y) = \exp(-ne^{-\lambda y})$$

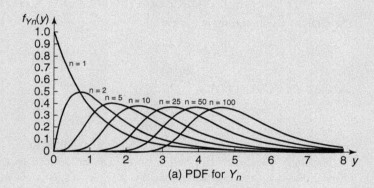

(a) PDF for Y_n

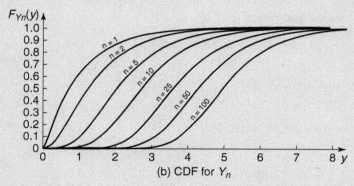

(b) CDF for Y_n

Figure E4.17   PDFs and CDFs of Y_n for different n ($\lambda = 1.0$).   ◀

The Asymptotic Distributions—In Example 4.17, we observed that in the case of the exponential initial distribution, the CDF of the largest value from samples of size n approaches the double exponential distribution as n increases; in this case, this double exponential distribution is the *asymptotic distribution* of the largest value. This characteristic is shown also in Fig. 4.2, illustrating the convergence of the exact distribution of Y_n to the asymptotic double exponential distribution as $n \to \infty$.

The characteristics illustrated in Fig. 4.2 for the initial exponential distribution actually apply also to other initial distributions; i.e., the distribution of an extreme value converges asymptotically in distribution as n increases. According to Gumbel (1954, 1960), there are three types of such asymptotic distributions (although not exhaustive) depending on the tail behavior of the initial PDFs; namely, as follows:

> ***Type I***: The double exponential form.
>
> ***Type II***: The single exponential form.
>
> ***Type III***: The exponential form with an upper (or lower) bound.

The extreme value from an initial distribution with an exponentially decaying tail (in the direction of the extreme) will converge asymptotically to the *Type I* limiting form. This was illustrated earlier in Example 4.17 and demonstrated graphically in Fig. 4.2. For an initial variate with a PDF that decays with a polynomial tail, the distribution of its extreme value will converge to the *Type II* limiting form, whereas if the extreme value is limited, the

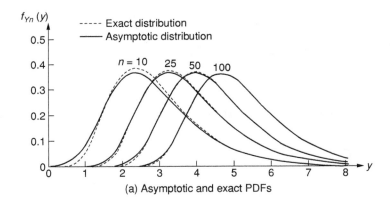

(a) Asymptotic and exact PDFs

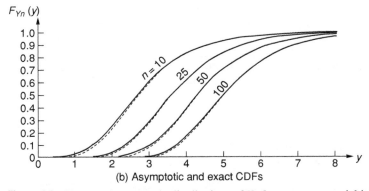

(b) Asymptotic and exact CDFs

Figure 4.2 Exact and asymptotic distributions of Y_n from an exponential initial distribution.

corresponding extremal distribution will converge asymptotically to the *Type III* asymptotic form.

Parameters of the Asymptotic Distributions—Even though the forms of the asymptotic distributions do not depend on the distributions of the initial variates, the parameters of the asymptotic distributions, such as u_n and α_n of the *Type I* asymptote, will depend on the distribution of the initial variate. We shall limit our illustrations below to the parameters of the *Type I* asymptotic form (a more complete discussion of the other asymptotic forms can be found in Ang and Tang, Vol. 2, 1984).

The Gumbel Distribution—The CDF of the *Type I* asymptotic form for the largest value, well known as the Gumbel distribution (Gumbel, 1958), is

$$F_{Y_n}(y) = \exp\left[-e^{-\alpha_n(y-u_n)}\right] \tag{4.29a}$$

and its PDF is

$$f_{Y_n}(y) = \alpha_n e^{-\alpha_n(y-u_n)}\exp\left[-e^{-\alpha_n(y-u_n)}\right] \tag{4.29b}$$

in which

$u_n =$ the most probable value of Y_n

$\alpha_n =$ an inverse measure of the dispersion of values of Y_n

Moreover, the mean and variance of the largest value, Y_n, and smallest value, Y_1, are related to the respective parameters as follows (see Ang and Tang, Vol. 2, 1984):

$$\mu_{Y_n} = u_n + \frac{\gamma}{\alpha_n} \tag{4.30a}$$

in which $\gamma = 0.577216$ (the Euler number); and

$$\sigma_{Y_n}^2 = \frac{\pi^2}{6\alpha_n^2} \tag{4.30b}$$

whereas for the smallest value, the corresponding mean and variance are

$$\mu_{Y_1} = u_1 - \frac{\gamma}{\alpha_1} \quad \text{and} \quad \sigma_{Y_1}^2 = \frac{\pi^2}{6\alpha_1^2}$$

▶ **EXAMPLE 4.18**

Consider an initial variate X with the standard normal distribution, i.e., $N(0,1)$, with the PDF

$$f_X(x) = \frac{1}{\sqrt{2\pi}}e^{-x^2/2}$$

In this case, the tail of the PDF is clearly exponential; hence, the asymptotic distribution of the largest value is of the double exponential form (*Type I*); specifically, the CDF of Y_n is

$$F_{Y_n}(y) = \exp[-e^{-\alpha_n(y-u_n)}]$$

and its PDF is

$$f_{Y_n}(y) = \alpha_n e^{-\alpha_n(y-u_n)}\exp[-e^{-\alpha_n(y-u_n)}]$$

with parameters

$$u_n = \sqrt{2 \ln n} - \frac{\ln \ln n + \ln 4\pi}{2\sqrt{2 \ln n}}$$

and

$$\alpha_n = \sqrt{2 \ln n}$$

Explanation of the derivation of the above parameters may be found in Ang and Tang, Vol. 2 (1984). The mean and standard deviation of the largest value can then be obtained, respectively, from Eqs. 4.30a and 4.30b.

If the initial variate X has a general Gaussian distribution, $N(\mu, \sigma)$, we observe from Example 4.2 that $(X - \mu)/\sigma$ will be $N(0,1)$. Then, the asymptotic distribution of the largest value from $(X - \mu)/\sigma$ will be of the double exponential form with the parameters u_n and α_n obtained above. Therefore, if Y_n' is the largest value from the initial Gaussian variate X, then it follows that

$$Y_n = \frac{Y_n' - \mu}{\sigma}$$

is the largest from $\dfrac{X - \mu}{\sigma}$, and the CDF of Y_n' is, therefore,

$$F_{Y_n'}(y') = F_{Y_n}\left(\frac{y' - \mu}{\sigma}\right) = \exp\left[-e^{-\alpha_n\left(\frac{y' - \mu - \sigma u_n}{\sigma}\right)}\right] = \exp\left[-e^{-\frac{\alpha_n}{\sigma}(y' - \mu - \sigma u_n)}\right]$$

Hence, the CDF of Y_n' is of the same double exponential form as Y_n, with the parameters

$$u_n' = \sigma u_n + \mu$$
$$\alpha_n' = \alpha_n/\sigma$$

◀

In the case of the smallest value from samples of size n, the corresponding distributions, PDF and CDF, would shift to the left as n increases. Similar to the largest value, the distribution of the smallest value will also converge (in distribution) to one of the three types of asymptotic distributions depending on the tail behavior (in the direction of the smallest value) of the PDF of the initial variate.

▶ **EXAMPLE 4.19**

Consider an initial variate with the standard Gaussian distribution $N(0, 1)$. The tail behavior in the lower end of this PDF is obviously also exponential; thus, the CDF of the smallest value from this initial variate will also converge to the *Type I* asymptotic form as follows:

$$F_{Y_1}(y) = 1 - \exp[-e^{\alpha_1(y - u_1)}]$$

and the corresponding PDF is

$$f_{Y_1}(y) = \alpha_1 e^{\alpha_1(y - u_1)} \exp[-e^{\alpha_1(y - u_1)}]$$

in which the parameters u_1 and α_1 are

$$u_1 = -\sqrt{2 \ln n} + \frac{\ln \ln n + \ln 4\pi}{2\sqrt{2 \ln n}} \; ; \quad \text{and}$$

$$\alpha_1 = \sqrt{2 \ln n}$$

◀

▶ **EXAMPLE 4.20**

If the initial variate X is described by the Rayleigh distribution with PDF,

$$f_X(x) = \frac{x}{\alpha^2} e^{-(1/2)(x/\alpha)^2} \qquad x \geq 0$$

in which α = the modal value. Although the tail behavior of this PDF may not be obvious, it is exponential (see Ang and Tang, Vol. 2, 1984), and thus the distribution of the largest value will converge asymptotically to the *Type I* form with the following parameters:

$$u_n = \alpha\sqrt{2\ln n}$$

and

$$\alpha_n = \frac{\sqrt{2\ln n}}{\alpha}$$

Then, according to Eqs. 4.30a and 4.30b, the mean and standard deviation of Y_n are, respectively,

$$\mu_{Y_n} = \alpha\sqrt{2\ln n} + 0.5772\left(\frac{\alpha}{\sqrt{2\ln n}}\right)$$

and

$$\sigma_{Y_n} = \frac{\pi}{\sqrt{6}}\frac{\alpha}{\sqrt{2\ln n}} = \frac{\pi\alpha}{2\sqrt{3\ln n}}$$

◀

The Fisher–Tippett Distribution—The *Type II* asymptotic distribution of the largest value is often referred to as the Fisher–Tippett distribution (Fisher and Tippett, 1928). Its CDF is

$$F_{Y_n}(y) = \exp\left[-\left(\frac{v_n}{y}\right)^k\right] \tag{4.31}$$

The corresponding PDF is

$$f_{Y_n}(y) = \frac{k}{v_n}\left(\frac{v_n}{y}\right)^{k+1}\exp\left[-\left(\frac{v_n}{y}\right)^k\right] \tag{4.32}$$

where the parameters

v_n = the most probable value of Y_n

k = the shape parameter, which is an inverse measure of the dispersion of values of Y_n

The mean and standard deviation of the variate Y_n are related to the above parameters as follows:

$$\mu_{Y_n} = v_n\Gamma\left(1 - \frac{1}{k}\right)$$

$$\sigma_{Y_n} = v_n\sqrt{\Gamma\left(1 - \frac{2}{k}\right) - \Gamma^2\left(1 - \frac{1}{k}\right)} \tag{4.33}$$

The *Type I* and *Type II* asymptotic forms are related through a logarithmic transformation as follows: If Y_n has the *Type II* asymptotic distribution of Eq. 4.31, with parameters v_n and k, the distribution of $\ln Y_n$ will have the *Type I* asymptotic form with parameters

$$u_n = \ln v_n \quad \text{and} \quad \alpha_n = k$$

The above logarithmic transformation applies also to the smallest value of Y_1 and $\ln Y_1$.

▶ **EXAMPLE 4.21**

If the initial variate X is lognormal, with parameters λ_X and ζ_X, $\ln X$ is normal with parameters $\mu = \lambda_X$ and $\sigma = \zeta_X$. Then, according to the results of Example 4.18, the largest value of $\ln X$ will converge asymptotically to the *Type I* distribution with parameters

$$u_n = \zeta_X\left(\sqrt{2\ln n} - \frac{\ln\ln n + \ln 4\pi}{2\sqrt{2\ln n}}\right) + \lambda_X \quad \text{and} \quad \alpha_n = \frac{\sqrt{2\ln n}}{\zeta_X}$$

Therefore, according to the above logarithmic transformation, the largest value of the initial lognormal variate X will converge to the *Type II* asymptotic form with parameters

$$v_n = e^{u_n} \quad \text{and} \quad k = \alpha_n$$

◄

The Weibull Distribution—For the *Type III* asymptotic form, the asymptotic distribution of the smallest value is of greater interest. In engineering, it is well known as the Weibull distribution (Weibull, 1951), which was discovered by Weibull for modeling the fracture strength of materials. Its CDF is given by

$$F_{Y_1}(y) = 1 - \exp\left[-\left(\frac{y - \varepsilon}{w_1 - \varepsilon}\right)^k\right]; \qquad y \geq \varepsilon \tag{4.34}$$

where

$w_1 = $ the most probable smallest value

$k = $ the shape parameter

$\varepsilon = $ the lower bound value of y

The mean and standard deviation of Y_1 is related to the above parameters as follows:

$$\mu_{Y_1} = \varepsilon + (w_1 - \varepsilon)\Gamma\left(1 + \frac{1}{k}\right);$$

and $\tag{4.35}$

$$\sigma_{Y_1} = (w_1 - \varepsilon)\sqrt{\Gamma\left(1 + \frac{2}{k}\right) - \Gamma^2\left(1 + \frac{1}{k}\right)}$$

► **EXAMPLE 4.22**

Suppose the lower-bound fracture strength of a welded joint is 4.0 ksi. If the actual strength of the joint Y_1 is modeled with a *Type III* smallest asymptotic distribution, Eq. 4.34, with parameters $w_1 = 15.0$ ksi and $k = 1.75$, the probability that the strength of the joint will be at least 16.5 ksi is

$$P(Y_1 \geq 16.5) = \exp\left[-\left(\frac{16.5 - 4}{15 - 4}\right)^{1.75}\right] = 0.286$$

The mean and standard deviation of the joint strength are, according to Eq. 4.35, respectively,

$$\mu_{Y_1} = 4 + (15 - 4)\Gamma\left(1 + \frac{1}{1.75}\right) = 4 + 11 \times \Gamma(1.5714) = 4 + 11 \times 0.8906$$

$$= 13.80 \text{ ksi};$$

and

$$\sigma_{Y_1} = (15 - 4)[\Gamma(1 + 2/1.75) - \Gamma^2(1 + 1/1.75)]^{1/2} = 11 \times [\Gamma(2.1429) - \Gamma^2(1.5714)]^{1/2}$$

$$= 11[1.069 - (0.8906)^2]^{1/2} = 5.78 \text{ ksi}$$

The values of the respective gamma functions indicated above were evaluated through *MATLAB*.

◄

▶ 4.3 MOMENTS OF FUNCTIONS OF RANDOM VARIABLES

In Sect. 4.2 we derived the probability distribution of a function of one or more random variables; theoretically, such a derived probability distribution of several random variables is analytically possible under certain conditions. For instance, the linear function of normal variates remains normal, whereas the product (or quotient) of lognormal variates will also be lognormal. However, in general, such derived probability distributions of the function (especially if nonlinear) may be very difficult (or even impossible) to derive analytically; in such general cases, Monte Carlo simulations would be necessary.

In these latter circumstances, we may use the moments, particularly the mean and the variance, of the function as an approximation of the probability distribution. This approximate approach may be sufficient for practical purposes even if the correct probability distribution must be left undetermined. Such moments are functionally related to the moments of the individual basic variates, and therefore may be derived approximately as functions of the moments of the basic variates.

4.3.1 Mathematical Expectations of a Function

The expected value of a function of random variables, called the *mathematical expectation*, can be derived as a generalization of Eq. 3.9b. That is, for a function of n variables, $Z = g(X_1, X_2, \ldots, X_n)$, its mathematical expectation is

$$E(Z) = \int_{-\infty}^{\infty} \cdots \int_{-\infty}^{\infty} g(x_1, x_2, \ldots, x_n) f_{X_1, X_2, \ldots X_n}(x_1, x_2, \ldots, x_n) dx_1 dx_2 \ldots dx_n$$

(4.36)

A similar generalization of Eq. 3.9a will yield the corresponding mathematical expectation $E(Z)$ of the function for discrete variates.

Below, we shall use Eq. 4.36 to derive the moments of linear functions of random variables, as well as the first-order approximate moments of nonlinear functions.

Mean and Variance of a Linear Function—Consider first the moments of the linear function

$$Y = aX + b$$

According to Eq. 3.9b, the mean value of Y is

$$E(Y) = E(aX + b) = \int_{-\infty}^{\infty} (ax + b) f_X(x)\, dx = a \int_{-\infty}^{\infty} x f_X(x)\, dx + b \int_{-\infty}^{\infty} f_X(x)\, dx$$

$$= aE(X) + b \tag{4.37}$$

whereas the variance is

$$\mathrm{Var}(Y) = E[(Y - \mu_Y)^2] = E[(aX + b - a\mu_X - b)^2] = a^2 \int_{-\infty}^{\infty} (x - \mu_X)^2 f_X(x)\, dx$$

$$= a^2 \mathrm{Var}(X) \tag{4.38}$$

For $Y = a_1 X_1 + a_2 X_2$, where a_1 and a_2 are constants

$$E(Y) = \int_{-\infty}^{\infty} \int_{-\infty}^{\infty} (a_1 x_1 + a_2 x_2) f_{X_1, X_2}(x_1, x_2) dx_1 dx_2$$

$$= a_1 \int_{-\infty}^{\infty} x_1 f_{X_1}(x_1) dx_1 + a_2 \int_{-\infty}^{\infty} x_2 f_{X_2}(x_2) dx_2$$

We can recognize that the last two integrals above are, respectively, $E(X_1)$ and $E(X_2)$; hence, we have for the sum of two random variables

$$E(Y) = a_1 E(X_1) + a_2 E(X_2) \tag{4.39}$$

and the corresponding variance is

$$
\begin{aligned}
\mathrm{Var}(Y) &= E[(a_1 X_1 + a_2 X_2) - (a_1 \mu_{X_1} + a_2 \mu_{X_2})]^2 \\
&= E[a_1(X_1 - \mu_{X_1}) + a_2(X_2 - \mu_{X_2})]^2 \\
&= E[a_1^2(X_1 - \mu_{X_1})^2 + 2a_1 a_2(X_1 - \mu_{X_1})(X_2 - \mu_{X_2}) + a_2^2(X_2 - \mu_{X_2})^2]
\end{aligned}
$$

We may recognize that the expected values of the first and third terms within the brackets are variances of X_1 and X_2, respectively, whereas the middle term is the covariance between X_1 and X_2. Hence, we have

$$\mathrm{Var}(Y) = a_1^2 \mathrm{Var}(X_1) + a_2^2 \mathrm{Var}(X_2) + 2a_1 a_2 \mathrm{Cov}(X_1, X_2) \tag{4.40}$$

Similarly, if $Y = a_1 X_1 - a_2 X_2$, the results would be

$$E(Y) = a_1 \mu_{X_1} - a_2 \mu_{X_2} \tag{4.41}$$

whereas the corresponding variance would be

$$\mathrm{Var}(Y) = a_1^2 \mathrm{Var}(X_1) + a_2^2 \mathrm{Var}(X_2) - 2a_1 a_2 \mathrm{Cov}(X_1, X_2) \tag{4.42}$$

If the variates X_1 and X_2 are statistically independent, $\mathrm{Cov}(X_1, X_2) = 0$; thus, Eqs. 4.40 and 4.42 become

$$\mathrm{Var}(Y) = a_1^2 \mathrm{Var}(X_1) + a_2^2 \mathrm{Var}(X_2)$$

The results we obtained above in Eqs. 4.39 through 4.42 can be extended to a general linear function of n random variables, such as

$$Y = \sum_{i=1}^{n} a_i X_i$$

in which the a_i's are constants. For this general case, we obtain the following: The mean value of Y,

$$E(Y) = \sum_{i=1}^{n} a_i E(X_i) = \sum_{i=1}^{n} a_i \mu_{X_i} \tag{4.43}$$

whereas the corresponding variances are

$$
\begin{aligned}
\mathrm{Var}(Y) &= \sum_{i=1}^{n} a_i^2 \mathrm{Var}(X_i) + \sum_{\substack{i,j=1 \\ i \neq j}}^{n} \sum a_i a_j \mathrm{Cov}(X_i, X_j) \\
&= \sum_{i=1}^{n} a_i^2 \sigma_{X_i}^2 + \sum_{\substack{i,j=1 \\ i \neq j}}^{n} \sum a_i a_j \rho_{ij} \sigma_{X_i} \sigma_{X_j}
\end{aligned} \tag{4.44}
$$

in which ρ_{ij} is the correlation coefficient between X_i and X_j, as defined in Eq. 3.72. Moreover, if we have another linear function of the X_i's,

$$Z = \sum_{i=1}^{n} b_i X_i$$

there may be a covariance between Y and Z, which is

$$\text{Cov}(Y, Z) = \sum_{i=1}^{n} a_i b_i \text{Var}(X_i) + \sum_{i, j=1\, i \neq j}^{n} \sum^{n} a_i b_j \text{Cov}(X_i, X_j)$$

$$= \sum_{i=1}^{n} a_i b_i \sigma_{X_i}^2 + \sum_{i, j=1\, i \neq j}^{n} \sum^{n} a_i b_j \rho_{ij} \sigma_{X_i} \sigma_{X_j}$$

(4.45)

▶ **EXAMPLE 4.23**

The maximum load on a column of a high-rise reinforced concrete building may be composed of the dead load (D), the live load (L), and the earthquake-induced load (E). The total maximum load carried by the column would be $T = D + L + E$. Suppose the statistics of the individual load components are as follows:

$$\mu_D = 2000 \text{ tons}, \quad \sigma_D = 210 \text{ tons}$$
$$\mu_L = 1500 \text{ tons}, \quad \sigma_L = 350 \text{ tons}$$
$$\mu_E = 2500 \text{ tons}, \quad \sigma_E = 450 \text{ tons}$$

If the three loads are statistically independent, i.e., $\rho_{ij} = 0$, the mean and variance of the total load T, according to Eqs. 4.43 and 4.44, are

$$\mu_T = 2000 + 1500 + 2500 = 6000 \text{ tons}$$

and

$$\sigma_T^2 = 210^2 + 350^2 + 450^2 = 369{,}100 \text{ tons}^2$$

Hence, the standard deviation is

$$\sigma_T = 607.5 \text{ tons}$$

However, the dead load, D, and the earthquake load, E, may be correlated, say with a correlation coefficient of $\rho_{ij} = 0.5$, whereas the live load L is uncorrelated with D and E. Then, the corresponding variance would be, according to Eq. 4.44,

$$\sigma_T^2 = 210^2 + 350^2 + 450^2 + 2(0.50)(210)(450) = 463{,}600 \text{ tons}^2$$

and the standard deviation becomes $\sigma_T = 680.88$ tons.

Now, suppose the mean and standard deviation of the load-carrying capacity of the column C are $\mu_C = 10{,}000$ tons and $\sigma_C = 1500$ tons. The probability that the column will be overloaded is then,

$$P(C < T) = P(C - T < 0)$$

But the mean value of $C - T = 10{,}000 - 6000 = 4000$ tons, and its standard deviation is

$$\sigma_{C-T} = \sqrt{607.5^2 + 1500^2} = 1618 \text{ tons}$$

Assuming that all the variables are Gaussian, and therefore, the difference $C - T$ is also Gaussian, the probability of overloading the column would be

$$P(C - T < 0) = \Phi\left(\frac{0 - 4000}{1618}\right) = \Phi(-2.47) = 1 - \Phi(2.47) = 1 - 0.9932 = 0.007$$ ◀

▶ **EXAMPLE 4.24**

In the event of an earthquake of intensity $I = 5$, the average economic losses in the central area of a city are estimated to be as follows, with respective standard deviations:

- Property losses—$\mu_P = \$2.5$ million; $\sigma_P = \$1.5$ million
- Loss of business—$\mu_B = \$6.0$ million; $\sigma_B = \$2.5$ million
- Cost of injuries—$\mu_J = \$4.0$ million; $\sigma_J = \$2.0$ million

Assume that the loss of lives is negligible, and among the three cost items above the loss of business is positively correlated with the property losses, with a correlation coefficient of 0.70, i.e., $\rho_{BP} = 0.70$, whereas the cost of injuries is uncorrelated with the other losses.

Assume also that each of the average earthquake losses indicated above varies with I^2, and the possible intensities in the area during an earthquake are $I = 4$, 5, and 6 with relative likelihoods of 2:1:1, respectively, whereas the standard deviation is invariant with I.

The total loss during an earthquake is

$$T = P + B + J$$

During an earthquake of intensity $I = 5$, the mean or expected total loss is

$$E(T|I = 5) = 2.5 + 6.0 + 4.0 = 12.5 \text{ million}$$

Since the expected total loss is conditional on I^2, we also have

$$E(T|I = 4) = 12.5\left(\frac{4}{5}\right)^2 = 8.00$$

$$E(T|I = 6) = 12.5\left(\frac{6}{5}\right)^2 = 18.00$$

The unconditional expected total loss during an earthquake is, therefore,

$$\mu_T = 8.0(2/4) + 12.5(1/4) + 18.0(1/4) = \$11.625 \text{ million}$$

The variance of the total loss is

$$\text{Var}(T) = 1.5^2 + 2.5^2 + 2.0^2 + 2(0.70)(1.5 \times 2.5) = 17.75$$

Hence, the standard deviation of the total loss is, $\sigma_T = \sqrt{17.75} = \$4.21 \text{ million}$ ◄

4.3.2 Mean and Variance of a General Function

For a general function of a single random variable X,

$$Y = g(X)$$

the exact moments of Y may be obtained as the mathematical expectations of $g(X)$; in particular, according to Eq. 3.9b, the mean and variance, respectively, would be

$$E(Y) = \int_{-\infty}^{\infty} g(X)f_X(x)\,dx$$

and

$$\text{Var}(Y) = \int_{-\infty}^{\infty} [g(x) - \mu_Y]^2 f_X(x)\,dx$$

Obviously, to obtain the mean and variance of the function Y with the above relations would require information on the PDF $f_X(x)$. In many applications, however, the PDF of X may not be available. In such cases, we seek approximate mean and variance of the function Y as follows.

We may expand the function $g(X)$ in a Taylor series about the mean value of X, that is,

$$g(X) = g(\mu_X) + (X - \mu_X)\frac{dg}{dX} + \frac{1}{2}(X - \mu_X)^2\frac{d^2g}{dX^2} + \cdots$$

where the derivatives are evaluated at μ_X.

Now, if we truncate the above series at the linear terms, i.e.,

$$g(X) \simeq g(\mu_X) + (X - \mu_X)\frac{dg}{dX}$$

we obtain the first-order approximate mean and variance of Y as

$$E(Y) \simeq g(\mu_X) \tag{4.46}$$

and

$$\text{Var}(Y) \simeq \text{Var}(X - \mu_X)\left(\frac{dg}{dX}\right)^2 = \text{Var}(X)\left(\frac{dg}{dX}\right)^2 \tag{4.47}$$

We should observe that if the function $g(X)$ is approximately linear (i.e., not highly nonlinear) for the entire range of X, Eqs. 4.46 and 4.47 should yield good approximations of the exact mean and variance of $g(X)$ (Hald, 1952). Moreover, when $\text{Var}(X)$ is small relative to $g(\mu_X)$, the above approximations should be adequate even when the function $g(X)$ is nonlinear.

The above first-order approximations may be successively improved by including the higher-order terms of the Taylor series expansion; for example, if we include the second-order term in the series, we can show that the corresponding *second-order* approximations are as follows:

$$E(Y) \simeq g(\mu_X) + \frac{1}{2}\text{Var}(X)\frac{d^2g}{dX^2} \tag{4.48}$$

and

$$\text{Var}(Y) \simeq \sigma_X^2\left(\frac{dg}{dX}\right)^2 - \frac{1}{4}\sigma_X^4\left(\frac{d^2g}{dX^2}\right)^2 + E(X - \mu_X)^3\frac{dg}{dX}\frac{d^2g}{dX^2} \\ + \frac{1}{4}E(X - \mu_X)^4\left(\frac{d^2g}{dX^2}\right)^2 \tag{4.49}$$

We observe that the second-order approximate variance of Y as indicated in Eq. 4.49 involves the third and fourth central moments of the original variate X, which are seldom evaluated in practical applications.

For practical purposes, the mean value is of first importance; thus, we may use the second-order mean of the function Y, Eq. 4.48, with its first-order variance of Eq. 4.47. In this way, we obtain an improved mean value of Y without involving more than the mean and variance of X.

▶ **EXAMPLE 4.25**

The maximum impact pressure (in psf) of ocean waves on coastal structures may be determined by

$$p_m = 2.7\frac{\rho K U^2}{D}$$

where U is the random horizontal velocity of the advancing wave, with a mean of 4.5 fps and a c.o.v. of 20%. The other parameters are all constants as follows:

$$\rho = 1.96 \text{ slugs/cu ft, the density of sea water}$$
$$K = \text{length of hypothetical piston}$$
$$D = \text{thickness of air cushion}$$

Assume a ratio of $K/D = 35$

The first-order mean and variance of p_m, according to Eqs. 4.46 and 4.47, are

$$E(p_m) \simeq 2.7(1.96)(35)(4.5)^2 = 3750.7 \text{ psf} = 26.05 \text{ psi}; \quad \text{and}$$

$$\text{Var}(p_m) \simeq \text{Var}(U)\left(2.7\rho\frac{K}{D}\right)^2 (2\mu_U)^2 = (0.20 \times 4.5)^2(2.7 \times 1.96 \times 35)^2 (2 \times 4.5)^2$$

Therefore, the standard deviation of p_m is

$$\sigma_{p_m} \simeq (0.20 \times 4.5)(2.7 \times 1.96 \times 35)(2 \times 4.5) = 1500.3 \text{ psf} = 10.42 \text{ psi}$$

For an improved mean value, we evaluate the second-order mean with Eq. 4.48 as follows:

$$E(Y) \simeq 3750.7 + \frac{1}{2}(0.20 \times 4.5)^2\left(2.7\rho\frac{K}{D}\right)(2)$$

$$= 3750.7 + \frac{1}{2}(0.20 \times 4.5)^2 (2.7 \times 1.96 \times 35 \times 2)$$

$$= 3750.7 + 150.0 = 3900.7 \text{ psf} = 27.09 \text{ psi}$$

This shows that for this case the first-order mean is about 4% less than the second-order mean. ◄

► **EXAMPLE 4.26**

Hypothetically, suppose the average number of commercial airplanes arriving over the Chicago O'Hare Airport during the peak hour from various major cities in the United States are as follows:

Originating City	Average No. of Arrivals	Standard Deviation
New York	10	4
Miami	6	2
Los Angeles	10	5
Washington, DC	12	4
San Francisco	8	4
Dallas	10	3
Seattle	5	2
Other U.S. cities	15	6

The total average number of arrivals is 76 planes, and the standard deviation of the total arrivals (assuming the arrivals from the different cities are statistically independent) is 11.22 planes.

Now, suppose that the holding time, T (in minutes), is an empirical function of the total arrivals as follows:

$$T = 3\sqrt{N_A}, \text{ in minutes}$$

in which N_A is the total number of arrivals during the peak hour. Then, by first-order approximation, the mean holding time would be

$$\mu_T \simeq 3\sqrt{\mu_{N_A}} = 3\sqrt{76} = 26.15 \text{ min}$$

and the variance is

$$\sigma_T^2 \simeq \sigma_{N_A}^2\left(\frac{3}{2}\mu_{N_A}^{-1/2}\right)^2 = (11.22)^2\left(\frac{3}{2}(76)^{-1/2}\right)^2 = 3.73$$

Therefore, the standard deviation is

$$\sigma_T = 1.93 \text{ min}$$

With Eq. 4.48, we obtain the corresponding second-order mean holding time:

$$\mu_T \simeq 3\sqrt{\mu_{N_A}} + \frac{1}{2}\sigma_{N_A}^2 \left(-\frac{3}{4}\mu_{N_A}^{-3/2}\right) = 26.15 - \frac{3}{8}(11.22)^2(76)^{-3/2} = 26.08 \text{ min}$$

In this case, the first-order mean is fairly accurate; it is almost the same as the second-order mean. ◀

Function of Multiple Random Variables—If Y is a function of several random variables,

$$Y = g(X_1, X_2, \ldots, X_n)$$

we obtain the approximate mean and variance of Y similarly as follows:

Expand the function $g(X_1, X_2, \ldots, X_n)$ in a Taylor series about the mean values $(\mu_{X_1}, \mu_{X_2}, \ldots, \mu_{X_n})$, yielding

$$Y = g(\mu_{X_1}, \mu_{X_2}, \ldots, \mu_{X_n}) + \sum_{i=1}^{n} (X_i - \mu_{X_i})\frac{\partial g}{\partial X_i}$$
$$+ \frac{1}{2}\sum_{i=1}^{n}\sum_{j=1}^{n}(X_i - \mu_{X_i})(X_j - \mu_{X_j})\frac{\partial^2 g}{\partial X_i \partial X_j} + \cdots$$

where the derivatives are all evaluated at $\mu_{X_1}, \mu_{X_2}, \ldots, \mu_{X_n}$.

If we truncate the above series at the linear terms, i.e.,

$$Y \simeq g(\mu_{X_1}, \mu_{X_2}, \ldots, \mu_{X_n}) + \sum_{i=1}^{n}(X_i - \mu_{X_i})\frac{\partial g}{\partial X_i}$$

We obtain the first-order mean and variance of Y, respectively, as follows:

$$E(Y) \simeq g(\mu_{X_1}, \mu_{X_2}, \ldots, \mu_{X_n}) \tag{4.50}$$

and

$$\text{Var}(Y) \simeq \sum_{i=1}^{n}\sigma_{X_i}^2\left(\frac{\partial g}{\partial X_i}\right)^2 + \sum_{\substack{i,j=1 \\ i\neq j}}^{n}\rho_{ij}\sigma_{X_i}\sigma_{X_j}\frac{\partial g}{\partial X_i}\frac{\partial g}{\partial X_j} \tag{4.51}$$

We observe that if X_i and X_j are uncorrelated (or statistically independent) for all i and j, i.e., $\rho_{ij} = 0$, then Eq. 4.51 becomes

$$\text{Var}(Y) \simeq \sum_{i=1}^{n}\sigma_{X_i}^2\left(\frac{\partial g}{\partial X_i}\right)^2 \tag{4.51a}$$

Equation 4.51 or 4.51a may be called the "propagation of uncertainty." We observe that it is a function of both the variances of the independent variables and of the sensitivity coefficients as represented by the partial derivatives.

The above first-order approximate mean and variance may also be improved by including the higher-order terms in the Taylor series expansion of $g(X_1, X_2, \ldots, X_n)$. In particular, the second-order approximate mean of Y would be

$$E(Y) \simeq g(\mu_{X_1}, \mu_{X_2}, \ldots, \mu_{X_n}) + \frac{1}{2}\sum_{i=1}^{n}\sum_{j=1}^{n}\rho_{ij}\sigma_{X_i}\sigma_{X_j}\left(\frac{\partial^2 g}{\partial X_i \partial X_j}\right) \tag{4.52}$$

where the derivatives are evaluated at the mean values of the X_i's. Again, if X_i and X_j are uncorrelated, Eq. 4.52 becomes

$$E(Y) \simeq g(\mu_{X_1}, \mu_{X_2}, \ldots, \mu_{X_n}) + \frac{1}{2} \sum_{i=1}^{n} \sigma_{X_i}^2 \left(\frac{\partial^2 g}{\partial X_i^2} \right) \qquad (4.52a)$$

▶ **EXAMPLE 4.27**

According to the Manning equation, the velocity of uniform flow, in fps, in an open channel is

$$V = \frac{1.49}{n} R^{2/3} S^{1/2}$$

where:

$S = $ slope of the energy line, in %

$R = $ the hydraulic radius, in ft

$n = $ the roughness coefficient of the channel

For a rectangular open channel with concrete surface, assume the following mean values and corresponding c.o.v.s:

Variable	Mean Value	c.o.v.
S	1%	0.10
R	2 ft	0.05
n	0.013	0.30

Assuming that the above random variables are statistically independent, the first-order mean and variance of the velocity V are, respectively,

$$\mu_V \simeq \frac{1.49}{0.013}(2)^{2/3}(1)^{1/2} = 182 \text{ fps}; \quad \text{and}$$

$$\sigma_V^2 \simeq \sigma_S^2 \left(\frac{1.49}{2\mu_n} \mu_R^{2/3} \mu_S^{-1/2} \right)^2 + \sigma_R^2 \left(\frac{2 \times 1.49}{3\mu_n} \mu_S^{1/2} \mu_R^{-1/3} \right)^2 + \sigma_n^2 \left(-1.49 \mu_R^{2/3} \mu_S^{1/2} \mu_n^{-2} \right)^2$$

$$= (0.10 \times 1)^2 \left(\frac{1.49}{2 \times 0.013}(2)^{2/3}(1)^{-1/2} \right)^2 + (0.05 \times 2)^2 \left(\frac{2 \times 1.49}{3 \times 0.013}(1)^{1/2}(2)^{-1/3} \right)^2$$

$$+ (0.30 \times 0.013)^2 \left(-1.49(2)^{2/3}(1)^{1/2}(0.013)^{-2} \right)^2 = 82.75 + 36.80 + 2979.2 = 3098.7$$

yielding the standard deviation

$$\sigma_V = 55.7 \text{ fps}$$

The corresponding second-order mean velocity would be, according to Eq. 4.52a,

$$\mu_V \simeq 182 + \frac{1}{2} \left[\sigma_S^2 \left(-\frac{1.49}{4\mu_n} \mu_R^{2/3} \mu_S^{-3/2} \right) + \sigma_R^2 \left(-\frac{2 \times 1.49}{9\mu_n} \mu_S^{1/2} \mu_R^{-4/3} \right) + \sigma_n^2 \left(\frac{2 \times 1.49}{\mu_n^3} \mu_R^{2/3} \mu_S^{1/2} \right) \right]$$

$$= 182 + \frac{1}{2} \left[\begin{array}{l} -(0.1)^2 \left(\frac{1.49}{4 \times 0.013}(2)^{2/3}(1)^{-3/2} \right) - (0.05 \times 2)^2 \left(\frac{2 \times 1.49}{9 \times 0.013} \right)(1)^{1/2}(2)^{-4/3} \\ +(0.30 \times 0.013)^2 \left(\frac{2 \times 1.49}{(0.013)^3}(2)^{2/3}(1)^{1/2} \right) \end{array} \right]$$

$$= 182 + \frac{1}{2}(-0.46 - 0.10 + 32.76) = 198.10 \text{ fps}$$

The first-order approximate mean velocity is about 8% lower than the corresponding second-order mean velocity. ◀

▶ **EXAMPLE 4.28**

The applied stress, S, in a beam is calculated as

$$S = \frac{M}{Z} + \frac{P}{A}$$

where:

$M = $ applied bending moment

$P = $ applied axial force

$A = $ cross-sectional area of the beam

$Z = $ section modulus of the beam

$M, Z,$ and P are random variables with respective means and c.o.v.s as follows:

$$\mu_M = 45{,}000 \text{ in--lb}; \qquad \delta_M = 0.10$$
$$\mu_Z = 100 \text{ in}^3; \qquad \delta_Z = 0.20$$
$$\mu_P = 5000 \text{ lb}; \qquad \delta_P = 0.10$$
$$A = 50 \text{ in}^2$$

Assume that M and P are correlated with a correlation coefficient of $\rho_{M,P} = 0.75$, whereas Z is statistically independent of M and P. We determine the mean and standard deviation of the applied stress S in the beam by first-order approximation as follows:

Mean value: $\quad \mu_S \simeq \dfrac{\mu_M}{\mu_Z} + \dfrac{\mu_P}{A} = \dfrac{45{,}000}{100} + \dfrac{5000}{50} = 550 \text{ psi}$

and variance:

$$\sigma_S^2 \simeq \sigma_M^2 \left(\frac{1}{\mu_Z}\right)^2 + \sigma_Z^2 \left(\frac{-\mu_M}{\mu_Z^2}\right)^2 + \sigma_P^2 \left(\frac{1}{A}\right)^2 + 2\rho_{M,P}\sigma_M\sigma_P\left(\frac{1}{\mu_Z}\right)\left(\frac{1}{A}\right)$$

$$= 4500^2\left(\frac{1}{100}\right)^2 + 20^2\left(\frac{-45{,}000}{100^2}\right)^2 + 500^2\left(\frac{1}{50}\right)^2 + 2 \times 0.75 \times 4500 \times 500\left(\frac{1}{100}\right)\left(\frac{1}{50}\right)$$

$$= 10{,}900.00$$

from which we obtain the standard deviation of S, $\sigma_S = 104.40$ psi.

Based on test data, the strength capacity of the beam, S_c, was estimated to have a mean strength of 800 psi and a standard deviation of 110 psi. Assuming that S and S_c are lognormal variates with the respective means and standard deviations determined above, we evaluate the parameters of the respective lognormal distributions as follows:

$$\zeta_S^2 = \ln\left(1 + \frac{104.40^2}{550^2}\right) = 0.0354; \qquad \zeta_{S_c}^2 = \left(\frac{110}{800}\right)^2 = (0.14)^2$$

and

$$\lambda_S = \ln 550 - \frac{1}{2}(0.0354) = 6.29; \qquad \lambda_{S_c} = \ln 800 - \frac{1}{2}(0.14)^2 = 6.67$$

The safety factor of the beam is defined as $\theta = S_c/S$. As S_c and S are both lognormal variates, the safety factor θ is also lognormal with the parameters

$$\lambda_\theta = \lambda_{S_c} - \lambda_S = 6.67 - 6.29 = 0.38$$

and

$$\zeta_\theta = \sqrt{\zeta_{S_c}^2 + \zeta_S^2} = \sqrt{(0.14)^2 + 0.0354} = 0.23$$

The beam will be overstressed when $\theta < 1.0$; therefore, the probability of this event is

$$P(\theta < 1.0) = \Phi\left(\frac{\ln 1.0 - \lambda_\theta}{\zeta_\theta}\right) = \Phi\left(\frac{0 - 0.38}{0.23}\right) = 1 - \Phi(1.65) = 1 - 0.950 = 0.050$$

That is, there is a 5% chance that the beam will be overstressed under the applied load. ◀

▶ **EXAMPLE 4.29**

A two-span bridge across a 300-m-wide river is to be constructed with a midspan pier about 150 m from one bank of the river. In order to locate the center position of the pier, a base line B is established along one bank as shown in Fig. E4.29. The position of the pier is to be determined by intersecting the lines of sight from Stations a and b, with the angle at Station a fixed at 90 degrees.

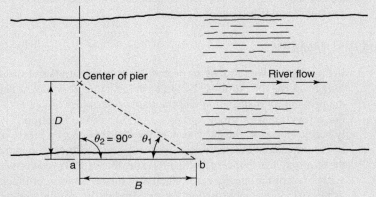

Figure E4.29 Locating position of midspan pier.

The proposed position of the center point of the pier is to be located 150 m from the base line, which has been measured to have a mean distance of $\overline{B} = 200$ m and a standard error of $\sigma_{\overline{B}} = 2$ cm. Also, the angle θ_1 measured from Station b has a mean measurement of $36°52'$ and a standard error of $2'$.

The first-order approximate mean and standard error of the required distance D from the base line to the pier location are as follows (in measurement theory, the standard deviation of the mean measurement is called the "standard error"):

$$D = \overline{B} \tan \overline{\theta}_1 = 200 \tan 36°52' = 149.98 \text{ m}$$

and

$$\sigma_D^2 = \sigma_{\overline{B}}^2 (\tan \overline{\theta}_1)^2 + (\overline{B} \sec^2 \overline{\theta}_1)^2 \sigma_{\theta_1}^2 = (0.02)^2 (\tan 36°52')^2 + (200 \sec^2 36°52')^2 (5.818 \times 10^{-4})^2$$
$$= 0.000225 + 0.033050 = 0.033275$$

Therefore, we find the standard error of the measured distance D to be

$$\sigma_D = 0.1824 \text{ m} = 18.24 \text{ cm}$$

whereas the second-order mean of the distance D is

$$D = 149.98 + \frac{1}{2}[\sigma_{\overline{B}}^2(0) + \sigma_{\overline{\theta}_1}^2(2\overline{B} \sec^2 \overline{\theta}_1 \tan \overline{\theta}_1)]$$
$$= 149.98 + 200(5.818 \times 10^{-4})^2 [(\sec^2 36°52')(\tan 36°52')]$$
$$= 149.98 + 0.00008 = 149.98 \text{ m}$$

Therefore, in this case, the second-order mean gives the same result as the first-order mean; i.e., the second-order approximation gives no improvement. ◀

Role of Monte Carlo Simulations—As we have observed in Sect. 4.2, the PDF of the function of a single variable can be derived analytically without much difficulty. However, in the case of a function of multiple random variables, the derivation of its probability distribution can be a formidable task, with some exceptions, such as the sum of multiple Gaussian variates or the product/quotient of multiple lognormal variates, as we saw in Sect. 4.2.2. Except for these special cases, we may have to resort to Monte Carlo simulations or other numerical methods to generate, approximately at least, the probability distribution of the function (see Chapter 5 for the basics of Monte Carlo simulations). Take, for example, the sum of two lognormal random variables; the distribution of the sum is neither normal nor lognormal. Also, the probability distribution of the product of two normal random variables is neither lognormal nor normal. In such cases, if information on the probability distribution of the function is required, a numerical method or Monte Carlo simulation is clearly necessary and provides a practical tool. The basics of Monte Carlo simulation are presented and illustrated in Chapter 5.

▸ 4.4 CONCLUDING SUMMARY

In this chapter, we saw that the probabilistic characteristics of a function of random variables may be derived from those of the basic constituent variates. These include, in particular, the probability distribution and the main descriptors (mean and variance) of the function. For a function of a single variable, the PDF of the function can be readily obtained analytically. However, the derivation of the distribution of a function of multiple variables can be complicated mathematically, especially for nonlinear functions. Therefore, even though the required distribution of a function may theoretically be derived, it is often impractical to apply, except for special cases, such as a linear function of independent Gaussian variates or the strictly product/quotient of independent lognormal variates. In this light, it is often necessary, in many applications, to describe the probabilistic characteristics of a function approximately in terms only of its mean and variance. Even then, the mean and variance of linear functions are easily amenable to exact evaluation; however, for a general nonlinear function, we must often resort to first-order (or second-order) approximations. Furthermore, when the probability distribution of a general function is required, we may need to resort to Monte Carlo simulations or other numerical methods.

▸ PROBLEMS

4.1 Suppose an engineering variable Y is an exponential function of a random variable X as follows:

$$Y = e^X$$

and X is normally distributed as N(2, 0.4). Derive the PDF of Y and show that it follows a lognormal distribution.

4.2 The absolute velocity (X) of particles in a gas follows a Maxwell distribution, with the PDF

$$f_X(x) = \begin{cases} \dfrac{4x^2}{a^3\sqrt{\pi}}\exp\left(-\dfrac{x^2}{a^2}\right) & \text{if } x > 0 \\ 0 & \text{otherwise} \end{cases}$$

where a is a constant. Determine the PDF $f_Y(y)$ for the particle kinetic energy $Y = \dfrac{1}{2}mX^2$, where m is the mass of a particle.

$$\text{Ans. } f_Y(y) = \begin{cases} \dfrac{4}{a^3}\sqrt{\dfrac{2y}{\pi m^3}}\exp\left(-\dfrac{2y}{ma^2}\right) & y \geq 0 \\ 0 & y < 0 \end{cases}$$

4.3 The sources of electrical power for a region are nuclear, fossil, and hydroelectric. The respective generating capacities of these sources can be described as independent Gaussian random variables as follows (in megawatt power):

Nuclear: N(100, 15)

Fossil: N(200, 40)

Hydro: N(400, 100)

(a) Determine the total power supply for the region; i.e., define the probability distribution of the power supply with the corresponding mean and standard deviation.

(b) The power demand of the region during normal weather is 400 megawatt, whereas during extreme weather the demand would be 600 megawatt. In any given year, normal weather is twice as likely to occur as extreme weather. With the generating capacity of (a), what is the probability that power shortage for the region (i.e., demand exceeding supply) will occur during the year?

(c) If power shortage should occur during a given year, what is the probability that it will occur during normal weather?

(d) Suppose that the three power sources are assigned to supply the following respective percentages of the total power demand:

Nuclear = 15% of demand; Fossil = 30% of demand; and Hydro = 55% of demand

During normal weather, what is the probability that at least one of the three sources will be unable to supply their respective allocations? Assume statistical independence.

4.4 Bob and John are traveling from city A to city D. Bob decides to take the upper route (through B), whereas John takes the lower route (through C) as shown in the following figure:

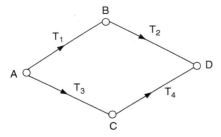

The travel times (in hours) between the cities indicated are normally distributed as follows:

$$T_1 \sim N(6, 2)$$
$$T_2 \sim N(4, 1)$$
$$T_3 \sim N(5, 3)$$
$$T_4 \sim N(4, 1)$$

Although the travel times can generally be assumed to be statistically independent, T_3 and T_4 are dependent with a correlation coefficient of 0.8.

(a) What is the probability that John will not arrive in city D within 10 hours? (*Ans. 0.397*)

(b) What is the probability that Bob will arrive in city D earlier than John by at least 1 hour? (*Ans. 0.327*)

(c) Which route (upper or lower) should be taken if one wishes to minimize the expected travel time from A to D? Justify. (*Ans. lower*)

4.5 A structure is resting on a soil stratum (see accompanying figure) in which the compressibility of each of the three layers of soil is well established. The settlement (in cm) of the building may be evaluated from

$$S = 0.3\,A + 0.2\,B + 0.1\,C$$

where A, B, and C are, respectively, the thickness (in m) of the three layers of soil. Suppose A, B, and C are modeled as independent normal random variables as follows:

$$A \sim N(5, 1)$$
$$B \sim N(8, 2)$$
$$C \sim N(7, 1)$$

(a) Determine the probability that the settlement will exceed 4 cm. (*Ans. 0.347*)

(b) If the total thickness of the three layers is known exactly as 20 m, and furthermore, thickness A and B are correlated with a correlation coefficient equal to 0.5, determine the probability that the settlement will exceed 4 cm. (*Ans. 0.282*)

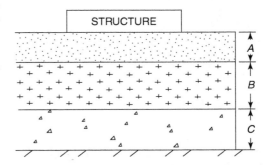

4.6 A friction pile is driven through three soil layers as shown in the figure below. The total bearing capacity of the friction pile (in tons) is obtained from

$$Q = 4A + B + 2C$$

where A, B, and C are penetration lengths (in meters) through each of the three soil layers, respectively.

Suppose A is N (5, 3); B is N (8, 2); A and B are negatively correlated with coefficient $\rho = -0.5$. The total length of the pile is 30 m. Determine the probability that the pile will fail to support the 40-ton load, i.e., the event that the capacity Q is less than 40 tons.

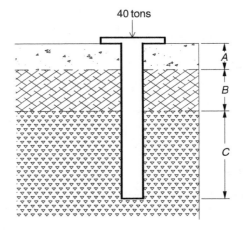

4.7 A city is located downstream of the confluence point of two rivers as shown below. The annual maximum flood peak in River-1 has an average of 35 m³/sec with the standard deviation of 10 m³/sec, whereas in River-2 the mean peak flow rate is 25 m³/sec and the standard deviation is 10 m³/sec. The annual maximum peak flow rates in both rivers are normally distributed with a correlation coefficient of 0.5. Presently, the channel which runs through the city can accommodate up to 100 m³/sec flow rate without flooding the city.

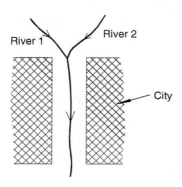

Answer the following questions.

(a) What are the mean and standard deviation of the annual maximum peak discharge passing through the city? (*Ans. 60 and 17.32 (m³/sec)*)

(b) What is the annual probability that the city will experience flooding based on the existing channel capacity? What is the corresponding return period? (*Ans. 0.0105; 96 years*)

(c) Calculate the probability that the city will experience flooding over a 10-year period. (*Ans. 0.10*)

(d) If the desired flooding probability over a 10-year period is to be reduced by half, by how much should the present channel capacity be enlarged? (*Ans. 104.5 m³/sec*)

4.8 Fibers may be embedded in concrete to increase its strength. Consider a cracked section as shown in the following diagram:

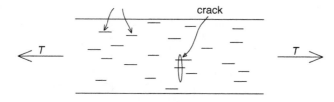

Suppose the total strength T of the cracked section is given by the following expression:

$$T = C + (F_1 + F_2 + \cdots + F_N)$$

where C is the strength contributed by the cement; F_1, F_2, etc. are the strength of each of the fibers across the crack; and N is

the total number of fibers across the crack. Suppose $C = N(30, 5)$; each F_i is $N(5, 3)$ and N is a discrete random variable with the following PMF.

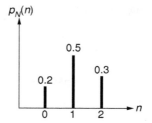

Assume C and the F_i's are statistically independent. What is the probability that the total strength T will be less than 30?

4.9 A shuttle bus operates from a shopping center, travels to Towns A and B sequentially, and then returns to the shopping center as shown in the figure below.

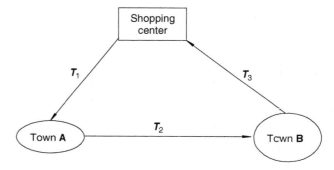

Assume that the respective travel times (under normal traffic) are independent Gaussian random variables with the following statistics:

Travel Time	Mean (min)	c.o.v.
T_1	30	0.30
T_2	20	0.20
T_3	40	0.30

However, during rush hours (8:00–10:00 AM and 4:00–6:00 PM), the mean travel time between Town A and Town B will increase by 50% with the same c.o.v.

(a) The scheduled time for one round trip of the shuttle bus is 2 hours. Determine the probability that a round trip will not be completed on schedule under normal traffic.

(b) What is the probability that a passenger, originating in Town A, will arrive at the shopping center within 1 hour under normal traffic?

(c) The number of passengers originating from Town B is twice that of the number originating from Town A. What percentage

of the passengers will arrive at the shopping center during rush hours in less than 1 hour?
(d) A passenger starting from Town B has an appointment at 3:00 PM at the shopping center. If the bus left Town B at 2:00 PM but has not arrived at the shopping center at 2:45 PM, what is the probability that he or she will arrive on time for his/her appointment?

4.10 The settlement of each footing shown in the figure below follows a normal distribution with a mean of 2 in. and a coefficient of variation of 30%. Suppose the settlements between two adjacent footings are correlated with a correlation coefficient of 0.7; that is, the differential settlement is

$$D = |S_1 - S_2|$$

where S_1 and S_2 are the settlements of footings 1 and 2, respectively.
(a) Determine the *mean* and *variance* of D.
(b) What is the probability that the magnitude of the differential settlement, i.e. $|D|$, will be less than 0.5 in.?

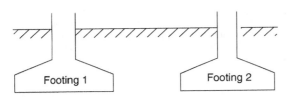

4.11 The system of pipes shown in the figure below is supposed to carry the storm runoffs X_1 and X_2, and the municipal waste water, X_3. The statistics of the flow rates (all in units of cubic feet per second, cfs), are as follows:

	Mean	c.o.v.	Distribution
X_1	10	0.30	Normal
X_2	15	0.20	Normal
X_3	20	0.00	—

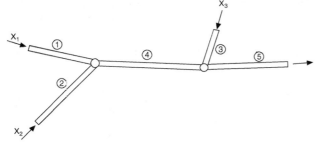

Because of proximity, X_1 and X_2 are dependent. Assume that the coefficient of correlation $\rho_{1,2} = 0.6$.
(a) Determine the mean value and standard deviation of the total rate of inflow to pipe 5.
(b) What is the probability that during a 1-minute interval, the volume of water flowing into pipe 2 exceeds that into pipe 1 by

at least 400 ft³? (*Hint:* assume the inflow rate into each pipe is constant during that minute.)
(c) Suppose the municipal waste water is projected to increase at a rate of 3 cfs per year, and pipe 5 has a capacity of 70 cfs. If the design criterion is that the probability of overflow at pipe 5 (total inflow exceeds capacity) after a storm should be less than 0.05, how many years will the current pipe 5 remain adequate, i.e., before a larger size pipe is needed?

4.12 The flows in two tributaries 1 and 2 combine to form stream 3 as shown in the figure below. Suppose the concentration of pollutants in tributary 1 is X, which is normally distributed as N(20, 4) parts per unit volume (puv), whereas that in tributary 2 is Y which is N(15, 3) puv. The flow rate in tributary 1 is 600 cubic feet per sec (cfs) and that in 2 is 400 cfs. Hence, the concentration of pollutants in the stream is

$$Z = \frac{600X + 400Y}{600 + 400} = 0.6X + 0.4Y$$

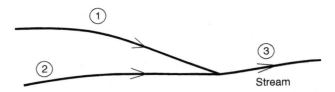

Pollution occurs if pollutant concentration exceeds 20 puv.
(a) What is the probability that *at least one* of the two tributaries will be polluted?
(b) Assume X and Y are statistically independent. Determine the probability of a polluted stream.
(c) Suppose the same industrial plant dumps the pollutants into the two tributaries; hence X and Y are correlated, say with $\rho = 0.8$. What is the probability of a polluted stream?

4.13 A contractor submits bids to three highway jobs and two building jobs. The probability of winning each job is 0.6. The profit from a highway job is normally distributed as N(100, 40) in thousand dollars, whereas that for a building job is N(80, 20) in thousand dollars. Assume that winning each job is an independent event and the profits between jobs are also statistically independent.
(a) What is the probability that the contractor will win at least two jobs?
(b) What is the probability that he will win exactly one highway job but none of the building jobs?
(c) Suppose he has won two highway jobs and one building job. What is the probability that the *total* profit will exceed 300 thousand dollars?
(d) How would the answer in part (c) change if the profit from the two highway jobs are correlated with a correlation coefficient of 0.8?

4.14 During the next month, the amount of water available for a city is lognormally distributed with a mean of 1 million gallons and a c.o.v. of 40%, whereas the total demand is expected to follow a lognormal distribution with a mean of 1.5 million gallons with a c.o.v. of 10%. What is the probability of a water shortage in this city over the next month?

4.15 A large parabolic antenna is designed against wind load. During a wind storm, the maximum wind-induced pressure on the antenna, P, is computed as

$$P = \frac{1}{2}CRV^2$$

where C = drag coefficient; R = air mass density in slugs/ft^3; V = maximum wind speed in ft/sec; and P = pressure in lb/ft^2. C, R, and V are statistically independent lognormal variates with the following respective means and c.o.v.'s:

$$\mu_C = 1.80; \qquad\qquad \delta_C = 0.20$$
$$\mu_R = 2.3 \times 10^{-3} \text{ slugs/ft}^3; \qquad \delta_R = 0.10$$
$$\mu_V = 120 \text{ ft/sec}; \qquad\qquad \delta_V = 0.45$$

(a) Determine the probability distribution of the maximum wind pressure P and evaluate its parameters.
(b) What is the probability that the maximum wind pressure will exceed 30 lb/ft^2?
(c) The actual wind resistance capacity of the antenna is also a lognormal random variable with a mean of 90 lb/ft^2 and a c.o.v. of 0.15. Failure in the antenna will occur whenever the maximum applied wind pressure exceeds its wind resistance capacity. During a wind storm, what is the probability of failure of the antenna?
(d) If the occurrences of wind storms in (c) constitute a Poisson process with a mean occurrence rate of once every 5 years, what is the probability of failure of the antenna in 25 years?
(e) Suppose five antennas were built and installed in a given region. What is the probability that at least two of the five antennas will not fail in 25 years? Assume that failures between antennas are statistically independent.

4.16 A pile is designed to have a mean capacity of 20 tons; however, because of variabilities, the pile capacity is lognormally distributed with a coefficient of variation (c.o.v.) of 20%. Suppose the pile is subject to a maximum lifetime load that is also lognormally distributed with a mean of 10 tons and a c.o.v. of 30%. Assume that pile load and capacity are statistically independent.
(a) Determine the probability of failure of the pile.
(b) A number of piles may be tied together to form a pile group to resist an external load. Suppose the capacity of the pile group is the sum of the capacities of the individual piles. Consider a pile group that consists of two single piles as described above. Because of proximity, the capacities between the two piles are correlated with a correlation coefficient of 0.8. Let T denote the capacity of this pile group.
  (i) Determine the mean value and c.o.v. of T.

(ii) Will the random variable T follow a normal, or lognormal, or some other probability distribution? Provide an explanation.

4.17 A corner column of a building is supported on a pile group consisting of two piles. The capacity of each pile is the sum of two independent contributions, namely, the frictional resistance, F, along the pile length and the end bearing resistance, B, at the pile tip. Suppose F and B are both normally distributed with mean values of 20 and 30 tons and coefficients of variation of 0.2 and 0.3, respectively. The capacities between the two piles are correlated with a correlation coefficient of $\rho = 0.8$.
(a) Determine the mean and c.o.v. of the capacity of one pile.
(b) Determine the mean and c.o.v. of the total capacity of the pile group.
(c) If the maximum load applied on the pile group is also normally distributed with a mean value of 50 tons and a c.o.v. of 0.3, what is the probability of failure of this pile group?

4.18 Consider a tower that has leaned an inclination of 18°. The tower continues to lean at an annual increment A, which is normally distributed with a mean value of 0.1° and a c.o.v. of 30%.
(a) Assume that the incremental amounts of additional leaning between years are statistically independent. Determine the probability that the final inclination of the tower after 16 years will exceed 20°.
(b) Suppose it is believed that the maximum inclination, M, before the tower collapses (i.e., the leaning capacity of the tower) is itself a normal random variable with a mean of 20° and a c.o.v. of 1%.
  (i) What is the probability that the tower will not collapse within the next 16 years?
  (ii) An alternative assumption, instead of statistical independence between years as in part (a), is that the tower will have the same incremental leaning in each year in the future. What then is the probability that the tower will not collapse after 16 years?

4.19 Five cylindrical tanks are used to store oil; each of the tanks is shown below. The weight, W, of each tank and its contents is 200 kips. When subjected to an earthquake, the horizontal inertial force may be calculated as

$$F = \frac{W}{g}a$$

where:

g = acceleration of gravity = 32.2 ft/sec^2

a = maximum horizontal acceleration of the earthquake ground motion

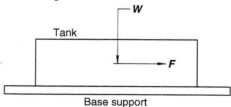

During an earthquake, the frictional force at the base of a tank keeps it from sliding. The coefficient of friction between the tank and the base support is a lognormal variate with a median value of 0.40 and a c.o.v. of 0.20. Also, during an earthquake assume that the maximum ground acceleration is a lognormal variate with a mean of 0.30g and a c.o.v. of 0.25.
(a) What is the probability that during an earthquake a tank will slide from its base support?
(b) Of the five tanks, what is the probability that none of them will slide during an earthquake, assuming that the conditions between the tanks are statistically independent?

4.20 The time T in minutes for each car to clear the toll station at Lion Rock Tunnel is exponentially distributed with a mean value of 5 sec. What is the probability that a line of 50 cars waiting to pay toll at this tunnel can be completely served in less than 3.5 min?

4.21 Consider a long cantilever structure shown in the figure below, which is loaded by a force F at a distance X from the fixed end A. Suppose F and X are independent lognormal random variables with mean values of 0.2 kips and 10 ft., respectively. The c.o.v.s are 20% and 30%, respectively.
(a) Determine the probability that the induced bending moment at A will exceed 3 ft–kip. (*Ans. 0.093*)
(b) Suppose there are 50 forces, each acting at a random location along the beam. Each force has a lognormal distribution with a mean of 0.2 kip and c.o.v. of 20%, and the location of each force is also lognormal with a mean of 10 ft. from A and a c.o.v. of 30%. Assume statistical independence between individual values of the 50 forces and also between individual locations of the 50 forces. Determine the probability that the overall induced bending moment at A will exceed 120 ft–kip. (*Ans. 5.5 × 10⁻⁵*)

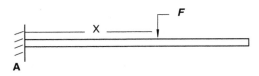

4.22 The salaries of Assistant Engineers (AEs) at a large engineering firm are uniformly distributed between $30,000 and $50,000 per year.
(a) What is the probability that a randomly selected AE at this firm will have an annual salary exceeding $40,000?
(b) If 50 AEs are selected at random from this firm, find the probability that their average salary will exceed $40,000.
(c) Why do the answers from (a) and (b) differ significantly? Elaborate.

4.23 Suppose the traffic on a long-span bridge consists of two types of vehicles: cars and trucks. The weight of each car has a mean of 4 kips and standard deviation of 1 kip, whereas the weight of a truck has a mean of 20 kips and standard deviation of 5 kips. Assume that the weight of each vehicle is lognormally distributed and the weights between vehicles are statistically independent. There are currently 100 cars and 30 trucks on the bridge.
(a) Determine the mean and c.o.v. of the total vehicle weight on the bridge.
(b) Estimate the probability that the total vehicle weight exceeds 1200 kips. State any assumption that you use.
(c) Suppose the total dead load of the bridge is normally distributed with a mean of 1200 kips and a c.o.v. of 10%.
 (i) What is the probability that the total vehicle load will exceed the total dead load?
 (ii) What is the probability that the sum of total dead and vehicle load will exceed 2500 kips?

4.24 "5-star" brand cement is shipped in batches, in which each batch contains 40 bags. Previous records show that the weight of a randomly selected bag of this brand of cement has a mean of 2.5 kg and a standard deviation of 0.1 kg, but its exact PDF is unknown.
(a) What is the mean weight of one batch of 5-star brand cement?
(b) Suppose the shipping company charges a penalty if a batch exceeds its mean weight by more than 1 kg. What is the probability that a batch of 5-star brand cement will receive a penalty?
(c) Suppose the standard deviation of each bag is changed to 1 kg but all other parameters remain the same. What is the probability of a penalty now? Comment on whether or not the increased standard deviation is desirable.

4.25 From an extensive survey of live floor loads of office buildings, it was determined that the sustained EUDL (equivalent uniformly distributed load) intensity may be modeled with a lognormal distribution with a mean EUDL of 12 psf and a c.o.v. of 30%. Over the life of a building, the live floor load may fluctuate because of changes in occupancy and use of the floor space.
(a) Assuming an average rate of occupancy change of once every 2 years, determine the exact distribution of the lifetime maximum live load EUDL for office buildings over a life of 50 years.
(b) Develop the corresponding asymptotic form for the distribution of the lifetime maximum EUDL, and evaluate the parameters for the 50-year life.

4.26 The PDF of an initial variate with a gamma distribution is

$$f_X(x) = \frac{v(vx)^{k-1}}{\Gamma(k)}e^{-vx}; \qquad x \geq 0; \qquad k > 1$$

where v and k are the parameters and $\Gamma(k)$ is the gamma function.
(a) Determine the CDF and PDF of the largest value from a sample of size n.
(b) Determine the appropriate asymptotic form for the distribution of the largest value as $n \to \infty$.

4.27 Axle loads of trailer trucks higher than 18 tons may be modeled with a shifted exponential distribution as follows:

$$f_X(x) = \frac{1}{3.2}e^{-(x-18)/3.2}; \qquad x \geq 0$$

Suppose that a culvert is subjected to 1355 of the above axle loads (> 18 tons) in a year. Assume that the axle loads are statistically independent and that the average traffic volume will remain constant over the years.

(a) Determine the mean and c.o.v. of the maximum axle load on the culvert over periods of 1, 5, 10, and 25 years.

(b) For a culvert life of 20 years, determine the probability that it will be subjected to an axle load of over 80 tons.

(c) Determine the "design axle load" corresponding to an allowable exceedance probability of 10% over the life of 20 years.

4.28 The daily level of dissolved oxygen (DO) concentration in a stream is assumed to be Gaussian distributed with a mean of 3.00 mg/1 and a standard deviation of 0.50 mg/l. The DO concentrations between days may be assumed to be statistically independent.

(a) Determine the asymptotic distribution of the minimum daily DO in a month (30 days); also in a year (365 days).

(b) Evaluate the probability that the daily DO concentration will be less than 0.5 mg/l in a month; in a year.

(c) Determine the mean and c.o.v. of the minimum daily DO concentration over a period of a month; over a period of a year.

4.29 A structure is designed with a capacity to withstand a wind velocity of 150 kph; i.e., any wind velocity up to 150 kph will not cause damage to the structure. In a hurricane-prone area, the maximum wind velocity during a hurricane may be modeled as an extreme *Type I* asymptotic distribution with a mean velocity of 100 kph and a c.o.v. of 0.45.

(a) For a structure with the above wind-resistant capacity, what would be the probability of wind damage during a hurricane?

(b) If the permissible damage probability were to be reduced to 1/10 of the original design, as described in (a) above, what should be the wind-resistant capacity of the revised design (in kph)?

(c) If the occurrence of hurricanes in the area is modeled as a Poisson process with a return period of 200 years, what is the probability of damage to the original structure over a life of 100 years? Also, what would be the corresponding damage probability to the revised structure? Assume that damages between hurricanes are statistically independent.

(d) Suppose that three structures with the original design were built in the area, each with the same wind damage probability. What is the probability that at least one of the structures will be damaged over the life of 100 years?

4.30 The wind gust at a building site is of concern during the erection of a steel building. Suppose the daily maximum wind velocity at a particular site can be modeled as a Gaussian random variable with a mean velocity of 40 mph and a c.o.v. of 25%.

(a) If the erection of a building can be completed in a period of 3 months, what is the most probable maximum wind velocity that can be expected during the erection of the building?

(b) If the shoring system used in the erection can withstand a wind gust velocity of 70 mph, what is the probability that the shoring system will be inadequate to withstand the maximum wind velocity during the erection of the building?

(c) What would be the mean and c.o.v. of the maximum wind velocity during the erection of the building?

4.31 In Problem 4.30, suppose the daily maximum wind at the building site is lognormally distributed (instead of normally) with the same mean of 40 mph and c.o.v. of 25%. In this case, if the erection of the building requires 3 months to complete:

(a) What is the most probable maximum wind velocity that can be expected during the erection of the building?

(b) What would be the probability that the shoring system will be inadequate to withstand the maximum wind velocity during the erection of the building? The capacity of the shoring system is 70 mph.

4.32 The hydraulic head loss in a pipe may be determined by the Darcy–Weisbach equation as follows:

$$h = \frac{fLV^2}{2Dg}$$

where:

L = length of a pipe

V = flow velocity of water in a pipe

D = pipe diameter

f = coefficient of friction

g = gravitational acceleration = 32.2 ft/sec^2

Suppose a pipe has the following properties:

Variable	Mean Value	c.o.v.
L	100 ft	0.05
D	12 in.	0.15
f	0.02	0.25
V	9.0 fps	0.2

(a) By first-order approximation, determine the mean and standard deviation of the hydraulic head loss of the pipe.

(b) Evaluate also the corresponding mean head loss by second-order approximation.

4.33 By simple beam theory, the maximum deflection of a prismatic cantilever beam under a concentrated load P applied at the end of the beam is given by

$$D = \frac{PL^3}{3EI}$$

where:

L = length of the beam

E = modulus of elasticity of the material

I = moment of inertia of the cross section of the beam

Suppose a 15-ft rectangular wood beam with nominal cross-sectional dimensions of 6″ × 12″ (actual mean dimensions are

5.5″ × 11.5″) is subjected to a concentrated load P with a mean of 500 lb and a c.o.v. of 20%. The modulus of elasticity E of the wood has a mean value of 3,000,000 psi and a c.o.v. of 0.25. We may assume that the c.o.v. of each cross-sectional dimension is 0.05. The variables P, E, and I are statistically independent.
(a) Determine the first-order mean and standard deviation of the end deflection D of the cantilever wood beam. Observe that the moment of inertia of a rectangular cross section is $I = \frac{1}{12}bh^3$, in which b and h are correlated with $\rho_{bh} = 0.80$
(b) Evaluate also the mean deflection of the beam by second-order approximation.

4.34 The following is an engineering formula, for Z:

$$Z = X\,Y^2\,W^{1/2}$$

where:

    X: uniformly distributed between 2 and 4

    Y: beta distribution between 0 and 3 with parameters $q=1$ and $r=2$

    W: exponentially distributed with a median of 1

and X, Y, and W are statistically independent.
(a) Determine the mean values and variances of X, Y, and W, respectively. (*Ans. 3, 1/3; 1, 1/2; 1.443, 2.081*)
(b) Estimate the mean, variance, and c.o.v. of Z using the first-order approximation. (*Ans. 3.6, 29.7, 1.51*)

4.35 A clay liner is constructed to reduce the amount of pollutants migrating from the landfill to the surrounding environment, as shown in the figure below.

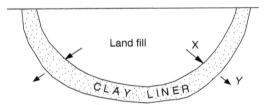

Suppose X is the flow rate of pollutants from the landfill without the installation of the clay liner; L is the effectiveness of the clay liner, such that the flow rate of pollutants ejected to the environment is given by

$$Y = (1 - L)\,X$$

X follows a lognormal distribution with a mean of 100 units per year and a c.o.v. of 20%. Acceptable performance requires that Y should not exceed 8 units per year.
(a) Suppose the quality of construction of the clay liner can be assured to have an effectiveness of 0.95. What is the probability of acceptable performance? (*Ans. 0.993*)
(b) Suppose the contractor cannot guarantee the effectiveness to be 0.95. Instead, he offers to provide a liner with a mean effectiveness of 0.96 but with a standard deviation of 0.01.

(i) Estimate the mean and variance of Y using the first-order approximate method. (*Ans. 4, 1.64*)
(ii) Determine if this offer from the contractor will yield a larger probability of acceptable performance. You may assume Y to be lognormal with values of mean and variances estimated in part (i).

4.36 Manning's formula frequently used for determining the flow capacity of a circular storm sewer is

$$Q_c = 0.463\,n^{-1}D^{2.67}S^{0.5}$$

in which Q_c is the flow rate (in ft³/sec); n is the roughness coefficient; D is the sewer diameter (in ft.); and S is the pipe slope (ft/ft).

Due to manufacturing imprecision and construction error, the sewer diameter and pipe slope, together with the roughness coefficient, are subject to uncertainty. Consider a section of sewer in a storm sewer system. The statistical properties of n, D, and S in the above Manning's formula are given in the following table.

Variable	Mean	Coefficient of Variation
n	0.015	0.10
D (ft.)	3.0	0.02
S (ft/ft)	0.005	0.05

Suppose that all three random variables are statistically independent.
(a) Using the first-order approximate formula, estimate the mean and variance of the flow carrying capacity of the sewer.
(b) Calculate the percentage of contribution of each random variable to the total uncertainty of the sewer flow capacity and identify the relative importance of each random variable. (*Ans. n: 74.2%, D: 21.2%; S: 4.6%*)
(c) Determine also the second-order mean of the sewer flow capacity.
(d) Assuming that the sewer flow capacity follows the lognormal distribution with the mean and variance obtained from part (a), determine the reliability that the sewer capacity can accommodate an inflow of 30 ft³/sec. (*Ans. 0.996*)

4.37 The amount of rainfall R in each rainstorm is independently lognormally distributed with a mean of 1 in. and a c.o.v. of 100%.
(a) Suppose 50 rainstorms have occurred last year, and other contributions to rainfall are negligible. Estimate the probability that last year's annual rainfall exceeded 60 in.?
(b) Suppose the amount of water collected from a single rainstorm is predicted by the following equation:

$$V = NR\ln(R + 1)$$

where V is the volume of water in thousand gallons and N is a random variable (independent of R), with a mean value of 1 and a c.o.v. of 50% representing the error of the prediction model.

(i) Determine the mean and c.o.v. of V using the first-order approximate formulas.

(ii) Are the approximations good in this case? Why?

(iii) Evaluate also the second-order mean of V.

4.38 The natural frequency, ω, of a single-degree-of-freedom spring-mass mechanical system is

$$\omega = \sqrt{\frac{K}{M}}$$

where K is the stiffness and M is the mass of the system.

(a) If the mean value and standard deviation of K are, respectively, 400 and 200, and $M = 100$, determine the first-order approximate mean value and standard deviation of the natural frequency of the system.

(b) Now, if M is also a random variable with a standard deviation of 20, determine the corresponding mean and standard deviation of the natural frequency.

(c) Determine the second-order approximate mean value of the natural frequency.

▷ REFERENCES

Ang, A. H-S., and W.H., Tang. *Probability Concepts in Engineering Planning and Design*, Vol. II, Decision, Risk and Reliability, John Wiley & Sons, Inc., New York, 1984.

Fisher, R.A., and L.H.C., Tippett. "Limiting Forms of the Frequency Distribution of the Largest or Smallest Number of a Sample," *Proc. Cambridge Philosophical Soc.*, XXIV, Part II, 1928, pp. 180–190.

Gumbel, E., *Statistics of Extremes*, Columbia University Press, New York, 1958.

Hald, A., *Statistical Theory and Engineering Applications*, John Wiley and Sons, Inc., New York, 1952.

Weibull, W., "A Statistical Distribution of Wide Applicability," *Journal of Applied Mechanics*, ASME, Vol. 18, 1951.

CHAPTER 5

Computer-Based Numerical and Simulation Methods in Probability

▶ 5.1 INTRODUCTION

The main thrust of this chapter is to present numerical and simulation methods for solving probabilistic problems that are otherwise difficult or practically impossible by analytical methods. These numerical methods must necessarily rely on high-speed computers for their applications. The objective here is to expand the usefulness and utility of probabilistic modeling of engineering problems containing uncertainties. As in the previous three chapters, Chapters 2 through 4, the probability concepts are also illustrated with numerical examples in the present chapter. However, the pertinent solutions would now require computer programs that may be coded using commercial softwares (such as *MATLAB*); also, there are no preprogrammed software packages that can be used for the required solutions. Some of the coded programs for solving the example problems may not be optimal but are intended only to be illustrative.

In many practical engineering situations, the problems (even deterministic ones) may be complicated and not amenable to analytic solutions. In such situations, numerical methods are necessary and often provide the only practical and effective approach. When the problems involve random variables, or require consideration of probability, the numerical process may include repeated simulations through Monte Carlo sampling techniques. When we consider problems involving aleatory and epistemic uncertainties separately, there is even greater need for Monte Carlo simulations.

For example, when the distributions of the individual basic variables are not all Gaussian, the probability distribution of the sum of two or more basic variables may not be Gaussian as we can observe from Sect. 4.2. Similarly, in the case of products or quotients of several random variables, unless all the random variables are lognormally distributed, the probability distribution of the function (product or quotient) may not be lognormal, as in Sect. 4.2. Moreover, when the function is a nonlinear function of several variates, regardless of distributions, the distribution of the function is often difficult or practically impossible to determine analytically. In these, and in other cases, we have to resort to the application of numerical methods to obtain practical solutions, even if approximations are involved.

We must emphasize that the purpose of this chapter is to illustrate the solution process of available numerical tools to enhance the application of probabilistic and statistical models in engineering. The descriptions of the numerical tools, which are generally well established, however, are beyond the scope of the chapter.

▶ 5.2 NUMERICAL AND SIMULATION METHODS

As we indicated above, numerical solutions to engineering problems are often required (even if approximations are involved); these include the use of numerical quadrature for integrating a complex function and the use of finite elements or finite differences to approximately discretize a continuous system. When random variables are involved, such numerical methods are often also necessary to obtain solutions to the probabilistic problems. These would include Monte Carlo simulation, which is an effective tool for complementing the probabilistic analytical methods and often may be the only practical means of finding a solution to a complex probabilistic problem. For specialized applications, there are also approximate numerical methods for finding solutions to a probabilistic problem, e.g., the methods of FORM (Cornell, 1969) and SORM (Der Kiureghian et al., 1987) for the reliability analysis of engineering systems. In general, however, Monte Carlo simulation is the numerical process that has wide applicability to problems involving probability.

Monte Carlo simulation (MCS) is a numerical process of repeatedly calculating a mathematical or empirical operator in which the variables within the operator are random or contain uncertainty with prescribed probability distributions. The numerical result from each repetition of the numerical process may be considered a sample of the true solution, analogous to an observed sample from a physical experiment. When random variables are involved, the values of the different variables are sampled from the respective probability distributions in each repetition. An essential component of the process, therefore, is the generation of the values of the random variables, each with its prescribed probability distribution, known as *random number generators* (see, for example, Rubinstein, 1981; Ang and Tang, 1984). Only the basic Monte Carlo sampling technique is used and illustrated here. More sophisticated Monte Carlo sampling methods, such as variance-reduction and the "smart" Monte Carlo methods, are available. However, they are beyond the scope of this book.

Monte Carlo simulation and numerical integration methods are practical tools for solving probability problems only with the availability of computers. The availability of commercial software such as *MATLAB* (2004), *MATHCAD* (2002), and *MATHEMATICA* (2002) should facilitate the application of these numerical methods. Also, MS Excel + *VISUAL BASIC* (e.g., Halverson, 2003; MacDonald, 2003) or similar spreadsheet software may be used for certain problems. The examples illustrated in this chapter were solved by Monte Carlo simulation or numerical procedures using *MATLAB* or *MATHCAD*. However, any of the other software mentioned above may be equally effective for performing the required numerical process; the choice of the software will depend on the familiarity and preference of the user or reader.

It should be emphasized that the numerical solution obtained by one repetition (or run) of a Monte Carlo simulation, with a given sample size, may be slightly different when repeated by another repetition of the same sample size; i.e., the numerical results obtained from different runs of the same simulation for a particular problem may not be exactly the same unless very large sample size is used in each run.

5.2.1 Essentials of Monte Carlo Simulation

A Monte Carlo simulation starts with the generation of random numbers with respective prescribed probability distributions. Methods for generating a set of random numbers with well-known distributions are widely available (e.g., Rubinstein, 1981; Ang and Tang, 1984).

Also, random number generators for a number of distributions are available in *MATLAB* with its companion Statistics Toolbox; *MATHCAD* and *MATHEMATICA* have similar random number generators for essentially the same distributions.

In any Monte Carlo simulation, the sample size, n, or number of random numbers generated for each distribution, must be prescribed. The accuracy of the solution obtained through Monte Carlo simulation will improve with the sample size. This accuracy is often important and may be measured by the coefficient of variation (c.o.v.) of the solution.

Later, in Eq. 6.27 of Chapter 6, we shall show that in estimating the probability p with a sample of size n, the sample mean $\overline{P}$ is an unbiased estimate of the probability p, and the sample variance of $\overline{P}$ is given by Eq. 6.28a as

$$\sigma_{\overline{p}}^2 = \frac{\overline{P}\left(1 - \overline{P}\right)}{n}$$

Therefore, the c.o.v. of $\overline{P}$ estimated with a sample of size n in a Monte Carlo simulation is

$$\text{c.o.v.}\,(\overline{P}) = \sqrt{\frac{(1 - \overline{P})}{n\overline{P}}} \tag{5.1}$$

Shooman (1968) has shown, based on an approximation, the same result in determining the probability of system performance through MCS.

To obtain the actual c.o.v. of a Monte Carlo solution, we have to repeatedly perform the same simulation process, say with N repeated simulations, each with the same sample size n, and obtain the mean and standard deviation of the N simulations from which the c.o.v. of the solution can be estimated. This is illustrated specifically in Example 5.5.

Using Eq. 5.1, we can also evaluate the error (in percent) of a Monte Carlo solution with a given sample size n,

$$\text{error (in \%)} = 200\sqrt{\frac{1 - \overline{P}}{n\overline{P}}} \tag{5.2}$$

Conversely, Eq. 5.2 may be used to determine the sample size, n, required in a Monte Carlo simulation for a specified tolerable % error in the estimated $\overline{P}$.

5.2.2 Numerical Examples

A number of examples are illustrated below, many of which involve solutions that are not practically possible or feasible by analytical means, and therefore require numerical procedures or Monte Carlo simulation (MCS). These include problems involving aleatory and epistemic uncertainties (see Section 5.2.3) that require repeated iterations; the solution processes are performed automatically through appropriate numerical methods with the use of personal computers.

The numerical solutions for each of the example problems are presented with the accompanying computer codes or programs in *MATHCAD* and/or *MATLAB*; in the case of *MATLAB*, the companion *Statistics* and *Symbolic Math Toolboxes* are necessary. However, any other commercial software, including *MATHEMATICA*, that is widely available may be equally convenient and effective for the examples illustrated here. In all cases, personal computers are necessary.

In order to understand the computer-based solutions, and the companion computer codes, to the examples illustrated below, some familiarity with *MATLAB* or *MATH-CAD* statements and programming is necessary. The appropriate *MATLAB* or *MATHCAD* programs developed for some of the examples are intended to be illustrative and may not necessarily be optimal.

▶ **EXAMPLE 5.1**

Suppose employees of a company have to travel by air between two major cities; the air travel time (in hours) between the cities is estimated to require a minimum of 4 hr and a maximum of 8 hr, depending on the weather and traffic conditions. From data based on experiences of its company employees, a beta distribution with parameters $q = 2.50$ and $r = 5.50$ is deemed to be appropriate to model the probability distribution of the actual travel time between the two cities.

The probability that the actual travel time T will be less than 6 hr, $P(T < 6)$, may be calculated conveniently through *MATLAB* using the incomplete beta function of the Statistics Toolbox. Therefore, according to Eq. 3.52, with

$$u = (6 - 4)/(8 - 4) = 0.5$$

the incomplete beta function from *MATLAB* gives

$$P(T < 6) = betainc(0.5, 2.5, 5.5) = 0.872$$

Note: $betainc(x, a, b)$ is the *MATLAB* syntax, or statement, for the incomplete beta function ratio.

The probability that the travel time will be between 4.5 hr and 5.5 hr would be

$$P(4.5 < T \leq 5.5) = betainc(0.375, 2.5, 5.5) - betainc(0.125, 2.5, 5.5)$$
$$= 0.6753 - 0.1058 = 0.570$$

Instead of the beta distribution, the gamma distribution was also suggested as a plausible alternative distribution for the travel time between the two cities. Assuming that the gamma distribution will have the same mean and standard deviation as the above beta distribution, which are $\mu_T = 5.32$ hr and $\sigma_T = 0.50$ hr, the corresponding parameters of the gamma distribution are $v = 21.73$ and $k = 115.60$. Then using *MATLAB* again, we obtain the probability that the actual travel time will be less than 6 hr as (see Eq. 3.43a)

$$P(T < 6) = gammainc(21.73 \times 6, 115.60) = 0.912$$

Note: $gammainc(vx, k)$ is the *MATLAB* syntax, or statement, for the incomplete gamma function ratio whereas the probability that the travel time will be between 4.5 hr and 5.5 hr is now

$$P(4.5 < T \leq 5.5) = gammainc(21.73 \times 5.5, 115.60) - gammainc(21.73 \times 4.5, 115.60)$$
$$= 0.6523 - 0.0428 = 0.610$$

Solution by Numerical Integration—Alternatively, the probability $P(T < 6)$ may be evaluated by numerical integration. This can be accomplished conveniently by quadrature integration using *MATLAB* as follows:

Create an "M-File": function f=bfunc(x)

```
f=((beta(2.5,5.5)).^-1).*(((x-4).^1.5).*(8-x).^4.5)./4.^7
```

and save it. Then perform the quadrature integration between 4 and 6 in the command window of *MATLAB* with the command,

```
quad(@bfunc,4,6)
```

obtaining a result of 0.872 for the probability $P(T < 6)$. The probability of T between 4.5 and 5.5 hours would be obtained with the command,

```
quad(@bfunc,4.5,5.5)
```

giving a result of 0.570.

We observe that these results are the same as those obtained above for the beta distribution. By replacing beta with gamma in the above M-File, similar results for the gamma distribution can be obtained. ◄

► **EXAMPLE 5.2**

This problem illustrates the ease of generating the CDF of the binomial distribution using MATLAB or MATHCAD.

In a car assembly plant, suppose that 50 cars are produced daily. There are 15 parts in a car that are critical for a vehicle to function properly. The parts are randomly selected for inspection, and according to the inspection record, the average percentage of defective parts is 6%. The conditions of the 15 parts in a car may be assumed to be statistically independent.

If three or more of the critical parts installed in a car are defective, the car will malfunction within a month or 1000 miles. The probability that a customer will purchase one of the malfunctioning cars, therefore, is

$$P(X \geq 3) = 1 - P(X < 3) = 1 - P(X \leq 2)$$
$$= 1 - binocdf(2,15,0.06) = 0.0571$$

where $binocdf(2,15,0.06)$ is the *MATLAB* statement for the CDF of the binomial distribution with parameters $n = 15$, $x = 2$, and $p = 0.06$.

Also, the probability that this assembly plant will produce no more than 4 defective vehicles per day, among the 50, is

$$P(X \leq 4) = binocdf(4,50,0.0571) = 0.8445$$

We might observe that the parameters in these problems are beyond the limits of Table A.2. *MATHCAD* may be used in place of *MATLAB* to obtain the same results. ◄

► **EXAMPLE 5.3**

Consider the sum of three random variables X_1, X_2, and X_3 that are statistically independent. All three variables are individually lognormally distributed with the following respective means and c.o.v.s:

$$\mu_{X_1} = 500 \text{ units}; \qquad \mu_{X_2} = 600 \text{ units}; \qquad \mu_{X_3} = 700 \text{ units}$$
$$\delta_{X_1} = 0.50; \qquad \delta_{X_2} = 0.60; \qquad \delta_{X_3} = 0.70$$

The corresponding parameters of the respective lognormal distributions, therefore, are

$$\zeta_{X_1} = \sqrt{\ln(1 + 0.50^2)} = 0.47; \quad \zeta_{X_2} = \sqrt{\ln(1 + 0.60^2)} = 0.55; \quad \zeta_{X_3} = \sqrt{\ln(1 + 0.70^2)} = 0.63$$

$$\lambda_{X_1} = \ln 500 - \frac{1}{2}(0.47)^2 = 6.10; \quad \lambda_{X_2} = \ln 600 - \frac{1}{2}(0.55)^2 = 6.25;$$

$$\lambda_{X_3} = \ln 700 - \frac{1}{2}(0.63)^2 = 6.35$$

In this case, the probability distribution of the sum $S = X_1 + X_2 + X_3$ will neither be normal nor lognormal. In fact, its distribution would be difficult to determine analytically. Therefore, if the probability distribution of S is required, Monte Carlo simulation would be a practical solution. With 10,000 repetitions (i.e., sample size, $n = 10,000$), we generate, using *MATHCAD*, the histogram of values of S as shown below, with the following statistics of S:

The mean value, the standard deviation, and the skewness of S are, respectively,

$$\text{mean(S)} = 1.799 \times 10^3 \qquad \text{Stdev(S)} = 644.869 \qquad \text{skew(S)} = 1.316$$

As the skewness coefficient is not zero, the distribution of the sum, S, is clearly not normal. Also, the 75-percentile and 90-percentile values of S are, respectively,

$$S_{75} = 2.121 \times 10^3 \qquad \text{and} \qquad S_{90} = 2.628 \times 10^3$$

MATHCAD STATEMENTS FOR THE ABOVE MCS SOLUTIONS, WITH A SAMPLE SIZE OF 10,000:

```
X₁:=rlnorm(10000,6.10,0.47)
X₂:=rlnorm(10000,6.25,0.55)
X₃:=rlnorm(10000,6.35,0.63)
S:=(X₁+X₂+X₃)
v:=sort(S)
S₉₀:=v₉₀₀₀
```

CONSTRUCTION OF HISTOGRAM FOR *S*:

```
n:=4C
j:=0..4C
intⱼ:=100+4000 j/n
h:=hist(int,S)
```

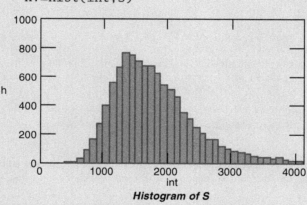

Histogram of S

Similarly, using *MATLAB*, we obtain by MCS with 10,000 repetitions, the following histogram for S:

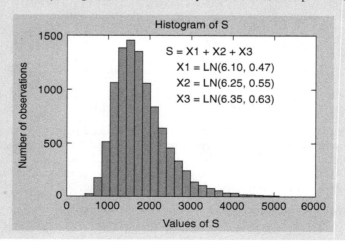

The corresponding mean value, standard deviation, and skewness coefficient of S are, respectively,

$$1.797E + 3, \qquad 651.1972, \qquad \text{and} \qquad 1.3767$$

whereas the 50% (median), 75%, and 90% values of S are, respectively,

$$1.680E + 3, \qquad 2.1212E + 3, \qquad \text{and} \qquad 2.6389E + 3$$

MATLAB STATEMENTS FOR EXAMPLE 5.3

```
X1=lognrnd(6.10,0.47,10000,1);
X2=lognrnd(6.25,0.55,10000,1);
X3=lognrnd(6.35,0.63,10000,1);
S=X1+X2+X3;
hist(S,40)                        % plots the histogram of S
mean(S);
std(S);
skewness(S)
prctile(S,50)                     % the 50% value of S
prctile(S,75);
prctile(S,90)
```

The solution for this example may also be performed using Microsoft EXCEL; the simulation can be performed in a spreadsheet through the following steps:

1. Go to Tools, select Add-in, then select Analysis Tool Pak and Analysis Tool Pak-VBA.

2. Since Excel does not have a built-in random number generator for lognormal distribution, one can first generate values for a normal random variable Y and then obtain the corresponding values of the lognormal random variable from the relation $X=\exp(Y)$.

3. Create three columns of normally distributed random numbers using Tools-Data/Analysis-Random Number Generation. Choose Normal and input the mean and standard deviation of (6.10, 0.47); (6.25, 0.55); (6.35, 0.63), respectively, from previously calculated parameters. Call these Y1, Y2, and Y3.

4. Create another set of three columns of X1, X2, and X3 by mapping the respective columns of Y1, Y2, and Y3 through the function $X = \exp(Y)$.

5. Create a fourth column of random numbers as X1+X2+X3 using EXCEL's ordinary arithmetic formulas. Copy the formula from the first cell to the last.

6. Use EXCEL's Tools-Data Analysis-Descriptive Statistics on the data created in step 5 to find the mean, standard deviation and, other pertinent statistics.

7. Use EXCEL's Tools-Data Analysis-Histogram on the data created in Step 5 to construct the histogram. Select the appropriate interval (bin) as needed.

The respective statistics of the sum of the three lognormal random variables are

$$\text{Mean} = 1795.4 \qquad \text{Standard deviation} = 671.5 \qquad \text{Skewness} = 2.623$$

The histogram of the sum S based on a sample size of 10,000 is shown in the following.

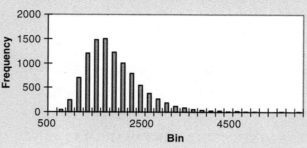

▶ **EXAMPLE 5.4**

Suppose the random variables in Example 5.3 are now Gaussian with the following respective means and standard deviations:

$$X_1 = N(500, 75); \qquad X_2 = N(600, 120); \qquad X_3 = N(700, 210)$$

and we are interested, this time, in the probability distribution of the product and quotient of the three variables, i.e.,

$$P = \frac{X_1 X_3}{X_2}$$

The distribution of the above function is clearly neither lognormal nor normal. To determine the probability distribution of P, we resort again to Monte Carlo simulation to determine the random values of the function P. With *MATHCAD*, we generate 100,000 sample values of each random variable.

The mean, median, standard deviation, and skewness of P are, respectively,

```
mean(P)=610.456   median(P)=575.856   Stdev(P)=256.239
skew(P)=1.16
```

Also, the 75-percentile and 90-percentile values of P are, respectively,

```
P75=744.586   and   P90=933.762
```

MATHCAD STATEMENTS FOR THE MCS SOLUTIONS FOR P:

```
X₁:=rnorm(100000,500,75)
X₂:=rnorm(100000,600,120)
X₃:=rnorm(100000,700,210)
```
$$P: \frac{\overrightarrow{X_1 X_3}}{X_2}$$
```
v:=sort(P)
P75:=v₇₅₀₀₀
P90=v₉₀₀₀₀
```

CONSTRUCTION OF HISTOGRAM FOR P:

```
h=hist(int,P)
n:=4C
j:=0..40
```
$$\text{int}_j := -500 + 3000 \cdot \frac{j}{40}$$

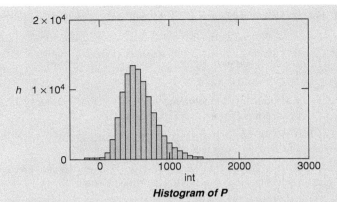

Histogram of P

Alternatively, using *MATLAB*, we also obtain through MCS with 10,000 repetitions the following histogram of *P*:

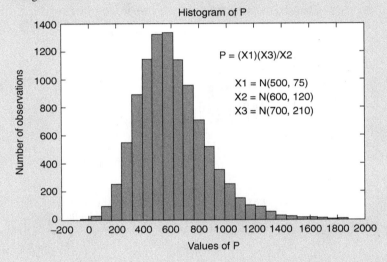

The corresponding mean value, standard deviation, and skewness coefficient of *P* are, respectively,

$$\text{mean(P)} = 609.76, \quad \text{std(P)} = 255.64, \quad \text{and} \quad \text{skewness(P)} = 1.17.$$

MATLAB STATEMENTS FOR EXAMPLE 5.4:

```
X1=normrnd(500,75,10000,1);
X2=normrnd(600,120,10000,1);
X3=normrnd(700,210,10000,1);
u=X1.*X3;
P=u./X2;
hist(P,40)                    % construct histogram of product P
mean(P)
std(P)
skewness(P)
```

For simulation using the EXCEL spreadsheet, the steps are as follows:

1. Go to <u>Tools</u>, select <u>Add-in</u>, then select <u>Analysis Tool Pak</u> and <u>Analysis Tool Pak-VBA</u>

2. Generate 3 columns of normally distributed random numbers using <u>Tools</u>—<u>Data Analysis</u>—<u>Random Number Generation</u>—(choose <u>Normal</u> and input the <u>parameters</u>: <u>mean</u> & <u>standard deviation</u>). Call these X1, X2, & X3.

3. Create a 4th column of random numbers as X1*X3/X2 using EXCEL's ordinary arithmetic formulas. Copy the formula from the first cell to the last.

4. Use EXCEL's <u>Tools</u>—<u>Data Analysis</u>—<u>Descriptive Statistics</u> on the data created in step 3 to find the mean, standard deviation and other pertinent statistics.

5. Use EXCEL's <u>Tools</u>—<u>DataAnalysis</u>—<u>Histogram</u> on the data created in step 3 to construct the histogram. Select the appropriate interval (bin) as needed.

The statistics of the product X1*X3/X2 based on 10,000 repetitions are as follows:

$$\text{Mean}=612.9 \quad \text{Standard deviation}=261.3 \quad \text{Skewness}=1.155$$

The corresponding histogram is shown in the following:

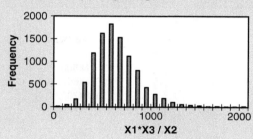

▶ **EXAMPLE 5.5**

In this example, we will consider a problem involving two independent random variables with different probability distributions. As a specific problem, consider the capacity of a pile group, R, and the applied load, S, that are, respectively, Gaussian and lognormal variates as follows:

$$R = N(50, 15)\,\text{tons}; \qquad S = LN(30, 0.33)\,\text{tons}$$

i.e., R is Gaussian with a mean of 50 tons and a standard deviation of 15 tons, whereas S is lognormal with a median of 30 tons and a c.o.v. of 0.33.

The probability of failure of the pile group under the applied load is

$$p_F = (R - S \leq 0)$$

Because the distribution of $(R - S)$ is clearly neither normal nor lognormal, Monte Carlo simulation is an effective means to evaluate the above probability of failure. For this purpose, we generate n sample values of R from the given Gaussian distribution and the same n sample values of S from the lognormal distribution, and we calculate for each ith pair of sample values the difference $(R_i - S_i)$. The ratio of the number of $(R_i - S_i) \leq 0$ to the total sample size n can then be used to represent the probability of failure.

That is,

$$p_F = \frac{\sum_{i=1}^{n}(R_i - S_i) \leq 0}{n}$$

The result obtained with *MATHCAD* for $n = 100,000$ yields the following failure probability:

$$p_F = 0.155$$

MATHCAD PROGRAM FOR MCS SOLUTION OF EXAMPLE 5.5:

```
R:=rnorm(100000,50,15)
S:=rlnorm(100000,ln(30),0.33)
```

$$v:=\overrightarrow{[(R-S)\leq 0]}$$

$$PF:=\frac{\sum_{i=1}^{99999} v_i}{100000} \qquad\qquad PF=0.155$$

According to Eq. 5.1, the c.o.v. of the above p_F obtained with a sample size of $n = 100,000$ would be 0.0074 or 0.74%.

Using *MATLAB* with the same sample size of $n = 100,000$, we obtain the failure probability of

$$p_F = 0.1566$$

MATLAB PROGRAM FOR EXAMPLE 5.5:

```
R=normrnd(50,15,100000,1);
S=lognrnd(log(30),0.33,100000,1);
x=(R-S);
y=sum(x<0);
pF=y/100000     yielding pF=0.156
```

Let us use this example to illustrate the c.o.v.s of MCS results as a function of the sample size n. For this purpose, it is convenient to use *MATLAB* with the program described below. For a given sample size n, the *MATLAB* program effectively performs the MCS 100 times giving the respective estimated c.o.v.s of the failure probability, p_F, as follows:

$$
\begin{array}{lll}
\text{For} & n = 100 & \text{c.o.v.}(pF) = 23\% \\
& n = 1000 & \text{c.o.v.}(pF) = 7\% \\
& n = 10{,}000: & \text{c.o.v.}(pF) = 2\% \\
& n = 100{,}000: & \text{c.o.v.}(pF) = 0.7\%
\end{array}
$$

Clearly, the above results show that the c.o.v. decreases with the sample size n.

MATLAB PROGRAM FOR CALCULATING pF 100 TIMES, EACH WITH SPECIFIED SAMPLE SIZE n:

```
R=normrnd(50,15,1000,100);  % generates 100 vectors each of
                               size n=1000
S=lognrnd(log(30),0.33,1000,100);
x=R-S;
y=sum(x<0);
p=y/1000;                    % a row vector of 100 values of p
pF=p'                        % a vector of 100 values of pF
```

From the vector of 100 pF's generated above, we estimate the mean and standard deviation of pF, and the corresponding c.o.v. (pF). ◄

► **EXAMPLE 5.6**

This problem is the same as that of Example 4.12, except that the individual work durations are now beta-distributed instead of Gaussian as in Example 4.12. Figure E5.6 shows the same activity network as that of Fig. E4.12.

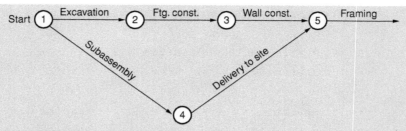

Figure E5.6 House framing activity network.

The framing of a house may be done by subassembling the components in a plant and then delivering them to the building site for framing. Simultaneously, while this subassembly of the components is being fabricated, the preparation of the site, which includes the excavation of the foundation through the construction of the foundation walls, can proceed at the same time. The different activities and their respective sequences may be represented with the activity network shown in Fig. E5.6; the required work durations for the respective activities are shown in the table below, with the respective distribution parameters.

		Completion Time (days)		Parameters of Beta Distribution			
Activity	Variable	Mean	Std. Dev.	q	r	a	b
1–2	X_1	2	1	2.00	3.00	0	5.00
2–3	X_2	1	$1/2$	2.00	3.00	0	2.50
3–5	X_3	3	1	2.00	3.00	1.00	6.00
1–4	X_4	5	1	2.00	3.00	3.00	8.00
4–5	X_5	2	$1/2$	2.00	3.00	1.00	3.50

Assume that the required completion times for the different activities are statistically independent beta-distributed variates, with the respective means and standard deviations as indicated in columns 3 and 4 in the above table. With the shape parameters of $q = 2.00$ and $r = 3.00$, the corresponding lower and upper bound values of the respective beta distributions are as indicated in the above table.

Clearly, to start framing the house, the foundation walls must be completed and the assembled components must also be delivered to the site. The probability that framing of the house can start within 8 days after work started on the job can be determined as follows.

With the durations of the respective activities listed in the table above, the total durations for the two sequences of activities are

$$T_1 = X_1 + X_2 + X_3$$

and

$$T_2 = X_4 + X_5$$

in which T_1 and T_2 are also statistically independent. As the individual variates $X_1, X_2, X_3, X_4,$ and X_5 are beta-distributed, the above sums for T_1 and T_2 are difficult to determine analytically. MCS provides a practical alternative. The required MCS results by *MATHCAD* using 100,000 repetitions are shown below, yielding the following:

$$P(T_1 < 8) = 0.899$$

and

$$P(T_2 \leq 8) = 0.802$$

As T_1 and T_2 are statistically independent, the probability that framing can start within 8 days is, therefore,

$$P(\text{framing}) = 0.899 \times 0.802 = 0.721$$

Both *MATHCAD* and *MATLAB* generate random numbers only for the standard beta distribution for x between 0 and 1 with shape parameters q and r. The corresponding beta-distributed random numbers for X between a and b with the same parameters can be obtained as $X = (b - a)x + a$.

MATHCAD STATEMENTS FOR MCS SOLUTION OF EXAMPLE 5.6:

```
x1:=rbeta(100000,2.0,3.0)          generate standard
x2:=rbeta(100000,2.0,3.0)            beta-distributed values
x3:=rbeta(100000,2.0,3.0)            x1, x2, x3, x4, x5
x4:=rbeta(100000,2.0,3.0)
x5:=rbeta(100000,2.0,3.0)

X1=5x1

X2=2.5x2

X3=(6-1)x3+1.00
X4:=(8-3)x4+3.00
X5:=(3.50-1.00)x5+1.00
```

$$T1 := \overrightarrow{(X1 + X2 + X3)}$$
$$T2 := \overrightarrow{(X4 + X5)}$$
$$s1 := \overrightarrow{(T1 \leq 8)} \quad s2 := \overrightarrow{(T2 \leq 8)}$$

$$p1 := \frac{\sum_{i=1}^{99999} s1_i}{100000} \qquad\qquad p2 := \frac{\sum_{j1}^{99999} s2_j}{100000}$$

$$P(T1 < 8) := p1 \quad p1 = 0.899 \qquad\qquad P(T2 < 8) := p2 \quad p2 = 0.802$$

With *MATLAB*, we determine that the corresponding probability that framing can start within 8 days is

$$P(\text{framing}) = 0.902 \times 0.802 = 0.723$$

MATLAB STATEMENTS FOR THE MCS SOLUTION OF EXAMPLE 5.6:

```
x1=betarnd(2,3,100000,1);      % generate standard beta-
x2=betarnd(2,3,100000,1)         distributed random numbers
x3=betarnd(2,3,100000,1)         x1, x2, x3, x4, x5
x4=betarnd(2,3,100000,1)
x5=betarnd(2,3,100000,1)

X1=5.*x;

X2=2.5.*x;

X3=(6-1).*x+1;

X4=(8-3).*x+3;

X5=(3.5-1).*x+1;

T1=X1+X2+X3;

T2=X4+X5;

y=sum(T1<8);
p1=y./100000                    % gives p1=0.902
z=sum(T2<8);
p2=z./100000                    % gives p2=0.802
```

The above results, obtained by *MATHCAD* or *MATLAB* may be compared with the result of Example 4.12, which gives a probability of 0.74 that framing can start within 8 days; again, in the case of Example 4.12, the durations for the individual activities are all Gaussian. We see that the result here is different from that of Example 4.12. ◀

▶ **EXAMPLE 5.7**

In Example 4.25 of Ang and Tang (1984), it was shown that the annual floods (maximum discharges), F, of the Wabash River in Mount Carmel, Illinois, may be modeled by the *Type I* largest asymptotic distribution with the following parameters: $u_n = 120 \times 10^3$ cfs and $\alpha_n = 0.015 \times 10^{-3}$.

Suppose that a bridge is planned to be constructed across the river with a center pier. The pier foundation is designed against scouring during high floods. The flood capacity of the pier foundation against scouring, R, may be assumed to be a lognormal random variable with a median capacity of 400×10^3 cfs and a c.o.v. of 0.30. The probability of failure of the center pier due to scouring during high floods is

$$p_F = P(R \leq F)$$

The required probability of failure, p_F, can be evaluated effectively by MCS. We describe below the *MATHCAD* statements to perform the MCS with 10,000 repetitions, showing the failure probability of the pier to be 0.037.

Special note: A *MATHCAD* program is developed to generate the required random numbers with a *Type 1* asymptotic distribution of largest values. The *Type 1* largest random numbers x with parameters α and β can be generated as follows:

$$F_X(x) = \exp(-e^{-\alpha(x-\beta)})$$

At probability $F_X(x) = u$, where $u =$ value of uniformly distributed random number,

$$x = \beta - (1/\alpha)\ln[\ln(1/u)]$$

MATHCAD PROGRAM FOR THE ABOVE PURPOSE AND FOR CALCULATING THE FAILURE PROBABILITY:

```
u:=runif(10000,0,1)     with β:=120·10³;   α:=0.015·10⁻³
```

$$F:=120\cdot10^3 - \frac{1}{0.015\cdot10^{-3}}\left(\ln\left(\ln\left(\frac{1}{u}\right)\right)\right)$$

```
R:=rlnorm(10000,ln(400·10³),0.30)
x:=[(R-F)≤0]
```

$$pF:=\frac{\displaystyle\sum_{i=1}^{9999} x_i}{10000} \qquad pF=0.037$$

CONSTRUCTION OF HISTOGRAM FOR F:

```
N:=4C
j:=1..N
```
$$int_j:=0+5\cdot10^5\cdot\frac{j}{N}$$
```
h:=hist(int,F)
```

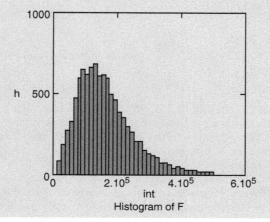

Histogram of F

Similarly, we obtain the solution of the same problem with the following *MATLAB* program, obtaining the failure probability of the pier as $p_F = 0.0359$.

MATLAB PROGRAM AND MCS RESULTS FOR EXAMPLE 5.7:

```
M-File:                          Explanations (indicated by %):
u=unifrnd(0,1,10000,1)
F=120*10^3-(1/(0.15*10^-3)).     % generates a vector of
  *(log(log(1/u)))                 10,000 Type I random numbers
R=lognrnd(log(400*10^3),0.3,10000,1)
x=(R-F)
y=sum(x<=0)
pF=y/10000                       % calculates the failure
                                   probability

end
Execution of this program yields pF=0.0359.                        ◀
```

▶ **EXAMPLE 5.8**   Suppose the load effect of heavy vehicles, each weighing X tons, on a bridge support is $X^{1.33}$ tons (including impact). If the number of heavy vehicles on the bridge is a random variable N with a Poisson distribution with a mean of 25, and the average vehicle weight on the bridge is a Gaussian random variable $N(12,4)$ tons, the total vehicle load effect on the bridge support, Y, would be given by

$$Y = N(X)^{1.33}$$

where N = a Poisson random variable with parameter $\lambda = 25$; i.e.,

$$P(N = n) = \frac{\lambda^n}{n!} e^{-\lambda} \quad \text{and} \quad X = N(12, 4)$$

Y is also a random variable; its distribution would be difficult to determine analytically. We resort to MCS with 1000 repetitions using *MATHCAD* yielding the following histogram for the vehicle load effect, Y:

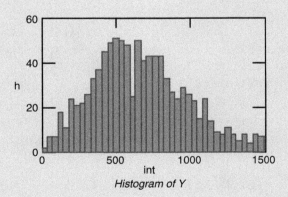

Histogram of Y

From the 1000 sample values, the corresponding mean, standard deviation, and skewness coefficient of Y are estimated to be as follows:

$$\text{mean}(Y) = 691.3 \text{ tons} \qquad \text{Stdev}(Y) = 346.365 \text{ tons} \qquad \text{skew}(Y) = 0.786$$

MATHCAD STATEMENTS FOR EXAMPLE 5.8:

```
N:=rpois(1000,25)
X:=rnorm(1000,12,4)
```
$$Y:=\overrightarrow{(N \cdot X)^{1.33}}$$

Using *MATLAB*, we also obtain through MCS with 1000 repetitions, the following histogram.

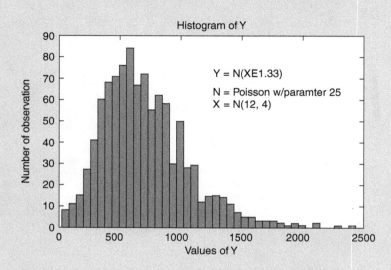

Histogram of Y

Y = N(XE1.33)

N = Poisson w/paramter 25

X = N(12, 4)

The corresponding mean, standard deviation, and skewness coefficient of Y are, respectively,

$$\text{mean}(Y) = 689.23 \text{ tons} \quad \text{Stdev}(Y) = 348.41 \text{ tons} \quad \text{skew}(Y) = 0.666$$

MATLAB STATEMENTS FOR THE SOLUTION OF EXAMPLE 5.8:

```
N=poissrnd(25,1000,1);
X=normrnd(12,4,1000,1);
Y=N.*X,^1.33;
hist(Y,40)                    % generates the histogram for Y
mean(Y)=689.23
std(Y)=348.41
skewness(Y)=0.666
```

It is of interest to observe that by first-order approximation (see Chapter 4), the corresponding mean and standard deviation for this example would be;

$$\text{mean}(Y) = 681.16 \text{ tons}, \qquad \text{Stdev}(Y) = 331.29 \text{ tons} \quad ◀$$

▶ **EXAMPLE 5.9**

Consider the problem that was illustrated in Example 4.14 involving a high-rise building consisting of a structure-foundation system designed to withstand wind-induced pressures. However, let us modify the problem by changing the distributions for some of the variables as follows:

Assume that the wind-induced pressure capacities of the superstructure and foundation are, respectively, normal variates:

$$R_s = N(70, 15) \text{ psf} \qquad \text{and} \qquad R_f^* = N(60, 20) \text{ psf}$$

The peak wind-induced pressure, P_w, on the building during a wind storm is

$$P_w = 1.165 \times 10^{-3}\, CV^2, \text{ in psf}$$

where:

$C =$ the drag coefficient, a Gaussian variate $N(1.80, 0.50)$.

$V =$ maximum wind speed, a *Type I* extremal variate with a modal speed of 100 fps, and a c.o.v. of 30%; i.e., the equivalent extremal parameters are $\alpha = 0.037$ and $u = 100$.

The probability distribution of P_w is clearly not any of the standard distribution; for this determination we resort to Monte Carlo simulation. With 1000 repetitions (or sample size) we generate the histogram of P_w with the following *MATHCAD* program:

MATHCAD STATEMENTS TO GENERATE V AND P_w:

```
C:=rnorm(1000,1.8,0.5)
u:=runif(1000,0,1)
```
$$V := 100 - \frac{1}{0.037}\left(\ln\left(\ln\left(\frac{1}{u}\right)\right)\right)$$
$$P := \overrightarrow{(1.165 \cdot 10^{-3} \cdot C \cdot V^2)}$$

CONSTRUCTION OF HISTOGRAM FOR P_w:

$$h := \text{hist(int,P)}: \quad N := 4C: \quad j := 1..N: \quad \text{int}_j := 0 + 100 \cdot \frac{j}{N}$$

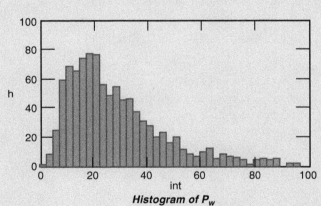

Histogram of P_w

With the results of the above simulations, obtained with a sample of size $n = 1000$, we estimate the sample mean and sample standard deviation of P_w using Eqs. 6.1 and 6.3 of Chapter 6, yielding, respectively.

```
mean(P) = 31.09
Stdev(P) = 24.18
skew(P) = 2.73
```

Moreover, by Monte Carlo simulation (with 1000 repetitions), we also obtain the failure probabilities of the superstructure and the foundation, using MATHCAD, respectively, as follows:

$$P_{F_s} = P(R_s - P_w \leq 0) = 0.072$$

and

$$P_{F_f} = P(R_f - P_w \leq 0) = 0.138$$

MATHCAD PROGRAM TO CALCULATE RESPECTIVE FAILURE PROBABILITIES OF SUPERSTRUCTURE AND FOUNDATION:

```
S:=rnorm(1000,70,15)
```

$$v:=\overrightarrow{(S-P)} \leq 0$$

$$PF:=\frac{\displaystyle\sum_{i=1}^{999} v_i}{1000} \qquad\qquad PF = 0.072$$

```
F:=rnorm(1000,60,20)
```

$$v:=\overrightarrow{(F-P)} \leq 0$$

$$Pf:=\frac{\displaystyle\sum_{i=1}^{999} v_i}{1000} \qquad\qquad Pf = 0.138$$

The failure of the structure-foundation system is the union of the failure of the superstructure and of the foundation; we observe that the failures of these two components are not statistically independent because of the common load P_w. By MCS, the failure probability of the system is found to be 0.144.

MATHCAD PROGRAM TO CALCULATE FAILURE PROBABILITY OF THE STRUCTURE-FOUNDATION SYSTEM:

$$pF := \begin{vmatrix} C \leftarrow rnorm(1000,1.80,0.50) \\ u \leftarrow runif(1000,0,1) \\ V \leftarrow 100 - \dfrac{1}{0.037} \cdot \left(\ln\left(\ln\left(\dfrac{1}{u}\right)\right)\right) \\ P \leftarrow \overrightarrow{(1.165\cdot 10^{-3}\cdot C\cdot V^2)} \\ S \leftarrow rnorm(1000,70,15) \\ F \leftarrow rnorm(1000,60,20) \\ \text{for } i \in 1..1000 \\ \quad \begin{vmatrix} x \leftarrow \overrightarrow{(S \leq P)} \\ y \leftarrow \overrightarrow{(F \leq P)} \\ pF \leftarrow \dfrac{\displaystyle\sum_{i=1}^{999}(x_i \vee y_i)}{1000} \\ pF \end{vmatrix} \\ pF \end{vmatrix} \qquad\qquad pF = 0.144$$

With *MATLAB*, the corresponding results (for sample size $n = 1,000$) of Example 5.9 are as follows:

For the superstructure, the failure probability is $\qquad\qquad p_{F_s} = 0.072$;

whereas for the foundation, the failure probability is $\qquad\qquad p_{F_f} = 0.127$;

and the probability of failure of the superstructure-foundation system is $\qquad p_F = 0.149$.

MATLAB STATEMENTS FOR GENERATING RANDOM NUMBERS OF *V* AND P_w:

```
u=unifrnd(0,1,1000,1);
V=100-(log(log(1./u)))/0.037;

hist(V,40)                              % plots histogram of V
```

```
mean(V)=115.3406
std(V)=35.1357

C=normrnd(1.80,0.50,1000,1);
Pw=1.165*10^-3.*C.*V.^2;

hist(Pw,40)                          % plots histogram of Pw

mean(Pw)=30.1943
std(Pw)=22.4987
skewness(Pw)=2.793
```

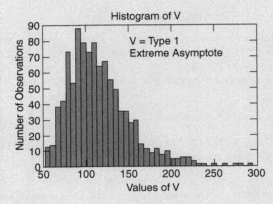

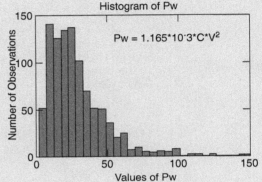

MATLAB STATEMENTS TO CALCULATE FAILURE PROBABILITY OF SUPERSTRUCTURE:

```
Rs=normrnd(70,15,1000,1);
n=(Rs-Pw)<0;
pFs=sum(n)/1000;        pFs=0.072
```

MATLAB STATEMENTS TO CALCULATE FAILURE PROBABILITY OF FOUNDATION:

```
Rf=normrnd(60,20,1000,1);
m=(Rf-Pw)<0;
pFf=sum(m)/1000         pFf=0.127
```

MATLAB PROGRAM TO CALCULATE FAILURE PROBABILITY OF STRUCTURE-FOUNDATION SYSTEM:

```
x=(Rs-Pw)<=0;
y=(Rf-Pw)<=0;
z=x|y;                                        % x ∪ y
PFss=sum(z)/1000        PFss=0.149
```

We observe that the results obtained by *MATLAB* and *MATHCAD* are slightly different; this is to be expected for a sample size of 1000 in the MCS. ◀

▶ **EXAMPLE 5.10**

In Example 4.9, suppose the storm drain of the city is a normal variate with a mean capacity of 1.5 million gallons per day (mgd) and a standard deviation of 0.3 mgd; i.e., the capacity is $N(1.5, 0.30)$. The storm drain serves two independent drainage sources within the city that are gamma distributed (instead of normal as in Example 4.9) with the same respective means and standard deviations as in Example 4.9; i.e., drainage source $A = $ gamma-distributed with $k = 12$ and $v = 17.5$ and drainage source $B = $ gamma-distributed with $k = 11$ and $v = 22.22$. These parameter values are consistent with those of Chapter 3, which defined mean $= k/v$ and variance $= k/v^2$. In the present case, with the indicated parameter values, the means and standard deviations of drainage sources A and B are, respectively, the same as those of Example 4.9; namely, $\mu_A = 0.70$ mgd, $\mu_B = 0.50$ mgd, and $\sigma_A = 0.20$ mgd, $\sigma_B = 0.15$ mgd.

The probability that the storm drain capacity will be exceeded during a storm can be determined as follows:

$$P(D < A + B) = P\{[D - (A + B)] < 0\}$$

Because D is a normal variate but A and B are independent gamma-distributed variates, the distribution of the function $S = [D - (A + B)]$ is difficult to determine analytically. Its distribution may be determined by Monte Carlo simulation as indicated below.

The probability that the drain capacity will be exceeded is found by MCS with the *MATHCAD* program below to be

$$P(S < 0) = 0.202$$

MATHCAD PROGRAMS TO GENERATE TWO-PARAMETER GAMMA-DISTRIBUTED RANDOM NUMBERS FOR *A* AND *B*:

Special note: MATHCAD (Version 11) generates random numbers for the one-parameter gamma distribution only. The following programs generate the two-parameter gamma-distributed random numbers.

$$A := \begin{vmatrix} \text{for } i \in 1..9999 \\ \quad u \leftarrow \text{runif}(10000, 0, 1) \\ \quad A_i \leftarrow \dfrac{-1}{17.5} \left(\sum_{j=1}^{12} \ln(u_j) \right) \\ \quad i \leftarrow i+1 \\ \quad A \\ A \end{vmatrix} \qquad B := \begin{vmatrix} \text{for } i \in 1..9999 \\ \quad u \leftarrow \text{runif}(10000, 0, 1) \\ \quad B_i \leftarrow \dfrac{-1}{22.22} \left(\sum_{j=1}^{11} \ln(u_j) \right) \\ \quad i \leftarrow i+1 \\ \quad B \\ B \end{vmatrix}$$

yielding the following:

```
mean(A)=0.685          mean(B)=0.495
Stdev(A)=0.197         Stdev(B)=0.149
skew(A)=0.605          skew(B)=0.653
```

MATHCAD PROGRAM TO CALCULATE PROBABILITY OF EXCEEDANCE:

$$D := \text{rnorm}(10000, 1.50, 0.30)$$

$$S := \overrightarrow{[D - (A+B)]}$$

$$x := S \leq 0$$

$$pF := \dfrac{\sum_{j=1}^{9999} x_j}{10000}$$

$$pF = 0.202$$

MATLAB also generates the one-parameter gamma distribution. The two-parameter gamma-distributed random numbers can be generated with the following *MATLAB* program.

MATLAB PROGRAM TO GENERATE TWO-PARAMETER GAMMA-DISTRIBUTED RANDOM NUMBERS FOR *A* AND *B*:

```
u=unifrnd(0,1,1000,12);
X=log(u);
Y=X';
Z=sum(Y(:,1:1000));
A=-(1/17.5)*Z;

hist(A,20)

mean(A)=0.6847
std(A)=0.1944
skewness(A)=0.4493
```

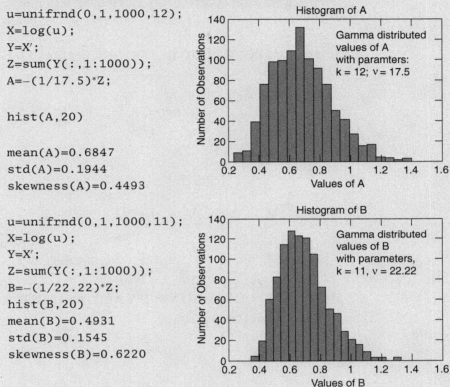

```
u=unifrnd(0,1,1000,11);
X=log(u);
Y=X';
Z=sum(Y(:,1:1000));
B=-(1/22.22)*Z;
hist(B,20)
mean(B)=0.4931
std(B)=0.1545
skewness(B)=0.6220
```

Using the *MATLAB* program below, the probability that the storm drain capacity will be exceeded is

    pF=0.196

MATLAB PROGRAM TO CALCULATE PROBABILITY OF EXCEEDANCE:

```
D=normrnd(1.50,0.30,1000,1);
E=D-(A'+B');
F=(E<=0);
G=sum(F);
pF=G/1000=0.1960
```

Clearly, the above result for this example (obtained by *MATHCAD* or *MATLAB*) is different from that of Example 4.9, which is 0.221. This is because the distributions of the two drainage sources are gamma distributed instead of normal as in Example 4.9.

Next, if another drainage source C = gamma distributed with parameters $k = 16$ and $v = 20$, equivalent to a mean of 0.80 mgd and standard deviation of 0.20 mgd, is to be added to the existing storm drain, how much must the mean capacity of the current storm drain be increased in order to maintain the original probability of exceedance (i.e., of 0.202) of the existing system? Assume that the standard deviation of the new storm drain will remain at 0.3 mgd.

In this case, a repeated Monte Carlo simulation may be performed as shown below. A *MATHCAD* program was developed below to automatically calculate the failure probability repeatedly for different assumed mean values of D between $\mu_D = 1.70$ and 3.0 until the failure probability was ≤ 0.202. The results obtained with this program indicate that the required upgraded mean capacity should be $\mu_D = 2.35$ mgd.

Hence, the mean capacity of the existing storm drain must be increased by $2.35 - 1.5 = 0.85$ mgd.

MATHCAD PROGRAM TO DETERMINE UPGRADED CAPACITY OF STORM DRAIN:

$$V := \begin{vmatrix} \text{for } v \in 1.70, 1.71..3.0 \\ \quad \begin{vmatrix} E \leftarrow \text{rnorm}(10000, v, 0.30) \\ g \leftarrow \overrightarrow{[E-(A+B+C)]} \leq 0 \\ p \leftarrow \dfrac{\sum\limits_{j=1}^{9999} g_j}{10000} \\ v \text{ if } p > 0.202 \\ (\text{break}) \text{ if } p \leq 0.202 \\ v \end{vmatrix} \\ V \leftarrow v \\ V \end{vmatrix}$$

$$\text{yielding } V = 2.35$$

MATHCAD PROGRAM TO VERIFY ABOVE RESULT:

```
D:=rnorm(10000,2.35,0.3)
```
$$g := \overrightarrow{[D-(A+B+C)]} \leq 0$$

$$p := \dfrac{\sum\limits_{j=1}^{9999} g_j}{10000} \qquad\qquad p=0.201$$

By *MATLAB*, the corresponding upgraded storm capacity is determined to be $V = 2.36$ mgd. This is obtained with the following *MATLAB* program.

MATLAB PROGRAM TO ITERATIVELY DETERMINE THE UPGRADED STORM DRAIN CAPACITY:

```
M-File                          Explanations (indicated by %)
u=unifrnd(0,1,1000,12)
X=log(u)
Y=X'
Z=sum(Y(:,1:1000))
A=-(1/17.5)*Z                   % generates vector of 1000 gamma-
                                  distributed values for A
v=unifrnd(0,1,1000,11)
x=log(v)
y=x'
z=sum(y(:,1:1000))
```

```
          B=-(1/22.22)*z                % generates vector of 1000 gamma-
                                            distributed values for B
          w=unifrnd(0,1,1000,16)
          a=log(w)
          b=a'
          c=sum(b(:,1:1000))
          C=-(1/20)*c                    % generates vector of 1000 gamma-
                                            distributed values for C

          for v=1.7:0.01:3.0             % designates range of V
          D=normrnd(n,0.30,1000,1)
          E=D-(A'+B'+C')
          F=(E<=0)
          G=sum(F)
          pF=(1/1000)*G                  % calculates failure probability
                                            of upgraded capacity

          if pF>0.196
            v=v+0.1
          elseif pF<=0.196
            V=v
            break
          end
```

Execution of the above program yields the required upgraded drain capacity $V = 2.36$ mgd. The corresponding calculated failure probability is $p_F = 0.192$, which is less than 0.196 of the original system. ◀

▶ **EXAMPLE 5.11**

In an apartment building, the number of occupants in the building during a fire would be highly variable. Suppose the total number of occupants living in an apartment is 150; however, during a fire the number of people in the building may be as few as 50. Therefore, during a fire the actual number would be a random variable that may be described with a beta distribution ranging between 50 and 150, with parameters $q = 1.50$ and $r = 3.00$. Assume also that during a fire the probability of serious injury to an individual in the building is 0.001. Therefore, assuming that injuries between individuals are statistically independent, the probability of x number of injuries during a fire is given by the binomial distribution (see Sect. 3.2) as follows:

$$P(X = x) = \binom{n}{x}(0.001)^x(1 - 0.001)^{n-x}$$

However, the number of occupants exposed to the fire, n, is a beta-distributed random variable as described above. Therefore, the required probability may be evaluated as

$$P(X = x) = \sum_{n=50}^{150} P(X = x|N = n)P(N = n) = \sum_{n=50}^{150}\left\{\left[\binom{n}{x}(0.001)^x(1 - 0.001)^{n-x}\right]P(N = n)\right\}$$

in which N is beta-distributed within the range of (50, 150) and parameters $q = 1.50$ and $r = 3.00$. Clearly, numerical solution is necessary to evaluate the above probability for given x. In this case, the solution involves the summation of discrete probabilities (including discretization of the beta PDF). For this purpose, a *MATLAB* program is developed below to perform the required solution automatically, yielding the following results:

$$P(X = 0) = 0.91890$$
$$P(X = 1) = 0.07632$$
$$P(X = 2) = 0.00332$$
$$P(X = 3) = 1.0033 \times 10^{-4}$$
$$P(X = 4) = 2.37 \times 10^{-6}$$

We may observe, for verification of the results, that the sum of the above probabilities is approximately equal to 1.00.

MATLAB PROGRAM FOR EXAMPLE 5.11:

```
M-File                    Explanations (indicated by %)

x = 1                     % illustrated for x = 1
Pinj = 0
for n=50:1:150
p=((factorial(n)./(factorial(x).*factorial(n-x))).*
   (.001).^x.*(.999).^(n-x))
f=(1./(beta(1.5,3.0))).*(((n-50).^(0.5)).*(150-n).^2)./
   (100).^3.5
P=p*f
if n<150
  n=n+1
  Pinj=Pinj+P             % calculates probability of injury with
                             n occupants
elseif n>=150
  n=n
  break
end
```

This example involves the CDF of the binomial distribution; however, the required CDF is much beyond the limits of most tables of this CDF, including Table A.2. For problems such as this example, the tools afforded by computer-based numerical solutions are clearly needed. ◀

▶ **EXAMPLE 5.12**

In environmental engineering, waste treatment is a necessary process of maintaining the ultimate BOD level of a waste. If a waste is not treated, the ultimate BOD level of the waste at a given stage may be predicted from the BOD measured at an earlier stage of degradation. Suppose that at time $t = 5$ days, the BOD was measured to be 0.22 mg/L. On this basis, the ultimate BOD level of the waste will eventually reach,

$$Lo = (BOD)_5 / [1 - \exp(-5k)]$$

if no further treatment is performed on the waste beyond $t = 5$ days. The parameter k is the reaction rate constant; it depends on the temperature, type of waste, and the degradation process. Assume that k may be modeled as a normal variate, $N(0.22, 0.03)$ per day.

Moreover, because of variability in the measurement of the BOD level, $(BOD)_5$ may be assumed to be $N(200, 10)$ in mg/L. Clearly, the distribution of Lo will not be normal; moreover, its distribution will not be any of the standard distributions. We may determine the distribution and statistics of the ultimate level of BOD, *Lo*, through MCS as shown below using *MATHCAD*. Through this process, we obtain the results as follows:

$$\text{mean(Lo)} = 303.72 \quad \text{Stdev(Lo)} = 29.2 \quad \text{skew(Lo)} = 0.79$$

The corresponding histogram is as shown below.

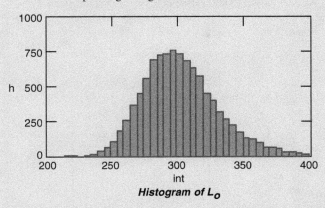

Histogram of L_O

MATHCAD STATEMENTS FOR EXAMPLE 5.12:

```
BOD₅:=rnorm(10000,200,10)
k:=rnorm(10000,0.22,0.03)
```

$$\text{Lo} := \frac{\overrightarrow{\text{BOD}_5}}{1 - e^{-5k}}$$

◄

5.2.3 Problems Involving Aleatory and Epistemic Uncertainties

In Chapter 1, Sect. 1.2, we discussed the two types of uncertainty, namely, the aleatory and the epistemic uncertainties. We emphasized that the significances between these two types of uncertainty are distinctly different, and ought to be evaluated and represented separately. Namely, the aleatory uncertainty is associated with the inherent variability of a random event, and its significance should be expressed in terms of the probability of occurrence of the event, whereas the epistemic uncertainty is associated with the imperfection in our calculation of the occurrence probability and thus may be represented by a range of possible errors in the calculated probability. That is, rationally, the epistemic uncertainty should not be included in the calculation of the probability, but rather should be reflected as an uncertainty in the calculated probability.

In this regard, the need for Monte Carlo simulation becomes even more essential, as illustrated in the following examples.

► **EXAMPLE 5.13**

Suppose an engineering system has been designed with a median safety factor, $\bar{\theta}$. The underlying probability of failure due to inherent variability, δ_θ, is

$$p_F = \Phi\left(-\frac{\ln \bar{\theta}}{\delta_\theta}\right)$$

where:

$\bar{\theta}$ = the median safety factor; assume $\bar{\theta} = 2.5$

δ_θ = the c.o.v. of θ; assume $\delta_\theta = 0.25$

However, because of epistemic uncertainty, $\overline{\theta}$ is also a random variable; suppose its distribution is lognormal with distribution $LN(2.5, 0.15)$, i.e., with median of 2.5 and c.o.v. of 0.15. Because $\overline{\theta}$ is a random variable, P_F is also a random variable; its distribution will not be one of the standard forms. We may generate the values and histogram of P_F through Monte Carlo simulation. The *MATHCAD* statements for the MCS with a sample size of 10,000 are shown below.

For each of the 10,000 values of $\overline{\theta}$, we calculate the above P_F. The result is a vector of 10,000 values of P_F, from which we can construct the histogram of P_F, as shown below for $\ln(pF)$, as well as estimate its mean, median, standard deviation, and skewness, and determine also its 90% value as follows:

$$\text{mean(pF)} = 8.414 \times 10^{-4} \quad \text{median(pF)} = 1.218 \times 10^{-4} \quad \text{Stdev(pF)} = 2.833 \times 10^{-3}$$

$$\text{skew(pF)} = 10.578 \quad\quad PF_{9000} = 1.842 \times 10^{-3}$$

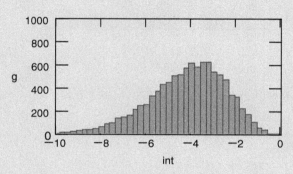

Histogram of log(pF)

We may emphasize that the aleatory uncertainty gives rise to the probability of failure p_F, whereas the epistemic uncertainty is reflected in the range and distribution of p_F. The distribution of p_F provides a decision maker the option to select conservative values of p_F, such as the 90% value $p_{F_{90}} = 1.842 \times 10^{-3}$.

MATHCAD STATEMENTS FOR EXAMPLE 5.13:

```
x:= rlnorm(1000,ln(2.5),0.15)
```
$$pF := cnorm\left(\frac{-\ln(x)}{0.25}\right)$$
```
mean(pF)=8.414×10⁻⁴
median(pF)=1.218×10⁻⁴
Stdev(pF)=2.833×10⁻³
skew(pF)=10.578
```

Using *MATLAB*, the corresponding results are as follows:

$$\text{mean} = 8.46e - 004, \quad \text{standard deviation} = 0.0029, \quad \text{and} \quad \text{skewness coefficient} = 17.61.$$

Also, the median and 90% values of PF are, respectively,

$$\text{median} = 1.228e - 004, \quad \text{and} \quad PF_{90} = 0.0019$$

and the corresponding histogram for log(PF) is as follows:

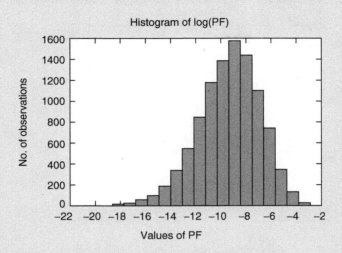

Histogram of log(PF)

MATLAB STATEMENTS FOR EXAMPLE 5.13:

```
A=lognrnd(log(2.5),0.15,10000,1)
X=-log(A)./0.25
pf=normcdf(X)
```

◄

► **EXAMPLE 5.14**

Suppose the probability of failure of an engineering sytem due to inherent variability (aleatory uncertainty) is

$$p_F = P(R - S \leq 0)$$

where:

$$R = N(40,15)$$
$$S = LN(25, 0.35), \text{ i.e., median of 25 and c.o.v. of } 0.35$$

Clearly, p_F can be easily evaluated by MCS but would be difficult to evaluate analytically, as $(R - S)$ is not of any standard distribution. Moreover, because of possible errors in the estimation of μ_R (mean of R) and s_m (median of S), there are epistemic uncertainties in the estimates of these parameters; assuming no bias and respective c.o.v.s of 0.10 and 0.15, we may denote the distributions of these parameters to be $N(1.0, 0.10)$ and $N(1.0, 0.15)$, respectively. Because of these epistemic uncertainties in μ_R and s_m, the true failure probability will be a random variable.

For each pair of values of μ_R and s_m, we calculate the above p_F, and for the ranges of the random values of μ_R and s_m, we obtain a vector of p_F.

The solution requires repeated iterations by a double-loop MCS; the inner loop calculates p_F for specified values of μ_R and s_m, whereas the outer loop generates the random vectors of R and S for the given pair of μ_R and s_m. This requires programming to perform the repeated iterations automatically; the results obtained using *MATHCAD* with a sample size of $n = 10,000$ are as follows:

The mean, standard deviation, and skewness coefficient of the failure probability are, respectively,

$$\text{mean(pF)} = 0.235 \qquad \text{Stdev(pF)} = 0.094 \qquad \text{skew(pF)} = 0.437$$

Also, the 90% value of pF is $pF_{90} = 0.366$. Again, for a conservative value of the failure probability, we may use $pF_{90} = 0.366$, instead of the mean pF. The corresponding histogram for pF is as follows:

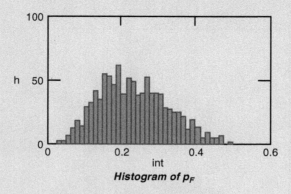

Histogram of p_F

The median, 75%, and 90% values are, respectively,

$$pF_{50} = 0.226$$
$$PF_{75} = 0.298$$
$$PF_{90} = 0.366$$

MATHCAD PROGRAM FOR EXAMPLE 5.14:

```
U := rnorm(1000,1.0,0.10)
V := rnorm(1000,1.0,0.15)
pF := | for i ∈ 0..999
      |     R ← rnorm(10000,40U_i,15)
      |     S ← rlnorm(10000,ln(25V_i),0.35)
      |     x ← (R-S) ≤ 0
      |              9999
      |              Σ x_j
      |              j=1
      |     pF_i ← ─────────
      |             10000
      |     pF
      | pF
```

The following *MATLAB* program also yields the solutions for this example.

MATLAB PROGRAM FOR EXAMPLE 5.14:

```
M-File:                          Explanations (indicated by %)
for i=1:1000
  u=normrnd(1.0,0.1,1000,1)      % generates a vector of μ_R
                                   of size 1000
  R=normrnd((40.*u(i,1)),        % generates a vector of R
           15,1000,1)              of size 1000 with ith value
                                   of μ_R.
  v=normrnd(1.0,0.15,1000,1)     % generates a vector of s̄_m
                                   of size 1000.
  S=lognrnd(log(25.*v(i,1)),     % generates a vector of S of
           0.35,1000,1)            size 1000 with ith value of s_m.
```

```
x=R-S
y=(x<=0)
n=sum(y)
pf=n/1000
pF(i,1)=pf'    % a vector of 1000 values of pF.
```

Execution of the above program yields a vector of 1000 values of pF, with the corresponding histogram as follows:

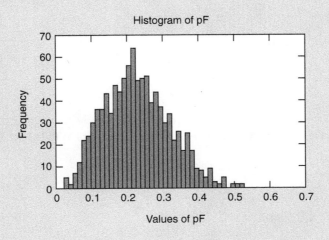

Histogram of pF

The mean, standard deviation, and skewness coefficient are as follows:

Mean of pF $= 0.2269$ Std. dev. of pF $= 0.0960$ Skewness of pF $= 0.5673$

Also, the 50%, 75%, and 90% values of pF are, respectively, 0.2140, 0.2870, and 0.3570.   ◄

► **EXAMPLE 5.15**

In Example 5.9, we can expect significant uncertainties (epistemic) in the specification of the mean value of the drag coefficient, μ_C, and in the estimation of the mean wind velocity, μ_V. Moreover, there may also be imperfection (consisting of bias and random error) in the equation for calculating the wind-induced pressure P_w. Let us suppose the following epistemic uncertainties, expressed in terms of the respective c.o.v.s.

$$\mu_C \text{ with c.o.v. of } \Delta_C = 0.30$$
$$\mu_V \text{ with c.o.v. of } \Delta_V = 0.15$$

And there is no bias in the equation for P_w, but there is uncertainty (in terms of c.o.v.) of $\Delta P_w = 0.20$. Therefore, the total epistemic uncertainty in the calculated wind-induced pressure P_w may be aggregated as

$$\Omega_{P_w} = \sqrt{0.30^2 + (2 \times 0.15)^2 + 0.20^2} = 0.47$$

In light of the above epistemic uncertainties, the wind-induced pressure P_w is recalculated by MCS using *MATHCAD* with a sample size of 10,000, yielding the following results:

mean(P) $= 34.208$ Stdev(P) $= 26.424$ skew(P) $= 3.255$

and the corresponding histogram for P_w is as follows:

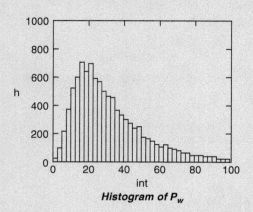

Histogram of P_w

MATHCAD PROGRAM FOR GENERATING WIND PRESSURE P_w:

$$P := \begin{vmatrix} N \leftarrow rnorm(1000,1.0,0.30) \\ M \leftarrow rnorm(1000,1.0,0.15) \\ L \leftarrow rnorm(1000,1.0,0.20) \\ \text{for } i \in 0..999 \\ \begin{bmatrix} C \leftarrow rnorm(1000,1.8 \cdot N_i,0.50) \\ u \leftarrow runif(1000,0,1) \\ V \leftarrow (100)M_i - \left[\dfrac{1}{0.037} \cdot \left(\ln \left(\ln \left(\dfrac{1}{u} \right) \right) \right) \right] \\ P \leftarrow 1.165 \cdot 10^{-3} \cdot \overrightarrow{\left(C \cdot V^2 \right)} \cdot L_i \\ \text{continue} \end{bmatrix} \\ P \end{vmatrix}$$

Clearly, therefore, the calculated failure probability of the superstructure (as well as of the foundation) will depend on the value of the mean wind pressure. Assuming that there are no epistemic uncertainties in R_s and R_f, we perform repeated MCS for 1000 sample values of the mean wind pressures (see *MATHCAD* program below), obtaining the probability distribution of the failure probabilities of the superstructure, and the foundation with the following respective statistics:

$$\text{mean(pFs)} = 0.092 \qquad \text{Stdev(pFs)} = 0.092 \qquad \text{skew(pFs)} = 2.264$$
$$\text{mean(pFf)} = 0.156 \qquad \text{Stdev(pFf)} = 0.121 \qquad \text{skew(pFf)} = 1.615$$

Also, from the same MCS, we obtain the histograms of the failure probabilities of the superstructure and foundation, respectively, as shown below.

On the basis of the respective histograms, we may determine the corresponding 90-percentile failure probabilities of the superstructure and foundation; these may be referred to as the values with a 10% probability of error.

For the superstructure, $\text{pF}_{90} = 0.244$ and for the foundation, $\text{pF}_{90} = 0.312$

As in Example 5.9, the failures of the superstructure and of the foundation are not statistically independent because of the common load P_w. By MCS, the failure probability of the structure-foundation

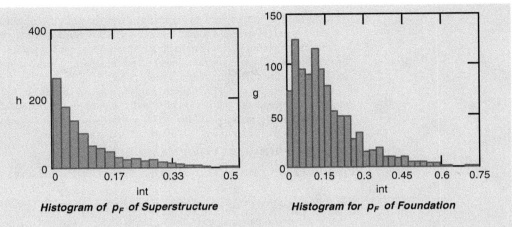

Histogram of p_F of Superstructure   **Histogram for p_F of Foundation**

system, which is the union of the superstructure and foundation failures, is found to be as follows:

$$\text{mean(pFSS)} = 0.173 \qquad \text{Stdev(pFSS)} = 0.129 \qquad \text{skew(pFSS)} = 1.445$$

The corresponding 90% value of the failure probability of the structure-foundation system is 0.361.

MATHCAD PROGRAM FOR CALCULATING EACH OF THE THREE FAILURE PROBABILITIES:

$$
\text{pFSS} := \Bigg|
\begin{aligned}
&N \leftarrow \text{rnorm}(1000,1.0,0.3) \\
&M \leftarrow \text{rnorm}(1000,1.0,0.15) \\
&L \leftarrow \text{rnorm}(1000,1.00,0.20) \\
&\text{for } j \in 0..999 \\
&\qquad \begin{bmatrix}
C \leftarrow \text{rnorm}(1000,1.8 \cdot N_j,0.5) \\
u \leftarrow \text{runif}(1000,0,1) \\
V \leftarrow \left[100 \cdot (M_j) - \dfrac{1}{0.037} \cdot \left(\ln\left(\ln\left(\frac{1}{u}\right)\right)\right)\right] \\
P \leftarrow 1.165 \times 10^{-3} \overrightarrow{\left(C \cdot V^2 \cdot L_j\right)} \\
S \leftarrow \text{rnorm}(1000,70,15) \\
F \leftarrow \text{rnorm}(1000,60,20) \\
x \leftarrow \overrightarrow{(S \leq P)} \\
\text{pFs}_j \leftarrow \sum_{i=1}^{999} \dfrac{x_i}{1000} \\
y \leftarrow \overrightarrow{(F \leq P)} \\
\text{pFf}_j \leftarrow \sum_{i=1}^{999} \dfrac{y_i}{1000} \\
\text{pFSS}_j \leftarrow \sum_{i=1}^{999} \dfrac{x_i \vee y_i}{1000} \\
\text{continue}
\end{bmatrix} \\
&\text{pFSS}
\end{aligned}
$$

The *MATHCAD* program for calculating the three failure probabilities (for the superstructure, the foundation, and the structure-foundation system) is described above. In this program, we should point out that the proper input/output must be specified prior to each execution of the program; i.e., pFs for superstructure, pFf for foundation, and pFSS for the structure-foundation system failure probabilities. As illustrated above, the program will yield pFSS.

The MCS solutions for Example 5.15 are obtained also by *MATLAB*. The corresponding *MATLAB* program to generate the distribution of P_w and calculate the failure probabilities of the superstructure and the foundation, as well as of the structure-foundation system, is as follows:

MATLAB PROGRAM FOR EXAMPLE 5.15:

M-File	Explanations (indicated in %)
`for i=1:1000`	
`    c=normrnd(1.0,0.3,1000,1)`	
`    C=normrnd(1.8.*c(i,1),0.50,`	`% generates vector of 1000`
`       1000,1)`	`  values of C`
`    v=normrnd(1.0,0.15,1000,1)`	
`    u=unifrnd(0,1,1000,1)`	
`    V=(100.*v(i,1))-(1/0.037).*`	`% generates vector of 1000`
`       (log(log(1./u)))`	`  values of V`
`    p=normrnd(1.0,0.20,1000,1)`	
`    Pw=(1.165*(10^-3)).*C.`	`% generates vector of 1,000`
`       *(V.^2).*p(i,1)`	`  values of Pw`
`    Rs=normrnd(70,15,1000,1)`	`% generates vector of 1000`
	`  values of Rs`
`    a=(Rs<=Pw)`	
`    Rf=normrnd(60,20,1000,1)`	`% generates vector of 1000`
	`  values of Rf`
`    x=(Rf<=Pw)`	
`    Ps(i,1)=sum(a)/1000`	`% vector of 1000 pF's of`
	`  superstructure`
`    Pf(i.1)=sum(x)/1000`	`% vector of 1000 pF's of`
	`  foundation`
`    U=a|x`	`% union of structure-`
	`  foundation system failure`
`    PSS(i,1)=sum(U)/1000`	`% vector of 1000 pF's of`
	`  struct-fdn system`
`end`	

The results obtained with a sample size of 1000 can be summarized as follows:

```
mean(Pw)=32.38;    std.dev.(Pw)=29.93;   skewness(Pw)=2.55
mean(Ps)=0.0900;   std.dev.(Ps)=0.0879;  skewness(Ps)=1.84;
  90%(Ps)=0.204
mean(Pf)=0.1531;   std.dev.(Pf)=0.1147;  skewness(Pf)=1.33;
  90%(Pf)=0.308
mean(PSS)=0.1729;  std.dev.(PSS)=0.1287; skewness(PSS)=1.28;
  90%(PSS)=0.348
```

and the respective histograms are shown below:

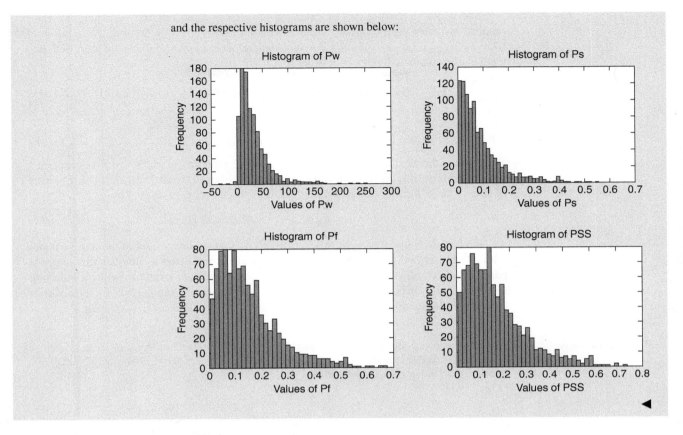

5.2.4 MCS Involving Correlated Random Variables

Monte Carlo simulation can also be applied to generate dependent or correlated random numbers. We shall limit our illustrations to the generation of correlated random numbers for two random variables, say X and Y, with known bivariate distributions. The procedure illustrated for bivariate distributions can be extended to multivariate distributions.

In general, for two dependent random variables X and Y with a general bivariate PDF $f_{X,Y}(x,y)$ or CDF $F_{X,Y}(x,y)$, the corresponding bivariate random numbers can be generated as follows:

The joint CDF can be expressed as

$$F_{X,Y}(x,y) = F_X(x)F_{Y|X}(y|x)$$

To generate the bivariate random numbers, we generate two vectors U_1 and U_2 of random numbers each of sample size n, and each with uniform distribution between 0 and 1. Then, the random numbers for X is generated as (see Ang and Tang, 1984)

$$\underline{X} = F_X^{-1}(U_1) \tag{5.3}$$

which is a vector of random numbers of X of the same sample size n as U_1.

The corresponding random numbers for Y given $X = x$ are then generated as

$$\underline{Y} = F_{Y|X}^{-1}(U_2|\underline{X}) \tag{5.4}$$

which is a vector of the same size as U_2 or $\underline{X}$.

The above procedure can be extended to three or more random variables, say for k random variables $X_1, X_2, \ldots, X_k$ with known joint PDF $f_{X_1,X_2,\ldots,X_k}(x_1, x_2, \ldots, x_k)$ or joint CDF $F_{X_1,X_2,\ldots,X_k}(x_1, x_2, \ldots, x_k)$. For such dependent random variables, the joint PDF and CDF can be expressed, respectively, as

$$f_{X_1,X_2,\ldots,X_k}(x_1, x_2, \ldots, x_k) = f_{X_1}(x_1) \cdot f_{X_2|X_1}(x_2|x_1) \cdots f_{X_k|X_1,\ldots,X_{k-1}}(x_k|x_1, \ldots, x_{k-1})$$

and

$$F_{X_1,X_2,\ldots,X_k}(x_1, x_2, \ldots, x_k) = F_{X_1}(x_1) \cdot F_{X_2|X_1}(x_2|x_1) \cdots F_{X_k|X_1,\ldots,X_{k-1}}(x_k|x_1, \ldots, x_{k-1})$$

To generate the random numbers for each of the k random variables, we first generate k vectors of uniformly distributed random numbers between 0 and 1, $U_1, U_2, \ldots, U_k$, each of size n. Then, the kth vector of n random numbers for X_k is generated by,

$$\underline{X_k} = F^{-1}_{X_k|X_1,\ldots,X_{k-1}}(U_k|x_1, \ldots, x_{k-1}) \tag{5.5}$$

Besides the general procedure described above that applies to a set of random variables with any prescribed joint distribution, there are also specialized techniques that apply to particular types of joint distributions, such as the multivariate joint normal distribution as illustrated below in Example 5.16 for the bivariate normal and Example 5.17 for the bivariate lognormal distributions.

▶ **EXAMPLE 5.16**

Consider the generation of correlated random numbers for two random variables, X and Y, with a bivariate normal distribution. Following Example 5.10 in Ang and Tang (1984), the joint PDF of a bivariate distribution can be expressed as

$$f_{X,Y}(x, y) = f_{Y|X}(y|x)f_X(x)$$

where $f_{Y|X}(y|x)$ is the conditional normal PDF of Y given X as defined in Chapter 3 with a mean and standard deviation of μ_Y and σ_Y, and $f_X(x)$ is the marginal normal PDF of X with mean and standard deviation of μ_X and σ_X. And X and Y are correlated with a correlation coefficient of ρ.

Suppose the parameters of the two random variables are defined as follows:

$$\mu_X = 150; \quad \sigma_X = 20$$
$$\mu_Y = 120; \quad \sigma_Y = 25; \quad \text{and} \quad \rho = 0.75$$

The following *MATLAB* program generates the correlated bi-variate normal random numbers with a sample size of $n = 100{,}000$:

MATLAB PROGRAM FOR EXAMPLE 5.16:

```
M-File                          Explanations (indicated by %)

x=normrnd(150,20,100000,1)      % generates vector of 100,000
                                  values of X

uy=120+0.75*(25/20).*(x-150)    % calculates the conditional
                                  mean of Y for each value of X

sy=25*sqrt(1-(.75)^2)

y=normrnd(uy,sy)                % generates vector of 100,000
                                  values of Y|x

end
```

The statistics of the results are summarized as follows:

$$\text{mean(X)} = 150.08; \quad \text{std(X)} = 20.4; \quad \text{mean(Y|x)} = 120.12;$$
$$\text{std(Y|x)} = 25.05 \quad \text{corrcoef(X,Y)} = 0.750$$

The generated random numbers, therefore, reproduce the statistics of the original or prescribed bivariate normal distribution.

Also, the respective skewness of the marginal histograms is as follows, with both close to zero:

$$\text{skewness(X)} = -0.015 \qquad \text{skewness(Y|x)} = 0.005$$

confirming that the random numbers generated for X and for $Y|x$ are, respectively, very close to being normally distributed.

The marginal histograms for X and $Y|x$ are, respectively:

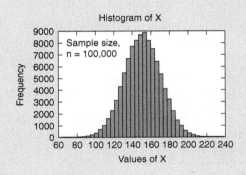

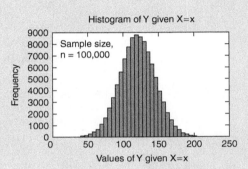

To further confirm the above MCS results, we calculate the probability $P(X < Y)$ analytically and by MCS, yielding the following for the above joint distribution:

$$\text{Exact probability,} \qquad P(X < Y) = 0.03522;$$
$$\text{MCS with sample size, } n = 100,000 \qquad P(X < Y) = 0.03544 \qquad ◀$$

▶ **EXAMPLE 5.17**

Next, consider the generation of random numbers of two random variables, X and Y, that are jointly lognormally distributed. As we saw in Example 4.3, if X is a lognormal variate, $\ln(X)$ will be normally distributed. Therefore, we can generate random numbers for $\ln(X)$ and $\ln(Y)$ with a bivariate normal distribution as described in Example 5.16, and then determine the respective lognormal random numbers for X and Y:

We illustrate this numerically as follows. Assume the following:

$$X \text{ is lognormally distributed with median, } x_m = 150; \text{ and c.o.v., } \delta_X = 0.13;$$
$$Y \text{ is lognormally distributed with median, } y_m = 120; \text{ and c.o.v., } \delta_Y = 0.21;$$
$$\text{and the correlation between } \ln(X) \text{ and } \ln(Y) \text{ is } 0.75.$$

For a sample size of $n = 10,000$, the following *MATLAB* statements generate the bivariate lognormal random numbers with the properties prescribed above.

MATLAB STATEMENTS FOR EXAMPLE 5.17:

M-File	*Explanations (indicated by %)*
`X=lognrnd(log(150),0.13, 10000,1)`	% generates vector of 10,000 values of lognormal *X*
`A=log(X)`	
`uB=log(120)+0.75.*(0.21./0.13). *(A-log(150))`	
`sB=(0.21).*sqrt(1-0.75^2)`	
`B=normrnd(uB,sB)`	% vector of 10,000 values of log(*Y*)
`Y=exp(B)`	% vector of 10,000 values of lognormal *Y*|*x*

`end`

The statistics of the results can be summarized as follows:

median(X) = 150.19   c.o.v.(X) = 0.131   skewness(X) = 0.396;   skewness(A) = −0.009

median(Y|x) = 120.35   c.o.v.(Y|x) = 0.213   skewness(Y|x) = 0.618;   skewness(B) = −0.031;

and corrcoef(A, B) = 0.749

The values of the above statistics can be observed to agree with those prescribed in the problem. Also, observe that the skewnesses of $A = \ln(X)$ and $B = \ln(Y|x)$ are both close to zero indicating that these are, respectively, normals.

The marginal histograms of *X* and *Y* are, respectively, as follows:

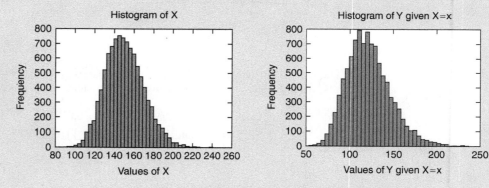

Let us now suppose that *X* is the supply of construction material and *Y* is the demand of the material in a construction project. It is reasonable to assume that the supply and demand are positively correlated; i.e., the supply will be higher or lower depending on the demand. We assume a correlation coefficient of 0.75 between *X* and *Y*. In this case, we may be interested in the probability of shortage of material for the project (supply will be less than the demand); i.e., $P(X < Y)$. We do this through MCS with the following *MATLAB* statements for a sample size of $n = 10,000$:

$$f = (X < Y);$$
$$s = \text{sum}(f);$$
$$P = s/10000$$

yielding the pertinent probability as $P(X < Y) = 0.0563$

It might be of interest to compare this probability with the corresponding probability if the demand and supply were uncorrelated or statistically independent. In this latter case, we can show that the probability would be $P(X < Y) = 0.1901$, showing, therefore, that the assumed correlation of 75% between X and Y significantly affects the probability of shortage. ◀

▶ **EXAMPLE 5.18**

Let us now illustrate the general procedure described earlier for the bivariate normal distribution of Example 5.16, in which the statistics of X and Y are as follows:

$$\mu_X = 150; \quad \sigma_X = 20$$
$$\mu_Y = 120; \quad \sigma_Y = 25; \quad \text{and} \quad \rho = 0.75$$

In this case, according to Eq. 5.3, each value of the vector $\underline{X}$ is $x = \Phi^{-1}(u_1) \cdot \sigma_X + \mu_X$, where u_1 is a value of U_1, whereas, according to Eq. 5.4, each value of $\underline{Y}$ given $X = x$ is

$$y|x = \Phi^{-1}(u_2) \cdot \sigma_Y \sqrt{1 - \rho^2} + \left[\mu_Y + \rho \left(\frac{\sigma_Y}{\sigma_X} \right)(x - \mu_X) \right], \text{ in which } u_2 \text{ is a value of } U_2.$$

Using *MATLAB*, we generate the corresponding bivariate random numbers of size 10,000 as follows.

MATLAB **STATEMENTS FOR EXAMPLE 5.18:**

```
u1=unifrnd(0,1,10000,1)        % generates vector of 10,000
                                 values of U₁

u2=unifrnd(0,1,10000,1)        % generates vector of 10,000
                                 values of U₂

x=norminv(u1).*20+150          % vector of 10,000 x's
y=(norminv(u2)).*(25).          % vector of 10,000 y|x.
  *(sqrt(1.-0.75.^2))+((0.75).
  *(1.25).*(x-150))+(120)
```

Execution of the above *MATLAB* statements yields the following statistics for X and $Y|x$.

```
mean(X) = 149.72;  std(X) = 19.88; mean(Y|x) = 119.52;  std(Y|x) = 24.91;
  corrcoef(X,Y) = 0.753
```

which are very close to those of Example 5.16. The respective histograms of X and Y are as follows:

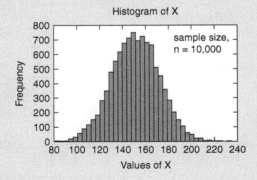

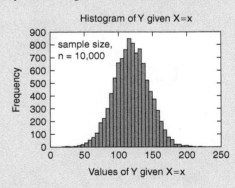

Also, the skewness of the marginal histograms are, respectively,

$$\text{skewness(X)} = -0.007 \quad \text{and} \quad \text{skewness(Y)} = -0.012$$

which are both close to zero. As expected, we observe that the above statistics agree closely with those of Example 5.16. ◀

▶ **EXAMPLE 5.19**   For a general bivariate distribution, consider the following PDF of X and Y:

$$f_{X,Y}(x, y) = \frac{1}{24}(x + y) \qquad 0 \leq x \leq 2; \quad 0 \leq y \leq 4$$

and the joint CDF of X and Y is

$$F_{X,Y}(x, y) = \frac{1}{48}(x^2 y + xy^2)$$

The corresponding marginal PDFs and CDFs are as follows:

$$f_X(x) = \frac{1}{6}(x + 2); \qquad f_Y(y) = \frac{1}{12}(y + 1)$$

$$F_X(x) = \frac{1}{6}\left(\frac{x^2}{2} + 2x\right); \qquad F_Y(y) = \frac{1}{12}\left(\frac{y^2}{2} + y\right)$$

whereas the conditional CDF of Y given $X = x$ is

$$F_{Y|X}(y|x) = \frac{(xy + y^2)}{4(x + 4)}$$

Because the product of the marginal PDFs of X and Y is not equal to the joint bi-variate PDF, the two variates are not statistically independent.

The above results can be confirmed using the Symbolic Math Toolbox of the following *MATLAB* program; also, the program generates the required inverse functions of the respective CDFs. The appropriate program for these purposes is as follows.

MATLAB PROGRAM WITH SYMBOLIC MATH TOOLBOX TO OBTAIN THE INVERSE FUNCTIONS, ETC.:

```
syms x y                         Explanations (indicated by %)
fXY=(1/24)*(x+y)                 % joint PDF of X and Y
FXy=int(fXY,y)                   % integrates joint PDF of X and
                                   Y wrto y
FXY=int(FXy,x)                   % integrates wrto x to obtain
                                   joint CDF of X & Y
fX=(1/6)*(x+2)                   % PDF of X
FX=int(fX,x)                     % integrates wrto x to obtain
                                   CDF of X
IFX=finverse(FX,x)               % inverse of CDF of X
fY=(1/2)*(y+1)                   % PDF of Y
FY=int(fY,y)                     % integrates wrto y to obtain
                                   CDF of Y
IFY=finverse(FY,y)
FYx=(1/4)*(x*y+y.^2)/(x+4)       % CDF of Y given X=x
IYx=finverse(FYx,y)              % inverse of conditional
                                   CDF of Y|x
```

With the inverse functions generated in the program above, the following *MATLAB* program will then generate the random numbers of X, Y, and $Y|x$:

MATLAB PROGRAM TO GENERATE THE BIVARIATE RANDOM NUMBERS OF X AND Y AND $Y|x$:

```
u1=unifrnd(0,1,10000,1)
x=-2+2.*(1+3.*u1).^(1/2)      % generates a vector of 10,000
                                values of X

u2=unifrnd(0,1,10000,1)
y=-1+(1+24.*u2).^(1/2)        % generates a vector of 10,000
                                values of Y

Yx=-1/2*x+1/2.*(x.^2+64*u2+   % vector of 10,000 values of Y|x
   16*u2.*x).^(1/2)
```

Executing the second program above, we obtain the following statistics for X, Y, and $Y|x$ as follows:

```
mean(X)=1.113;      std(X)=0.565;      skewness(X)=-0.231
mean(Y)=2.429       std(Y)=1.062;      skewness(Y)=-0.413
mean(Y|x)=2.513;    std(Y|x)=1.023;    skewness(Y|x)=-0.480
   corrcoef(X,Y)=-0.054
```

From these statistics, we can see that the statistics of $Y|x$ are different from those of Y, indicating that there is some dependency between X and Y. The correlation coefficient of -0.054 indicates that there is a weak correlation between X and Y. The corresponding histograms can be displayed as shown below:

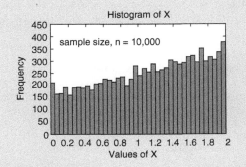

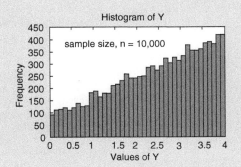

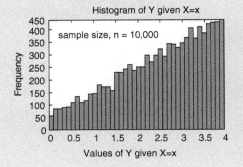

▶ **EXAMPLE 5.20**   Consider next a more complicated bivariate joint PDF as follows:

$$f_{X,Y}(x, y) = \frac{1}{3}\left(\frac{x}{y^2} + 1\right); \qquad 0 \le x \le 2; \quad 1 \le y \le 2$$

We can easily show that the corresponding marginal distributions are

$$f_X(x) = \frac{1}{3}\left(\frac{x}{2} + 1\right); \qquad F_X(x) = \frac{1}{3}\left(\frac{x^2}{4} + x\right)$$

$$f_Y(y) = \frac{2}{3}\left(\frac{1}{y^2} + 1\right); \qquad F_y(y) = \frac{2}{3}\left(y - \frac{1}{y}\right)$$

and the conditional CDF of $Y|x$ is

$$F_{Y|x}(y|x) = \frac{x/2 + y - x/2y - 1}{1 + x/4}$$

Using the *MATLAB* Symbolic Math Toolbox, we obtain the respective inverse functions of the above CDFs with the following program:

```
syms x y                    Explanations (indicated by %)
fXY=(x/y^2)+1               % joint PDF of X and Y
FXy=int(fXY,y)              % integrates joint PDF of X & Y wrto y
FXY=int(FXy,x)              % integrates wrto x to obtain joint
                              CDF of X & Y
fX=(1/3)*(1+x/2)            % PDF of X
FX=int(fX,x)                % integrates wrto x to obtain CDF of X
IFX=finverse(FX,x)          % inverse of CDF of X
fY=(2/3)*(1+1/y^2)          % PDF of Y
FY=int(fY,y)                % integrates wrto y to obtain CDF of Y
IFY=finverse(FY,y)          % inverse of CDF of Y
FYx=(-x/(2*y)+y             % CDF of Y|x
    +x/2-1)/(1+x/4)
IYx=finverse(FYx,y)         % inverse of conditional CDF of Y|x
```

Then, using the respective inverse functions obtained above, the following *MATLAB* program generates the vectors of X, Y, and $Y|x$.

```
u=unifrnd(0,1,10000,1)
x=-2+2*(1+3.*u).^(1/2)          % generates a vector of 10000
                                  values of X
v=unifrnd(0,1,10000,1)
y=(3/4).*v+(1/4).                % generates a vector of 10000
  *(9.*v.^2+16).^(1/2)             values of Y
Yx=-(1/4).*x+1/2+(1/2).          % generates a vector of 10,000
   *v+(1/8).*v.*x+(1/8).            values of Y|x
   *(4.*x.^2+16.*x-8.*v.
   *x-4.*v.*x.^2+16+32.*v+16.
   *v.^2+8.*v.^2.*x+v.^2.
   *x.^2).^(1/2)
```

The results are as follows:

```
mean(X)=1.111;      std(X)=0.565;      skewness(X)=-0.228
mean(Y)=1.462;      std(Y)=0.291;      skewness(Y)=0.154
mean(Y|x)=1.477;    std(Y|x)=0.291;    skewness(Y|x)=0.094
         corrcoef(X,Y)=-0.036
```

In this case, we also find the correlation coefficient between X and Y to be -0.036, indicating that there is some slight dependency between X and Y.

The respective histograms are also portrayed below.

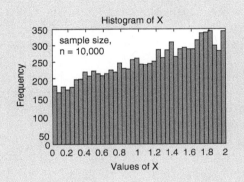

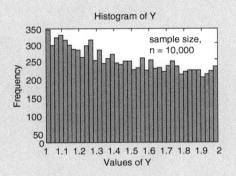

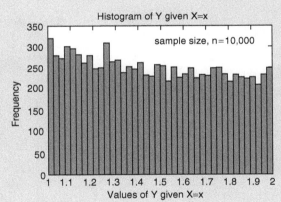

▶ EXAMPLE 5.21

Finally, consider the following joint PDF for two random variables X and Y:

$$f_{X,Y}(x, y) = \frac{1}{54}(x^2 + y^2); \qquad 0 \le x \le 3; \qquad 0 \le y \le 3$$

The corresponding joint CDF would be

$$F_{X,Y}(x, y) = \frac{1}{162}(yx^3 + xy^3)$$

and the associated marginal distributions for X, Y, and the conditional distribution of $Y|x$ are, respectively.

$$f_X(x) = \frac{1}{18}(x^2 + 3); \qquad f_Y(y) = \frac{1}{18}(y^2 + 3);$$

$$F_X(x) = \frac{1}{54}(x^3 + 9x); \qquad F_Y(y) = \frac{1}{54}(y^3 + 9y); \qquad F_{Y|x}(y|x) = \frac{(yx^2 + y^3)}{3(x^2 + 9)}$$

USING THE *MATLAB* SYMBOLIC MATH TOOLBOX, WE OBTAIN THE RESPECTIVE INVERSE FUNCTIONS OF THE ABOVE CDFS WITH THE FOLLOWING PROGRAM:

```
syms x y                    Explanations (indicated by %)
fXY=(1/54)*(x^2+y^2)        % joint PDF of X and Y
FXy=int(fXY,y)              % integrates joint PDF of X & Y wrto y
FXY=int(FXy,x)              % integrates wrto x to obtain joint
                             CDF of X & Y
fX=(1/18)*(3+x^2)           % PDF of X
FX=int(fX,x)                % integrates wrto x to obtain CDF of X
IFX=finverse(FX,x)          % inverse of CDF of X
fY=(1/18)*(3+y^2)           % PDF of Y
FY=int(fY,y)                % integrates wrto y to obtain CDF of Y
IFY=finverse(FY,y)          % inverse of CDF of Y
FYx=(1/3)*(y*x^2+y^3)/      % CDF of Y|x
   (9+x^2)
IYx=finverse(FYx,y)         % generates the inverese of the
                             CDF of Y|x
```

Then, using the respective inverse functions obtained above, the following *MATLAB* program generates the vectors of X, Y, and $Y|x$.

```
u=unifrnd(0,1,10000,1)
x=(27*u+3*(3+81*u.^2).^(1/2)).^(1/3)     % generates vector x
   -3./(27*u+3*(3+81*u.^2)
   .^(1/2)).^(1/3)
v=unifrnd(0,1,10000,1)
y=(27*v+3*(3+81*v.^2).^(1/2)).          % generates vector y
   ^(1/3)-3./(27*v+3*(3+81*v.^2).
   ^(1/2)).^(1/3)
Yx=1/6*(2916*v+324*.x^2.                 % generates vector Y|x
   *v+12*(12*.x.^6+59049
   *v.^2+ 13122*v.^2*x.
   ^2+729*v.^2.*x.^4).
   ^(1/2)).^(1/3)-2*x.^2/
   (2916*v+324*x.^2.*v+12*(12*x.^6+
   59049*v.^2.+13122*v.^2.*x.^2+729*v.
   ^2.*x.^4).^(1/2)).^(1/3)
```

obtaining the following statistics for X, Y, and $Y|x$:

```
mean(X)=1.871;    std(X)=0.826;    skewness(X)=-0.557;
mean(Y)=1.869;    std(Y)=0.823;    skewness(Y)=-0.541
mean(Y|x)=2.033;  std(Y|x)=0.747;  skewness(Y|x)=-0.783;
   corrcoef(X,Y)=-0.150
```

On the basis of the above statistics, we can see that there is dependency between X and Y; specifically, the statistics for Y are different from those of $Y|x$. Also, there is a correlation coefficient of -0.150 between X and Y indicating some dependency.

The corresponding histograms are displayed, respectively, below:

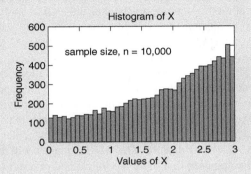

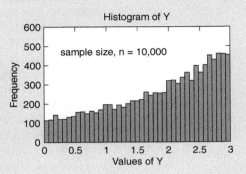

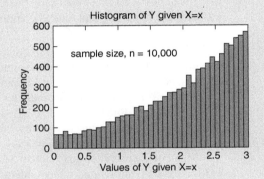

With the results generated above, let us examine the following. Suppose

X = the transportation cost per 100 ton of cargo shipped, in $1000

Y = the total tonnage of cargo shipped, in unit of 100 tons

In this case, it is reasonable to expect that the total tonnage shipped will be negatively correlated with the price of shipping; as we assumed earlier, the correlation coefficient between X and Y is -0.150. As the owner of a shipping company, we might be interested in the probability that the revenue from shipping will be below a certain threshold amount. Suppose that this threshold is $3000. The total revenue is XY; therefore, we need to determine the probability $P(XY < 3)$. The following *MATLAB* program calculates this probability.

```
f=(x.*Yx<3);    % calculates (XY < 3) for each pair of
                       values of X and Y/x
S=sum(f);
P=S/10000       % yielding P(X·Y|x< 3) = 0.412
```

whereas if X and Y were statistically independent, the corresponding *MATLAB* program would be

```
g=(x.*y<3);
s=sum(g);
p=g/10000       yielding P(XY<3)=0.454
```

The above results, therefore, indicate that the correlation affects the threshold probability. ◂

▶ 5.3   CONCLUDING SUMMARY

The main point of this chapter is to illustrate, through specific numerical examples, the application of computer-based numerical procedures and Monte Carlo simulation for formulating and solving probabilistic problems of engineering significance. These are problems for which analytical methods are impractical or difficult to apply. For this purpose, appropriate computer programs must be coded for the solution of each of the example problems using *MATLAB* and/or *MATHCAD*, although other commercial softwares such as *MATHEMATICA* can be equally effective. These programs are, therefore, part of the solution process of the examples. The principal objective of these illustrations is to expand the usefulness and utility of probabilistic models in engineering beyond the capability of purely analytical methods. However, there is no intention to describe the underlying numerical methods nor the principles of Monte Carlo simulation, which are widely available elsewhere.

To understand the coded programs for the solutions of the example problems, some familiarity with *MATLAB* and/or *MATHCAD* languages will be necessary and helpful. Finally we should emphasize that there are no canned programs that can be used to solve the type of problems illustrated here. Each of the problems is unique and its solution requires probabilistic modeling and related computer programs.

▶ PROBLEMS

The problems in this chapter require solutions by Monte Carlo simulation or numerical procedures. For this purpose, personal computers (PCs) with appropriate software are necessary. Commercially available software includes *MATHCAD, MATLAB*, and *MATHEMATICA. MS Excel + VISUAL BASIC* may also be used. To understand and apply the software languages used in this chapter, some familiarity with one or more of these softwares (particularly *MATHCAD* or *MATLAB*) will be helpful.

Many of the problems in Chapters 3 and 4 may also be solved by numerical procedures or Monte Carlo simulations, either with the probability distributions as specified in Chapters 3 and 4 or with revised distributions as appropriate. Specifically, consider the following:

5.1 This problem is similar to that of Example 4.11 of Chapter 4. The integrity of the columns in a high-rise building is essential for the safety of the building. The total load acting on the columns may include the effects of the dead load D (primarily the weight of the structure), the live load L (that includes human occupancy, furniture, movable equipment, etc.), and the wind load W.

These individual load effects on the building columns may be assumed to be statistically independent variates with the following respective probability distributions and corresponding parameters:

Distribution of D is normal with $\mu_D = 4.2$ tons; and $\sigma_D = 0.3$ ton

Distribution of L is lognormal with $\mu_L = 6.5$ tons; and $\sigma_L = 0.8$ ton

Distribution of W is *Type I* extreme with $\mu_W = 3.4$ tons; and $\sigma_W = 0.7$ ton

The total combined load effect S on each of the columns is

$$S = D + L + W$$

The individual columns were designed with a mean strength that is equal to 1.5 times the total mean load that it carries, and may be assumed to be a lognormal variate with a c.o.v. of 15%. The strength of each column R is clearly independent of the applied load. A column will be over-stressed when the applied load S exceeds the strength R; determine the probability of occurrence of this event $(R < S)$.

5.2 The annual operational cost for a waste treatment plant is a function of the weight of solid waste, W, the unit cost factor, F, and an efficiency coefficient, E, as follows:

$$C = \frac{WF}{\sqrt{E}}$$

where W, F, and E are statistically independent variates. Assume that the respective probability distributions are as follows, with the following medians and coefficients of variation (c.o.v.):

Variable	Distribution	Median	c.o.v.
W	lognormal	2000 tons/per year	20%
F	beta (from 0 to 50)	$20 per ton	15%
E	normal	1.6	12.5%

As C is a function of the product and quotient of the three variates with the respective distributions indicated above, its probability distribution is difficult to determine analytically. By Monte Carlo simulation, with a sample size of $n = 10,000$, determine the probability that the annual cost of operating the waste treatment plant will exceed $35,000.

The result may be compared with that of Example 4.13 in which the random variables were all assumed to be lognormally distributed.

5.3 In Problem 4.14, the amount of water supply available for a city in the next month is lognormally distributed with a mean of 1 million gallons and a c.o.v. of 40%, whereas the total demand is expected to follow a lognormal distribution with a mean of 1.5 million gallons with a c.o.v. of 10%.

(a) Evaluate by MCS the probability of a water shortage in this city over the next month. The result should be the same as, or close to, that of Example 4.15.

(b) Suppose the available water supply for the city in the next month is beta distributed with parameters $q = 2.00$ and $r = 4.00$, whereas the total demand is still lognormally distributed. The means and c.o.v.s remain the same. Evaluate the end values a and b of the beta distribution, and estimate by MCS the probability of a water shortage in the city in the next month.

(c) In part (a), suppose the mean supply of 1 million gallons can actually vary between 0.80 and 1.1 million gallons in the next month (equivalent to an epistemic uncertainty with a c.o.v. of 0.09). Determine by MCS the statistics of the probability of water shortage in the city over the next month.

5.4 A shuttle bus operates from a shopping center, travels to Towns A and B sequentially, and then returns to the shopping center as shown in the figure below.

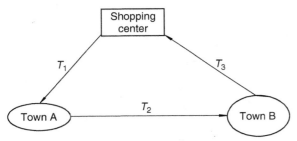

Assume that the respective travel times (under normal traffic) are independent random variables with the following respective distributions and statistics:

Travel Time	Distribution	Mean (min)	c.o.v.
T_1	Gaussian	30	0.30
T_2	lognormal	20	0.20
T_3	lognormal	40	0.30

However, during rush hours (8:00–10:00 AM and 4:00–6:00 PM), the mean travel time between Town A and Town B will increase by 50% with the same c.o.v.

(a) The scheduled time for one round trip of the shuttle bus is 2 hours. Determine by MCS the probability that a round trip will not be completed on schedule under normal traffic.

(b) What is the probability that a passenger, originating in Town A, will arrive at the shopping center within 1 hour under normal traffic?

(c) The number of passengers originating from Town B is twice that of the number originating from Town A. What percentage of the passengers will arrive at the shopping center during rush hours in less than 1 hour?

(d) A passenger starting from Town B has an appointment at 3:00 PM at the shopping center. If the bus left Town B at 2:00 PM but has not arrived at the shopping center at 2:45 PM, what is the probability that he or she will arrive on time for his or her appointment?

In Problem 4.9, the distributions of the travel times were all Gaussian random variables. The results obtained above may be compared with those of Problem 4.9.

5.5 The settlement of each footing shown in the figure below follows a normal distribution with a mean of 2 in. and a coefficient of variation of 30%. Suppose the settlements between two adjacent footings are correlated with a correlation coefficient of 0.7; that is, the differential settlement is

$$D = |S_1 - S_2|$$

where S_1 and S_2 are the settlements of footings 1 and 2, respectively.

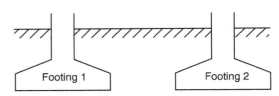

(a) Determine the mean and variance of D.

(b) What is the probability that the magnitude of the differential settlement will be less than 0.5 in.?

By Monte Carlo simulation, determine the solutions to (a) and (b). The results may be compared with those of P4.10 of Chapter 4.

(c) Next, suppose that the distributions of S_1 and S_2 are lognormally distributed with the same mean and c.o.v. as in the original problem In this case, determine the solutions to part (a) and part (b).

5.6 A city is located downstream of the confluence point of two rivers as shown below. The annual maximum flood peak in River 1 has an average of 35 m³/sec with a standard deviation of 10 m³/sec, whereas in River 2 the mean peak flow rate is 25 m³/sec and a standard deviation of 10 m³/sec. The annual maximum peak flow rates between the two rivers are statistically independent lognormal random variables.

Presently, the channel which runs through the city can accommodate up to 100 m³/sec flow rate without flooding the city.

Perform Monte Carlo simulations to answer the following questions.

(a) What are the mean and standard deviation of the annual maximum peak discharge passing through the city?

(b) What is the annual risk that the city will experience flooding based on the existing channel capacity? What is the corresponding return period?

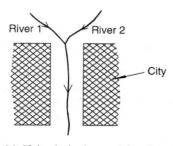

(c) If the desired annual flooding risk is to be reduced by half, by how much should the present channel capacity be enlarged?
(d) Repeat parts (a) to (c) if the annual maximum flood peaks of Rivers 1 and 2 are correlated with a correlation coefficient of 0.8.

5.7 Five cylindrical tanks are used to store oil; each of the tanks is shown below. The weight, W, of each tank and its contents is 200 kips. When subjected to an earthquake, the horizontal inertial force may be calculated as

$$F = \frac{W}{g}a$$

where:

g = acceleration of gravity = 32.2 ft/sec^2

a = maximum horizontal acceleration of the earthquake ground motion

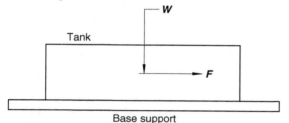

During an earthquake, the frictional force at the base of a tank keeps it from sliding. The coefficient of friction between the

tank and the base support is a beta-distributed variate with shape parameters $q = r = 3.00$, and a median value of 0.40 and a c.o.v. of 0.20. Also, during an earthquake assume that the maximum ground acceleration is a *Type I* extremal variate with a mean of 0.30 g and a c.o.v. of 0.35%.

(a) By Monte Carlo simulation, determine the probability that during an earthquake, a tank will slide from its base support?
(b) In order to reduce the probability of sliding to 30% of that in (a), each of the tanks will need to be anchored to its base. For this purpose, what should be the sliding shear resistance of the anchors for each tank?
(c) There can be epistemic uncertainties in specifying or estimating the parameters of the relevant distributions. Suppose the median value of the frictional coefficient has a c.o.v. of 25%, and the mean maximum ground acceleration has a c.o.v. of 45%. In this case, determine the statistics and histogram of the probability of sliding of a tank during an earthquake. Select the 90% value for a conservative probability of sliding.

5.8 The last part of Problem 4.16 is difficult to determine analytically. Let us perform MCS to obtain the required results.

A pile is designed to have a mean capacity of 20 tons; however, because of variabilities, the pile capacity is lognormally distributed with a coefficient of variation (c.o.v.) of 20%. Suppose the pile is subject to a maximum lifetime load that is also lognormally distributed with a mean of 10 tons and a c.o.v. of 30%.

A number of piles may be tied together to form a pile group to resist an external load. Suppose the capacity of the pile group is the sum of the capacities of the individual piles. Consider a pile group that consists of two single piles as described above. Because of proximity, the capacities between the two piles are correlated with a correlation coefficient of $\rho = 0.8$. Let T denote the capacity of this pile group.

Determine the mean value and c.o.v. of T, and calculate the probability of failure of the pile group by MCS.

5.9 Determine the solutions to Problem 3.58 by MCS.

5.10 Determine the solutions to Problem 3.59 by MCS.

As we indicated earlier, many of the problems in Chapters 3 and 4 may be solved by numerical methods or MCS. In particular, if the probability distributions in many of the problems in Chapter 4 are changed, appropriate numerical methods or MCS may be required to obtain the pertinent solutions.

▶ REFERENCES AND SOFTWARE

Ang, A. H-S. and W.H. Tang. *Probability Concepts in Engineering Planning and Design*, Vol. II—Decision, Risk, and Reliability, John Wiley & Sons, Inc. New York, 1984.

Cornell, C. A., "A Probability-Based Structural Code," *Jour. of the American Concrete Institute*, 1969, pp. 974–985.

Der Kiureghian, A., H.-Z. Lin, and S.-J. Hwang, "Second-order reliability approximations," *Jour. of Engineering Mech.*, ASCE, 113(8), 1987, pp. 1208–1225.

Halvorson, M., *Microsoft VISUAL BASIC Step by Step*, Microsoft Press, Redmond, WA, 2003.

MacDonald, M., *Microsoft VISUAL BASIC. Net Programmer's Handbook*, Microsoft Press, Redmond, WA, 2003.

MATLAB, Version 7, The Mathworks, Inc., Natick, MA, 2004.

MATHCAD. 11, Mathsoft Engineering and Education, Inc., Cambridge, MA, 2002.

MATHEMATICA, Version 4.2, Wolfram Research, Inc., Champaign, IL, 2002.

Rubinstein, R.Y., *Simulation and the Monte Carlo Method.* John Wiley and Sons, Inc., New York, 1981.

Shooman, M.L., *Probabilistic Reliability: An Engineering Approach*, McGraw-Hill Book, Co., New York, NY 1968.

CHAPTER 6

Statistical Inferences from Observational Data

▶ ## 6.1 ROLE OF STATISTICAL INFERENCE IN ENGINEERING

In the previous chapters, particularly in Chapters 3 and 4, we observed that once we know (or can specify) the probability distribution function (the PDF or CDF) of a random variable, and the values of its parameters, we can calculate the probability of an event defined by values of the random variable. The calculated probability clearly depends on the values of the parameters as well as on the form of the distribution model (PDF or CDF). Naturally, therefore, methods for determining the parameters and for selecting the appropriate form of distribution are of practical importance.

For the above purposes, the required information must be based on real-world data. For example, in determining the maximum wind speed for the wind-resistant design of a high-rise building, past records of measured wind velocities in the neighborhood of the building site are useful and important; similarly, in properly designing an improved left-turn lane at a highway crossing, repeated counts of vehicles making left turns at the intersection would provide the proper and useful information. On the basis of the observational data, the parameters may be estimated statistically, and information on the appropriate form of distribution may also be inferred.

In many geographical regions of the world, observational data on natural processes, such as rainfall intensities, flood levels of rivers, wind velocities, earthquake magnitudes and frequencies, traffic volumes, pollutant concentrations, and ocean wave heights and forces, have been and continue to be recorded and catalogued. Also, field and laboratory data on the variabilities of concrete strength, yield and ultimate strengths of steel, fatigue lives of materials, shear strength of soils, efficiency of construction crews and equipment, measurement errors in surveying, and many others continue to be acquired as needed for engineering purposes. All these statistical data provide valuable information from which the appropriate probability model and the corresponding parameters required for engineering applications may be developed or evaluated.

The techniques for deriving the appropriate probabilistic model and for estimating the parameter values from available observational data are embodied in the methods of *statistical inference*, in which information obtained from sampled data is used to represent (approximately at least) the corresponding information regarding the population from which the samples are derived. Inferential methods of statistics, therefore, provide a link between the real world and the idealized probability models prescribed or assumed in a probabilistic analysis. The role of statistical inference in the decision-making process is schematically illustrated in Fig. 6.1. The present chapter is devoted to statistical methods for estimating

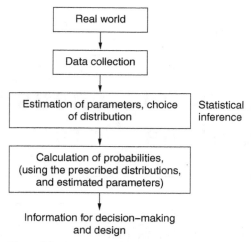

Figure 6.1   Role of statistical inference in decision-making process.

the parameters of a probability model (distribution). The subject of empirically determining or verifying the form of a probability distribution is covered in the following chapter (Chapter 7).

Although there are other inferential methods of statistics, including more esoteric and specialized techniques, we shall limit our consideration to those concepts and methods that are most basic and useful in engineering applications. These include principally the methods of estimation (point and interval estimations) and the testing of hypotheses in this chapter; the empirical determination and validation of probability distributions are treated in Chapter 7; regression and correlation analyses involving two or more random variables are the subjects of Chapter 8; and finally, the Bayesian approach to statistical estimation is covered as part of Chapter 9.

▶ 6.2 STATISTICAL ESTIMATION OF PARAMETERS

Classical estimation of parameters consists of two types: (1) *point* and (2) *interval* estimations. Point estimation is aimed at calculating a single number, from a set of observational data, to represent the parameter of the underlying population, whereas interval estimation determines an interval within which the true parameter lies, with a statement of "confidence" represented by a probability that the true parameter will be in the interval.

6.2.1 Random Sampling and Point Estimation

As alluded to earlier, the parameters of a probability model or distribution may be evaluated or estimated on the basis of a set of observational data obtained from the corresponding "population"; such a data set represents a *sample* of the population, and thus the value of a parameter evaluated on the basis of the sample values is necessarily an *estimate* of the true parameter. In other words, the exact values of the population parameters are generally unknown; the best that we can do is to estimate their values based on finite *samples* of the population.

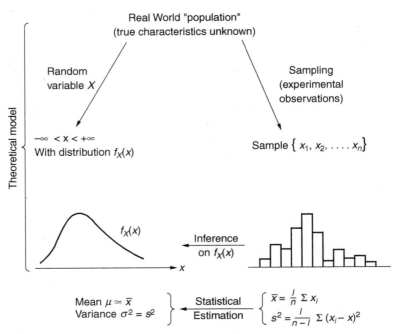

Figure 6.2 Role of sampling in statistical inference.

The role of sampling in statistical inference is illustrated graphically in Fig. 6.2. The real-world population may be modeled by a random variable X, with PDF $f_X(x)$ and associated parameters, such as μ and σ in the case of the normal distribution. A finite number of observations from the population would be taken, constituting a sample, from which the parameters may be estimated; the same set of sample observations may be used to empirically determine the most appropriate distribution model (see Chapter 7).

Given a set of sample observations from a population, there are different methods for estimating the parameters; among these are the *method of moments* and the *method of maximum likelihood*. Irrespective of the method used, the estimation of the parameters is necessarily based on a finite set of sample values $x_1, x_2, \ldots, x_n$, called a *sample of size n* from the population X. Usually, such a sample is assumed to constitute a *random sample;* this means that the successive sample values are statistically independent and the underlying population remains the same from one sample observation to another.

There are certain properties that are desirable of a point estimator; these are, namely, the properties of *unbiasedness, consistency, efficiency,* and *sufficiency,* defined, respectively, as follows:

- If the expected value of an estimator is equal to the parameter, the estimator is said to be *unbiased.* Unbiasedness, therefore, implies that the average value of the estimator (if repeated estimates were to be made) will be equal to the parameter.

- *Consistency* implies that as $n \rightarrow \infty$, the estimator approaches the value of the parameter. Consistency, therefore, is an asymptotic property—practically, it means that the error in the estimator decreases as the sample size increases.

- *Efficiency* refers to the variance of the estimator—for a given set of data, an estimator θ_1 is said to be more efficient than another θ_2 if the variance of θ_1 is smaller than that of θ_2.

- Finally, an estimator is said to be *sufficient* if it utilizes all the information in a sample that is pertinent to the estimation of the parameter.

In practice, however, it is seldom possible to require all or many of the above properties of an estimator. In fact, there are very few, if any, estimators that possess all the desirable properties. In the following sequel, we shall refer to one or more of these properties in connection with specific estimators.

The Method of Moments

In Chapter 3, we saw that the mean and variance, which are the first and second moments of a random variable, are the main descriptors of the random variable. These moments are related to the parameters of the distribution of the random variable as indicated in Table 6.1 for a number of common distributions. For example, in the case of the Gaussian distribution, the parameters μ and σ are, respectively, also the mean and standard deviation of the variate, whereas in the case of the gamma distribution, the parameters v and k are related to the mean and variance of the variate as follows: $E(X) = k/v$ and $\text{Var}(X) = k/v^2$.

Therefore, on the basis of the relationships between the moments of a random variable and the parameters of the corresponding distribution, such as those summarized in Table 6.1, it follows that the parameters of a distribution may be determined by first estimating the mean and variance (and higher moments if there are more than two parameters) of the random variable. This process, in essence, is the basis of the *method of moments*.

Given a sample of size n, with sample values $(x_1, x_2, \ldots, x_n)$, we may use the *sample moments* to estimate the respective moments of the random variable. In this regard, just as the mean and variance are the weighted averages of X and $(X-\mu)^2$, the *sample mean* and *sample variance* may be defined as the respective averages of the sample values; namely,

for the sample mean,
$$\bar{x} = \frac{1}{n}\sum_{i=1}^{n} x_i \tag{6.1}$$

and for the sample variance,
$$s^2 = \frac{1}{n-1}\sum_{i=1}^{n}(x_i - \bar{x})^2 \tag{6.2}$$

Accordingly, $\bar{x}$ and s^2 are, respectively, the *point estimates* of the population mean μ and population variance σ^2.

We might point out that the sample variance of Eq. 6.2 is an "unbiased" estimate of σ^2; observe, however, that this is not quite the average of the sample values of $(x_i - \bar{x})^2$ for $i = 1, 2, \ldots, n$. The average would require division by n, instead of $(n-1)$, which would be a biased estimator of σ^2.

By expanding the squared terms of Eq. 6.2, it can be shown that Eq. 6.2 may be expressed also as

$$s^2 = \frac{1}{n-1}\left(\sum_{i=1}^{n} x_i^2 - n\bar{x}^2\right) \tag{6.3}$$

TABLE 6.1 Common Distributions and Their Parameters

Distribution	PDF or PMF	Parameters	Relations to Moments
Binomial	$p_X(x) = \binom{n}{x} p^x (1-p)^{n-x};$ $x = 0, 1, 2, \ldots, n$	p	$\mu_X = np$ $\sigma_X^2 = np(1-p)$
Geometric	$p_X(x) = p(1-p)^{x-1};$ $x = 1, 2, \ldots$	p	$\mu_X = 1/p$ $\sigma_X^2 = (1-p)/p^2$
Poisson	$P_X(x) = \dfrac{(vt)^x}{x!} e^{-vt};$ $x = 0.1, 2, \ldots$	v	$\mu_X = vt$ $\sigma_X^2 = vt$
Exponential	$f_X(x) = \lambda e^{-\lambda x};$ $x \geq 0$	λ	$\mu_X = 1/\lambda$ $\sigma_X^2 = 1/\lambda^2$
Gamma	$f_X(x) = \dfrac{v(vx)^{k-1}}{\Gamma(k)} e^{-vx};$ $x \geq 0$	v, k	$\mu_X = k/v$ $\sigma_X^2 = k/v^2$
3-P Gamma *(Shifted gamma)*	$f_x(x) = \dfrac{v[v(x-\gamma)]^{k-1}}{\Gamma(k)} \exp[-v(x-\gamma)];$ $x \geq \gamma$	$v, \gamma, k \geq 1.0$	$\mu_X = k/v + \gamma$ $\sigma_X^2 = k/v^2$ $\theta_X = 2/\sqrt{k}, \ v > 0$ $= -2\sqrt{k}, \ v < 0$
Gaussian *(normal)*	$f_X(x) = \dfrac{1}{\sqrt{2\pi}\sigma} \exp\left[-\dfrac{1}{2}\left(\dfrac{x-\mu}{\sigma}\right)^2\right];$ $-\infty < x < \infty$	μ, σ	$\mu_X = \mu$ $\sigma_X^2 = \sigma^2$
Lognormal	$f_X(x) = \dfrac{1}{\sqrt{2\pi}\zeta x} \exp\left[-\dfrac{1}{2}\left(\dfrac{\ln x - \lambda}{\zeta}\right)^2\right];$ $x \geq 0$	λ, ζ	$\mu_X = \exp\left(\lambda + \dfrac{1}{2}\zeta^2\right)$ $\sigma_X^2 = \mu_X^2\left(e^{\zeta^2} - 1\right)$
Rayleigh	$f_X(x) \dfrac{x}{\alpha^2} \exp\left[-\dfrac{1}{2}\left(\dfrac{x}{\alpha}\right)^2\right];$ $x \geq 0$	α	$\mu_X = \sqrt{\dfrac{\pi}{2}}\alpha$ $\sigma_X^2 = \left(2 - \dfrac{\pi}{2}\right)\alpha^2$
Uniform	$f_X(x) = \dfrac{1}{(b-a)}; \quad a \leq x \leq b$	a, b	$\mu_X = \dfrac{a+b}{2}$ $\sigma_X^2 = \dfrac{1}{12}(b-a)^2$
Triangular	$f_X(x) = \dfrac{2}{b-a}\left(\dfrac{x-a}{u-a}\right); \quad a \leq x \leq u$ $= \dfrac{2}{b-a}\left(\dfrac{b-x}{b-u}\right); \quad u \leq x \leq b$	a, b, u	$\mu_X = \dfrac{1}{3}(a+b+u)$ $\sigma_X^2 = \dfrac{1}{18}(a^2+b^2+u^2-ab-au-bu)$

(Continued)

TABLE 6.1 (*Continued*)

Distribution	PDF or PMF	Parameters	Relations to Moments
Beta	$f_X(x) = \dfrac{1}{B(q, r)} \dfrac{(x - a)^{q-1}(b - x)^{r-1}}{(b - a)^{q+r-1}};$ $a \leq x \leq b$	q, r	$\mu_X = a + \dfrac{q}{q + r}(b - a)$ $\sigma_X^2 = \dfrac{qr(b - a)^2}{(q + r)^2(q + r + 1)}$
Gumbel (*Type* I largest)	$f_X(x) = \exp(-e^{-\alpha(x-u)});$	u, α	$\mu_X = u + \dfrac{0.5772}{\alpha}$ $\sigma_X^2 = \dfrac{\pi^2}{6\alpha^2}$
Weibull (*Type* III *smallest*)	$f_X(x) = \dfrac{k}{w - \varepsilon}\left(\dfrac{x - \varepsilon}{w - \varepsilon}\right)^{k-1} e^{-\left(\frac{x-\varepsilon}{w-\varepsilon}\right)^k};$ $x \geq \varepsilon$	k, w	$\mu_X = \varepsilon + (x - \varepsilon)\Gamma\left(1 + \dfrac{1}{k}\right)$ $\sigma_X^2 = (w - \varepsilon)^2\left[\Gamma\left(1 + \dfrac{2}{k}\right) - \Gamma^2\left(1 + \dfrac{1}{k}\right)\right]$

▶ **EXAMPLE 6.1**

Consider the data of the crushing strength of concrete tested for 25 cylinders listed in Table E6.1. Applying Eqs. 6.1 and 6.3, we obtain the sample mean and sample variance, respectively,

TABLE E6.1 **Sample Mean and Sample Variance of Example 6.1**

Cylinder #	Crushing Strength, x_i	x_i^2
1	5.6 ksi	31.36
2	5.3	28.09
3	4.0	16.00
4	4.4	19.36
5	5.5	30.25
6	5.7	32.49
7	6.0	36.00
8	5.6	31.36
9	7.1	50.41
10	4.7	22.09
11	5.5	30.25
12	5.9	34.81
13	6.4	40.96
14	5.8	33.64
15	6.7	44.89
16	5.4	29.16
17	5.0	25.00
18	5.8	33.64
19	6.2	38.44
20	5.6	31.36
21	5.7	32.49
22	5.9	34.81
23	5.4	29.16
24	5.1	26.01
25	5.7	32.49
	$\sum = 140.00$	$\sum = 794.52$

as follows (using the results in Table E6.1):

$$\bar{x} = \frac{1}{25}\sum_{i=1}^{25} x_i = \frac{140}{25} = 5.6\,\text{ksi}$$

and

$$s^2 = \frac{1}{24}\left[\sum_{i=1}^{25} x_i^2 - 25(5.6)^2\right] = \frac{1}{24}(794.52 - 784) = 0.44$$

Therefore, the estimates of the population mean and standard deviation are $\bar{x} = 5.6\,\text{ksi}$ and $s = \sqrt{0.44} = 0.66\,\text{ksi}$.

Now, if the probability distribution of the crushing strength is modeled with the Gaussian distribution, then its parameters (see Table 6.1) would be $\hat{\mu} = 5.6\,\text{ksi}$ and $\hat{\sigma} = \sqrt{0.44} = 0.66\,\text{ksi}$.

However, if the crushing strength of concrete is assumed to be gamma distributed, then according to Table 6.1, the parameters v and k would be

$$\hat{k}/\hat{v} = 5.6 \qquad \text{and} \qquad \hat{k}/\hat{v}^2 = 0.44$$

from which we obtain the corresponding parameters:

$$\hat{v} = 12.73 \qquad \text{and} \qquad \hat{k} = 71.29 \qquad\qquad ◀$$

▶ **EXAMPLE 6.2**

Data for the fatigue life of 75 S-T aluminum yielded the sample mean and sample variance for the fatigue life (in terms of loading cycles) as follows:

$$\bar{x} = 26.75\,\text{million cycles}$$
$$s^2 = 360.0\,(\text{million cycles})^2$$

According to the relationships given in Table 6.1, the mean and variance of a lognormal distribution are given by

$$\mu_X = \exp\left(\lambda + \frac{1}{2}\zeta^2\right)$$
$$\sigma_X^2 = \mu_X^2(e^{\zeta^2} - 1)$$

Substituting the sample mean and sample variance, respectively, for the true mean and variance, we obtain the estimated values of the parameters λ and ζ as follows (denoting $\hat{\lambda}$ and $\hat{\zeta}$ as the respective estimators):

$$\exp\left(\hat{\lambda} + \frac{1}{2}\hat{\zeta}^2\right) = 26.75$$

and,

$$(26.75)^2(e^{\hat{\zeta}^2} - 1) = 360.0$$

from which we obtain, $\hat{\lambda} = 3.08$ and $\hat{\zeta} = 0.64$. ◀

The Method of Maximum Likelihood

Another method of point estimation is the method of *maximum likelihood*. In contrast to the method of moments, the maximum likelihood method derives the point estimator of a parameter directly.

Consider a random variable X with PDF $f(x; \theta)$, in which θ is the parameter. If the observed sample values are $x_1, x_2, \ldots, x_n$, then we may inquire, "what is the most likely value of θ that will produce the particular set of observations?" In other words, among the possible values of θ, what is the value that will maximize the likelihood of obtaining the observed set of sample values $x_1, x_2, \ldots, x_n$? This is the rationale underlying the maximum likelihood method of point estimation.

It is reasonable to assume that the likelihood of obtaining a particular sample value x_i is proportional to the value of the PDF at x_i. Then, assuming random sampling, the likelihood of obtaining a set of n independent observations $x_1, x_2, \ldots, x_n$ is

$$L(x_1, x_2, \ldots, x_n; \theta) = f(x_1; \theta) f(x_2; \theta) \cdots f(x_n; \theta) \qquad (6.4)$$

which is the *likelihood function* of observing the set of sample values $x_1, x_2, \ldots, x_n$. We may then define the *maximum likelihood estimator* $\hat{\theta}$ as the value of θ that maximizes the likelihood function of Eq. 6.4, and thus may be obtained from the following equation (if the likelihood function, L, is differentiable with respect to the parameter θ),

$$\frac{\partial L(x_1, x_2, \ldots, x_n; \theta)}{\partial \theta} = 0 \qquad (6.5)$$

Because the likelihood function of Eq. 6.4 is a product function, it is more convenient to maximize the logarithm of the likelihood function; i.e.,

$$\frac{\partial \log L(x_1, x_2, \ldots, x_n; \theta)}{\partial \theta} = 0 \qquad (6.6)$$

The solution for $\hat{\theta}$ from Eq. 6.6, of course, is the same as that obtained through Eq. 6.5. For PDFs that are defined with two or more parameters, the likelihood function becomes

$$L(x_1, x_2, \ldots, x_n; \theta_1, \ldots, \theta_m) = \prod_{i=1}^{n} f(x_i; \theta_i, \ldots, \theta_m) \qquad (6.7)$$

where $\theta_1, \ldots, \theta_m$ are the m parameters to be estimated. In this case, the maximum likelihood estimators of these parameters may be obtained from the solution to the following set of simultaneous equations:

$$\frac{\partial \log L(x_1, x_2, \ldots, x_n; \theta_1, \ldots, \theta_m)}{\partial \theta_j} = 0; \qquad j = 1, 2, \ldots, m \qquad (6.8)$$

The maximum likelihood estimate (MLE) of a parameter possesses many of the desirable properties of an estimator described earlier. In particular, for large sample size n, the maximum likelihood estimator is often considered to yield the "best" estimate of a parameter, in the sense that asymptotically it has the minimum variance (Hoel, 1962).

▶ **EXAMPLE 6.3**

The times between successive arrivals of vehicles at a toll booth were observed as follows:

$$1.2, 3.0, 6.3, 10.1, 5.2, 2.4, \text{ and } 7.1 \text{ sec}$$

If the inter-arrival times of vehicles at the toll booth are modeled with an exponential distribution, with a PDF of

$$f_T(t) = \frac{1}{\lambda} e^{-t/\lambda}$$

in which λ is the parameter of the distribution and is also the mean inter-arrival time.

According to Eq. 6.4, the likelihood function for observing the seven inter-arrival times is

$$L(t_1, t_2, \ldots, t_7; \lambda) = \prod_{i=1}^{7} \frac{1}{\lambda} \exp(-t_i/\lambda) = (\lambda)^{-7} \exp\left(-\frac{1}{\lambda} \sum_{i=1}^{7} t_i\right)$$

and the corresponding logarithm is

$$\text{Log } L = -7 \text{ Log } \lambda - \frac{1}{\lambda} \sum_{i=1}^{7} t_i; \text{ and for the present example, Log } L = -7 \text{ Log } \lambda - \frac{35.30}{\lambda}$$

Thus, with Eq. 6.6, we obtain

$$\frac{\partial \text{ Log } L}{\partial \lambda} = -\frac{7}{\lambda} + \frac{35.30}{\lambda^2} = 0$$

from which we obtain the MLE of λ as

$$\hat{\lambda} = \frac{35.30}{7} = 5.04 \text{ sec}$$

From the above result, we may infer generally that the MLE for the parameter λ, from a sample of size n, is

$$\hat{\lambda} = \frac{1}{n} \sum_{i=1}^{n} t_i$$

◀

▶ **EXAMPLE 6.4**

Tri-axial laboratory tests are performed on five specimens of saturated sand. Each specimen is subjected to cyclic vertical loads with stress amplitudes of ± 200 psf. The number of load cycles applied until each of the specimens failed were observed to be as follows: 25, 20, 28, 33, and 26 cycles.

If the number of load cycles to failure of saturated sand is modeled with a lognormal distribution, the two parameters λ and ζ may be estimated by the maximum likelihood method as follows.

According to Eq. 6.7, the likelihood function for the case of a sample of size n would be

$$L(x_1, x_2, \ldots, x_n; \lambda, \zeta) = \prod_{i=1}^{n} \left\{ \frac{1}{\sqrt{2\pi} \zeta x_i} \exp\left[-\frac{1}{2} \left(\frac{\ln x_i - \lambda}{\zeta} \right)^2 \right] \right\}$$

$$= \left(\frac{1}{\sqrt{2\pi} \zeta} \right)^n \left(\prod_{i=1}^{n} \frac{1}{x_i} \right) \exp\left[-\frac{1}{2\zeta^2} \sum_{i=1}^{n} (\ln x_i - \lambda)^2 \right]$$

Taking the natural logarithm of both sides, the above becomes

$$\ln L(x_1, \ldots, x_n; \lambda, \zeta) = -n \ln\sqrt{2\pi} - n \ln \zeta - \sum_{i=1}^{n} \ln x_i - \frac{1}{2\zeta^2} \sum_{i=1}^{n} (\ln x_i - \lambda)^2$$

To obtain the maximum likelihood function, we require

$$\frac{\partial \ln L}{\partial \lambda} = 0; \quad \text{yielding} \quad \frac{1}{\zeta^2} \sum_{i=1}^{n} (\ln x_i - \lambda) = 0$$

and

$$\frac{\partial \ln L}{\partial \zeta} = 0; \quad \text{yielding} \quad -\frac{n}{\zeta} + \frac{1}{\zeta^3} \sum_{i=1}^{n} (\ln x_i - \lambda)^2 = 0$$

From the two equations above, we obtain the MLE for the two parameters, respectively, as

$$\hat{\lambda} = \frac{1}{n} \sum_{i=1}^{n} \ln x_i \quad \text{and} \quad \hat{\zeta}^2 = \frac{1}{n} \sum_{i=1}^{n} (\ln x_i - \lambda)^2$$

Applied to the present example, with the observed number of load cycles for the five specimens, we obtain the corresponding MLE as follows:

$$\hat{\lambda} = \frac{1}{5}(\ln 25 + \ln 20 + \ln 28 + \ln 33 + \ln 26) = 3.26$$

and

$$\hat{\zeta}^2 = \frac{1}{5}[(\ln 25 - 3.26)^2 + (\ln 20 - 3.26)^2 + (\ln 28 - 3.26)^2$$
$$+ (\ln 33 - 3.26)^2 + (\ln 26 - 3.26)^2] = 0.027$$

or,

$$\hat{\zeta} = 0.164$$

For this example, the same parameters estimated by the method of moments would be as follows:

The sample mean is $\quad \bar{x} = \frac{1}{5}(25 + 20 + 28 + 33 + 26) = 26.40 \text{ cycles}$

and the sample variance is

$$s^2 = \frac{1}{4}[(25 - 26.4)^2 + (20 - 26.4)^2 + (28 - 26.4)^2 + (33 - 26.4)^2 + (26 - 26.4)^2] = 22.30$$

Thus, the sample standard deviation is $s = 4.72$ cycles.

Using the relationships given in Table 6.1 for the lognormal distribution, we have

$$26.40 = \exp\left(\lambda + \frac{1}{2}\zeta^2\right)$$

and

$$22.30 = (26.40)^2\left(e^{\zeta^2} - 1\right)$$

From which we obtain the corresponding estimates of the two parameters as

$$\hat{\lambda} = 3.26 \quad \text{and} \quad \hat{\zeta} = 0.178$$

We see that the two methods yield the same estimates for the parameter λ, but the estimates for the parameter ζ are different by about 8% between the two methods. ◄

Estimation of Proportion

In many engineering problems requiring the probability of occurrence or nonoccurrence of an event, the appropriate probability measure may be estimated as the proportion of experimental or field observations of the event—for example, the annual occurrence probability of hurricane-intensity wind in a coastal region, the proportion of vehicular traffic making left turns at a major intersection, or the proportion of embankment material meeting a specified compaction standard.

In such situations, the required probability may be estimated as the proportion of occurrences (of an event) in a Bernoulli sequence. Suppose that we have a sequence of n independent trials or random variables $X_1, X_2, \ldots, X_n$, where every X_i is a two-valued random variable; i.e., $X_i = 0$ or 1 denoting the occurrence or nonoccurrence of an event in the ith trial. Then, the sequence $X_1, X_2, \ldots, X_n$ constitutes a *random sample* of size n.

The probability of occurrence of an event, p, in a trial is the parameter of the binomial PMF associated with the Bernoulli sequence. The *maximum likelihood estimator* of the parameter p can be shown to be

$$\hat{P} = \frac{1}{n}\sum_{i=1}^{n}X_i \tag{6.9}$$

In other words, the estimate of p is the proportion of occurrences of an event among the sequence of n trials.

6.2.2 Sampling Distributions

Distribution of the Sample Mean with Known Variance

Thus far, we have used the sample mean $\overline{x}$ to estimate the population mean μ; naturally, questions about possible inaccuracy in the estimate can be raised. We examine such questions on the basis of the distribution of the sample mean.

First, for a sample of size n, the specific set of sample values $x_1, x_2, \ldots, x_n$ may be conceived as a single realization from a set of statistically independent sample random variables $X_1, X_2, \ldots, X_n$. Moreover, in random sampling, the PDFs of these sample random variables are assumed to be identically distributed as that of the population X; i.e.,

$$f_{X_1}(x_1) = f_{X_2}(x_2) = \cdots = f_{X_n}(x_n) = f_X(x)$$

Therefore, in reality, the sample mean is also a random variable,

$$\overline{X} = \frac{1}{n}\sum_{i=1}^{n}X_i \tag{6.10}$$

Its expected value is

$$\mu_{\overline{X}} = E\left(\frac{1}{n}\sum_{i=1}^{n}X_i\right) = \frac{1}{n}(n\mu) = \mu \tag{6.11}$$

which means that the expected value of the sample mean is equal to the population mean, and therefore $\overline{X}$ is an unbiased estimator of the population mean μ.

Observing that $X_1, X_2, \ldots, X_n$ are statistically independent and identically distributed as X, the variance of $\overline{X}$ is, therefore,

$$\text{Var}(\overline{X}) = \text{Var}\left(\frac{1}{n}\sum_{i=1}^{n}X_i\right) = \frac{1}{n^2}\text{Var}\left(\sum_{i=1}^{n}X_i\right) = \frac{1}{n^2}(n\sigma^2) = \frac{\sigma^2}{n} \tag{6.12}$$

or the standard deviation of the sample mean is $\sigma_{\overline{X}} = \dfrac{\sigma}{\sqrt{n}}$.

Therefore, the sample mean $\overline{X}$ has a mean value μ and standard deviation $\sigma/\sqrt{n}$; the latter is the *sampling error* underlying $\overline{X}$ as the estimator of μ and constitutes an epistemic uncertainty as defined in Chapter 1. According to the results of Chapter 4, if the underlying population X is Gaussian, the sample mean $\overline{X}$ is also Gaussian. Moreover, if the sample size n is sufficiently large, the sample mean $\overline{X}$ will be approximately Gaussian, by virtue of the central limit theorem, even if the underlying population is non-Gaussian. Therefore, for most practical applications, $\overline{X}$ may be assumed to be a normal variate with distribution $N(\mu, \sigma/\sqrt{n})$. Clearly, as the sample size n increases, the distribution of the sample mean $\overline{X}$ becomes narrower, as illustrated in Fig. 6.3; thus, the estimator $\overline{x}$ for the population mean μ improves with the sample size n.

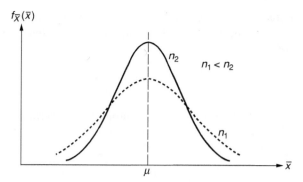

Figure 6.3 PDF of sample mean $\overline{X}$ with varying n.

It follows, therefore, that the distribution of $\dfrac{\overline{X} - \mu}{\sigma/\sqrt{n}}$ is $N(0,1)$.

Distribution of the Sample Mean with Unknown Variance

In general, the standard deviation of the population, σ, is unknown and may need to be estimated with the sample variance of Eq. 6.2 or 6.3. Therefore, instead of σ, we have to use s. In such cases, the random variable $\dfrac{\overline{X} - \mu}{s/\sqrt{n}}$ will not be normal, particularly if the sample size n is small. More accurately, the random variable $\dfrac{\overline{X} - \mu}{s/\sqrt{n}}$ will have the Student's t-distribution (Freund, 1962) with $(n-1)$ degrees-of-freedom. The PDF of the t-distribution is

$$f_T(t) = \frac{\Gamma[(f+1)/2]}{\sqrt{\pi f}\,\Gamma(f/2)}\left(1 + \frac{t^2}{f}\right)^{-\frac{1}{2}(f+1)}; \qquad -\infty < t < \infty \qquad (6.13)$$

in which f is the number of degrees-of-freedom (d.o.f.). A number of PDFs of the t-distribution is shown in Fig. 6.4 for various values of $f =$ d.o.f. We can observe that each of the PDFs has the bell-shape curve similar to the Gaussian PDF and is symmetrical about the origin. For small values of f, the PDFs of the t-distribution have wider spread than the standard normal PDF; however, as f increases, it tends toward the standard normal distribution as can be seen in Fig. 6.4.

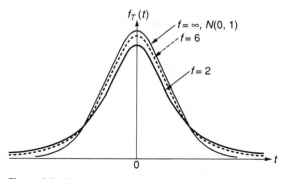

Figure 6.4 PDFs of the student's t-distribution for varying d.o.f.

Distribution of the Sample Variance

From Eq. 6.2, we may infer that the sample variance of the sample random variables is

$$S^2 = \frac{1}{n-1}\sum_{i=1}^{n}(X_i - \overline{X})^2 \tag{6.14}$$

which is also a random variable with expected value

$$E(S^2) = \frac{1}{n-1}E\left[\sum_{i=1}^{n}(X_i - \overline{X})^2\right] = \frac{1}{n-1}E\left\{\sum_{i=1}^{n}[(X_i - \mu) - (\overline{X} - \mu)]^2\right\}$$

$$= \frac{1}{n-1}\left\{\sum_{i=1}^{n}E(X_i - \mu)^2 - nE(\overline{X} - \mu)^2\right\}$$

but since $f_{X_i}(x) = f_X(x)$, then $\qquad E(X_i - \mu)^2 = \sigma^2$

and $\qquad\qquad\qquad\qquad E(\overline{X} - \mu)^2 = \dfrac{\sigma^2}{n}$

Hence, we have

$$E(S^2) = \frac{1}{n-1}\left(\sum_{i=1}^{n}\sigma^2 - n\cdot\frac{\sigma^2}{n}\right) = \frac{1}{n-1}(n\sigma^2 - \sigma^2) = \sigma^2 \tag{6.15}$$

On the basis of Eq. 6.15, the sample variance of Eq. 6.14 is, therefore, an unbiased estimator of the population variance, σ^2, as we asserted earlier for Eq. 6.2.

The variance of S^2 is given by (see Hald, 1952):

$$\mathrm{Var}(S^2) = \frac{\sigma^4}{n}\left(\frac{\mu_4}{\sigma^4} - \frac{n-3}{n-1}\right) \tag{6.16}$$

where $\mu_4 = E(X - \mu)^4$ is the *fourth central moment* or *kurtosis* of the population X. We may observe that as n increases, the variance of S^2 decreases.

For sampling in Gaussian populations, i.e., X is a Gaussian random variable, the distribution of the sample variance can be determined as follows.

We may rewrite Eq. 6.14 as

$$(n-1)S^2 = \sum_{i=1}^{n}[(X_i - \mu) - (\overline{X} - \mu)]^2 = \sum_{i=1}^{n}(X_i - \mu)^2 - n(\overline{X} - \mu)^2$$

Dividing both sides by σ^2, we obtain

$$\frac{(n-1)S^2}{\sigma^2} = \sum_{i=1}^{n}\left(\frac{X_i - \mu}{\sigma}\right)^2 - \left(\frac{\overline{X} - \mu}{\sigma/\sqrt{n}}\right)^2 \tag{6.17}$$

where X_i and $\overline{X}$ are both Gaussian. We may recognize that the first term on the right-hand side of Eq. 6.17 is the sum of squares of n independent standard normal variates; it can be shown, by generalizing the result of Example 4.8, that this has a chi-square distribution with n degrees-of-freedom (denoted as χ_n^2). Similarly, the second term on the right-hand side of Eq. 6.17 is also a chi-square distribution with one degree-of-freedom. Moreover, according to Hoel (1962), the sum of two chi-square distributions, respectively, with degrees-of-freedom p and q is also a chi-square distribution with $(p + q)$ degrees-of-freedom. On these bases, the distribution of $(n - 1)S^2/\sigma^2$ is a chi-square distribution with $(n - 1)$ degrees-of-freedom (d.o.f.); i.e., χ_{n-1}^2.

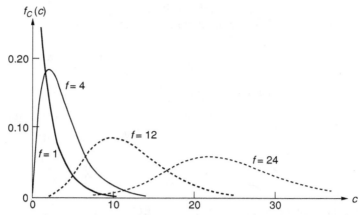

Figure 6.5 Chi-Square PDFs with varying d.o.f $= f$.

In general, the PDF of a chi-square distribution with f degrees-of-freedom (d.o.f.) is given by

$$f_C(c) = \frac{1}{2^{f/2}\Gamma\left(\dfrac{f}{2}\right)} c^{\left(\frac{f}{2}-1\right)} e^{-c/2} \tag{6.18}$$

The PDFs of Eq. 6.18 are shown in Fig. 6.5 for different degrees-of-freedom f. We may observe from this figure that as f increases, the χ_f^2 distribution tends to approach the normal distribution by virtue of the central limit theorem.

The mean and variance of the χ_f^2 distribution are, respectively,

$$\mu_C = f$$

and

$$\sigma_C^2 = 2f; \quad \text{or} \quad \sigma_C = \sqrt{2f}$$

The information developed in Sect. 6.2.2 will be useful in our presentation of hypothesis testing and in determining the confidence intervals, which are the topics of Sects. 6.3 and 6.4, respectively.

▶ 6.3 TESTING OF HYPOTHESES

6.3.1 Introduction

Hypothesis testing is a statistical method for making inferential decisions about the population based on information provided by available sampled data. The inference may be with respect to one or more population parameters or to the distribution model; the latter is a goodness-of-fit test which is covered in Chapter 7. Such hypothesis testings may be useful in many engineering applications. For example, engineering specifications often specify certain minimum values, say μ_o, such as the minimum yield strength of rebars. In reinforced concrete construction, the engineer would be interested in whether the available rebars satisfy the required minimum strength. In order to form the proper decision, samples of rebars may be tested, and the results used to formulate the appropriate decision regarding compliance with the required minimum. Such is one of the purposes of *hypothesis testing*.

A hypothesis is a statement regarding a parameter (or parameters) of the population. There are usually two competing hypotheses; the first one is the *null hypothesis*, denoted by H_o, and the competing one is the *alternative hypothesis*, denoted by H_A. The null hypothesis is formulated as an equality, whereas the alternative hypothesis is normally an inequality; for instance,

$$\text{\textit{Null hypothesis:}} \qquad H_o\text{: } \mu = \mu_o$$
$$\text{\textit{Alternative hypothesis:}} \ H_A\text{: } \mu \neq \mu_o$$

where μ is a population parameter and μ_o is the specified or required standard, such as the specified minimum yield strength of rebars in a specification.

Then, based on the set of observed data, and appropriate statistical analysis, the null hypothesis is accepted or rejected; in the latter case (i.e., rejection of the null hypothesis), the alternative hypothesis must be accepted. Again, we might emphasize that hypothesis testing is to make statistical inferences on the population parameter and/or distribution. In other words, hypothesis testing is to determine whether or not a population parameter is equal to a specified value, within a specified statistical significance level.

6.3.2 Hypothesis Test Procedure

The major steps needed in hypothesis testing are as follows.

Step 1—Define the *null* and the *alternative* hypotheses.

We have described above one definition of the null and alternative hypotheses, which may be called the two-sided test, involving both the upper and lower tails of a distribution. There are also one-sided tests as follows:

$$H_o\text{: } \mu = \mu_o$$
$$H_A\text{: } \mu > \mu_o$$

that involve the upper tail of a distribution. Similarly, the one-sided test involving the lower tail of a distribution would be

$$H_o\text{: } \mu = \mu_o$$
$$H_A\text{: } \mu < \mu_o$$

Step 2—Identify the proper *test statistic* and its *distribution.*

The appropriate test statistic, and its probability distribution, will depend on the population parameter that is being tested.

Step 3—Based on a sample of observed data, *estimate* the test statistic.

Step 4—Specify the *level of significance.*

Because the test statistic is a random variable, and its value is estimated from a finite sample of observations, there is a probability of error that a wrong hypothesis may be selected. In this regard, there are two possible types of errors:

Type I error—rejecting the null hypothesis H_o when in fact it is true.

Type II error—accepting the null hypothesis H_o when in fact it is false.

The probability of a Type I error, usually denoted α, is called the *level of significance*. This level should be selected with some care, although its selection is largely subjective. In practice, values of α between 1% and 5% are often selected as the levels of significance.

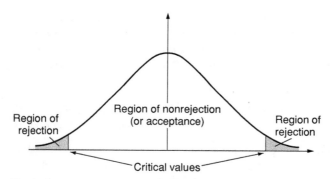

Figure 6.6 Regions of rejection and acceptance.

Although there is a probability of a Type II error, usually denoted β, this error is seldom used in practice. Therefore, we shall concentrate only on the Type I error.

Step 5—Define the *region for rejection* of the null hypothesis.

Based on the probability distribution of the test statistic, we can define the region for rejecting the null hypothesis corresponding to the specified level of significance α. In the case of a two-sided test, the region may be divided equally in the two-tail regions of the distribution, whereas in a one-sided test, the region will be in the upper tail region or the lower tail region, as illustrated in Fig. 6.6.

The region of rejection is defined by the appropriate *critical value(s)* of the test statistic corresponding to the specified level of significance α, as indicated in Fig. 6.6. Such critical values will depend on the appropriate probability distribution of the test statistic. The complementary region is the *region of nonrejection* of the null hypothesis (for practical purposes, this is equivalent to the region of acceptance of the null hypothesis).

▶ **EXAMPLE 6.5**
Test of the Mean with
Known Variance

Suppose the specification for the yield strength of rebars required a mean value of 38 psi. It is, therefore, essential that the population of rebars to be used in the construction of a reinforced concrete structure has the required mean strength. From the rebars delivered to the construction site by the supplier, the engineer ordered that a sample of 25 rebars be randomly selected and tested for yield strengths. The sample mean from the 25 tests yielded a value of 37.5 psi. It is known that the standard deviation of rebar strength from the supplier is 3.0 psi.

Since the engineer would be concerned only with rebars having mean yield strength lower than 38 psi, a one-sided test is appropriate. Specifically, the appropriate null and alternative hypotheses are as follows:

$$H_o: \mu_Y = 38 \text{ psi}$$

and

$$H_A: \mu_Y < 38 \text{ psi}$$

In this case, we may use the sample mean $\overline{X}$ to formulate the test statistic

$$Z = \frac{\overline{X} - \mu}{\sigma/\sqrt{n}}$$

As $\overline{X}$ is normal with a mean μ and a standard deviation of $\sigma/\sqrt{n}$, the distribution of Z is $N(0, 1)$. From the 25 sampled data, the estimated value of Z is

$$z = \frac{37.5 - 38.0}{3.0/\sqrt{25}} = -0.833$$

At a 5% level of significance, $\alpha = 5\%$, the critical value of z is found in Table A.1, yielding $z_\alpha = \Phi^{-1}(0.05) = -\Phi^{-1}(0.95) = -1.95$. Since the estimated statistic is -0.833, which is outside the region of rejection, or is within the region of acceptance, the null hypothesis is accepted. Therefore, the rebars from the supplier satisfy the required yield strength of the specification and are acceptable. ◀

▶ **EXAMPLE 6.6**
Test of the Mean with Unknown Variance

In Example 6.5, the standard deviation, σ, of the population of rebars is known to be 3.0 psi. In many cases, the information on the population standard deviation may not be known and must be estimated also from the available sampled data. In Example 6.5, suppose the 25 tests yielded the following results:

$$\overline{x} = 37.5\,\text{psi} \qquad \text{and} \qquad s = 3.50\,\text{psi}$$

The null and alternative hypotheses will remain the same as in Example 6.5. However, the test statistic will now be

$$T = \frac{\overline{X} - \mu}{s/\sqrt{n}}$$

From the test results, we obtain the estimated value of the test statistic to be

$$t = \frac{37.5 - 38}{3.5/\sqrt{25}} = -0.714$$

With $f = 25 - 1 = 24$ d.o.f., we obtain the critical value of t from Table A.3 at the 5% significance level to be $t_\alpha = -1.711$. Therefore, the value of the test statistic is $-0.714 > -1.711$ which is outside the region of rejection; hence, the null hypothesis is accepted, and the rebars from the supplier remain acceptable for yield strength. ◀

▶ **EXAMPLE 6.7**
Test of the Variance

Continuing with the same Example 6.5, let us now test the population variance of $\sigma^2 = 9.0$. For this purpose, let us suppose that the sample size was increased to 41 and the tests yielded the results: $\overline{x} = 37.60$ psi and $s = 3.75$ psi. The null and alternative hypotheses for this case are

$$H_o: \sigma^2 = 9.0$$
$$H_A: \sigma^2 > 9.0$$

For this test, the appropriate test statistic would be

$$C = \frac{(n-1)S^2}{\sigma^2}$$

which has the chi-square distribution of Eq. 6.17 with $(n-1)$ d.o.f.
With the results for the 41 sampled test data, the estimated value of the test statistic is

$$c = \frac{40(3.75)^2}{9} = 62.50$$

For a significance level of $\alpha = 2.5\%$, we obtain from Table A.4 with $f = (41-1) = 40$ d.o.f. the critical value of $c_{0.975} = 59.34$. Since $c = 62.50$ which is $> c_{0.975}$, and therefore is in the region of rejection, the null hypothesis is rejected and the alternative hypothesis must be accepted, which implies that $\sigma > 3.0$. ◀

▶ 6.4 CONFIDENCE INTERVALS (INTERVAL ESTIMATION)

In Sect. 6.2, we discussed the point estimation of the population parameters. The estimated values from a particular sample of size n may be considered the "best estimate" of the parameter of interest. However, there is no information about the accuracy of the estimate, except that the estimate will improve with the sample size n. In Sect. 6.3, we introduced the testing of hypotheses, which can be used to determine whether or not a target value of a population parameter (a minimum or maximum value), such as specified or required in a standard, is satisfied based on the statistics of the sampled data. Again, there is no quantitative measure of the accuracy of the estimated parameters. In order to provide some quantitative measures of accuracy of an estimator, we may use the *confidence intervals*. A confidence interval for a parameter defines a range (with lower and upper limits, or simply an upper limit or a single lower limit) within which the true value of the parameter will lie with a prescribed probability.

6.4.1 Confidence Interval of the Mean

With Known Variance

When the variance of the population is known, we saw in Sect. 6.2.2 that the probability distribution of the sample mean $\overline{X}$ is Gaussian with a distribution $N(\mu, \sigma/\sqrt{n})$. It follows that $K = \dfrac{\overline{X} - \mu}{\sigma/\sqrt{n}}$ is a standard normal random variable with the distribution $N(0, 1)$. On this basis, we may form the statement

$$P\left(k_{\alpha/2} < \frac{\overline{X} - \mu}{\sigma/\sqrt{n}} \leq k_{(1-\alpha/2)}\right) = (1 - \alpha)$$

in which $(1 - \alpha)$ is a specified probability, and

$$k_{\alpha/2} = -\Phi^{-1}(1 - \alpha/2) \qquad \text{and} \qquad k_{(1-\alpha/2)} = \Phi^{-1}(1 - \alpha/2)$$

are the lower and upper critical values, as shown in Fig. 6.7.

The above probability statement may be rewritten as

$$P\left(\overline{x} + k_{\alpha/2}\frac{\sigma}{\sqrt{n}} < \mu \leq \overline{x} + k_{(1-\alpha/2)}\frac{\sigma}{\sqrt{n}}\right) = (1 - \alpha)$$

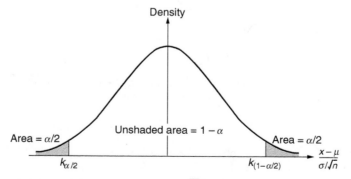

Figure 6.7 Standard normal PDF of $\dfrac{\overline{X} - \mu}{\sigma/\sqrt{n}}$ and the critical values.

From which we obtain the $(1 - \alpha)$ confidence interval for the population mean, μ, as

$$\langle \mu \rangle_{1-\alpha} = \left(\overline{x} + k_{\alpha/2} \frac{\sigma}{\sqrt{n}} ; \overline{x} + k_{(1-\alpha/2)} \frac{\sigma}{\sqrt{n}} \right) \tag{6.19}$$

in which $k_{\alpha/2} = -\Phi(1 - \alpha/2)$ and $k_{(1-\alpha/2)} = \Phi(1 - \alpha/2)$

▶ **EXAMPLE 6.8**

Let us consider the yield strength of rebars that was described in Example 6.5. However, we now wish to determine the 95% confidence interval of the true mean yield strength of rebars. We recall that the essential information is as follows: The standard deviation of the yield strength is known to be 3.0 psi; the number of rebars that were tested is 25; and the estimated sample mean is 37.5 psi.

We determine the 95% confidence interval of the mean yield strength μ as follows:

First, we determine the lower critical value $k_{\alpha/2} = k_{0.025} = -\Phi^{-1}(0.975) = -1.96$; and the upper critical value $k_{(1-\alpha/2)} = k_{0.975} = \Phi^{-1}(0.975) = 1.96$. Thus,

$$\langle \mu \rangle_{0.95} = \left(37.5 - 1.96 \frac{3.0}{\sqrt{25}} ; 37.5 + 1.96 \frac{3.0}{\sqrt{25}} \right) = (36.32; 38.68) \text{ psi}$$

The above interval shows that with 95% probability, the value of the true mean μ will be between 36.32 psi and 38.68 psi. ◂

With Unknown Variance

In most practical situations, the population variance may not be known. In such cases, we must also estimate the sample variance s^2, besides the sample mean $\overline{x}$, from a sample of size n. Then, with reference to Sect. 6.2.2, the distribution of the statistic $T = \dfrac{\overline{X} - \mu}{s/\sqrt{n}}$ has a Student's t-distribution with $(n - 1)$ d.o.f. as shown in Fig. 6.8. On this basis, we can form the following probability statement:

$$P \left(t_{\frac{\alpha}{2}, n-1} < \frac{\overline{X} - \mu}{s/\sqrt{n}} \leq t_{(1-\frac{\alpha}{2}), n-1} \right) = (1 - \alpha)$$

where $t_{\frac{\alpha}{2}, n-1}$ and $t_{(1-\frac{\alpha}{2}), n-1}$ are, respectively, the lower and upper critical values at probabilities of $\alpha/2$ and $(1 - \alpha/2)$ of the t-distribution with $(n - 1)$ d.o.f. as indicated in Fig. 6.8. These critical values are tabulated in Table A.3 for given d.o.f. Observe that

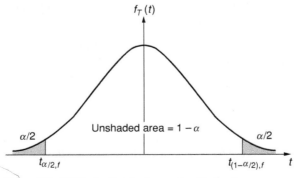

Figure 6.8 PDF of t-distribution and critical values.

because of symmetry of the t-distribution about the origin, $t_{\alpha/2, n-1} = -t_{1-\alpha/2, n-1}$, from which we determine the $(1 - \alpha)$ confidence interval for the population mean as

$$\langle \mu \rangle_{1-\alpha} = \left(\overline{x} + t_{\frac{\alpha}{2}, n-1} \frac{s}{\sqrt{n}}; \; \overline{x} + t_{(1-\frac{\alpha}{2}), n-1} \frac{s}{\sqrt{n}} \right) \tag{6.20}$$

We might point out that for large sample size n, e.g., > 50, the t-distribution approaches the standard normal distribution, as we saw earlier in Sect. 6.2.2. Therefore, in such cases, the confidence interval obtained with Eq. 6.20 can be expected to be close to that obtained with Eq. 6.19.

▶ **EXAMPLE 6.9**

In Example 6.6, the sample of 25 rebars tested for yield strengths yielded the following information: $\overline{x} = 37.5$ psi, and $s = 3.5$ psi. In this case, since the population variance is unknown and must be estimated also from the sampled data, the appropriate statistic is $T = \dfrac{\overline{X} - \mu}{s/\sqrt{n}}$ which has the t-distribution. For a 95% confidence interval, we obtain the lower and upper critical values from Table A.3 to be $t_{0.025,24} = -2.064$, and $t_{0.975,24} = 2.064$, respectively. Therefore, with Eq. 6.20, the 95% confidence interval for the mean yield strength is

$$\langle \mu \rangle_{0.95} = \left(37.5 - 2.064 \frac{3.5}{\sqrt{25}}; \; 37.5 + 2.064 \frac{3.5}{\sqrt{25}} \right) = (36.06; 38.94) \text{ psi}$$

Comparing this result with that of Example 6.8, we see that this 95% confidence interval is wider than that of Example 6.8. However, if the sample size n is increased to 120 from 25, the corresponding 95% confidence interval would become,

$$\langle \mu \rangle_{0.95} = \left(37.5 - 1.980 \frac{3.5}{\sqrt{120}}; \; 37.5 + 1.980 \frac{3.5}{\sqrt{120}} \right) = (36.87; 38.13) \text{ psi}$$

◀

One-Sided Confidence Limit of the Mean

The confidence interval we have established above is called a *two-sided confidence interval* because it defines the lower and upper bound values of the population mean μ. However, in practice, there are situations in which only the lower bound value or the upper bound value of the mean value is pertinent. In such cases, we would be interested in the *one-sided confidence limit* of the population mean. For example, in the case of material strength, or the traffic capacity of a highway, or the capacity of a flood control channel, the lower limit of the mean μ would be of main concern.

For such purposes, the $(1 - \alpha)$ lower confidence limit, which will be denoted $\langle \mu \rangle_{1-\alpha}$, means that the population mean μ is larger than this limit with a probability of $(1 - \alpha)$. Again, the determination of the lower confidence limit will depend on whether the population standard deviation, σ, is known or unknown. If prior information on σ is available, the lower limit is obtained based on the standard normal distribution of the statistic $K = \dfrac{\overline{X} - \mu}{\sigma/\sqrt{n}}$ as follows:

$$P \left(\frac{\overline{X} - \mu}{\alpha/\sqrt{n}} \leq k_{(1-\alpha)} \right) = (1 - \alpha)$$

where $(1 - \alpha)$ is the specified confidence level and $k_{(1-\alpha)}$ is the critical value given by $k_{(1-\alpha)} = \Phi^{-1}(1 - \alpha)$. Rearranging the terms in the above equation, we have

$$P \left(\mu \geq \overline{X} - k_{(1-\alpha)} \frac{\sigma}{\sqrt{n}} \right) = (1 - \alpha)$$

from which we obtain the $(1-\alpha)$ lower confidence limit for the population mean μ as

$$\langle\mu\rangle_{1-\alpha} = \left(\bar{x} - k_{(1-\alpha)}\frac{\alpha}{\sqrt{n}}\right) \tag{6.21}$$

However, if the population variance, σ^2, is not known, and the sample variance s^2 must be used in its place, then the appropriate statistic would be $T = \dfrac{\overline{X} - \mu}{s/\sqrt{n}}$ which has a t-distribution with $(n-1)$ d.o.f. Then the $(1-\alpha)$ lower confidence limit would be

$$\langle\mu\rangle_{1-\alpha} = \left(\bar{x} - t_{1-\alpha,\,n-1}\frac{s}{\sqrt{n}}\right) \tag{6.22}$$

in which $t_{1-\alpha,n-1}$ is the critical value at the probability of $(1-\alpha)$ of the t-distribution.

▶ **EXAMPLE 6.10**

Test results for 100 randomly selected specimens of 1-cm diameter A36 steel yielded the following sample mean and sample standard deviation of the yield strength of the material: $\bar{x} = 2200\,\text{kgf}$ (kilogram force) and $s = 220\,\text{kgf}$. For specification purposes, the manufacturer is required to specify the 95% lower confidence limit of the mean yield strength. As the sample size $n = 100$ is reasonably large, we may assume that the population standard deviation, σ, is fairly well represented by the sample standard deviation $s = 220\,\text{kgf}$.

Then, with $(1-\alpha) = 0.95$, $\alpha = 0.05$; we have from Table A.1,

$$k_{0.95} = \Phi^{-1}(0.95) = 1.65$$

Therefore, the 95% lower confidence limit is

$$\langle\mu\rangle_{0.95} = 2200 - 1.65\frac{220}{\sqrt{100}} = 2164\,\text{kgf}$$

In other words, the manufacturer of the steel can be 95% confident that the mean yield strength is at least 2164 kgf. ◀

▶ **EXAMPLE 6.11**

In Example 6.10, if the same sample mean and sample standard deviation were obtained from a sample of only 15 specimens, i.e., $n = 15$, the corresponding 95% lower confidence limit may be determined with Eq. 6.22 as follows:

From Table A.3, we obtain the critical value at the probability of $(1-\alpha) = 0.95$, and with d.o.f. $f = 14$, $T_{0.95,14} = 1.761$. Thus, the corresponding 95% lower confidence limit of the population mean would be

$$\langle\mu\rangle_{0.95} = \left(2200 - 1.761\frac{220}{\sqrt{15}}\right) = 2100\,\text{kgf} \qquad ◀$$

Conversely, there are situations in which the upper confidence limits of the population mean are of interest. For example, in determining the wind load for the design of a building, the upper limit of the mean wind force would be relevant; similarly, in assessing the flood of a river, the upper limit of the mean maximum flow of the river would be of concern. In such cases, the upper confidence limit on μ is of interest.

On the same basis as that of Eq. 6.21, we can show that the $(1-\alpha)$ *upper confidence limit* (if the population variance σ^2 is known) is

$$\langle\mu\rangle_{1-\alpha} = \left(\bar{x} + k_{1-\alpha}\frac{\alpha}{\sqrt{n}}\right) \tag{6.23}$$

whereas if the population variance is unknown, the corresponding $(1-\alpha)$ upper confidence limit must be based on the t-distribution as in Eq. 6.22, and thus would be

$$\langle\mu\rangle_{1-\alpha} = \left(\bar{x} + t_{1-\alpha,\,n-1}\frac{s}{\sqrt{n}}\right) \tag{6.24}$$

We may emphasize that the confidence intervals given in Eqs. 6.19 and 6.20 and the one-sided confidence limits of Eqs. 6.21 through 6.24 for the population mean μ are exact if the underlying population is Gaussian. However, for practical purposes, these results may be applied also to non-Gaussian populations especially if the sample sizes are reasonably large (e.g., $n > 20$) by virtue of the central limit theorem; for this reason, irrespective of the distribution of the underlying populations, the preceding equations may be used to determine (approximately, at least) the confidence intervals or one-sided confidence limits of the population mean μ.

▶ **EXAMPLE 6.12**

Recorded data (from Linsley and Franzini, 1964) of 25 storms and associated runoffs on the Monocacy River at Jug Bridge, Maryland, are presented in Table E6.12.
Denoting

$$X = \text{precipitation}$$
$$Y = \text{runoff}$$

we obtain the respective sample means and sample variances as follows;

$$\bar{x} = \frac{53.89}{25} = 2.16 \text{ in.} \quad \text{and} \quad s_X^2 = \frac{1}{24}[153.39 - 25(2.16)^2] = 1.53 \quad \text{or} \quad s_X = 1.24 \text{ in.}$$

and

$$\bar{y} = \frac{20.05}{25} = 0.80 \text{ in.} \quad \text{and} \quad s_Y^2 = \frac{1}{24}[24.68 - 25(0.80)^2] = 0.36 \quad \text{or} \quad s_Y = 0.60 \text{ in.}$$

TABLE E6.12   Precipitation and Runoff Data

Storm No.	X = Precipitation (in.)	Y = Runoff (in.)
1	1.11	0.52
2	1.17	0.40
3	1.79	0.97
4	5.62	2.92
5	1.13	0.17
6	1.54	0.19
7	3.19	0.76
8	1.73	0.66
9	2.09	0.78
10	2.75	1.24
11	1.20	0.39
12	1.01	0.30
13	1.64	0.70
14	1.57	0.77
15	1.54	0.59
16	2.09	0.95
17	3.54	1.02
18	1.17	0.39
19	1.15	0.23
20	2.57	0.45
21	3.57	1.59
22	5.11	1.74
23	1.52	0.56
24	2.93	1.12
25	1.16	0.64

With the above sample information, we obtain the 99% confidence interval for the mean precipitation μ_X as follows.

From Table A.3, with $f = 24$ d.o.f., we obtain the critical values $t_{0.005,24} = -2.797$ and $t_{0.995,24} = 2.797$, obtaining the confidence interval for the mean precipitation as

$$\langle \mu_X \rangle_{0.99} = \left(2.16 - 2.797 \frac{1.24}{\sqrt{25}}; \ 2.16 + 2.797 \frac{1.24}{\sqrt{25}} \right) = (1.47; \ 2.85) \text{ in.}$$

For the runoff, we would be more interested in the upper confidence limit of the mean runoff μ_Y. Thus, with Eq. 6.24, we obtain the 99% upper confidence limit as follows:

From Table A.3, with $f = 24$ d.o.f., we obtain the critical value $t_{0.99,24} = 2.492$, and the upper confidence limit for the mean runoff as

$$\langle \mu_Y \rangle_{0.99} = \left(0.80 + 2.492 \frac{0.60}{\sqrt{25}} \right) = 1.10 \text{ in.} \qquad ◀$$

On the Determination of Sample Size

The size of the sample data, n, is clearly of fundamental importance in hypothesis testing and in the determination of confidence intervals. From Eq. 6.19, we observe that for a $(1 - \alpha)$ confidence interval of the population mean μ, the half width of the interval is $k_{(1-\alpha/2)} \frac{\sigma}{\sqrt{n}}$.

We also observe that $\frac{\sigma}{\sqrt{n}}$ is the standard deviation of the sample mean; therefore, the half width $k_{(1-\alpha/2)} \frac{\sigma}{\sqrt{n}}$ is the number of sample standard deviations on either side of the sample mean $\bar{x}$. Hence, if the population variance σ^2 is known, and with a prescribed half width length w, the sample size n required for a $(1 - \alpha)$ confidence interval of the population mean can be obtained from

$$k_{(1-\alpha/2)} \frac{\sigma}{\sqrt{n}} = w$$

from which the required sample size is

$$n = \frac{1}{w^2} \left(\sigma k_{(1-\alpha/2)} \right)^2 \qquad (6.25)$$

in which w is the prescribed length of the half width expressed in terms of the number of standard deviations of the sample mean.

We might observe from Eq. 6.25 that the sample size n increases inversely with the prescribed half width, w; it also increases with the specified level of confidence $(1 - \alpha)$.

▶ **EXAMPLE 6.13**

In a traffic survey, the speeds of vehicles are measured by laser guns for the purpose of determining the mean vehicle speed on a particular city street. It is known that the standard deviation of vehicle speeds on city streets with the same posted speed limit is 3.58 kph. If we wish to determine the mean vehicle speed to within ±1 kph (kilometer per hour) with a 99% confidence, what should be the sample size of our observations?

If we prescribe $w = 1.0$, and from Table A.1 we obtain $k_{0.995} = \Phi^{-1}(0.995) = 2.58$, then with Eq. 6.25 the required sample size is

$$n = \frac{1}{(1.0)^2} (3.58 \times 2.58)^2 = 85.3 \qquad \text{or} \qquad 86$$

However, if we wish to reduce the half width to $w = 0.50$, then the required sample size for the same level of confidence would be increased to

$$n = \frac{1}{(0.5)^2}(3.58 \times 2.58)^2 = 341.2 \quad \text{or} \quad 342$$ ◀

In Eq. 6.25, we assumed that the population variance is known. If the population variance is unknown, it must be estimated with s^2 and the appropriate statistic would be

$$T = \frac{\overline{X} - \mu}{s/\sqrt{n}}$$

which has the t-distribution. In this case, the sample size n necessary for a $(1 - \alpha)$ confidence interval with a prescribed half width of w can be shown to be

$$n = \frac{1}{w^2}\left(s \cdot t_{1-\alpha/2, n-1}\right)^2 \tag{6.26}$$

in which $t_{1-\alpha/2, n-1}$ is the value of the variate at the probability of $(1 - \alpha/2)$ in the t-distribution of Table A.3 with $(n - 1)$ d.o.f.

From Eq. 6.26, we see that $t_{1-\alpha/2, n-1}$ is a function of n; hence, in this case, the solution for n will require trial and error.

6.4.2 Confidence Interval of the Proportion

The confidence interval for the proportion p (the occurrence probability) may be developed as follows: Referring to Eq. 6.9, we first observe that

$$E(\hat{P}) = E\left(\frac{1}{n}\sum_{i=1}^{n}X_i\right) = \frac{1}{n}\sum_{i=1}^{n}E(X_i)$$

but

$$E(X_i) = 1(p) + 0(1 - p) = p$$

Therefore,

$$E(\hat{P}) = \frac{1}{n}(np) = p \tag{6.27}$$

showing that the sample mean $\hat{p}$ is an unbiased estimate of p. And the variance of $\hat{P}$ is

$$\text{Var}(\hat{P}) = \frac{1}{n^2}\sum_{i=1}^{n}\text{Var}(X_i) = \frac{1}{n^2}\sum_{i=1}^{n}\left[E(X_i^2) - E^2(X_i)\right]$$

where $E(X_i^2) = 1^2(p) + 0^2(1 - p) = p$; therefore,

$$\text{Var}(\hat{P}) = \frac{1}{n^2} \cdot n(p - p^2) = \frac{p(1 - p)}{n} \tag{6.28}$$

For large n, $\hat{P}$ will be approximately Gaussian by virtue of the central limit theorem. Furthermore, the variance of Eq. 6.28 may be approximated by

$$\text{Var}(\hat{P}) = \frac{\hat{p}(1 - \hat{p})}{n} \tag{6.28a}$$

where $\hat{p}$ is the sample estimate of p. On the basis of the above, we can form the following statement:

$$P\left(k_{\alpha/2} < \frac{\hat{p} - p}{\sqrt{\hat{p}(1 - \hat{p})/n}} \leq k_{(1-\alpha/2)}\right) = (1 - \alpha)$$

from which we obtain the $(1 - \alpha)$ confidence interval of p as

$$\langle p \rangle_{1-\alpha} = \left(\hat{p} + k_{\alpha/2}\sqrt{\frac{\hat{p}(1 - \hat{p})}{n}}; \ \hat{p} + k_{(1-\alpha/2)}\sqrt{\frac{\hat{p}(1 - \hat{p})}{n}} \right) \tag{6.29}$$

where $k_{\alpha/2} = -\Phi^{-1}(1 - \alpha/2)$, and $k_{(1-\alpha/2)} = \Phi^{-1}(1 - \alpha/2)$.

► **EXAMPLE 6.14**

In order to control the quality of the compaction of the subgrade in a highway pavement project, 50 soil specimens were prepared and tested. Three out of the sample of 50 specimens were found to have compaction below the CBR requirement.

On the basis of the test results, we estimate the proportion of the subgrade to have satisfactory compaction (i.e., satisfying the CBR requirement) to be

$$\hat{p} = \frac{47}{50} = 0.94$$

That is, 94% of the subgrade for the highway pavement constructed may be expected to be well compacted. Furthermore, the 95% confidence interval of the proportion of well-compacted pavement is, according to Eq. 6.29, with $k_{0.025} = -\Phi^{-1}(0.975) = -1.96$ and $k_{0.975} = \Phi^{-1}(0.975) = 1.96$

$$\langle p \rangle_{0.95} = \left(0.94 - 1.96\sqrt{\frac{0.94(1 - 0.94)}{50}}; \ 0.94 + 1.96\sqrt{\frac{0.94(1 - 0.94)}{50}} \right)$$

$$= (0.938; 0.942) \qquad \blacktriangleleft$$

6.4.3 Confidence Interval of the Variance

In Sect. 6.2.2, we saw that if the population is Gaussian, the random variable $\dfrac{(n - 1)S^2}{\sigma^2}$ has a chi-square distribution with $f = (n - 1)$ d.o.f. as described in Eq. 6.18. On this basis, the $(1 - \alpha)$ confidence interval for the population variance σ^2 can be formulated with the following probability statement:

$$P\left[c_{\alpha/2, n-1} < \frac{(n - 1)S^2}{\sigma^2} \leq c_{1-\alpha/2, n-1} \right] = (1 - \alpha)$$

from which we obtain the $(1 - \alpha)$ confidence interval for σ^2 as

$$\langle \sigma^2 \rangle_{1-\alpha} = \left[\frac{(n - 1)s^2}{c_{1-\alpha/2, n-1}}; \ \frac{(n - 1)s^2}{c_{\alpha/2, n-1}} \right] \tag{6.30}$$

where s^2 is the sample variance of Eq. 6.3 and $c_{\alpha/2, n-1}$ and $c_{1-\alpha/2, n-1}$ are, respectively, the values of the χ^2 random variable at probabilities of $\alpha/2$ and $(1 - \alpha/2)$ with $(n - 1)$ d.o.f. which are tabulated in Table A.4.

Equation 6.30 is a two-sided confidence interval. Corresponding one-sided confidence limits may also be developed, which are as follows. For the lower $(1 - \alpha)$ confidence limit,

$$\langle \sigma^2 \rangle_{1-\alpha} = \frac{(n - 1)s^2}{c_{1-\alpha, n-1}} \tag{6.31}$$

whereas the corresponding upper $(1 - \alpha)$ confidence limit would be

$$\langle \sigma^2 \rangle_{1-\alpha} = \frac{(n - 1)s^2}{c_{\alpha, n-1}} \tag{6.32}$$

For the population variance, we may be more concerned with either the lower bound or the upper bound value, depending on the situation at hand. Accordingly, one or the other of Eqs. 6.31 and 6.32 may be more useful than Eq. 6.30.

▶ **EXAMPLE 6.15**

In Example 6.12, we estimated the sample variance of the runoff on the Monocacy River, from 25 storms, to be 0.36 in^2. We determine the upper 95% confidence limit of the population variance for the runoff of the river to be (with $c_{0.05,24} = 13.848$ from Table A.4),

$$(\sigma^2)_{0.95} = \frac{(25-1)(0.36)}{13.848} = 0.624 \text{ in.}^2$$

Therefore, the upper 95% confidence limit of the standard deviation would be $\sqrt{0.624} = 0.790$ in. ◀

6.5 MEASUREMENT THEORY

One of the major applications of the techniques developed for parameter estimation and for determining the confidence intervals is in problems of measurement theory, such as in surveying and photogrammetry. Problems involving measurements require estimating a fixed (but unknown) quantity from a sample of repeated measurements of the same quantity; such problems are analogous to the estimation of the unknown population mean with the sample mean.

For instance, in measuring a physical distance, δ, between two points, several (say n) repeated measurements of the same distance may be taken constituting a sample of size n. The objective is to estimate the (unknown) true distance δ from the n sample measurements[*] $d_1, d_2, \ldots, d_n$. In this regard, the distance δ is analogous to the population mean μ. Therefore, the methods developed in Sects. 6.2 and 6.4 are directly applicable for problems of measurement theory, except for changes in notations and terminology. In particular, the point estimate of the true distance δ is the sample mean

$$\overline{d} = \frac{1}{n}\sum_{i=1}^{n} d_i \tag{6.33}$$

In this case, the series of measurements $d_1, d_2, \ldots, d_n$ are a specific set of sample values from the independent sample random variables $D_1, D_2, \ldots, D_n$, such that the estimator of δ is also a random variable,

$$\overline{D} = \frac{1}{n}\sum_{i=1}^{n} D_i \tag{6.34}$$

with expected value, $E(\overline{D}) = \delta$, indicating that Eq. 6.34 is an unbiased estimator of the true distance δ, whereas the variance of $\overline{D}$ is $\mathrm{Var}(\overline{D}) = \dfrac{s^2}{n}$, in which

$$s^2 = \frac{1}{n-1}\sum_{i=1}^{n} (d_i - \overline{d})^2$$

which is the sample variance of the set of sample measurements.

In measurement theory, the standard deviation of $\overline{D}$, i.e., $\sigma_{\overline{D}} = \dfrac{s}{\sqrt{n}}$, is known as the *standard error*.

---

[*]Observed measurements will generally contain *systematic* and *random errors*. It is assumed here that all measurements have been adjusted for systematic errors.

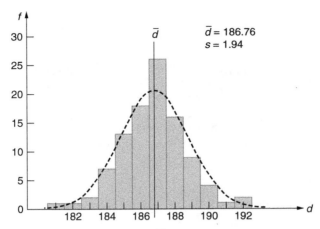

Figure 6.9 Histogram and $N(\overline{d}, s)$ distribution of measured distances.

The independent random variables $D_1, D_2, \ldots, D_n$ are assumed to be Gaussian; this is supported by observations of measured distances, such as, for example, Fig. 6.9. It follows that the random variable $\dfrac{\overline{D} - \delta}{s/\sqrt{n}}$ has a t-distribution with $(n-1)$ d.o.f.; therefore, the $(1 - \alpha)$ confidence interval for δ is

$$\langle \delta \rangle_{1-\alpha} = \left(\overline{d} + t_{\alpha/2,\, n-1} \frac{s}{\sqrt{n}};\ \overline{d} + t_{1-\alpha/2,\, n-1} \frac{s}{\sqrt{n}} \right) \tag{6.35}$$

in which $t_{1-\alpha/2,\, n-1}$ is the value of the random variable T with the PDF of Eq. 6.13 at the probability of $(1 - \alpha/2)$ with $(n-1)$ d.o.f., which can be found in Table A.3, and because of symmetry of the t-distribution $t_{\alpha/2} = -t_{1-\alpha/2}$.

► **EXAMPLE 6.16**

The straight-line distance between two geodetic stations A and B is measured with an electronic ranging instrument called a *tellerometer*. Ten repeated and independent measurements were taken of the distance as follows:

$$
\begin{array}{ll}
d_1 = 45{,}479.4 \text{ m} & d_6 = 45{,}479.2 \text{ m} \\
d_2 = 45{,}479.6 \text{ m} & d_7 = 45{,}479.6 \text{ m} \\
d_3 = 45{,}479.3 \text{ m} & d_8 = 45{,}479.5 \text{ m} \\
d_4 = 45{,}479.5 \text{ m} & d_9 = 45{,}479.3 \text{ m} \\
d_5 = 45{,}479.8 \text{ m} & d_{10} = 45{,}479.1 \text{ m}
\end{array}
$$

The estimated distance is, therefore,

$$\overline{d} = \frac{1}{10} \left(\sum_{i=1}^{10} d_i \right) = \frac{1}{10}(45{,}479.4 + \cdots + 45{,}479.1) = 45{,}479.43 \text{ m}$$

and the variance of the measured distances is

$$s^2 = \frac{1}{9} \sum_{i=1}^{10} \left[(d_i - \overline{d})^2 \right] = 0.0445 \text{ m}^2; \quad \text{or} \quad s = 0.21 \text{ m}$$

Therefore, the standard error of the estimated distance is

$$\sigma_{\overline{D}} = \frac{s}{\sqrt{n}} = \frac{0.21}{\sqrt{10}} = 0.0664 \text{ m}$$

We also determine the 90% confidence interval of the true distance δ with Eq. 6.35 as follows: With $f = n - 1 = 9$ d.o.f., we obtain from Table A.3, $t_{0.95,9} = 1.833$; then,

$$\langle\delta\rangle_{0.90} = \left(45,479.43 - 1.833\frac{0.21}{\sqrt{10}};\ 45,479.43 + 1.833\frac{0.21}{\sqrt{10}}\right)$$

$$= (45,479.31;\ 45,479.55)\ \text{m}$$

◀

For a function of one or more distances (or other geometrical dimensions), the function may be evaluated on the basis of the mean measurements; i.e., if ζ is a function of k distances $\delta_1, \delta_2, \ldots, \delta_k$,

$$\zeta = g(\delta_1, \delta_2, \ldots, \delta_k) \tag{6.36}$$

where the true distances $\delta_1, \delta_2, \ldots, \delta_k$ must necessarily be estimated by the respective mean measurements $\overline{d}_1, \overline{d}_2, \ldots, \overline{d}_k$. The estimator of ζ may then be obtained through first-order approximation (Eq. 4.50) as

$$\overline{\zeta} = g(\overline{D}_1, \overline{D}_2, \ldots, \overline{D}_k) \tag{6.37}$$

and its mean value is

$$\mu_{\overline{\zeta}} \simeq g(\overline{d}_1, \overline{d}_2, \ldots, \overline{d}_k) = \zeta \tag{6.38}$$

and based on Eq. 4.51a, the variance is

$$\sigma_{\overline{\zeta}}^2 \simeq \sum_{i=1}^{k}\left(\frac{\partial\overline{\zeta}}{\partial\overline{D}_i}\right)^2 \sigma_{\overline{D}_i}^2 \tag{6.39}$$

Assuming $\overline{\zeta}$ to be Gaussian with mean ζ according to Eq. 6.38, and standard error $\sigma_{\overline{\zeta}}$ as given by Eq. 6.39, we obtain the $(1 - \alpha)$ confidence interval for ζ as

$$\langle\zeta\rangle_{1-\alpha} = (\overline{\zeta} + k_{\alpha/2}\cdot\sigma_{\overline{\zeta}};\ \overline{\zeta} + k_{(1-\alpha/2)}\cdot\sigma_{\overline{\zeta}}) \tag{6.40}$$

in which $k_{(1-\alpha/2)} = \Phi^{-1}(1 - \alpha/2)$ and $k_{\alpha/2} = -k_{(1-\alpha/2)}$.

▶ **EXAMPLE 6.17**

Consider a rectangular tract of land as shown in Fig. E6.17. Each of the dimensions of the rectangle is measured several times as indicated below.

Dimension	No. of Independent Measurements	Mean Measurement, $\overline{d}_i$	Sample Variance, s_i^2
D	9	60 m	0.81 m^2
B	4	70 m	0.64 m^2
C	4	30 m	0.32 m^2

We determine the 95% confidence interval of the area of the tract of land as follows: The area of the rectangle is

$$A = (B + C)D$$

And according to Eq. 6.37, the area is estimated by

$$\overline{A} = (\overline{B} + \overline{C})\overline{D}$$

Using the respective mean measured dimensions, the estimated area is

$$\overline{A} = (70 + 30)60 = 6000\ \text{m}^2$$

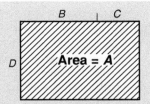

Figure E6.17 Rectangular track of land.

The variance of the estimator $\overline{A}$, according to Eq. 6.39, is

$$\sigma_{\overline{A}}^2 = \overline{D}^2\sigma_{\overline{B}}^2 + \overline{D}^2\sigma_{\overline{C}}^2 + (\overline{B} + \overline{C})^2\sigma_{\overline{D}}^2$$

$$= (60)^2\left(\frac{s_B^2}{4}\right) + (60)^2\left(\frac{s_C^2}{4}\right) + (100)^2\left(\frac{s_D^2}{9}\right)$$

$$= (60)^2\left(\frac{0.64}{4}\right) + (60)^2\left(\frac{0.32}{4}\right) + (100)^2\left(\frac{0.81}{9}\right)$$

$$= 3600(0.16) + 3600(0.08) + 10,000(0.09) = 1764\,\mathrm{m}^4$$

Thus, the standard error of the area A is $\sigma_{\overline{A}} = 42\,\mathrm{m}^2$.

Finally, using Eq. 6.40, we obtain the 95% confidence interval for the area A (with $k_{0.025} = -1.96$ and $k_{0.975} = 1.96$) as

$$\langle A\rangle_{0.95} = (6000 - 1.96 \times 42; 6000 + 1.96 \times 42)$$

$$= (5916.9; 6083.1)\,m^2 \qquad \blacktriangleleft$$

▶ 6.6 CONCLUDING SUMMARY

In modeling real-world phenomena, the form of the probability distribution of a random variable may be deduced theoretically on the basis of physical considerations or inferred empirically on the basis of observational data; the latter will be the subject of Chapter 7. However, the parameters of a distribution model, or the main descriptors (mean and variance) of the random variable must necessarily be related empirically to the real world; therefore, estimation based on factual data is essential and required for practical applications. The essential statistical methods of parameter estimation are presented in this chapter, which are of two types, namely, the *point* and *interval* estimations.

The common methods of point estimation are the *method of moments* and the *method of maximum likelihood*; the first method estimates a parameter indirectly by first evaluating the moments (such as the mean and variance) of the random variable through the corresponding sample moments, whereas the latter method derives the estimator of a parameter directly.

Clearly, when population parameters are estimated on the basis of limited or finite samples, errors of estimations are unavoidable. Such *sampling errors* constitute one source of epistemic uncertainty. Within the classical methods of estimation, such errors are not included in the point estimates of the parameters; the estimators may be tested for compliance with specified values or prescribed standards through *hypotheses testing*, or expressed in terms of appropriate *confidence intervals* that contain the true parameters within a prescribed probability (or confidence). An alternative consideration of these errors will be discussed in the Bayesian approach to estimation, which is the subject of Chapter 9.

▶ **PROBLEMS**

6.1 The foundation for a building is designed to rest on 100 piles based on the individual pile capacity of 80 tons. Nine test piles were driven at random locations into the supporting soil stratum and loaded until failure of each pile occurred. The results are as follows:

Test Piles	Pile Capacity (tons)
1	82
2	75
3	95
4	90
5	88
6	92
7	78
8	85
9	80

(a) Estimate the mean and standard deviation of the individual pile capacity to be used at the site.
(b) At the 5% significance level, should the piles be accepted based on the results of the nine test piles? That is, perform a one-sided hypothesis test, with the null hypothesis that the mean pile capacity is 80 tons.
(c) Establish the 98% confidence interval for the mean pile capacity, assuming that the standard deviation of the population is known, $\sigma = s$.
(d) Determine the 98% confidence interval for the mean pile capacity on the basis of unknown variance.

6.2 The average speed of vehicles on a highway is being studied.
(a) Suppose observations on 50 vehicles yielded a sample mean of 65 mph. Assume that the standard deviation of vehicle speed is known to be 6 mph. Determine the two-sided 99% confidence intervals of the mean speed.
(b) In part (a), how many additional vehicles' speed should be observed such that the mean speed can be estimated to within ± 1 mph with 99% confidence?
(c) Suppose John and Mary are assigned to collect data on the speed of vehicles on this highway. After each person has separately observed 10 vehicles, what is the probability that John's sample mean will exceed Mary's sample mean by 2 mph?
(d) Repeat part (c) if each person has separately observed 100 vehicles instead.

6.3 Suppose the annual maximum stream flow of a given river has been observed for 10 years yielding the following statistics:

$$\text{sample mean} = \overline{x} = 10,000 \text{ cfs}$$
$$\text{sample variance} = s^2 = 9 \times 10^6 (\text{cfs})^2$$

(a) Establish the two-sided 90% confidence interval on the mean annual maximum stream flow. Assume a normal population. *Ans. (8261, 11739)*

(b) If it is desired to estimate the mean annual maximum stream flow to within ±1,000 cfs with 90% confidence, how many additional years of observation will be required? Assume the sample (not the true value) variance based on the new set of data will be approximately 9×10^6 (cfs)^2. *(Ans. 17)*

6.4 Five piles have been load tested until failure; the load measured at failure denotes the actual capacity of the given pile. The following table summarizes the data from the load tests:

Pile Test No.	Actual Capacity A	Predicted Capacity P	N = A/P
1	20.5	13.6	to be determined
2	18.5	20.4	
3	10.0	8.8	
4	15.3	14.3	
5	26.2	22.8	

Observe that the capacity of each pile has also been predicted by a theoretical model as indicated in the table above. The factor N is simply the ratio of the actual pile capacity to the predicted pile capacity; i.e., $N = A/P$.
(a) Complete the table by calculating the respective value of N for each test pile.
(b) Determine the sample mean and variance of N. *(Ans. 1.154, 0.048)*
(c) Determine the 95% confidence interval of the mean value of N. *Ans. (0.881, 1.427)*
(d) In order to estimate the mean value of N to ± 0.02 with 90% confidence, how many additional piles should be tested? Assume that the variance of N is known and equal to 0.045 for this part. *(Ans. 300)*
(e) Assume N is a normal random variable whose mean value and variance are given exactly by the corresponding sample values from part (b). Consider a new site where a pile has been designed and its capacity is predicted by the model to be 15 tons. What is the probability that the pile will fail under a load of 12 tons? *(Ans. 0.0537)*

6.5 Concrete placed on a structure was subsequently cored after 28 days, and the following results were obtained of the compressive strengths from five test specimens:

$$4142, \; 3405, \; 3402, \; 4039, \; 3372 \text{ psi}$$

(a) Determine the 90% two-sided confidence interval of the mean concrete strength.
(b) Suppose the confidence interval established in part (a) is too wide, and the engineer would like to have a confidence interval to be ± 300 psi of the computed sample mean concrete strength. Generally, more specimens of concrete would be needed to keep the same confidence level. However, without additional samples,

what is the confidence level associated with the specified interval based on the five measurements given above?

(c) If the required minimum compressive strength is 3500 psi, perform a one-sided hypothesis test at the 2% significance level.

6.6 At a weigh station, the weights of trailer trucks were observed before crossing a highway bridge.

(a) Suppose observations on 30 trucks yielded a sample mean of 12.5 tons. Assume that the standard deviation of truck weights is known to be 3 tons. Determine the two-sided 99% confidence intervals of the mean weight of trailer trucks on the particular highway.

(b) In part (a), how many additional trucks should be observed such that the mean truck weight can be estimated to within ± 1.0 ton with 99% confidence?

6.7 The distribution of ocean wave height, H, may be modeled with the Rayleigh PDF as

$$f_H(h) = \frac{h}{\alpha^2} \cdot e^{-\frac{1}{2}(h/\alpha)^2} \qquad h \geq 0$$

in which α is the parameter of the distribution. Suppose that the following measurements of the wave height were observed:

1.50, 2.80, 2.50, 3.20, 1.90, 4.10, 3.60, 2.60, 2.90, 2.30 m

Estimate the parameter α by the method of maximum likelihood (MLE).

6.8 The daily dissolved oxygen concentration (DO) for a location A downstream from an industrial plant has been recorded for the past 10 consecutive days as tabulated below:

(a) Suppose the minimum concentration of DO required by the Environmental Protection Agency is 2.0 mg/l. Perform a hypothesis test to determine whether the stream quality satisfies the EPA standard at the significance level of 5%.

Day	DO (mg/l)
1	1.8
2	2
3	2.1
4	1.7
5	1.2
6	2.3
7	2.5
8	2.9
9	1.9
10	2.2

(b) Assume that the daily DO concentration has a normal distribution $N(\mu, \sigma)$; estimate the values of the parameters μ and σ.

(c) Determine the 95% confidence interval for the true mean DO concentration.

6.9 The height of a radio tower may be determined by measuring the horizontal distance L from the center of the tower base to the instrument, and the vertical angle β, as shown in the figure below.

(a) The distance L is measured three times with the following readings:

124.30, 124.20, and 124.40 ft

Determine the estimated distance and its standard error. *(Ans. 124.30 ft; 0.0577 ft)*

(b) The angle β is measured five times, with the following readings:

40°24.6′, 40°25.0′, 40°25.5′, 40°24.7′, 40°25.2′

Determine the estimated angle, and its standard error. *(Ans: 40°25′; 0.164′).*

(c) Estimate the height of the tower H. Assume that the instrument is 3 ft high with a standard deviation of 0.01 ft.

(d) Evaluate the standard error of the estimated height of the tower, $\sigma_{\overline{H}}$.

(e) Determine the 98% confidence interval of the actual height of the tower.

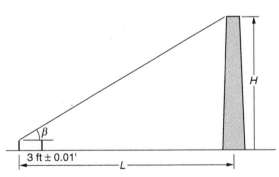

Determining height of tower.

6.10 Each of the inner and outer radii of a circular ring, as shown in the following figure was measured five times with the following readings:

outer radius, $r_o = 2.5, 2.4, 2.6, 2.6, 2.4$ cm

inner radius, $r_i = 1.6, 1.5, 1.6, 1.4, 1.4$ cm

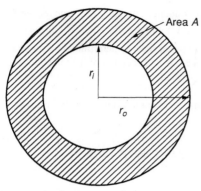

Area of circular ring.

(a) Determine the best estimates of the outer and inner radii, and the corresponding standard errors.

(b) The hatched area between the two concentric circles can be estimated based on the mean values of the estimated outer and inner radii; namely, the area is

$$\overline{A} = \pi(\overline{r}_o^2 - \overline{r}_i^2)$$

What is the estimated area? *Ans. 12.57 cm².*

(c) Determine the standard error of the estimated area. *(Ans. 0.819 cm²)*

(d) If it is desired to determine the sample mean of r_o to within ±0.07 cm with 99% confidence, how many additional independent measurements will be required? *(Ans. 12)*

6.11 The two distances b_1 and b_2, in meters, and the angle β, shown in the figure below were measured independently as shown in the table.

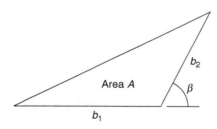

b_1, m	b_2, m	β in ° *and* ′
120.4	89.8	60° 20′
119.8	89.6	60° 10′
120.2	90.4	59° 45′
120.3	90.2	59° 35′
119.6	89.5	60° 5′
120.1		59° 50′
119.7		
119.4		

(a) Determine the estimated (mean) distances b_1 and b_2, and the angle β.

(b) Evaluate the respective standard errors.

(c) Estimate the area, A, of the triangular lot, and evaluate the associated standard error by first-order approximation.

(d) Determine the 90% confidence interval of the area.

6.12 The lengths a and b of the triangular lot, shown below, were measured independently as follows:

a, m	b, m
80.5	121.4
79.9	120.6
80.2	120.3
80.1	120.1
79.6	120.2
80.8	

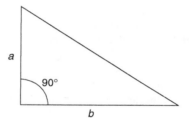

(a) What is the estimated area of the lot?

(b) Determine the standard error of the estimated lot area.

(c) What is the 95% confidence interval of the true area?

(d) From a regression analysis (see Chapter 8), suppose the expected cost of land is given by

$$E(C|A) = 10,000 + 15\,A \qquad \text{(in \$)}$$

in which A is in m², and the conditional standard deviation of C for given A is

$$s_{C|A} = 20,000 \qquad \text{(in \$)}$$

What would be the probability that the lot cost will exceed $90,000, assuming normal distribution?

6.13 The mileage of a certain make of car may not be exactly that rated by the manufacturer. Suppose ten cars of the same model were tested for combined city and highway mileages, with the following results:

Car No.	Observed Mileage
1	35 mpg
2	40
3	37
4	42
5	32
6	43
7	38
8	32
9	41
10	34

(a) Estimate the sample mean and sample standard deviation of the actual mileage of this particular make of car.

(b) Suppose that the stated mileage of this particular model of cars is 35 mpg; perform a hypothesis test to verify the stated mileage with a significance level of 2%.

(c) Determine the corresponding 95% confidence interval of the actual mileage.

► REFERENCES

Freund, J. E., *Mathematical Statistics*, Prentice-Hall, Englewood Cliffs, New Jersey, 1962.

Hald, A., *Stastistical Theory and Engineering Applications*, J. Wiley and Sons, Inc., New York, 1965.

Hoel, P. G., *Introduction to Mathematical Statistics*, 3rd Ed., J. Wiley and Sons, Inc., New York, 1962.

Linsley, R. K. and Franzini, J. B., *Water Resources Engineering*, McGraw-Hill Book Co., New York, 1964, p. 68.

CHAPTER 7

Determination of Probability Distribution Models

▶ 7.1   INTRODUCTION

The probability distribution model appropriate to describe a random phenomenon is generally not known; i.e., the functional form of the probability distribution is not defined. Under certain circumstances, the basis or properties of the physical process underlying the random phenomenon may suggest the form of the required distribution. For example, if a process is composed of the sum of many individual effects, the Gaussian distribution may be appropriate on the basis of the central limit theorem, whereas if the extremal conditions of a physical process are of interest, one of the asymptotic extreme-value distributions may be a suitable model (discussed in Sect. 7.4).

In many cases, the required probability distribution may need to be determined empirically based on the available observational data. For example, if the frequency diagram for a set of data can be constructed, the required distribution model may be inferred by visually comparing a particular PDF with the corresponding frequency diagram (see Chapter 1 for examples). Alternatively, the available data may be plotted on *probability papers* prepared for specific distributions (as described below in Sect. 7.2); if the data points plot approximately with a linear trend on one of these papers, the distribution associated with this paper may be an appropriate distribution model.

An assumed or prescribed prior probability distribution, perhaps determined empirically as described above or developed on theoretical grounds, may be verified, or disproved, in the light of available data using certain statistical tests, known as *goodness-of-fit* tests for distribution. Moreover, when two or more distributions appear to be plausible probability distribution models, such tests can be used to discriminate the relative validity of the different distributions. Two such goodness-of-fit tests are commonly used for these purposes—the *Chi-square* (χ^2) test and the *Kolmogorov–Smirnov* (K–S) test; a third test—the *Anderson–Darling* test—is particularly useful when the tails of a distribution are important.

In practice, the choice of the appropriate distribution model may be dictated by mathematical tractability or convenience. For example, because of the mathematical simplifications possible with the normal distribution and the wide availability of probability information (such as probability tables) for this distribution, the normal or lognormal distribution is frequently used to model nondeterministic problems—at times even when there is no clear-cut basis for such a model. Probabilistic information derived on the basis of such prescribed distribution models could be useful, particularly when the information is needed only for relative purposes. However, when the form of a distribution is important, particularly when ample data are available, the methods described in this chapter should provide the tools needed for its determination.

► 7.2 PROBABILITY PAPERS

Graph papers that are constructed for plotting observed data and their corresponding cumulative frequencies (or probabilities) are called *probability papers.* Probability papers are constructed such that a given probability paper is associated with a particular probability distribution; i.e., different probability papers correspond to different probability distributions.

Preferably, a probability paper should be constructed using a transformed probability scale in such a manner as to obtain a linear graph between the cumulative probabilities of the underlying distribution and the values of the random variable. For example, in the case of the uniform distribution, the cumulative probabilities are linearly related to the values of the variate; thus, the probability paper for this distribution would be constructed using arithmetic scales for both the values of the variate and the associated cumulative probabilities (between 0 and 1.0). For other distributions, however, special scales will be required for the cumulative probabilities in order to achieve the desired linear relationship.

7.2.1 Utility and Plotting Position

Observed data may be plotted on any probability paper. Each value of a set of data must be plotted at the appropriate cumulative probability; this data point consisting of an observed value and a cumulative probability is called a *plotting position* on a probability paper. The plotting position for each data point may be determined as follows:

> *For a set of N observations $x_1, x_2, \ldots, x_N$, arranged in increasing order, the mth value is plotted at the cumulative probability of $\dfrac{m}{N+1}$.*

This plotting position applies to all probability papers; its theoretical basis is discussed in Gumbel (1954). There are also other plotting positions, such as $(m-\frac{1}{2})/N$, which was advocated by Hazen (1930) and has been used widely; however, this plotting position has certain theoretical weaknesses. In particular, when there are N observations, this plotting position with probability of $(m-\frac{1}{2})/N$ would yield a return period of $2N$ for the largest observation instead of the correct value of N (Gumbel, 1954). Still other plotting positions have been suggested (e.g., Kimball, 1946); however, none seems to have the theoretical attributes and computational simplicity of $\dfrac{m}{N+1}$.

The utility of a probability paper is to provide a graphical portrayal of the frequency curve produced by a set of observational data. The linearity (or linear trend), or lack of a linear trend, of the set of sample data points plotted on a particular probability paper may be used as a basis for inferring or determining whether the distribution of the underlying population is the same as that of the probability paper. On this basis, therefore, probability papers may be used to establish or explore the possible distribution (s) of the underlying population. In the following sections, we shall describe the construction of several probability papers. In particular, in Sects. 7.2.2 and 7.2.3 we illustrate the construction and application of two commonly used probability papers—namely, the normal and the lognormal probability papers. In Sect. 7.2.4 we describe the construction of probability papers for other distributions.

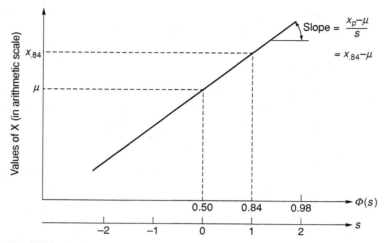

Figure 7.1 Construction of the normal probability paper.

7.2.2 The Normal Probability Paper

The *normal* (or *Gaussian*) probability paper is constructed on the basis of the standard normal CDF as follows:

- One axis, in arithmetic scale, represents the values of the random variable X, as illustrated in Fig. 7.1.
- On the other axis, perpendicular to the first, are two parallel scales; one in arithmetic scale represents values of the standard normal variate, s, whereas the other shows the cumulative probabilities, $\Phi(s)$, corresponding to the indicated values of s as shown also in Fig. 7.1.

A normal variate X with probability distribution $N(\mu, \sigma)$ is represented on this paper by a straight line passing through the point at $X = \mu$ and $\Phi(s) = 0.50$, with a slope of $(x_p - \mu)/s$, which is equal to the standard deviation σ, where x_p is the value of the variate at probability p. In particular, at $p = 0.84$, $s = 1$; hence, the slope is $(x_{0.84} - \mu)$.

Any set of data may be plotted on the normal probability paper; however, if the resulting graph of the data points, plotted with the plotting positions described earlier in Sect. 7.2.1, shows a lack of linearity (or linear trend), this would indicate that the underlying distribution of the variate is not Gaussian. Conversely, if the data points plotted on this normal probability paper show a linear trend, the straight line drawn through these data points represents a specific normal distribution for the data set, at least within the range of the observations.

▶ **EXAMPLE 7.1**

Data for the fracture toughness of steel plates are given in Table E7.1; the measured toughness has been rearranged in increasing order. The fracture toughness and corresponding cumulative probability of the data are plotted on the normal probability paper as shown in Fig. E7.1.

In Fig. E7.1, the data in Table E7.1 are plotted on a normal probability paper. Values of the fracture toughness K_{Ic} are plotted against the corresponding cumulative probabilities of $m/(N+1)$, with $N = 26$. The straight line drawn through the data points (by eye) as shown in Fig. E7.1 represents the normal distribution of the observed fracture toughness data, from which we obtain the mean value $\mu_{K_{Ic}} = 77$ ksi$\sqrt{\text{in}}$. From the same straight line, we also observe that the value of K_{Ic} at the probability of 84% is 81.6; thus, the standard deviation is $\sigma_{K_{Ic}} = 81.6 - 77 = 4.6$ ksi$\sqrt{\text{in}}$.

TABLE E7.1   Fracture Toughness of Steel Base Plates (after Kies et al., 1965)

m	K	$m/(N+1)$	m	K	$m/(N+1)$
1	69.5	0.0370	2	71.9	0.0741
3	72.6	0.1111	4	73.1	0.1418
5	73.3	0.1852	6	73.5	0.2222
7	74.1	0.2592	8	74.2	0.2963
9	75.3	0.3333	10	75.5	0.3704
11	75.7	0.4074	12	75.8	0.4444
13	76.1	0.4815	14	76.2	0.5185
15	76.2	0.5556	16	76.9	0.5926
17	77.0	0.6296	18	77.9	0.6667
19	78.1	0.7037	20	79.6	0.7407
21	79.7	0.7778	22	79.9	0.8148
23	80.1	0.8518	24	82.2	0.8889
25	83.7	0.9259	26	93.7	0.9630

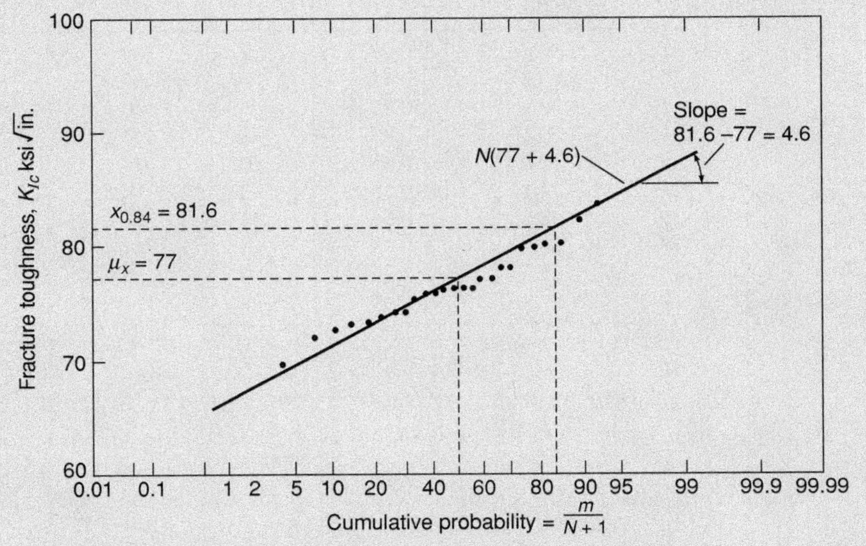

Figure E7.1   Fracture toughness of steel plotted on normal probability paper.   ◀

7.2.3   The Lognormal Probability Paper

The lognormal probability paper can be obtained from the normal probability paper by simply changing the arithmetic scale for values of the variate X on the normal probability paper to a logarithmic scale. The resulting paper would be as shown in Fig. E7.2. In this case, the standard normal variate becomes,

$$S = \frac{\ln X - \lambda}{\zeta} = \frac{\ln X - \ln x_m}{\zeta}$$

where x_m is the median value of X.

Observational data from a lognormal population should plot with a linear trend on the lognormal probability paper, so that a straight line can be drawn through the data points. From this straight line, the median x_m is simply the value of the variate on this line corresponding to the cumulative probability of 0.50, whereas the parameter ζ (or the c.o.v.) is given by the slope of the line, i.e.,

$$\zeta = \frac{\ln(x/x_m)}{s}$$

Conversely, if the data points of the mth value among N observations plotted at a cumulative probability of $m/(N+1)$ do not yield a linear trend on this probability paper, then the underlying population may not be lognormal.

▶ **EXAMPLE 7.2**

Data for the fracture toughness of MIG welds are shown in Table E7.2. Values of the toughness, arranged in increasing order, are shown in columns 2 and 5 in Table E7.2, and the corresponding plotting positions $m/(N+1)$ are shown in columns 3 and 6.

TABLE E7.2   Fracture Toughness of MIG Welds (data from Kies et al., 1965)

m	K	$\dfrac{m}{N+1}$	m	K	$\dfrac{m}{N+1}$
1	54.4	0.05	2	62.6	0.10
3	63.2	0.15	4	67.0	0.20
5	70.2	0.25	6	70.5	0.30
7	70.6	0.35	8	71.4	0.40
9	71.8	0.45	10	74.1	0.50
11	74.1	0.55	12	74.3	0.60
13	78.8	0.65	14	81.8	0.70
15	83.0	0.75	16	84.4	0.80
17	85.3	0.85	18	86.9	0.90
19	87.3	0.95			

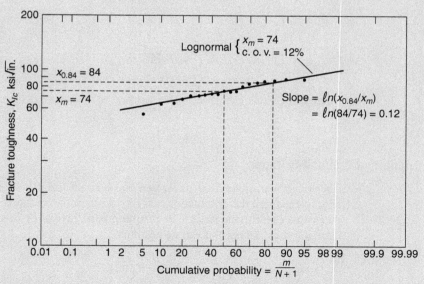

Figure E7.2   Fracture toughness of welds plotted on lognormal probability paper.

We can observe from Fig. E7.2 that the data points plotted on this probability paper yield a linear trend. Accordingly, the straight line drawn through the data points represents the lognormal distribution for the MIG welds with a median value of 74 ksi and a c.o.v. of 12% as indicated in Fig E7.2. ◀

▶ **EXAMPLE 7.3**

Measured data of the precipitation and runoff on the Monocacy River that were described earlier in Example 6.12 are rearranged in increasing order as shown in Table E7.3. The corresponding data points are plotted on the respective lognormal probability papers as shown in Fig. E7.3a for precipitation, X, and in Fig. E7.3b for the runoff, Y.

TABLE E7.3   Precipitation and Runoff Data on the Monocacy River

m	Precipitation X (in.)	Runoff Y (in.)	$\dfrac{m}{N+1}$	m	Precipitation X (in.)	Runoff Y (in.)	$\dfrac{m}{N+1}$
1	1.01	0.17	0.038	2	1.11	0.19	0.077
3	1.13	0.23	0.115	4	1.15	0.33	0.192
5	1.16	0.39	0.192	6	1.17	0.39	0.231
7	1.17	0.40	0.269	8	1.20	0.45	0.308
9	1.52	0.52	0.346	10	1.54	0.56	0.385
11	1.54	0.59	0.423	12	1.57	0.64	0.462
13	1.64	0.66	0.500	14	1.73	0.70	0.538
15	1.79	0.76	0.577	16	2.09	0.77	0.615
17	2.09	0.78	0.654	18	2.57	0.95	0.692
19	2.75	0.97	0.731	20	2.93	1.02	0.769
21	3.19	1.12	0.808	22	3.54	1.24	0.846
23	3.57	1.59	0.885	24	5.11	1.74	0.923
25	5.62	2.92	0.962				

On the basis of these two graphs, we may conclude from Fig. E7.3a that the distribution of the precipitation is not lognormal, whereas, from Fig. E7.3b, the runoff could reasonably be modeled with the lognormal distribution with parameters $\lambda_Y = \ln 0.66 = 0.42$ and $\zeta_Y = 0.79$.

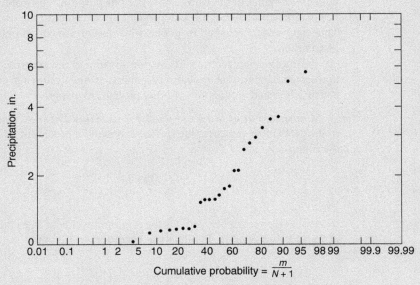

Figure E7.3a Precipitation plotted on lognormal probability paper.

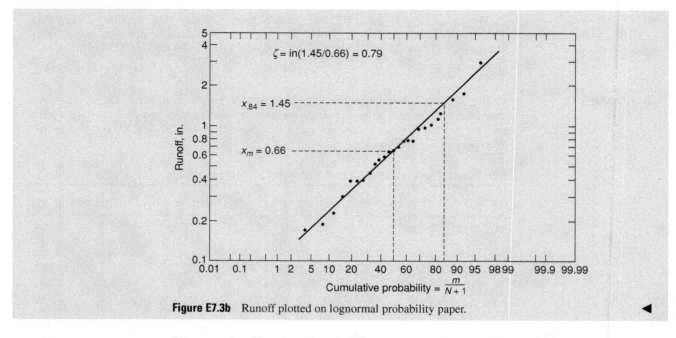

Figure E7.3b Runoff plotted on lognormal probability paper.

The normal and lognormal probability papers are commercially available.

7.2.4 Construction of General Probability Papers

A probability paper may be constructed for any probability distribution. For a given distribution, the corresponding probability paper should be such that values of a variate and the respective cumulative probabilities will yield a straight line on the appropriate paper. Conversely, any straight line on a specific probability paper represents a particular distribution with given values of the parameters. For this purpose, a probability paper, therefore, should be constructed such that it is independent of the values of the parameters of the distribution. This can be accomplished by defining a *standard variate*, if one exists, appropriate for the distribution.

In the last two sections above, we illustrated the construction and application of the normal and lognormal probability papers; in the following examples, we illustrate the construction and application of other probability papers.

Construction of the Exponential Probability Paper—Consider now the construction of the probability paper for the shifted exponential probability distribution. The PDF of this distribution is

$$f_X(x) = \lambda e^{-\lambda(x-a)}; \qquad x \geq a$$
$$= 0; \qquad\qquad x < a$$

where λ is the parameter, and a is the minimum value of X. In this case, the standard variate is $S = \lambda(X - a)$, and its PDF is then, according to Eq. 4.6,

$$f_S(s) = f_X\left(\frac{s}{\lambda} + a\right)\left|\frac{1}{\lambda}\right| = e^{-s}; \qquad s \geq 0$$
$$= 0 \qquad\qquad\qquad s < 0$$

TABLE 7.1   Grid Line Positions for Shifted Exponential Probability Paper

s	$F_S(s)$	s	$F_S(s)$	s	$F_S(s)$
0.11	0.10	1.20	0.70	3.00	0.95
0.22	0.20	1.39	0.75	3.10	0.955
0.36	0.30	1.61	0.80	3.22	0.96
0.51	0.40	1.90	0.85	3.35	0.965
0.69	0.50	2.30	0.90	3.51	0.97
0.80	0.55	2.53	0.92	3.69	0.975
0.92	0.60	2.66	0.93	3.91	0.98
1.05	0.65	2.81	0.94	4.20	0.985
				4.61	0.99

with corresponding CDF,

$$F_S(s) = 1 - e^{-s}; \qquad s \geq 0$$

On the basis of the above, we construct the exponential probability paper as follows:

- On one (e.g., the horizontal) axis, scale the values of the standard variate s (in arithmetic scale); on the same or a parallel axis, mark the corresponding cumulative probabilities according to the above $F_S(s)$.

- On the other (perpendicular) axis, mark the values of the original variate X (in arithmetic scale).

For illustration, specific values of s and the corresponding $F_S(s)$ are calculated as summarized in Table 7.1. Grid lines are then drawn for given $F_S(s)$ at the indicated values of s shown in Table 7.1. The resulting paper is shown in Fig. 7.2.

A straight line, with positive slope, on this paper represents a particular shifted exponential distribution, in which the intercept on the x-axis is the value of a, and its slope is $1/\lambda$.

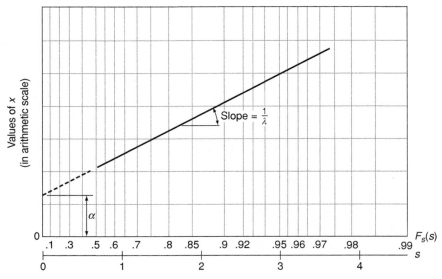

Figure 7.2 Construction of the exponential probability paper.

► **EXAMPLE 7.4**

Sample values from an exponential population should plot with a linear trend on the exponential probability paper. To illustrate this, consider the hypothetical set of data in Table E7.4 for a random variable X. The mth values of X and corresponding plotting positions $m/(N+1)$ are shown plotted in Fig. E7.4. From the straight line drawn visually through the data points in Fig. E7.4, we obtain estimates of the minimum value $a = 150$, and the slope $1/\lambda = 2000/2.69 = 743$; thus, the parameter $\lambda = 0.001346$.

TABLE E7.4   Sample Values from an Exponential Population

x	m	$\dfrac{m}{N+1}$	x	m	$\dfrac{m}{N+1}$	x	m	$\dfrac{m}{N+1}$
200	1	0.024	1635	37	0.902	952	30	0.756
201	2	0.049	559	21	0.512	1844	38	0.927
203	3	0.073	909	28	0.683	952	31	0.756
212	5	0.122	408	17	0.415	1427	36	0.878
248	7	0.171	2497	39	0.951	306	11	0.268
389	16	0.390	774	24	0.585	787	25	0.610
1331	35	0.854	946	29	0.707	254	8	0.195
1031	33	0.805	2781	40	0.976	772	23	0.561
208	4	0.098	308	12	0.293	842	27	0.659
226	6	0.146	274	9	0.220	981	32	0.780
289	10	0.244	531	19	0.463	1122	34	0.829
543	20	0.488	460	18	0.439	611	22	0.537
360	15	0.366	791	26	0.634	332	13	0.317
						343	14	0.341

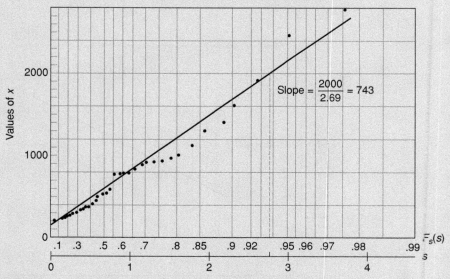

Figure E7.4   Sample values of X plotted on exponential probability paper.   ◄

In the case of an exponential distribution, we observe that

$$1 - F_X(x) = e^{-\lambda x} \qquad \text{or} \qquad \ln[1 - F_X(x)] = -\lambda x$$

Therefore, by scaling $[1 - F_X(x)]$ in logarithmic scale on one axis, and values of X in arithmetic scale on the other perpendicular axis, the graph between $[1 - F_X(x)]$ and x is a straight line on this semilogarithmic paper with a slope of λ. On this semilog paper, however, values of sample data must be plotted at the plotting positions of $[1 - m/(N + 1)]$.

The Gumbel Probability Paper—The *Type I* (Gumbel) distribution is one of the asymptotic distributions of extreme values (see Chapter 4). Its CDF for the distribution of the largest value is given by the double exponential form,

$$F_X(x) = \exp[-e^{-\alpha(x-u)}]; \qquad -\infty < x < \infty$$

where u and α are the parameters. For this distribution, the standard variate is

$$S = \alpha(X - u)$$

and its CDF is

$$F_S(s) = \exp(-e^{-s})$$

TABLE 7.2 Grid Line Positions for Constructing Gumbel Probability Paper

s	$F_S(s)$	s	$F_S(s)$
−2.0	0.0006	2.5	0.921
−1.5	0.011	3.0	0.951
−1.0	0.066	3.5	0.970
−0.5	0.192	4.0	0.982
0	0.368	4.5	0.989
0.5	0.545	5.0	0.993
1.0	0.692	5.5	0.996
1.5	0.800	6.0	0.998
2.0	0.873	7.0	0.999

Specific values of s and corresponding probabilities $F_S(s)$ are calculated as summarized in Table 7.2. These tabulated values and probabilities can then be used to construct the Gumbel probability paper as follows:

- Scale the values of s in arithmetic scale on one axis, and the associated probabilities $F_S(s)$ on the same (or a parallel) axis as shown in Fig. 7.3.

- The other perpendicular axis as shown in Fig. 7.3 are the values of the variate X in arithmetic scale.

The result is the *Type I* extremal probability paper, which is also known as the *Gumbel probability paper*. A straight line on this paper represents a particular *Type I* asymptotic extreme value distribution; the value of X on this line at $s = 0$ or $F_S(s) = 0.368$ is the value u, whereas the slope of the straight line is $1/\alpha$, as illustrated in Fig. 7.3.

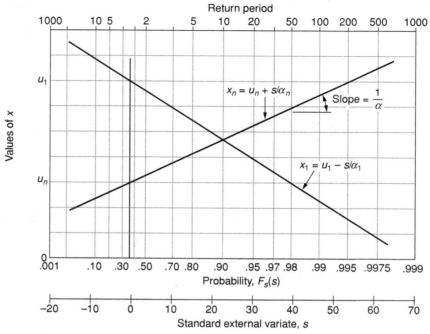

Figure 7.3 Construction of the Gumbel probability paper.

▶ **EXAMPLE 7.5**

In Table E7.5 we show data on the occurrence of the annual largest magnitude earthquakes for 31 years observed between 1932 and 1962 in California (Epstein and Lomnitz, 1966). Also shown in this table are the corresponding plotting positions that are used to plot the data on the Gumbel probability paper as shown in Fig. E7.5. From this figure, we can determine the parameters (at $s = 0$), $u = 5.7$ and the slope of the graph $1/\alpha = \frac{8-5.7}{4.6-0} = 0.50$ or $\alpha = 2.00$.

TABLE E7.5   Annual Maximum Earthquake Magnitudes in California (1932–1962)

m	mag, x	$\dfrac{m}{N+1}$	m	mag, x	$\dfrac{m}{N+1}$	m	mag, x	$\dfrac{m}{N+1}$
1	4.9	0.031	11	5.8	0.344	21	6.2	0.656
2	5.3	0.063	12	5.8	0.375	22	6.2	0.688
3	5.3	0.094	13	5.8	0.406	23	6.3	0.718
4	5.5	0.125	14	5.9	0.438	24	6.3	0.750
5	5.5	0.156	15	6.0	0.469	25	6.4	0.781
6	5.5	0.188	16	6.0	0.500	26	6.4	0.813
7	5.5	0.218	17	6.0	0.531	27	6.5	0.844
8	5.6	0.250	18	6.0	0.562	28	6.5	0.875
9	5.6	0.281	19	6.0	0.594	29	7.1	0.906
10	5.6	0.313	20	6.0	0.625	30	7.1	0.938
						31	7.7	0.969

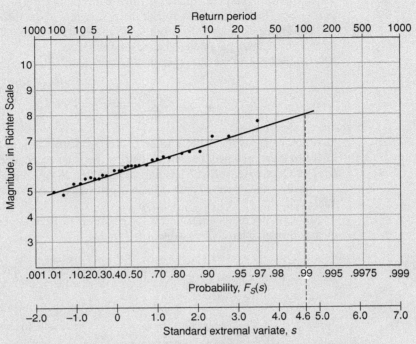

Figure E7.5 Annual largest earthquakes plotted on Gumbel probability paper.

▷ 7.3 TESTING GOODNESS-OF-FIT OF DISTRIBUTION MODELS

When a particular probability distribution has been specified to model a random phenomenon, determined perhaps on the basis of available data plotted on a given probability paper, or through visual inspection of the shape of the histogram, the validity of the specified or assumed distribution model may be verified or disproved statistically by *goodness-of-fit* tests. Three such tests for distribution are available and used widely—the *chi-square, the Kolmogorov–Smirnov* (or K–S), and the *Anderson–Darling* (or A–D) methods; one or the other of these methods may be used to test the validity of a specified or assumed distribution model. When two (or more) distributions appear to be plausible models, the same test may be used also to discriminate the relative superiority between (or among) the assumed distribution models. The A–D test is particularly useful when the tails of a distribution are of importance.

7.3.1 The Chi-Square Test for Goodness-of-Fit

Consider a sample of n observed values of a random variable and an assumed probability distribution for modeling the underlying population. The chi-square goodness-of-fit test compares the *observed frequencies* $n_1, n_2, \ldots, n_k$ of k values (or in k intervals) of the variate with the corresponding *theoretical frequencies* $e_1, e_2, \ldots, e_k$ calculated from the assumed theoretical distribution model. The basis for appraising the goodness of this comparison is the distribution of the quantity

$$\sum_{i=1}^{k} \frac{(n_i - e_i)^2}{e_i}$$

which approaches the chi-square (χ_f^2) distribution with $f = k - 1$ degrees-of-freedom (d.o.f.) as $n \to \infty$ (Hoel, 1962). However, if the parameters of the theoretical model are unknown and need to be estimated also from the available data, the d.o.f. f must be reduced by one for every unknown parameter that must be estimated.

On the basis of the above, if the assumed distribution yields

$$\sum_{i=1}^{k} \frac{(n_i - e_i)^2}{e_i} < c_{1-\alpha, f} \tag{7.1}$$

in which $c_{1-\alpha,f}$ is the critical value of the χ_f^2 distribution at the cumulative probability of $(1 - \alpha)$, the assumed theoretical distribution is an acceptable model, at the *significance level* α. Otherwise, if Eq. 7.1 is not satisfied, the assumed distribution model is not substantiated by the observed data at the α significance level. Values of $c_{1-\alpha,f}$ are tabulated in Table A.4.

In applying the chi-square test for goodness-of-fit, it is generally necessary for satisfactory results to have (if possible) $k \geq 5$ and $e_i \geq 5$.

► **EXAMPLE 7.6**

Severe thunderstorms have been recorded at a given station over a period of 66 years. During this period, the frequencies of severe thunderstorms observed are as follows:

20 years with zero storm	6 years with three storms
23 years with one storm	2 years with four storms
15 years with two storms	

The histogram of the annual number of rainstorms recorded at the station is displayed in Fig. E7.6. From the data, we estimate the mean annual occurrence rate of severe rainstorms to be 1.197 per year. In Fig. E7.6, we also display the Poisson distribution with an annual occurrence rate of $v = 1.197$, which appears to fit the observed histogram well.

Let us now apply the chi-square test to determine whether the Poisson distribution is a suitable model, at the 5% significance level. In this case, since four storms in a year were observed only twice, these data are combined with the data for three storms per year; thus, we have $k = 4$. Table E7.6 shows the calculations needed for the chi-square test.

Since the parameter v is estimated from the observed data, the d.o.f. for the χ_f^2 distribution is $f = 4-2 = 2$. From Table A.4, we obtain at $(1-\alpha) = 0.95$, $c_{0.95,2} = 5.99$. And from Table E7.6,

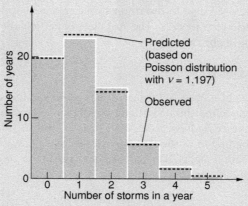

Figure E7.6 Histogram and the Poisson model for storm occurrences.

TABLE E7.6   Chi-Square Test of Poisson Model for Rainstorm Occurrences

No. of Storms per Year	Observed Frequencies, n_i	Theoretical Frequencies, e_i	$(n_i - e_i)^2$	$\dfrac{(n_i - e_i)^2}{e_i}$
0	20	19.94	0.0036	0.0002
1	23	23.87	0.7569	0.0317
2	15	14.29	0.5041	0.0353
≥3	8	7.90	0.0100	0.0013
$\sum$	66	66.00		0.0685

we have

$$\sum (n_i - e_i)^2/e_i = 0.068 < 5.99$$

Therefore, the Poisson distribution is suitable for modeling the annual occurrences of rainstorms at the station, at the 5% significance level. ◀

▶ **EXAMPLE 7.7**

Consider the histogram of the crushing strength of concrete cubes shown in Fig. E7.7. Also shown in the same figure are the normal and lognormal PDFs with the same mean and standard deviation as estimated from the observed data set. Visually, it appears that the two theoretical distributions are equally valid to model the crushing strength of concrete.

   In this case, the chi-square test can be used to discriminate the relative goodness of fit between the two candidate distributions. For this purpose, eight intervals of the crushing strength are considered as indicated in Table E7.7.

   The two parameters of both the normal and lognormal distributions were estimated from the sample data; consequently, the number of d.o.f. in both cases is $f = 8 - 3 = 5$. Therefore, at the significance level of $\alpha = 5\%$, we obtain from Table A.4, $c_{0.95,5} = 11.07$. Comparing this with the sum of the last two columns in Table E7.7, we see that both the normal and the lognormal are suitable for modeling the crushing strength of concrete; however, the lognormal model is superior to the normal model as indicated by the values of $\sum (n_i - e_i)^2/e_i$ for the two distributions, as shown in columns 5 and 6 of Table E7.7.

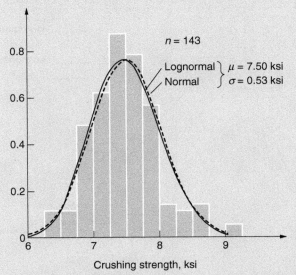

Figure E7.7   Histogram of crushing strength of concrete cubes. (Data from Cusens and Wettern, 1959.)

TABLE E7.7   Computations for Chi-Square Tests of Normal and Lognormal Distributions

Interval (ksi)	Observed Frequency, n_i	Theoretical Frequencies, e_i		$(n_i - e_i)^2/e_i$	
		Normal	Lognormal	Normal	Lognormal
< 6.75	9	11.1	9.9	0.40	0.09
6.75–7.00	17	13.2	14.0	1.09	0.92
7.00–7.25	22	21.1	22.1	0.04	0.00
7.25–7.50	31	26.1	26.9	0.92	0.62
7.50–7.75	28	26.1	25.6	0.14	0.23
7.75–8.00	20	21.0	19.8	0.05	0.00
8.00–8.50	9	20.2	19.4	6.22	5.57
> 8.50	7	4.2	5.3	1.87	0.54
$\sum$	143	143.0	143.0	10.73	7.97

◀

▶ **EXAMPLE 7.8**

From a sample of 320 observations, the histogram of residual stresses of wide flange steel beams is shown in Fig. E7.8. Superimposed on the same figure are the PDFs of three theoretical distribution models: the normal, lognormal, and the shifted (or 3-parameter) gamma distributions. From the data set, the first three moments were estimated to be, $\mu = 0.3561$, $\sigma = 0.1927$, and $\theta = 0.8230$ (skewness coefficient). Accordingly, the normal and lognormal distributions are assumed to have the estimated values for μ and σ, whereas the shifted gamma is assumed with the three estimated moments.

By visual inspection of Fig. E7.8, we can observe that the normal distribution is clearly not suitable. However, the lognormal and the shifted gamma distributions appear to fit the histogram well. In order to verify the relative validities of the three distribution models, we perform the chi-square test for goodness-of-fit with the calculations shown in Table E7.8.

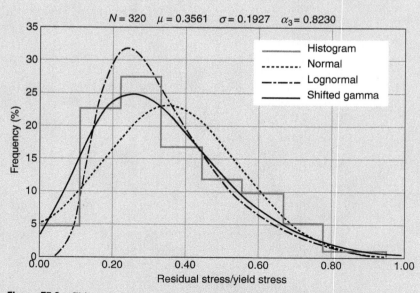

Figure E7.8   Chi-square tests to discriminate three distribution models for residual stresses.

TABLE E7.8   Computations for Chi-Square Tests of Three Distributions

Interval	Observed	Theoretical Frequencies			$\Sigma(n_i - e_i)^2/e_i$		
Res. Str./Yield Str.	Frequency	Norm	Lognorm	Shifted Gamma	Norm	Lognorm	Shifted Gamma
< 0.111	15	32.6	6.5	23.2	9.48	10.90	2.88
0.111–0.222	72	45.4	73.2	61.0	15.60	0.02	1.97
0.222–0.333	88	67.0	95.9	78.1	6.57	0.66	1.25
0.333–0.444	54	71.6	66.0	66.5	4.33	2.17	2.35
0.444–0.555	38	55.4	37.1	44.2	5.44	0.02	0.87
0.555–0.666	31	31.0	19.5	24.9	0.00	6.75	1.51
0.666–0.777	16	12.5	10.1	12.4	0.96	3.40	1.03
0.777–0.888	3	3.7	5.3	5.7	0.12	0.99	1.27
> 0.888	3	0.9	6.3	4.0	4.81	1.75	0.25
Σ	320	320	320	320	47.3	26.7	13.4

From Table E7.8, we see that among the three distributions, the shifted gamma distribution gives the lowest value for $\sum(n_i - e_i)^2/e_i$. Also, at the significance level of 1% and a d.o.f. of $f = 9 - 4 = 5$, we obtain from Table A.4 the critical value of $c_{0.99,5} = 15.09$ for the normal and lognormal distributions, whereas for the shifted gamma distribution $f = 9 - 5 = 4$ and $c_{0.99,4} = 13.28$. Therefore, according to the chi-square test, only the shifted gamma distribution (among the three distributions) is approximately valid at the 1% significance level for modeling the probability distribution of residual stresses in wide-flange beams. ◀

It may be emphasized that because there is some arbitrariness in the choice of the significance level α, the chi-square goodness-of-fit test (as well as the Kolmogorov–Smirnov and the Anderson–Darling methods, described subsequently in Sects. 7.3.2 and 7.3.3) may not provide absolute information on the validity of a specific distribution. For example, it is conceivable that a distribution acceptable at one significance level may be unacceptable at another significance level; this can be illustrated with the shifted gamma distribution of Example 7.8, in which the distribution is valid at the 1% significance level but will not be valid at the 5% level.

In spite of this arbitrariness in the selection of the significance level, however, such statistical goodness-of-fit tests remain useful, especially for determining the relative goodness-of-fit of two or more theoretical distribution models, as illustrated in Examples 7.7 and 7.8. Moreover, these tests should be used only to help verify the validity of a theoretical model that has been selected on the basis of other prior considerations, such as through the application of appropriate probability papers, or even visual inspection of an appropriate PDF with the available histogram.

7.3.2 The Kolmogorov–Smirnov (K–S) Test for Goodness-of-Fit

Another widely used goodness-of-fit test is the *Kolmogorov–Smirnov* or K–S test. The basic premise of this test is to compare the experimental cumulative frequency with the CDF of an assumed theoretical distribution. If the maximum discrepancy between the experimental and theoretical frequencies is larger than normally expected for a given sample size, the theoretical distribution is not acceptable for modeling the underlying population; conversely, if the discrepancy is less than a critical value, the theoretical distribution is acceptable at the prescribed significance level α.

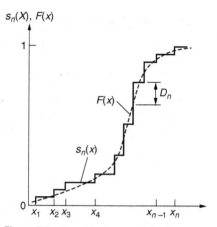

Figure 7.4 Empirical cumulative frequency versus theoretical CDF.

For a sample of size n, we rearrange the set of observed data in increasing order. From this ordered set of sample data, we develop a stepwise experimental cumulative frequency function as follows:

$$
\begin{aligned}
S_n(x) &= 0 & x < x_1 \\
&= \frac{k}{n} & x_k \leq x < x_{k+1} \\
&= 1 & x \geq x_n
\end{aligned}
\tag{7.2}
$$

where $x_1, x_2, \ldots, x_n$ are the observed values of the ordered set of data, and n is the sample size. Figure 7.4 shows a step-function plot of S_n and also an assumed theoretical CDF $F_X(x)$. In the K–S test, the maximum difference between $S_n(x)$ and $F_X(x)$ over the entire range of X is the measure of discrepancy between the assumed theoretical model and the observed data. Let this maximum difference be denoted by

$$
D_n = \max_x |F_X(x) - S_n(x)|
\tag{7.3}
$$

Theoretically, D_n is a random variable. For a significance level α, the K–S test compares the observed maximum difference, D_n of Eq. 7.3, with the critical value D_n^α which is defined for significance level α by

$$
P(D_n \leq D_n^\alpha) = 1 - \alpha
\tag{7.4}
$$

The critical values D_n^α at various significance levels α are tabulated in Table A.5 for various sample size n. If the observed D_n is less than the critical value D_n^α, the proposed theoretical distribution is acceptable at the specified significance level α; otherwise, the assumed theoretical distribution would be rejected.

The K–S test has some advantage over the chi-square test. With the K–S test, it is not necessary to divide the observed data into intervals; hence, the problem associated with small e_i and/or small number of intervals k in the chi-square test would not be an issue with the K–S test.

▶ **EXAMPLE 7.9**

The data for the fracture toughness of steel plates in Example 7.1 were plotted on a normal probability paper as shown in Fig. E7.1. The data appear to yield a linear trend producing a straight line corresponding to a normal distribution $N(77, 4.6)$. Let us now perform a K–S test to evaluate the

TABLE E7.9 Computations of S_n and $F_X(x)$ for Determining D

k	x	$S_n(x)$	$F_X(x) = \Phi\left(\dfrac{x-77}{4.6}\right)$	k	x	$S_n(x)$	$F_X(x) = \Phi\left(\dfrac{x-77}{4.6}\right)$
1	69.5	0.00	0.05	14	76.2	0.54	0.43
2	71.9	0.08	0.13	15	76.2	0.58	0.43
3	72.6	0.12	0.17	16	76.9	0.62	0.49
4	73.1	0.15	0.20	17	77.0	0.66	0.50
5	73.3	0.19	0.21	18	77.9	0.69	0.57
6	73.5	0.23	0.22	19	78.1	0.73	0.59
7	74.1	0.27	0.26	20	79.6	0.77	0.71
8	74.2	0.31	0.27	21	79.7	0.81	0.72
9	75.3	0.35	0.36	22	79.9	0.85	0.74
10	75.5	0.38	0.37	23	80.1	0.88	0.75
11	75.7	0.42	0.39	24	82.2	0.92	0.87
12	75.8	0.46	0.40	25	83.7	0.96	0.93
13	76.1	0.50	0.42	26	93.7	1.00	0.999

appropriateness of the proposed normal distribution model in light of the available data, at the 5% significance level.

With the tabulated data that have been rearranged in increasing order in Table E7.1, we illustrate the calculations of Eq. 7.2 for the empirical cumulative frequency and the corresponding theoretical CDF for the $N(77, 4.6)$ as shown in Table E7.9. The two cumulative frequencies are plotted as shown in Fig. E7.9. From Fig. E7.9, or Table E7.9, we can observe that the maximum discrepancy between the two cumulative frequencies is $D_n = 0.16$ and occurs at $x = K_{I_c} = 77$ ksi $\sqrt{\text{in}}$.

In this case, the sample size is $n = 26$; therefore, at the 5% significance level, we obtain the critical value of D_n^α from Table A.5 to be $D_{26}^{.05} = 0.265$. Since the maximum discrepancy $D_n = 0.16$ is less than 0.265, the normal distribution $N(77, 4.6)$ is verified as an acceptable model at the 5% significance level.

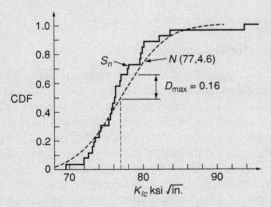

Figure E7.9 Cumulative frequencies for the K–S test of fracture toughness.

◀

▶ **EXAMPLE 7.10**

In Example 7.8, we tested the goodness-of-fit of three candidate distributions (the normal, lognormal, and the shifted gamma) for the residual stresses in wide flange steel beams using the chi-square test. With the same data, let us now use the K–S test to determine the goodness-of-fit of the same three distributions.

For the K–S test, we constructed Fig. E7.10 to display the empirical cumulative frequency function of the observed data and the respective CDFs of the normal, lognormal, and shifted gamma distributions. The respective maximum discrepancies for the three theoretical distributions, in accordance with Eq. 7.3, are found to be as follows:

For the normal:	$D_n = 0.1148$
For the lognormal:	$D_n = 0.0785$
For the shifted gamma:	$D_n = 0.0708$

With the sample size of $n = 320$ and at the significance level of 5%, we obtain from Table A.5 the critical value of $D_{320}^{.05} = 1.36/\sqrt{n} = 1.36/\sqrt{320} = 0.0760$. Comparing the above D_n values for the respective distributions with this critical value, we see that at the significance level of 5%, the shifted gamma distribution (with $D_n = 0.0708 < 0.0760$) is verified to be acceptable for modeling the residual stresses, whereas the normal and lognormal distributions must be rejected.

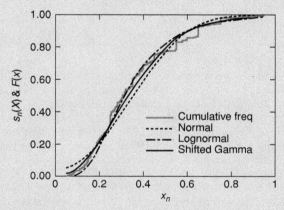

Figure E7.10   Empirical cumulative frequency and the CDFs of three distribution models ◀

7.3.3   The Anderson–Darling (A–D) Test for Goodness-of-Fit

In the K–S test, the probability scale is in an arithmetic scale. We may observe that both the proposed theoretical and the empirical CDFs are relatively flat at the tails of the probability distributions. Hence, the maximum deviation in the K–S test will seldom occur in the tails of a distribution, whereas in the chi-square test, the empirical frequencies at the tails must generally be grouped together. As a result, either test would not reveal any discrepancy between the empirical and theoretical frequencies at the tails of the proposed distribution. The Anderson–Darling (A–D) goodness-of-fit test was introduced by Anderson and Darling (1954) to place more weight or discriminating power at the tails of the distribution. This can be important when the tails of a selected theoretical distribution are of practical significance. The procedure for applying the A–D method can be described with the following steps:

1. Arrange the observed data in increasing order: $x_1, x_2, \ldots, x_i, \ldots, x_n$, with x_n as the largest value.

2. Evaluate the CDF of the proposed distribution $F_X(x_i)$ at x_i, for $i = 1, 2, \ldots n$.

3. Calculate the Anderson–Darling (A–D) statistic

$$A^2 = -\sum_{i=1}^{n}[(2i-1)\{\ln F_X(x_i) + \ln[1 - F_X(x_{n+1-i})]\}/n] - n \tag{7.5}$$

4. Compute the adjusted test statistic A^* to account for the effect of sample size n. This adjustment will depend on the selected form of the distribution; see Eqs. 7.7 through 7.10.

5. Select a significance level α and determine the corresponding critical value c_α for the appropriate distribution type. Values of c_α are given in Table A.6 (*a* through *d*) for four common distributions.

6. For a given distribution, compare A^* with the appropriate critical value c_α. If A^* is less than c_α, the proposed distribution is acceptable at the significance level α.

The test is valid for a sample size larger than 7; i.e, $n > 7$. Because the A–D statistic is expressed in terms of the logarithm of the probabilities, it would receive more contributions from the tails of a distribution. We emphasize that the critical value, c_α, in the A–D test for a given significance level α depends on the form of the proposed theoretical distribution. Also the adjusted A–D statistic, A^*, will depend on the sample size n. These are given and defined in Table A.6, specifically, as follows:

For the *normal distribution*, the critical value c_α is given by

$$c_\alpha = a_\alpha \left(1 + \frac{b_0}{n} + \frac{b_1}{n^2}\right) \tag{7.6}$$

in which values of a_α, b_0, and b_1 are given in Table A.6a for a prescribed significance level α, and the adjusted A–D statistic for a sample size n is

$$A^* = A^2 \left(1.0 + \frac{0.75}{n} + \frac{2.25}{n^2}\right) \tag{7.7}$$

For the *exponential distribution*, the critical value of c_α at a specified significance level α is given in Table A.6b, and the corresponding adjusted A–D statistic, A^*, is given by

$$A^* = A^2 \left(1.0 + \frac{0.6}{\sqrt{n}}\right) \tag{7.8}$$

For the *gamma distribution*, the critical value of c_α depends on the value of the parameter k as given in Table A.6c. Moreover, the adjusted A–D statistic also depends on the parameter k as follows:

For $k = 1$, $\quad A^* = A^2(1.0 + 0.6/n)$

For $k \geq 2$ $\quad A^* = A^2 + (0.2 + 0.3/k)/n \tag{7.9}$

For the *extremal distributions*, of the Gumbel and Weibull types, the critical values of c_α at specified significance levels α are given in Table A.6d. In this case, the adjusted A–D statistic is given by

$$A^* = A^2 \left(1.0 + \frac{0.2}{\sqrt{n}}\right) \tag{7.10}$$

► **EXAMPLE 7.11**

The steel toughness data of Example 7.1 was previously fitted with a normal distribution, and its goodness-of-fit was validated in Example 7.9 by the K–S Test. With the same data, we will now perform a validation of the normal distribution with the A–D Test at the 5% significance level. Table E7.11 summarizes the result of the calculations according to the procedure outlined earlier. The proposed normal model is N(76.99, 4.709) with the parameters, μ and σ, estimated from the sample of size 26. From the results in Table E7.11, we calculate the A–D statistic as

$$A^2 = -\frac{-699.476}{26} - 26 = 0.903$$

TABLE E7.11 Computations for the Anderson-Darling Test of the Normal Distribution

i	x_i	$F_X(x_i)$	$F_X(x_{27-i})$	$(2i\text{-}1)\{\ln F_X(x_i) + \ln[1 - F_X(x_{27-i})]\}$
1	69.5	0.055787	0.999806	−11.4342
2	71.9	0.139745	0.922853	−13.5899
3	72.6	0.175460	0.865630	−18.7375
4	73.1	0.204226	0.745369	−20.6953
5	73.3	0.216478	0.731552	−25.6083
6	73.5	0.229144	0.717368	−30.1072
7	74.1	0.269526	0.710143	−33.1429
8	74.2	0.276587	0.592990	−32.7622
9	75.3	0.359648	0.576430	−31.9883
10	75.5	0.375650	0.500652	−31.7974
11	75.7	0.391869	0.492180	−33.9035
12	75.8	0.400052	0.433188	−34.1294
13	76.1	0.424850	0.433188	−35.5937
14	76.2	0.433188	0.424850	−37.5221
15	76.2	0.433188	0.400052	−39.0774
16	76.9	0.492180	0.391869	−37.3946
17	77.0	0.500652	0.375650	−38.3753
18	77.9	0.576430	0.359648	−34.8824
19	78.1	0.592990	0.276587	−31.3151
20	79.6	0.710143	0.269526	−25.5977
21	79.7	0.717368	0.229144	−24.2892
22	79.9	0.731552	0.216478	−23.9313
23	80.1	0.745369	0.204226	−23.5042
24	82.2	0.865630	0.175460	−15.8497
25	83.7	0.922853	0.139745	−11.3098
26	93.7	0.999806	0.055787	−2.9375

$$\sum = -699.476$$

and the corresponding adjusted statistic becomes

$$A^* = 0.903 \left(1.0 + \frac{0.75}{26} + \frac{2.25}{26^2}\right) = 0.932$$

From Table A.6a, we obtain at the significance level of $\alpha = 0.05$, the values for a_α, b_0, and b_1 as $a_\alpha = 0.7514$; $b_0 = -0.795$; and $b_1 = -0.890$, yielding the critical value of

$$c_{0.05} = 0.7514 \left(1 + \frac{-0.795}{26} + \frac{-0.890}{26^2}\right) = 0.727$$

Since $A^* > 0.727$, the normal distribution is not acceptable at the 5% significance level. However, at the 1% significance level, the constants a_α, b_0, and b_1 from Table A.6a are 1.0348, −1.013, and −0.93 respectively. Hence, the critical value is $c_{0.01} = 0.994$. Therefore, at the 1% significance level, since $A^* < c_{0.01}$, the normal distribution would be acceptable. ◀

We might point out that the A–D test described above for the normal distribution is applicable also to the lognormal distribution. As the logarithm of a lognormal variate is normally distributed, we simply need to take the logarithms of the sample values of the variate in applying the same A–D test as for the normal distribution. That is, all the computations for a normal distribution would remain the same except that the sample values of the variate must be replaced by the respective logarithms of the variate; e.g., in Table E7.11, x_i must be replaced by the corresponding $\ln(x_i)$.

▶ **EXAMPLE 7.12**

As shown in Fig. E7.5, the annual largest earthquake magnitudes in California observed between 1932 and 1962 show a linear trend on the Gumbel probability paper. On this basis, the Gumbel distribution with the parameters $u = 5.7$ and $\alpha = 2.0$ is a viable model for the annual maximum earthquakes in California. We will now perform an A–D test for goodness-of-fit of this distribution at the 5% significance level.

The required calculations are summarized in Table E7.12.
Therefore, according to Eq. 7.5, the A–D statistic is

$$A^2 = -\frac{-976.86}{31} - 31 = 0.512$$

and the adjusted statistic according to Eq. 7.10 is

$$A^* = 0.512 \left(1.0 + \frac{0.2}{\sqrt{31}} \right) = 0.530$$

TABLE E7.12 Computations for the Anderson–Darling Test of the Gumbel Distribution

i	x_i	$F_X(x_i)$	$F_X(x_{32-i})$	$(2i-1)\{\ln F_X(x_i) + \ln[1 - F_X(x_{32-i})]\}$
1	4.9	0.00706	0.98185	−8.962
2	5.3	0.10801	0.94100	−15.167
3	5.3	0.10801	0.94100	−25.279
4	5.5	0.22496	0.81718	−22.338
5	5.5	0.22496	0.81718	−28.720
6	5.5	0.22496	0.78146	−33.138
7	5.5	0.22496	0.78146	−39.164
8	5.6	0.29482	0.73993	−38.523
9	5.6	0.29482	0.73993	−43.660
10	5.6	0.29482	0.69220	−45.594
11	5.8	0.44099	0.69220	−41.938
12	5.8	0.44099	0.57764	−38.654
13	5.8	0.44099	0.57764	−42.015
14	5.9	0.51154	0.57764	−41.370
15	6.0	0.57764	0.57764	−40.910
16	6.0	0.57764	0.57764	−43.732
17	6.0	0.57764	0.57764	−46.553
18	6.0	0.57764	0.51154	−44.286
19	6.0	0.57764	0.44099	−41.825
20	6.0	0.57764	0.44099	−44.086
21	6.2	0.69220	0.44099	−38.928
22	6.2	0.69220	0.29482	−30.839
23	6.3	0.73993	0.29482	−29.272
24	6.3	0.73993	0.29482	−30.573
25	6.4	0.78146	0.22496	−24.571
26	6.4	0.78146	0.22496	−25.573
27	6.5	0.81718	0.22496	−24.207
28	6.5	0.81718	0.22496	−25.121
29	7.1	0.94100	0.10801	−9.981
30	7.1	0.94100	0.10801	−10.331
31	7.7	0.98185	0.00706	−1.550
				$\sum = -976.86$

From Table A.6d, we obtain the critical value $c_\alpha = 0.757$ at the significance level of 5%. Therefore, according to the A–D test, since $A^* < c_\alpha$, the Gumbel distribution is an acceptable or valid distribution at the 5% significance level. ◀

▶ 7.4 INVARIANCE IN THE ASYMPTOTIC FORMS OF EXTREMAL DISTRIBUTIONS

In determining the appropriate probability distribution, it is well to emphasize that when extreme values are involved, one of the asymptotic forms of extremal distributions may be appropriate. In this regard, the invariance property of the asymptotic forms should also be of relevance; that is, the asymptotic form of an extreme value distribution will not change with the value of n.

We saw in Chapter 4, Sect. 4.2.3, that the largest and smallest values of samples of size n from a known initial distribution will converge to one of the three asymptotic forms of extreme value distributions as $n \to \infty$. This means that when our interest or concern is related to the extremes, one of the asymptotic forms of distributions may be selected. In this regard, we may wish to re-emphasize that the appropriate asymptotic form depends only on the tail behavior of the initial distribution. In particular, the following properties of the tail behavior are of significance:

- The largest (or smallest) value from an initial distribution with an exponential tail in the direction of the extreme, such as the exponential and normal distributions, will approach the *Type I* asymptotic form.

- The largest (or smallest) value from an initial distribution with a polynomial tail in the direction of the extreme, will converge asymptotically to the *Type II* asymptotic form. This includes, in particular, the lognormal distribution.

- The largest value from an initial distribution with an upper bound (or smallest value with a lower bound) value will approach the *Type III* asymptotic form.

Finally, we should also emphasize that if the initial distribution is one of the asymptotic extremal forms, the distribution for increasing n will remain of the same asymptotic form; i.e., the form of distribution of the extremes will remain invariant with increasing n, although the parameters will change with n. In particular, we observe, for example, that if Y is the annual maximum with a Gumbel distribution, the maximum in n years will also be Gumbel distributed; we show this as follows:
Suppose

$$F_Y(y) = \exp[-e^{-\alpha(y-u)}]$$

with parameters u and α.

Then, according to Eq. 4.23, the distribution of the maximum in n years, Y_n, would be

$$F_{Y_n}(y) = [F_Y(y)]^n = \exp\left[-ne^{-\alpha(y-u)}\right] = \exp\left[-e^{-\alpha\left[y-\left(u+\frac{\ln n}{\alpha}\right)\right]}\right] \tag{7.11}$$

Therefore, Y_n remains a Gumbel (*Type I*) distribution with corresponding parameters

$$u_n = u + \frac{\ln n}{\alpha} \qquad \text{and} \qquad \alpha_n = \alpha$$

▶ **EXAMPLE 7.13**

In Example 7.5, we observed that the occurrence of the annual largest magnitude earthquakes was determined to follow the Gumbel (*Type I*) extreme value distribution with parameters $u = 5.7$ and $\alpha = 2.00$. On the basis of Eq. 7.11, the distribution of the 10-year maximum magnitude will remain a Gumbel distribution with parameters

$$u_n = 5.7 + \frac{\ln 10}{2.0} = 6.85 \quad \text{and} \quad \alpha_n = 2.0$$

On this basis, we can say that the most probable largest magnitude earthquake in California in the next ten years will be 6.85 on the Richter scale. Similarly, in the next 25 years, the most probable earthquake will be of magnitude

$$u_n = 5.7 + \frac{\ln 25}{2.0} = 7.31$$

By the same token, extrapolating from 10 years to 25 years, we will also obtain

$$u_n = 6.85 + \frac{\ln 2.5}{2.00} = 7.31$$

◀

▶ **EXAMPLE 7.14**

In Example 4.21, the fracture strength of a welded joint was modeled by the Weibull distribution (the *Type III* asymptotic distribution of smallest value), as defined in Eq. 4.32, with parameters $w = 15$ ksi, $k = 1.75$, and the minimum strength is $\varepsilon = 4$ ksi. Now, suppose that we have a structural member with five welded joints of the same type. Then, according to Eq. 4.25, the CDF of the lowest fracture strength among the five welded joints in the structural member would be

$$F_{Y_1}(y) = 1 - \exp\left[-n\left(\frac{y - \varepsilon}{w - \varepsilon}\right)^k\right]$$

which remains a Weibull distribution. However, with $n = 5$, the probability that the lowest fracture strength of the structural member will be at least 16.5 ksi becomes

$$P(Y_1 \geq 16.5) = \exp\left[-5 \times \left(\frac{16.5 - 4}{15 - 4}\right)^{1.75}\right] = 0.002$$

which is a much lower probability than that of 0.286 for a single joint as shown in Example 4.21. ◀

▶ 7.5 CONCLUDING SUMMARY

In Chapter 6 we presented the statistical methods for estimating the parameters of a probability distribution from a sample of observed data, and related hypothesis test of the parameters. In this chapter we presented the empirical techniques for determining or selecting the appropriate distribution in light of the same data, and also methods for testing the goodness-of-fit of the selected or assumed distribution(s).

A probability paper constructed for a particular probability distribution may be used to select an appropriate probability distribution. The presence, or absence, of a linear trend in the data points plotted on a particular probability paper can be used to determine whether the distribution associated with the probability paper is appropriate to represent the underlying population.

The validity of an assumed, or selected, theoretical distribution may also be appraised by goodness-of-fit tests; the two commonly used tests are the *chi-square* and the *Kolmogorov–Smirnov* or *K–S* tests. A third test, the *Anderson–Darling* method may be a more accurate test for goodness-of-fit when the tail(s) of the assumed distribution is important. All such tests, however, depend on the prescribed level of significance, the choice of which is largely subjective. Nevertheless, these tests are statistically useful, particularly for discriminating the relative validities of two or more candidate distribution models.

Finally, it is well to emphasize that when extreme values are involved, one of the asymptotic forms of extremal distribution may be appropriate. The invariance property of the asymptotic forms is also of significance in such applications.

▶ PROBLEMS

7.1 The data for the fracture toughness of steel base plates have been compiled as shown in Table E7.1. In Example 7.1, these data were plotted on a normal probability paper.
(a) Plot the same set of data on a lognormal probability paper, and draw a straight line through the plotted data points if a linear trend can be observed.
(b) Estimate the median and c.o.v. of the fracture toughness on the basis of the straight line of Part (a).

7.2 The annual maximum wind velocity V at a town has been recorded for the past 20 years as follows:

Year	Wind Speed, V (kph)	Year	Wind Speed, V (kph)
1980	78.2	1990	78.4
1981	75.8	1991	76.4
1982	81.8	1992	72.9
1983	85.2	1993	76
1984	75.9	1994	79.3
1985	78.2	1995	77.4
1986	72.3	1996	77.1
1987	69.3	1997	80.8
1988	76.1	1998	70.6
1989	74.8	1999	73.5

(a) Construct a *Type I* extreme value (Gumbel) probability paper, and plot the above data for the maximum wind velocity on this paper.
(b) Draw a straight line through the data points in the above probability paper (if a linear trend is observed) and determine the parameters of the underlying extremal distribution.
(c) Perform a chi-square test for goodness-of-fit of the *Type I* distribution for modeling the annual maximum wind velocity at the 2% significance level.

7.3 The ultimate strains of 15 #5 steel reinforcing bars were measured, yielding the results shown below (data from Allen, 1972):

(a) Plot the data on both the normal and lognormal probability papers. From a visual inspection of the plotted data points, which of the two underlying distributions is a better model to represent the probability distribution of the ultimate strain of reinforcing bars?

Bar #	Ultimate Strain, U (in %)	Bar #	Ultimate Strain, U (in %)	Bar #	Ultimate Strain, U (in %)
1	19.4	6	17.9	11	16.1
2	16.0	7	17.8	12	16.8
3	16.6	8	18.8	13	17.0
4	17.3	9	20.1	14	18.1
5	18.4	10	19.1	15	18.6

(b) Perform chi-square tests for both the normal and lognormal distributions at a significance level of 2%, and on the basis of the test results, determine which of the two is a more suitable distribution to model the ultimate strains of steel reinforcing bars.
(c) Perform also the Kolmogorov–Smirnov tests for the same two distributions, and determine the relative validity of the two distributions on the basis of this test.

7.4 The exponential distribution is often used to model the time-to-failure (or malfunction) of mechanical equipment or machines. In order to verify the distribution of a certain type of diesel engine, a large number of units of the engine were tested, and the time-to-failure (in hours) of the units were recorded as tabulated below.

Time-to-failure (hr) of Diesel Engines

0.13	121.58	2959.47	2.34
0.78	672.87	124.09	393.37
3.55	62.09	85.28	84.09
14.29	656.04	380.00	646.01
54.85	735.89	298.58	412.03
216.40	895.80	678.13	813.00
1296.93	1057.57	861.93	39.10
952.65	470.97	1885.22	633.98
8.82	151.44	862.93	658.38
29.75	163.95	1407.52	855.95

(a) Construct the exponential probability paper, and plot on it the above time-to-failure data of the particular type of engine.
(b) If there is a linear trend in the data points on the above exponential paper, estimate the minimum and the mean time-to-failure of the particular type of diesel engine.

(c) Perform a chi-square test to evaluate the validity of the exponential distribution at the 5% significance level.
(d) Alternatively, perform also a Kolmogorov–Smirnov test of the same distribution at the same 5% significance level.

7.5 The number of vehicles arriving at a toll booth, per minute, were observed as follows:

0, 3, 1, 2, 0, 1, 1, 1, 2, 0, 1, 4, 3, 1, 1, 0, 0, 1, 0, 2, 2, 0, 1, 0, 0

(a) Assuming that the arrival rate of vehicles at the toll booth is a Poisson process, estimate the mean arrival rate.
(b) Perform a chi-square test to determine the validity of the Poisson distribution at the 1% significance level.

7.6 The PDF of the Rayleigh distribution of a random variable X is

$$f_X(x) = \frac{x}{\alpha^2} e^{-\frac{1}{2}(x/\alpha)^2}; \qquad x \geq 0$$
$$= 0 \qquad ; \qquad x < 0$$

in which the parameter α is the mode, or most probable value, of X.
(a) Construct the probability paper of the above Rayleigh distribution. What does the slope of a straight line on this paper represent?
(b) A set of data for strain range induced by vehicle loads on highway bridge members is tabulated below. Plot the set of data on the Rayleigh probability paper constructed in Part (a) above.
(c) In light of the results of Parts (a) and (b) above, what conclusion can you draw regarding the Rayleigh distribution as to its suitability for modeling the live-load stress-range in highway bridges? To verify or invalidate the Rayleigh distribution, perform a K–S test at the 1% significance level.
(d) Determine the most probable value of the strain-range (if possible) from Part (b).

Measured Strain Range (micro-in./in.)

48.4	52.7	42.4
47.1	44.5	146.2
49.5	84.8	115.2
116.0	52.6	43.0
84.1	53.6	103.6
99.3	33.5	64.7
108.1	43.8	69.8
47.3	56.3	44.0
93.7	34.5	36.2
36.3	62.8	50.6
122.5	180.5	167.0

(Data courtesy of W. H. Walker.)

7.7 Data on the rate of oxygenation, K, in streams have been obtained for the Cincinnati Pool, Ohio River, at 20°C and summarized as follows. (Data from Kothandaraman, 1968.)

Range of K/per day	No. of Observations
0.000–0.049	1
0.050–0.099	11
0.100–0.149	20
0.150–0.199	23
0.200–0.249	15
0.250–0.299	11
0.300–0.349	2

(a) If a normal distribution is proposed to model the oxygenation rate at the Cincinnati Pool, Ohio River, estimate the mean and standard deviation of the distribution.
(b) Perform a chi-square test for the goodness-of-fit of the proposed distribution at the 5% significance level.

7.8 A random variable X with a triangular distribution between a and $a + r$ is described by the following PDF:

$$f_X(x) = \frac{2(x - a)}{r^2}; \qquad a \leq x \leq a + r$$
$$= 0 \qquad ; \qquad \text{otherwise}$$

(a) Determine the appropriate standard variate S for the triangular distribution.
(b) Construct the corresponding probability paper. What do the values of X at $F_S(0)$ and $F_S(1.0)$ on this paper mean?
(c) Suppose the following sample values were observed for X:

36	32	34	71
18	69	45	66
56	71	53	58
64	50	55	53
72	28	62	48
			75

Plot the above set of sample values on the triangular probability paper constructed in Part (b) above, and from this plot estimate the minimum and maximum values of X.

7.9 The shear strength [in kips per square foot (ksf)] of 13 undisturbed samples of clay from the Chicago subway project are tabulated below (data from Peck, 1940):

Shear Strength of Clay, ksf

0.35	0.42	0.49	0.70	0.96
0.40	0.43	0.58	0.75	
0.41	0.48	0.68	0.87	

(a) Plot the above data on a lognormal probability paper, and draw a straight line through the data points (if a linear trend can be observed).
(b) Estimate the parameters of the distribution from the straight line in Part (a) above.
(c) Test the goodness-of-fit of the lognormal distribution using the Kolmogorov–Smirnov test at the 2% significance level. Do

the same with the Anderson–Darling test at the 2.5% significance level.

7.10 Passenger cars coming to an intersection must stop at a stop sign and must wait for a gap sufficiently long to cross or make a turn. The acceptance gap, G, measured in seconds, varies from driver to driver; some drivers are more alert or more risk-taking, whereas others are more careful or slow-moving. Observations at several similar intersections in a city were recorded as follows:

Acceptance Gap, G (sec)	Number of Observations
0.5–1.5	0
1.5–2.5	6
2.5–3.5	34
3.5–4.5	132
4.5–5.5	179
5.5–6.5	218
6.5–7.5	183
7.5–8.5	146
8.5–9.5	69
9.5–10.5	30
10.5–11.5	3
11.5–12.5	0

(a) Plot a histogram for the above data of the observed acceptance gap.
(b) From the above histogram, determine whether the normal or lognormal distribution is a better model for the acceptance gap, G.
(c) From the observed data, estimate the sample mean and sample standard deviation of the acceptance gap. For this purpose, the average gap length may be used for each of the intervals of G in the first column of the tabulated data.
(d) Perform chi-square goodness-of-fit tests for both the normal and lognormal distributions at the 1% significance level, and on this basis determine which of the two distributions is more appropriate to model the acceptance gap G.

7.11 Measured data for the annual rainfall intensity in a watershed area was presented in Table 1.1 of Chapter 1. These data are repeated below as follows:

Observed Rainfall Intensity, in.

43.30	54.49	58.71
53.02	47.38	42.96
63.52	40.78	55.77
45.93	45.05	41.31
48.26	50.37	58.83
50.51	54.91	48.21
49.57	51.28	44.67
43.93	39.91	67.72
46.77	53.29	43.11
59.12	67.59	

(a) Based on the histogram shown in Fig. 1.1 of Chapter 1, a lognormal or a gamma PDF may be a plausible distribution to model the rainfall intensity of the watershed area. Plot the data on a lognormal probability paper.
(b) Prescribe a gamma distribution and estimate its parameters by the method of moments.
(c) Determine which of the two distributions is better suited to model the pertinent rainfall intensity by performing chi-square goodness-of-fit tests for both distributions.

7.12 Data for the observed settlements of piles and the corresponding calculated settlements are compiled in Table E8.3 of Chapter 8. Based on these data, we can calculate the ratios of the observed to the corresponding calculated settlements; the results are as follows:

Ratios of Observed Settlement to Calculated Settlement

0.12	0.97	0.86	1.14	0.94
2.37	0.88	0.92	1.01	0.99
1.02	1.04	0.99	0.87	0.52
0.94	1.06	1.38	1.04	1.18
1.00	0.86	0.82	0.84	1.09

The ratio of the observed to the calculated settlements is a measure of the accuracy of the calculational method. From the above data, we can observe that this ratio has considerable variability.

(a) Assuming that the ratio is a Gaussian random variable, plot the above data on a normal probability paper and observe if there is a linear trend of the data points.
(b) If a linear trend is observed, draw a straight line through the data points and estimate the mean and standard deviation from the straight line. Perform a chi-square goodness-of-fit test for the normal distribution at the 5% significance level. Also, do the same with the Anderson–Darling test.
(c) Otherwise, if no linear trend can be observed in Part (b), plot the same data on another probability paper, such as the lognormal paper, and determine the suitability of this alternative distribution to model the relevant ratio, including a goodness-of-fit test at the 2% significance level to verify its suitability.

7.13 In Example 8.8 of Chapter 8, the data tabulated in Table E8.8 include the mean depth of glacier lakes in the Swiss Alps, which we summarized below as follows:

Mean Depth of Glacier Lakes, m

2.9	12.0	33.3
5.0	13.6	27.9
4.7	28.6	46.9
7.1	18.6	50.0
10.4	34.3	83.3

Obviously, the mean depths of glacier lakes are highly variable. Assuming that the above data are representative of glacier lakes in general, determine the appropriate distribution to model the mean depth of glacier lakes. Verify also the selected distribution with a goodness-of-fit test.

▶ REFERENCES

Allen, D. E., "Statistical Study of the Mechanical Properties of Reinforcing Bars," *Building Research Note*, No. 85, National Research Council, Ottawa, Canada, April 1972.

Anderson, T. W., and Darling, D. A. (1954), "A Test of Goodness-of-Fit," *Jour. of America Statistical Association*. 49, 765–769.

Cusens, A. R., and Wettern, J. H., "Quality Control in Factory-Made Precast Concrete," *Civil Engineering and Public Works Review*, Vol. 54, 1959.

D'Agostino, R. B., and Stephens, M. S., *Goodness-of-Fit Technique*. Marcel Dekker, Inc., New York and Basel, 1986.

Epstein, B., and Lomnitz, C., "A Model for the Occurrence of Large Earthquakes," *Nature*, August 1966, pp. 954–956.

Gumbel, E. J., "Statistical Theory of Extreme Values and Some Practical Applications," *Applied Mathematics Series 33*, National Bureau of Standards, Washington, D.C., February 1954.

Hazen, A., *Flood, Flows, A Study in Frequency and Magnitude*, J. Wiley & Sons, Inc., New York, 1930.

Hoel, P. G., *Introduction to Mathematical Statistics*, 3rd ed., J. Wiley & Sons, Inc., New York, 1962.

Kies, J. A., Smith, H. L., Romine, H. E., and Bernstein, M., "Fracture Testing of Weldments," ASTM *Special Publ. No. 381*, 1965, pp. 328–356.

Kimball, B. F., "Assignment of Frequencies to a Completely Ordered Set of Sample Data," *Transactions*, American Geophysical Union, Vol. 27, 1946, 843–846.

Kothandaraman, V., "A Probabilistic Analysis of Dissolved Oxygen-Biochemical Oxygen Demand Relationship in Streams," Ph.D. Dissertation, University of Illinois at Urbana-Champaign, 1968.

Lockhart, R. A., and Stephens, M. A. "Goodness-of-Fit Tests for the Gamma Distribution," Technical Report, Department of Mathematics and Statistics, Simon Fraser University, British Columbia, Canada, 1985.

Pearson, E. S., and Hartley, H. O., *Biometrika Tables for Statisticians*, Vol. 2. Cambridge University Press, New York, 1972.

Peck, R. B., "Sampling Methods and Laboratory Tests for Chicago Subway Soils," *Proc. Purdue Conf. on Soil Mechanics and Its Applications*, Lafayette, IN, 1940.

Petitt, A. N., "Testing the Normality of Several Independent Samples Using the Anderson-Darling Statistic." *Jour. Roy. Stat. Soc. C 26*, 156–161, 1977.

Stephens, M. A. "Goodness-of-Fit for the Extreme Value Distribution," *Biometrika*, Vol. 64, 1977, pp. 583–588.

Regression and Correlation Analyses

⊳ 8.1 INTRODUCTION

When there are two (or more) variables, there may be some relationship between (or among) the variables. In the presence of randomness, the relationship between the two variables will not be unique; given the value of one variable, there is a range of possible values of the other variable. The relationship between the variables, therefore, requires probabilistic description. If the probabilistic relationship between the variables is described in terms of the mean and variance of one random variable as a function of the value of the other variable, this requires what is known as *regression analysis*. When the analysis is limited to linear mean-value functions, it is called *linear regression*. More generally, however, regression may be nonlinear. The linear or nonlinear relationship obtained from a regression analysis does not necessarily represent any causal relation between the variables; i.e., there may not be any cause-and-effect relationship between the variables. Such a relationship, however, may be used to predict the value or statistics of one variable based on the value of the other *control* variable.

For linear regression, the degree of linearity in the relationship between two random variables may be measured by the *statistical correlation*, in particular, by the *correlation coefficient* as defined in Sect. 3.3.2. When the correlation coefficient is high, close to ±1.0, one can expect high confidence in being able to predict the value of one variable based on information about the value of the other (control) variable. Evaluation of the correlation coefficient from a set of observed data is through *correlation analysis*.

In this chapter, we start our discussion of linear regression with constant variance, including the determination of the conditional variance and confidence interval of a regression equation, and the related correlation analysis. This is then extended to linear regression with nonconstant variance, nonlinear regression, and multiple regression analyses, concluding with illustrations of engineering applications of regression analyses.

⊳ 8.2 FUNDAMENTALS OF LINEAR REGRESSION ANALYSIS

8.2.1 Regression with Constant Variance

When pair-wise data points for two variables, say X and Y, are plotted on a two-dimensional graph, the data points may show some scatter as illustrated in Fig. 8.1. Such a plot is called a *scattergram*.

In the case shown in Fig. 8.1, we can say that the range of possible values of Y tends to increase with increasing values of X, or vice versa. However, knowing the value of one

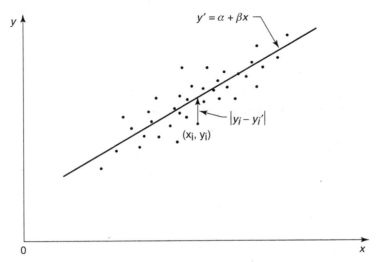

Figure 8.1 Scattergram of two variables X and Y.

variable, say $X = x$, does not give perfect information on the other variable Y. Conceivably, the range of values of Y is governed by a probability distribution. We may observe also from Fig. 8.1 that the mean value of Y increases with increasing values of X, and if this relationship is linear, we have a *linear regression*; i.e.,

$$E(Y|X = x) = \alpha + \beta x \tag{8.1}$$

where α and β are constants, known as the *regression coefficients*, which are the *intercept* and *slope*, respectively, of the straight line. Equation 8.1 is known as the *regression equation*, and it represents the regression of Y on X. The regression coefficients α and β must necessarily be estimated from the available data.

From the scatter of the data points in Fig. 8.1, we would expect a variance of Y that may depend on the value of X, i.e., a conditional variance of Y given $X = x$ or $\mathrm{Var}(Y|X = x)$. In general, this conditional variance may vary with x. However, let us first consider the case in which $\mathrm{Var}(Y|X = x)$ is constant.

We can observe from the scattergram of the plotted data points that there could conceivably be many straight lines, depending on the values of the regression coefficients, that might represent the mean-value of Y as a linear function of X. The "best" straight line may be the one that passes through the data points with the least cumulative error. To obtain this particular straight line, we observe from Fig. 8.1 that for each data point (x_i, y_i) the difference between the observed value y_i and the value from a candidate straight line $y_i' = \alpha + \beta x_i$ is $|y_i - y_i'|$. For a sample of observed data pairs of size n, i.e., $[(x_1, y_1), (x_2, y_2), \ldots, (x_n, y_n)]$, the total absolute error for all the data points can be represented by the total cumulative squared error; i.e.,

$$\Delta^2 = \sum_{i=1}^{n} \left(y_i - y_i'\right)^2 = \sum_{i=1}^{n} (y_i - \alpha - \beta x_i)^2 \tag{8.2}$$

Then, we may obtain the straight line with the least squared error by minimizing Δ^2 of Eq. 8.2, yielding the following equations for obtaining α and β. That is,

$$\frac{\partial \Delta^2}{\partial \alpha} = \sum_{i=1}^{n} -2(y_i - \alpha - \beta x_i) = 0$$

and

$$\frac{\partial \Delta^2}{\partial \beta} = \sum_{i=1}^{n} -2x_i (y_i - \alpha - \beta x_i) = 0$$

The above procedure is known as the *method of least squares*, from which we obtain the least-squares estimates of α and β from the sample of size n as follows:

$$\hat{\alpha} = \frac{1}{n} \sum y_i - \frac{\hat{\beta}}{n} \sum x_i = \bar{y} - \hat{\beta}\bar{x} \tag{8.3}$$

and

$$\hat{\beta} = \frac{\sum x_i y_i - n\bar{x}\,\bar{y}}{\sum x_i^2 - n\bar{x}^2} = \frac{\sum (x_i - \bar{x})(y_i - \bar{y})}{\sum (x_i - \bar{x})^2} \tag{8.4}$$

where

$$\sum = \sum_{i=1}^{n}$$

$\bar{x}, \bar{y} = $ the sample mean of X and Y, respectively

$n = $ the sample size

Therefore, the least-squares regression equation is

$$E(Y|X = x) = \hat{\alpha} + \hat{\beta}x \tag{8.5}$$

which may be written simply as $E(Y|x) = \hat{\alpha} + \hat{\beta}x$.

We may emphasize that, strictly speaking, the above regression line is valid only over the range of values of X for which data had been observed. Extrapolating beyond the range of observed data may lead to erroneous conclusions.

Equation 8.1 or 8.5 is referred to as the *regression of Y on X*, in which the regression line or equation can be used to predict the mean value of Y given a value of X. If both X and Y are random variables, we may also obtain the least-squares *regression of X on Y* using the same procedure; in this latter case, we would obtain the regression equation for $E(X|Y = y)$, which can be used to predict the mean value of X for a given value of Y. In general, this latter equation is a different linear equation from that of $E(Y|X = x)$; however, the two linear regression lines will always intersect at $(\bar{x}, \bar{y})$. For example, in Problem 8.4, Meadows, et al. (1972) compiled the per capita energy consumption Y versus the per capita GNP output X of different countries. If the interest is to predict the per capita energy consumption on the basis of the GNP output of a country, the regression of Y on X would be appropriate. On the other hand, the per capita GNP output of a country may be estimated on the basis of the per capita energy consumption of its populace; in this latter case, the regression of X on Y would be required.

8.2.2   Variance in Regression Analysis

The Conditional Variance

The regression equation of Eq. 8.5 predicts the mean value of Y as a function of X. Accordingly, the relevant variance of Y that is of interest must be conditional on a given value of X, i.e., the *conditional variance* $\text{Var}(Y|X = x)$. For the case where this conditional variance is assumed to be constant with x, an unbiased estimate of this variance from a sample of size n is

$$s_{Y|x}^2 = \frac{1}{n-2} \sum_{i=1}^{n} (y_i - y_i')^2$$

$$= \frac{1}{n-2} \left[\sum_{i=1}^{n} (y_i - \bar{y})^2 - \hat{\beta}^2 \sum_{i=1}^{n} (x_i - \bar{x})^2 \right] \qquad (8.6)$$

We observe from Eq. 8.2 that this is also

$$s_{Y|x}^2 = \frac{\Delta^2}{n-2} \qquad (8.6a)$$

The corresponding conditional standard deviation, therefore, is $s_{Y|x}$.

The ratio of the conditional variance of Eq. 8.6 relative to the original sample variance s_Y^2 is a measure of the reduction in the original variance of Y resulting from taking into account the variation of the variance with X. This reduction in the variance may be represented by

$$r^2 = 1 - \frac{s_{Y|x}^2}{s_Y^2} \qquad (8.7)$$

We shall see in Sect. 8.3.1, Eq. 8.12, that r^2 is closely related to the correlation coefficient ρ.

8.2.3 Confidence Intervals in Regression

As the regression equation (e.g., of Y on X) gives the predicted mean value of Y for given values of the control variable $X = x$, we might be interested in the *confidence interval* of the regression equation, which should provide some measure of the range of the true equation. Such a confidence interval is equivalent to the confidence intervals of the estimated mean values of Y, i.e., $E(Y|X = x_i)$, for varying values of x_i.

Since the regression coefficients $\hat{\alpha}$ and $\hat{\beta}$ are estimated from finite samples of size n, they are individually t-distributed with $(n-2)$ degrees-of-freedom (Hald, 1952); therefore, the mean value $\bar{y}_i = E(Y|X = x_i)$ estimated from the linear regression equation at $X = x_i$ will also have a t-distribution with $(n-2)$ d.o.f. Accordingly, the statistic

$$\frac{\bar{Y}_i - \mu_{Y|x_i}}{s_{Y|x} \sqrt{\dfrac{1}{n} + \dfrac{(x_i - \bar{x})^2}{\sum (x_i - \bar{x})^2}}}$$

will have the t-distribution with $(n-2)$ d.o.f (Hald, 1952). On this basis, we may obtain the $(1 - \alpha)$ confidence interval for the regression equation at several selected values of $X = x_i$ as

$$\langle \mu_{Y|x_i} \rangle_{1-\alpha} = \bar{y}_i \pm t_{(1-\frac{\alpha}{2}),n-2} \cdot s_{Y|x} \sqrt{\frac{1}{n} + \frac{(x_i - \bar{x})^2}{\sum (x_i - \bar{x})^2}} \qquad (8.8)$$

where $\bar{y}_i = E(Y|X = x_i)$, $s_{Y|x}$ is the conditional standard deviation of Y as given in Eq. 8.6, and $t_{(1-\frac{\alpha}{2}),n-2}$ is the value of the t-distributed variate at the probability of $(1 - \frac{\alpha}{2})$ ·

with $(n-2)$ d.o.f. as given in Table A.3. Among the confidence intervals of Eq. 8.8 at the selected discrete values of x_i, the interval will be minimum at $x_i = \bar{x}$, the mean value of X. Connecting these discrete points along the regression line should yield the appropriate confidence interval of the regression equation.

▶ **EXAMPLE 8.1**

Observed data for *blow counts, N*, and corresponding measured unconfined compressive strength, q, of very stiff clay are given in the first two columns of Table E8.1. The sample of 10 pairs of data is also plotted in Fig. E8.1.

On the basis of the calculations in Table E8.1 we obtain the respective sample means of N and q as

$$\overline{N} = \frac{187}{10} = 18.7 \quad \text{and} \quad \overline{q} = \frac{21.23}{10} = 2.123 \text{ tsf}$$

and corresponding sample variances

$$s_N^2 = \frac{1}{9}(4353 - 10 \times 18.7^2) = 95.12 \quad \text{and} \quad s_q^2 = \frac{1}{9}(56.09 - 10 \times 2.123^2) = 1.22$$

From the results of Table E8.1, we also obtain

$$\hat{\beta} = \frac{492.77 - 10 \times 18.7 \times 2.123}{4358 - 10 \times 18.7^2} = 0.112 \quad \text{and} \quad \hat{\alpha} = 2.123 - 0.112 \times 18.7 = 0.029$$

and the regression equation is

$$E(q|N = N_i) = 0.029 + 0.112 N_i$$

which is superimposed on the data points in Fig. E8.1.
The conditional variance of q for given N is

$$s_{q|N}^2 = \frac{\Delta^2}{n-2} = \frac{0.305}{8} = 0.038$$

and the corresponding conditional standard deviation is $s_{q|N} = 0.195$ tsf.

TABLE E8.1 Computations for Example 8.1*

Blow Counts, N_i	Compressive Strength (tsf), q_i	N_i^2	q_i^2	$N_i q_i$	$q_i' = \hat{\alpha} + \hat{\beta}N$	$(q_i - q_i')^2$
4	0.33	16	0.11	1.32	0.48	0.022
8	0.90	64	0.81	7.20	0.93	0.001
11	1.41	121	1.99	15.51	1.26	0.023
16	1.99	256	3.96	31.84	1.82	0.029
17	1.70	289	2.89	28.90	1.93	0.053
19	2.25	361	5.06	42.75	2.16	0.008
21	2.60	441	6.76	54.60	2.38	0.048
25	2.71	625	7.34	67.75	2.83	0.014
32	3.33	1024	11.09	106.56	3.61	0.078
34	4.01	1156	16.08	136.34	3.84	0.029
$\sum$ 187	21.23	4353	56.09	492.77		$\Delta^2 = 0.305$

*The tabulated results can be organized conveniently using spreadsheets as in Example 8.3.

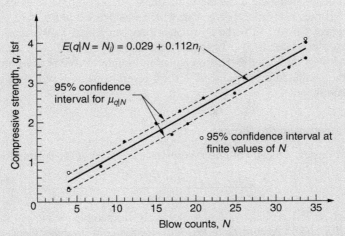

Figure E8.1 Unconfined compressive strength vs. blow counts of stiff clay.

Using the calculations in Table E8.1, the estimated correlation coefficient according to Eq. 8.9 is

$$\hat{\rho} = \frac{1}{9} \frac{492.77 - 10 \times 18.7 \times 2.123}{\sqrt{95.12}\sqrt{1.22}} = 0.99$$

To determine the 95% confidence interval for the regression equation of q on N, let us use the following selected values of $N_i = 4, 11, 19$, and 34, and with $t_{0.975,8} = 2.306$ from Table A.3, we obtain

$$\text{At } N_i = 4; \quad \langle \mu_{q|N} \rangle_{0.95} = 0.477 \pm 2.306 \times 0.195 \sqrt{\frac{1}{10} + \frac{(4 - 18.7)^2}{(4353 - 10 \times 18.7^2)}}$$
$$= (0.210 \rightarrow 0.744) \text{ tsf}$$

$$\text{At } N_i = 11 \quad \langle \mu_{q|N} \rangle_{0.95} = 1.261 \pm 2.306 \times 0.195 \sqrt{\frac{1}{10} + \frac{(11 - 18.7)^2}{(4353 - 10 \times 18.7^2)}}$$
$$= (1.076 \rightarrow 1.446) \text{ tsf}$$

$$\text{At } N_i = 19 \quad \langle \mu_{q|N} \rangle_{0.95} = 2.157 \pm 2.306 \times 0.195 \sqrt{\frac{1}{10} + \frac{(19 - 18.7)^2}{(4353 - 10 \times 18.7^2)}}$$
$$= (2.015 \rightarrow 2.299) \text{ tsf}$$

$$\text{At } N_i = 34 \quad \langle \mu_{q|N} \rangle_{0.95} = 3.837 \pm 2.306 \times 0.195 \sqrt{\frac{1}{10} + \frac{(34 - 18.7)^2}{(4353 - 10 \times 18.7^2)}}$$
$$= (3.562 \rightarrow 4.112) \text{ tsf}$$

The above confidence intervals at the selected discrete values of N are also displayed in Fig. E8.1. These may then be used to construct the 95% confidence interval for the linear regression equation of q on N, which are the lower and upper dash curves in Fig. E8.1. ◀

▶ 8.3 CORRELATION ANALYSIS

Intuitively, if the conditional standard deviation $s_{Y|x}$ is small (close to zero), we can say that the linear regression equation provides a good predictor of Y for given values of X. However, a better statistical measure of the linear relationship between two random variables X and Y is the *correlation coefficient*, which was defined in Eq. 3.72 as $\rho_{X,Y} = \text{Cov}(X, Y)/\sigma_X\sigma_Y$, where $\text{Cov}(X, Y)$ is the covariance between X and Y. We shall see later in Eq. 8.12 that $\rho_{X,Y}$ is related to $S_{Y|x}$. In essence, the correlation coefficient is a measure of the goodness-of-fit of the linear regression equation in light of the set of sampled data. The accuracy of the predicted mean value of Y for a given value of X will then depend on the correlation coefficient.

8.3.1 Estimation of the Correlation Coefficient

For a set of n pairs of observations, the estimate of the correlation coefficient is

$$\hat{\rho}_{X,Y} = \frac{1}{n-1} \frac{\sum_{i=1}^{n}(x_i - \overline{x})(y_i - \overline{y})}{s_X s_Y} = \frac{1}{n-1} \frac{\sum_{i=1}^{n} x_i y_i - n\overline{x}\,\overline{y}}{s_X s_Y} \tag{8.9}$$

where $\overline{x}$, $\overline{y}$, s_X, and s_Y are, respectively, the sample means and sample standard deviations of X and Y. According to Eq. 3.73, the value of ρ ranges from -1 to $+1$. If the estimated $\hat{\rho}_{X,Y}$ is large (close to ± 1.0), there is a strong linear relationship between X and Y; this is illustrated in Fig. 8.2, which shows the linear regression line between the *compression index* and *void ratio* of soils. Conversely, if $\hat{\rho}_{X,Y}$ is very small or close to zero (uncorrelated), this would indicate a lack of linear relationship between X and Y, as illustrated in Fig. 8.3 between the *modulus of rupture* and the *modulus of elasticity* of laminated wood.

From Eqs. 8.4 and 8.9, we can show that the estimated correlation coefficient is

$$\hat{\rho} = \frac{\sum(x_i - \overline{x})(y_i - \overline{y})}{\sum(x_i - \overline{x})^2}\left(\frac{s_X}{s_Y}\right) = \hat{\beta}\frac{s_X}{s_Y} \tag{8.10}$$

or

$$\hat{\beta} = \hat{\rho}\frac{s_Y}{s_X} \tag{8.10a}$$

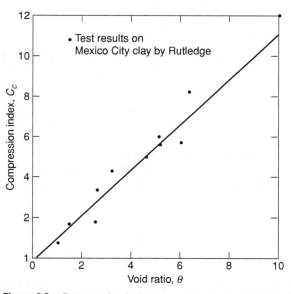

Figure 8.2 Compression index vs. void ratio of soil. (After Nishida, 1956.)

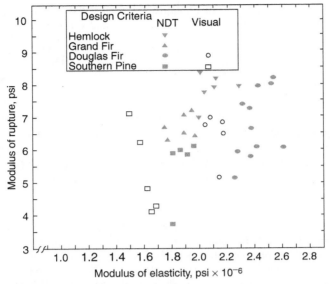

Figure 8.3 Modulus of rupture vs. modulus of elasticity of laminated wood. (After Galligan and Snodgrass, 1970.)

Equation 8.10a gives a useful relationship between the estimates of the correlation coefficient ρ and the slope of the regression line β. Furthermore, by substituting Eq. 8.10a into Eq. 8.6, we obtain

$$s^2_{Y|x} = \frac{1}{n-2}\left[\sum(y_i - \overline{y})^2 - \hat{\rho}^2\frac{s^2_Y}{s^2_X}\sum(x_i - \overline{x})^2\right] = \frac{n-1}{n-2}s^2_Y(1 - \hat{\rho}^2) \quad (8.11)$$

from which we also have

$$\hat{\rho}^2 = 1 - \frac{n-2}{n-1}\frac{s^2_{Y|x}}{s^2_Y} \quad (8.12)$$

which we see is equal to r^2 of Eq. 8.7 for large n. On this basis, therefore, we can conclude that a higher value of $|\hat{\rho}|$ means a greater reduction in the conditional variance associated with the linear regression equation, and hence a more accurate prediction of Y based on the regression of Y on X.

8.3.2 Regression of Normal Variates

Theoretically, the correlation coefficient ρ is defined only for the bivariate normal distribution of two correlated random variables. Moreover, the assumptions of linear model and constancy of variance underlying linear regression are, in fact, inherent properties of two populations that are jointly normal. From these standpoints, when we perform a linear regression of Y on X with constant variance, and estimate the regression coefficients and the correlation coefficient, we are implicitly assuming that the underlying populations are normal.

We may recall from Example 3.35 that if two random variables X and Y are jointly normally distributed, the conditional mean and variance of Y given $X = x$ are as follows:

$$E(Y|X = x) = \mu_Y + \rho\frac{\sigma_Y}{\sigma_X}(x - \mu_X) \tag{8.13}$$

and

$$\text{Var}(Y|X = x) = \sigma_Y^2(1 - \rho^2) \tag{8.14}$$

in which ρ is the correlation coefficient between the two variates. The results of Eqs. 8.13 and 8.14, therefore, clearly show that if two variates are jointly normal, the regression of Y on X, or vice versa the regression of X on Y, is linear with constant conditional variance; i.e., independent of x or y. Specifically, for the regression of Y on X, we see that the linear equation of Eq. 8.13 is of the form of Eq. 8.1 with a slope of $\beta = \rho\dfrac{\sigma_Y}{\sigma_X}$ and an intercept of $\alpha = \mu_Y - \beta\mu_X$.

▶ **EXAMPLE 8.2**

In Example 6.12 of Chapter 6, we presented the precipitation and corresponding runoff data recorded during the 25 rainstorms on the Monocacy River. In hydrology, it is of interest to be able to predict the runoff of the river on the basis of the precipitation; for this purpose, the regression of the runoff on precipitation, therefore, is relevant. Denoting Y for runoff and X for precipitation, we tabulate the calculations necessary to evaluate the regression coefficients in Table E8.2.

From the results of Table E8.2, we obtain the sample means and sample variances of X and Y, respectively, as follows:

$$\bar{x} = \frac{53.89}{25} = 2.16; \quad \bar{y} = \frac{20.05}{25} = 0.80$$

and corresponding sample variances

$$s_X^2 = \frac{1}{24}(153.44 - 25 \times 2.16^2) = 1.533; \quad s_Y^2 = \frac{1}{24}(24.68 - 25 \times 0.80^2) = 0.362$$

Thus, we obtain the following regression coefficients:

$$\hat{\beta} = \frac{59.24 - 25 \times 2.16 \times 0.80}{153.44 - 25(2.16)^2} = 0.435; \quad \text{and} \quad \hat{\alpha} = 0.80 - 0.435 \times 2.16 = -0.140$$

Graphically, the linear regression line is shown in Fig. E8.2.

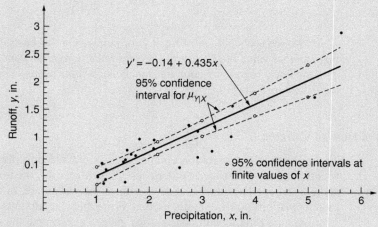

Figure E8.2 Data points and regression of runoff on precipitation.

TABLE E8.2 Computations for Example 8.2

Storm No.	Precip., x_i (in.)	Runoff, y_i (in.)	$x_i y_i$	x_i^2	y_i^2	$y_i' = \hat{\alpha} + \hat{\beta} x_i$	$(y_i - y_i')^2$
1	1.11	0.52	0.58	1.23	0.270	0.343	0.0313
2	1.17	0.40	0.47	1.37	0.160	0.369	0.0009
3	1.79	0.97	1.74	3.20	0.941	0.637	0.1110
4	5.62	2.92	16.40	31.60	8.526	2.280	0.4000
5	1.13	0.17	0.19	1.28	0.029	0.351	0.0328
6	1.54	0.19	0.29	2.37	0.036	0.530	0.1158
7	3.19	0.76	2.43	10.15	0.578	1.245	0.2360
8	1.73	0.66	1.14	2.99	0.436	0.612	0.0023
9	2.09	0.78	1.63	4.37	0.608	0.770	0.0001
10	2.75	1.24	3.41	7.55	1.538	1.059	0.0328
11	1.20	0.39	0.47	1.44	0.152	0.381	0.0001
12	1.01	0.30	0.30	1.02	0.090	0.299	0.0000
13	1.64	0.70	1.15	2.69	0.490	0.574	0.0158
14	1.57	0.77	1.21	2.46	0.593	0.544	0.0511
15	1.54	0.59	0.91	2.37	0.348	0.530	0.0036
16	2.09	0.95	1.99	4.36	0.902	0.770	0.0326
17	3.54	1.02	3.62	12.55	1.040	1.400	0.1442
18	1.17	0.39	0.46	1.37	0.152	0.368	0.0004
19	1.15	0.23	0.26	1.32	0.053	0.360	0.0169
20	2.57	0.45	1.16	6.60	0.202	0.980	0.2810
21	3.57	1.59	5.66	12.74	2.528	1.415	0.0306
22	5.11	1.74	8.90	26.18	3.028	2.084	0.1185
23	1.52	0.56	0.85	2.31	0.314	0.521	0.0015
24	2.93	1.12	3.28	8.58	1.254	1.135	0.0002
25	2.93	0.64	0.74	1.34	0.410	0.365	0.0755
	$\sum = 53.89$	20.05	59.24	153.44	24.678		$\Delta^2 = 1.7350$

The constant conditional variance of the runoff for a given precipitation is

$$s_{Y|x}^2 = \frac{\Delta^2}{n-2} = \frac{1.735}{25-2} = 0.075$$

and its conditional standard deviation is $s_{Y|x} = \sqrt{0.075} = 0.274$ in. In this case, using Eq. 8.9, we obtain the correlation coefficient as

$$\hat{\rho} = \frac{1}{24} \frac{59.24 - 25 \times 2.16 \times 0.80}{\sqrt{1.533}\sqrt{0.362}} = 0.898$$

Assuming that the runoff at a given precipitation is a normal variate, we may assess the probability of a specified runoff at a given precipitation. For instance, the probability that the runoff of the Monocacy River will exceed 2 in. during a rainstorm with 4 in. of precipitation would be as follows. For a precipitation of $X = 4$ in., the mean runoff would be

$$E(Y|X = 4) = -0.14 + 0.435 \times 4 = 1.6 \text{ in.}$$

Therefore, the normal distribution of the runoff Y when the precipitation is $X = 4$ in. is $N(1.6, 0.274)$ in. Hence, the probability that the runoff will exceed 2 in. is

$$P(Y > 2 | X = 4) = 1 - \Phi\left(\frac{2 - 1.6}{0.274}\right) = 1 - 0.928 = 0.072$$

We can also establish the 95% confidence interval as follows: For this purpose, let us select five discrete values of the precipitation at $x_i = 1.0$, 2.16, 3.0, 4.0, and 5.0 in. Then, according to Eq. 8.8, the respective confidence intervals at these five different values of x_i are, individually, with $t_{0.975,23} = 2.069$ from Table A.3.

At $x_i = 1.0$: $\quad (\mu_{Y|1.0})_{0.95} = 0.295 \pm 2.069 \times 0.274\sqrt{\dfrac{1}{25} + \dfrac{(1.0 - 2.16)^2}{153.44 - 25 \times 2.16^2}}$

$\qquad\qquad\qquad = (0.138 \rightarrow 0.452)\,\text{in.}$

At $x_i = 2.16$: $\quad (\mu_{Y|2.16})_{0.95} = 0.800 \pm 2.069 \times 0.274\sqrt{\dfrac{1}{25} + \dfrac{(2.16 - 2.16)^2}{153.44 - 25 \times 2.16^2}}$

$\qquad\qquad\qquad = (0.687 \rightarrow 0.913)\,\text{in.}$

At $x_i = 3.0$: $\quad (\mu_{Y|3.0})_{0.95} = 1.165 \pm 2.069 \times 0.274\sqrt{\dfrac{1}{25} + \dfrac{(3.0 - 2.16)^2}{153.44 - 25 \times 2.16^2}}$

$\qquad\qquad\qquad = (1.027 \rightarrow 1.303)\,\text{in.}$

At $x_i = 4.0$: $\quad (\mu_{Y|4.0})_{0.95} = 1.600 \pm 2.069 \times 0.274\sqrt{\dfrac{1}{25} + \dfrac{(4.0 - 2.16)^2}{153.44 - 25 \times 2.16^2}}$

$\qquad\qquad\qquad = (1.368 \rightarrow 1.832)\,\text{in.}$

At $x_i = 5.0$: $\quad (\mu_{Y|5.0})_{0.95} = 2.035 \pm 2.069 \times 0.274\sqrt{\dfrac{1}{25} + \dfrac{(5.0 - 2.16)^2}{153.44 - 25 \times 2.16^2}}$

$\qquad\qquad\qquad = (1.746 \rightarrow 2.324)\,\text{in.}$

which are also displayed in Fig. E8.2, therefore yielding the 95% confidence interval of the regression line of the runoff on precipitation. We can observe that the minimum range of the 95% confidence interval along the regression line occurs at the mean value of $\bar{x} = 2.16$ in. of precipitation. ◀

▶ **EXAMPLE 8.3**

Table E8.3 shows a set of data of observed settlements of pile groups (Col. 3), reported by Viggiani (2001), under the respective loads; also shown in the same table (Col. 4) are the corresponding calculated settlements using a nonlinear model proposed by Viggiani (2001). We may perform the regression of the observed settlement, Y, on the calculated settlement, X; the calculations are summarized in Table E8.3.

From Table E8.3, we obtain the sample means and sample standard deviations of Y and X, respectively, as follows:

$$\bar{y} = \frac{600.36}{25} = 24.014\,\text{mm} \qquad \bar{x} = \frac{563.34}{25} = 22.534\,\text{mm}$$

$$s_Y = \sqrt{\frac{1}{24}(48064.16 - 25 \times 24.014^2)} = 37.44\,\text{mm}$$

$$s_X = \sqrt{\frac{1}{24}(41185.44 - 25 \times 22.534^2)} = 34.45\,\text{mm}$$

According to Eqs. 8.3 and 8.4, we obtain the corresponding regression coefficients for the regression of Y on X as follows:

$$\hat{\beta} = \frac{43,842.89 - 25 \times 24.014 \times 22.534}{41,185.44 - 25 \times 22.534^2} = 1.064 \quad \text{and} \quad \hat{\alpha} = 24.014 - 1.064 \times 22.535 = 0.038$$

The data given in Columns 3 and 4 of Table E8.3 are plotted in Fig. E8.3.
Therefore, the linear regression equation of Y on X is

$$E(Y|X = x) = 0.038 + 1.064x$$

TABLE E8.3 Summary of Data and Calculations for Example 8.3

Case No.	Load, mn	Obs. S, y_i, mm	Cal. S, x_i, mm	y_i^2	x_i^2	$x_i y_i$	$y_i' = 0.038 + 1.064 x_i$	$(y_i' - y_i)^2$
1	0.22	0.7	5.9	0.49	34.81	4.13	6.316	31.539
2	1.2	64	27	4096	729	1728	28.728	1244.11
3	0.89	5	4.9	25	24.01	24.5	5.252	0.064
4	0.44	25	26.6	625	707.56	665	28.34	11.156
5	0.44	29.5	29.4	870.25	864.36	867.3	31.32	3.312
6	1.35	3.8	3.9	14.44	15.21	14.82	4.188	0.15
7	1.93	5.9	6.7	34.81	44.89	39.53	7.167	1.605
8	0.49	38.1	36.8	1452.61	1354.24	1402.08	39.193	1.195
9	1.3	185	175	34,225	30,625	32,375	186.238	1.533
10	5.09–8.9	28.1	32.7	789.61	1069.29	918.87	34.831	45.306
11	0.06–0.26	0.6	0.7	0.36	0.49	0.42	0.783	0.033
12	0.66	29.2	31.9	852.64	1017.61	931.48	33.98	22.848
13	0.76	32	32.4	1024	1049.76	1036.8	34.511	6.305
14	0.82	5.4	3.9	29.16	15.21	21.06	4.188	1.469
15	0–1.25	3.6	4.4	12.96	19.36	15.84	4.72	1.254
16	3.93	35.9	31.6	1288.81	998.56	1134.44	33.66	5.018
17	0.47	11.6	11.5	134.56	132.25	133.4	12.274	0.454
18	0.41	12.7	14.6	161.29	213.16	185.42	15.572	8.248
19	1.02	46	44.4	2116	1971.36	2042.4	47.28	1.638
20	0.24	3.8	4.5	14.44	20.25	17.1	4.826	1.053
21	0.105	4.9	5.2	24.01	27.04	25.48	5.571	0.45
22	1.65	11.7	11.8	136.89	139.24	138.06	12.593	0.797
23	0.598	1.83	3.49	3.35	12.18	6.39	3.751	3.489
24	0.68	9.43	8	88.92	64	75.44	8.55	0.744
25	0.95	6.6	6.05	43.56	36.6	39.93	6.475	0.016
Σ		600.36	563.34	48064.2	41185.4	43842.9		1393.79

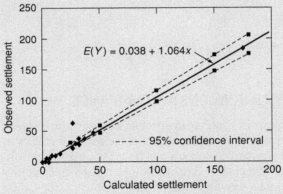

Figure E8.3 Data points and regression of observed settlement on calculated settlement.

and with Eq. 8.9, we obtain the correlation coefficient

$$\hat{\rho} = \frac{1}{24} \left(\frac{43842.89 - 25 \times 24.014 \times 22.534}{37.44 \times 34.45} \right) = 0.98$$

which shows a very high correlation between the observed and calculated settlements. The conditional standard deviation of Y given X, according to Eq. 8.6a, is

$$s_{Y|x} = \sqrt{\frac{1394}{23}} = 7.78$$

We may observe that this $s_{Y|x}$ is much smaller than the unconditional standard deviation $s_Y = 37.44$ mm.

We may also construct the 95% confidence interval of the regression line following Eq. 8.8. For this purpose, we first evaluate the corresponding 95% confidence intervals at the following selected discrete values of x_i: 25 mm, 50 mm, 100 mm, 150 mm, and 180 mm. From Table A.3, $t_{0.975,23} = 2.069$.

At $x_i = 25$ mm: $\quad \langle \mu_Y \rangle_{.95} = 26.600 \pm 2.069 \times 7.784 \sqrt{\dfrac{1}{25} + \dfrac{(25 - 22.534)^2}{(41,185.4 - 25 \times 22.534^2)}}$

$$= (23.38 \to 29.82) \text{ mm}$$

At $x_i = 50$ mm: $\quad \langle \mu_Y \rangle_{.95} = 53.238 \pm 2.069 \times 7.784 \sqrt{\dfrac{1}{25} + \dfrac{(50 - 22.534)^2}{(41,185.4 - 25 \times 22.534^2)}}$

$$= (49.08 \to 57.39) \text{ mm}$$

At $x_i = 100$ mm: $\quad \langle \mu_Y \rangle_{.95} = 106.44 \pm 2.069 \times 7.784 \sqrt{\dfrac{1}{25} + \dfrac{(100 - 22.534)^2}{(41,185.4 - 25 \times 22.534^2)}}$

$$= (98.38 \to 114.50) \text{ mm}$$

At $x_i = 150$ mm: $\quad \langle \mu_Y \rangle_{.95} = 159.64 \pm 2.069 \times 7.784 \sqrt{\dfrac{1}{25} + \dfrac{(150 - 22.534)^2}{(41,185.4 - 25 \times 22.534^2)}}$

$$= (147.06 \to 172.22) \text{ mm}$$

At $x_i = 180$ mm: $\quad \langle \mu_Y \rangle_{.95} = 191.56 \pm 2.069 \times 7.784 \sqrt{\dfrac{1}{25} + \dfrac{(180 - 22.534)^2}{(41,185.4 - 25 \times 22.534^2)}}$

$$= (176.19 \to 206.92) \text{ mm}$$

By connecting two lines through the respective lower-bound and upper-bound values, as calculated above, we obtain the 95% confidence interval of the regression line as shown in dash lines in Fig. E8.3. ◀

▶ 8.4 LINEAR REGRESSION WITH NONCONSTANT VARIANCE

The scattergram of observed data points may sometimes show a significant variation in the degree of scatter with varying values of the control variable; in such cases, the conditional variance or standard deviation about a regression equation will not be constant and may be a function of the control variable. The regression analysis presented in Sect. 8.2.1 can be extended to take account of the variation of the conditional variance. For this purpose, the conditional variance may be expressed as

$$\text{Var}(Y|X = x) = \sigma^2 g^2(x) \tag{8.15}$$

where σ is an unknown constant and $g(x)$ is a predetermined function of x. Again, for linear regression of Y on X, we have

$$E(Y|X = x) = \alpha + \beta x$$

in which the regression coefficients α and β may be different from those of Eq. 8.1. In this case, it would be reasonable to assume that data points in regions of small conditional variance should carry higher "weights" than those in regions of large conditional variances. On this premise, accordingly, we assign weights inversely proportional to the conditional variance, or

$$w'_i = \frac{1}{\text{Var}(Y|X = x_i)} = \frac{1}{\sigma^2 g^2(x_i)}$$

Then we can show that the total squared error is

$$\Delta^2 = \sum_{i=1}^{n} w_i (y_i - \alpha - \beta x_i)^2$$

from which the least-squares estimates of the regression coefficients α and β become

$$\hat{\alpha} = \frac{\sum w_i y_i - \hat{\beta} \sum w_i x_i}{\sum w_i} \tag{8.16}$$

and

$$\hat{\beta} = \frac{\sum w_i(\sum w_i x_i y_i) - (\sum w_i x_i)(\sum w_i y_i)}{\sum w_i(\sum w_i x_i^2) - (\sum w_i x_i)^2} \tag{8.17}$$

in which

$$w_i = \sigma^2 w'_i = \frac{1}{g^2(x_i)}$$

An unbiased estimate of σ^2 from a sample of size n is then

$$s^2 = \frac{\sum w_i(y_i - \hat{\alpha} - \hat{\beta} x_i)^2}{n - 2} = \frac{\sum w_i(y_i - \overline{y}_i)^2}{n - 2} \tag{8.18}$$

Hence, an estimate of the conditional variance of Y given $X = x$, according to Eq. 8.15, is

$$s^2_{Y|x} = s^2 \cdot g^2(x) \tag{8.19}$$

or

$$s_{Y|x} = sg(x) \tag{8.19a}$$

Correlation Coefficient and Confidence Interval

The correlation coefficient of Eq. 8.9 or 8.10 should still be valid for the present case; in particular, Eq. 8.10 is valid in which $\hat{\beta}$ is given by Eq. 8.17. That is, $\hat{\rho} = \hat{\beta} \frac{s_X}{s_Y}$, where s_X and s_Y are the respective sample standard deviations of X and Y. Also, the corresponding confidence interval may be determined with Eq. 8.8, where

$\overline{y}_i = E(Y|X = x_i)$ is given by the linear regression equation with the regression coefficients estimated with Eqs. 8.16 and 8.17; and

$$s_{Y|x} = g(x)\sqrt{\frac{\sum w_i(y_i - \overline{y}_i)^2}{n - 2}} \text{ of Eq. 8.19a and 8.18.}$$

▶ **EXAMPLE 8.4**

Data of the observed maximum settlements and the maximum differential settlements for 18 storage tanks in Libya are plotted as shown in Fig. E8.4. From this figure, we can observe that the scatter of the differential settlement appears to increase linearly with the maximum settlement. Accordingly, we may assume that the conditional standard deviation of the maximum differential settlement Y increases linearly with the maximum settlement X; that is, $g(x) = x$, $\text{Var}(Y|X = x) = \sigma^2 x^2$, and $w_i = 1/x_i^2$.

The details of the calculations are tabulated in Table E8.4, from which we obtain the sample means and sample standard deviations of X and Y, respectively, as follows:

$$\overline{x} = \frac{29.7}{18} = 1.65 \qquad \text{and} \qquad \overline{y} = \frac{19.9}{18} = 1.11$$

and

$$s_X = \sqrt{\frac{1}{17}(63.49 - 18 \times 1.65^2)} = 0.923; \qquad s_Y = \sqrt{\frac{1}{17}(28.87 - 18 \times 1.11^2)} = 0.627$$

Also, with Eqs. 8.16 and 8.17, we obtain the regression coefficients

$$\hat{\beta} = \frac{22.38 \times 12.42 - 15.84 \times 11.31}{22.38 \times 18 - 15.84^2} = 0.65 \qquad \text{and} \qquad \hat{\alpha} = \frac{11.31 - 0.65 \times 15.84}{22.38} = 0.045$$

Therefore, the linear regression of Y on X is $E(Y \mid x) = 0.045 + 0.65x$.

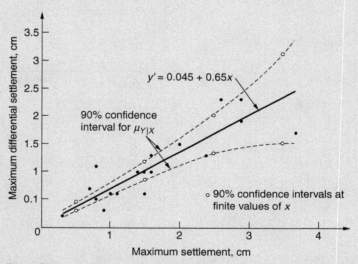

Figure E8.4 Settlements of tanks in Libya. (Data from Lambe and Whitman, 1969.)

According to Eq. 8.18, the estimate of σ^2 is

$$s^2 = \frac{0.9418}{18 - 2} = 0.0589$$

and from Eq. 8.19a, we obtain the conditional standard deviation of Y as a function of X to be

$$s_{Y|x} = x\sqrt{0.0589} = 0.243x$$

In this case, the estimated correlation coefficient, from Eq. 8.10, is

$$\hat{\rho} = \hat{\beta}\frac{s_X}{s_Y} = 0.65 \times \frac{0.923}{0.627} = 0.96$$

The corresponding confidence interval for the regression equation may be developed as follows. For this purpose, let us select the following discrete values of x_i: 0.5, 1.5, 2.5, and 3.5. Then, for a 90% confidence, we obtain with Eq. 8.8 and $t_{0.95,16} = 1.746$ from Table A.3, the following respective intervals at the selected values of x_i:

$$\text{At } x_i = 0.5: \quad \langle \mu_{Y|x} \rangle_{0.90} = 0.370 \pm 1.746(0.243 \times 0.5)\sqrt{\frac{1}{18} + \frac{(0.5 - 1.65)^2}{(63.49 - 18 \times 1.65^2)}}$$

$$= (0.288 \rightarrow 0.451)$$

$$\text{At } x_i = 1.5: \quad \langle \mu_{Y|x} \rangle_{0.90} = 0.975 \pm 1.746(0.243 \times 1.5)\sqrt{\frac{1}{18} + \frac{(1.5 - 1.65)^2}{(63.49 - 18 \times 1.65^2)}}$$

$$= (0.823 \rightarrow 1.127)$$

$$\text{At } x_i = 2.5: \quad \langle \mu_{Y|x} \rangle_{0.90} = 1.670 \pm 1.746(0.243 \times 2.5)\sqrt{\frac{1}{18} + \frac{(2.5 - 1.65)^2}{(63.49 - 18 \times 1.65^2)}}$$

$$= (1.326 \rightarrow 2.014)$$

$$\text{At } x_i = 3.5: \quad \langle \mu_{Y|x} \rangle_{0.90} = 2.320 \pm 1.746(0.243 \times 3.5)\sqrt{\frac{1}{18} + \frac{(3.5 - 1.65)^2}{(63.49 - 18 \times 1.65^2)}}$$

$$= (1.512 \rightarrow 3.122)$$

With the above intervals at the selected discrete values of x_i, we can construct the 90% confidence interval for the regression equation of the maximum differential settlement on the maximum settlement, as displayed in Fig. E8.4.

TABLE E8.4   Calculations for Example 8.4

Tank No., i	Max. Settlement x_i (cm)	Diff. Settlement y_i (cm)	w_i	$w_i x_i$	$w_i y_i$	$w_i x_i y_i$	$w_i x_i^2$	$w_i(y_i - \bar{y}_i)^2$
1	0.3	0.2	11.11	3.33	2.22	0.67	1.0	0.0178
2	0.7	0.7	2.04	1.43	1.43	1.00	1.0	0.0816
3	0.8	0.5	1.56	1.25	0.78	0.62	1.0	0.0066
4	0.8	1.1	1.56	1.25	1.72	1.37	1.0	0.4465
5	0.9	0.3	1.23	1.11	0.37	0.33	1.0	0.1339
6	1.0	0.6	1.00	1.00	0.60	0.60	1.0	0.0090
7	1.1	0.6	0.83	0.91	0.50	0.55	1.0	0.0212
8	1.4	1.0	0.51	0.71	0.51	0.71	1.0	0.0010
9	1.5	1.0	0.44	0.67	0.44	0.66	1.0	0.0002
10	1.6	1.0	0.39	0.63	0.39	0.62	1.0	0.0028
11	1.6	1.3	0.39	0.63	0.51	0.81	1.0	0.0180
12	2.0	1.5	0.25	0.50	0.38	0.75	1.0	0.0060
13	2.4	1.3	0.17	0.42	0.22	0.53	1.0	0.0158
14	2.6	2.3	0.15	0.38	0.35	0.90	1.0	0.0479
15	2.9	1.9	0.12	0.34	0.23	0.66	1.0	0.0001
16	2.9	2.3	0.12	0.34	0.28	0.80	1.0	0.0164
17	3.7	1.7	0.07	0.27	0.12	0.44	1.0	0.0394
18	1.5	0.6	0.44	0.67	0.26	0.40	1.0	0.0776
$\sum$	29.7	19.9	22.38	15.84	11.31	12.42	18.0	0.9418

◀

▶ 8.5   MULTIPLE LINEAR REGRESSION

A dependent variable may be a function of more than one independent or control variable. In such cases, if the variables are random, the mean and variance of the dependent variable will also be functions of the independent variables. When the mean-value function is assumed to be linear, the resulting analysis is called *multiple linear regression*. The basic analysis of multiple linear regression is simply a generalization of the regression analysis developed in Sect. 8.2 for two variables.

Suppose the dependent variable of interest is Y and that it is a linear function of k random variables $X_1, X_2, \ldots, X_k$. It follows that the mean value of Y at $X_1 = x_{i1}$, $X_2 = x_{i2}, \ldots, X_k = x_{ik}$; $i = 1, 2, \ldots, n$, is

$$y_i' = \beta_o + \beta_1 x_{i1} + \beta_2 x_{i2} + \cdots + \beta_k x_{ik} \tag{8.20}$$

for each i, where $\beta_0, \beta_1, \ldots, \beta_k$ are constant regression coefficients that must be estimated from the observed data $(x_{i1}, x_{i2}, \ldots, x_{ik})$, whereas the conditional variance of Y for given values $x_{i1}, x_{i2}, \ldots, x_{ik}$ is assumed to be constant for any i; i.e.,

$$\mathrm{Var}(Y|x_{i1}, x_{i2}, \ldots, x_{ik}) = \sigma^2 \tag{8.21}$$

or is a known function of $(x_{i1}, x_{i2}, \ldots, x_{ik})$; i.e.,

$$\mathrm{Var}(Y|x_{i1}, x_{i2}, \ldots, x_{ik}) = \sigma^2 g^2(x_{i1}, x_{i2}, \ldots, x_{ik}) \tag{8.22}$$

Regression analysis is then performed to estimate $\beta_o, \beta_1, \ldots, \beta_k$ and σ^2 based on a set of observed data of size n, namely, $(x_{i1}, x_{i2}, \ldots, x_{ik}, y_i)$, $i = 1, 2, \ldots, n$.

Equation 8.20 may be written conveniently in matrix form as follows:

$$\mathbf{y}' = \mathbf{X}\boldsymbol{\beta} \tag{8.20a}$$

where $\mathbf{y}'$ is a vector $\mathbf{y}' = \{y_1', y_2', \ldots, y_n'\}$, in which each y_i' is given by Eq. 8.20; $\boldsymbol{\beta}$ is a vector of the regression coefficients $\boldsymbol{\beta} = \{\beta_o, \beta_1, \ldots, \beta_k\}$, and $\mathbf{X}$ is an n by $k + 1$ matrix (i.e., n rows and $k + 1$ columns):

$$\mathbf{X} = \begin{bmatrix} 1 & x_{11} & x_{12} & \cdots & x_{1k} \\ 1 & x_{21} & x_{22} & \cdots & x_{2k} \\ \cdot & \cdot & \cdot & \cdots & \cdot \\ \cdot & \cdot & \cdot & \cdots & \cdot \\ \cdot & \cdot & \cdot & \cdots & \cdot \\ 1 & x_{n1} & x_{n2} & \cdots & x_{nk} \end{bmatrix}$$

Let us now concentrate our attention on the case in which the conditional variance is constant. The total squared error for a set of n data points is then

$$\Delta^2 = \sum_{i=1}^{n} (y_i - y_i')^2 = \sum_{i=1}^{n} [y_i - \beta_o - \beta_1 x_{i1} - \cdots - \beta_k x_{ik}]^2 \tag{8.23}$$

Then, on the basis of least squares, we minimize Δ^2 to obtain the following set of linear equations for determining the estimates of β_j, $j = 0, 1, 2, \ldots, k$:

$$\frac{\partial \Delta^2}{\partial \beta_o} = \sum_{i=1}^{n} [y_i - \hat{\beta}_o - \hat{\beta}_1 x_{i1} - \cdots - \hat{\beta}_k x_{ik}] = 0$$

$$\frac{\partial \Delta^2}{\partial \beta_1} = \sum_{i=1}^{n} x_{i1} [y_i - \hat{\beta}_o - \hat{\beta}_1 x_{i1} - \cdots - \hat{\beta}_k x_{ik}] = 0 \tag{8.24}$$

$$\vdots$$

$$\frac{\partial \Delta^2}{\partial \beta_k} = \sum_{i=1}^{n} x_{ik} [y_i - \hat{\beta}_o - \hat{\beta}_1 x_{i1} - \cdots - \hat{\beta}_k x_{ik}] = 0$$

Equation 8.24 consists of $k + 1$ equations with the $k + 1$ unknown regression coefficients. Again, in matrix notation, the set of equations of Eq. 8.24 may be written as

$$\mathbf{X}^T \mathbf{X} \hat{\boldsymbol{\beta}} = \mathbf{X}^T \mathbf{y} \tag{8.24a}$$

where $\mathbf{y} = \{y_1, y_2, \ldots, y_n\}$ is the vector of the observed values of Y and T is the transpose.

Premultiplying both sides of Eq. 8.24a by the inverse of the matrix $\mathbf{X}^T\mathbf{X}$, we obtain the solutions for the least-squares estimates of the regression coefficients as

$$\hat{\boldsymbol{\beta}} = \left(\mathbf{X}^T\mathbf{X}\right)^{-1}\mathbf{X}^T\mathbf{y} \tag{8.25}$$

where $\hat{\boldsymbol{\beta}}$ is a $(k+1)$ vector, $\hat{\boldsymbol{\beta}} = \{\beta_o, \beta_1, \ldots, \beta_k\}$, with which we obtain the multiple regression equations, in matrix form,

$$\mathbf{y}' = \mathbf{X}\hat{\boldsymbol{\beta}} \tag{8.26}$$

In scalar form, Eq. 8.26 represents a set of k multiple regression equations as follows:

$$y_i' = \hat{\beta}_o + \sum_{j=1}^{k} \hat{\beta}_j x_{ij}; \quad i = 1, 2, \ldots, n; \ j = 1, 2, \ldots, k \tag{8.26a}$$

where $\hat{\beta}_o$ and $\hat{\beta}_j$ are the components of $\hat{\boldsymbol{\beta}}$.

An unbiased estimate of the conditional variance of Y for given values of $X_1, X_2, \ldots, X_k$ is

$$s_{Y|x_1,x_2,\ldots,x_k}^2 = \frac{\Delta^2}{n-k-1} = \frac{\sum_{i=1}^{n}\left(y_i - y_i'\right)^2}{n-k-1} \tag{8.27}$$

in which y_i' is given by Eq. 8.26a. The reduction in the original variance of Y may also be evaluated with r of Eq. 8.7 by substituting $s_{Y|x_1,\ldots,x_k}$ for $s_{Y|x}$.

▶ **EXAMPLE 8.5**

An important factor in determining the frost depth for highway pavement design is the mean annual temperature for the site under consideration. The mean annual temperatures that have been recorded at 10 different weather stations in West Virginia are summarized in Table E8.5a.

At locations in the state where temperature records are not available, the mean annual temperature at a construction site in West Virginia may be predicted on the basis of its elevation and latitude based on the information recorded in Table E8.5a.

For the above purpose, the following multiple linear equation may be assumed:

$$E(Y|x_1, x_2) = \beta_o + \beta_1 x_1 + \beta_2 x_2$$

where

Y = the mean annual temperature, in °F

x_1 = elevation, in feet above sea level

x_2 = north latitude, in degrees

Table 8.5a Mean Annual Temperatures in West Virginia. (Data from Moulton and Schaub, 1969.)

Weather Station	Elevation, ft	North Latitude, deg.	Mean Annual Temperature, °F
Bayard	2375	39.27	47.5
Buckhannon	1459	39.00	52.3
Charleston	604	38.35	56.8
Flat Top	3242	37.58	48.4
Kearneysville	550	39.38	54.2
Madison	675	38.05	55.1
New Martinsville	635	39.65	54.4
Pickens	2727	38.66	48.8
Rainelle	2424	37.97	50.5
Wheeling	659	40.10	52.7

To evaluate the regression coefficients, we form the following matrix based on the data of Table E8.5a:

$$\mathbf{X} = \begin{bmatrix} 1 & 2375 & 39.27 \\ 1 & 1459 & 39.00 \\ 1 & 604 & 38.35 \\ 1 & 3242 & 37.58 \\ 1 & 550 & 39.38 \\ 1 & 675 & 38.05 \\ 1 & 635 & 39.65 \\ 1 & 2727 & 38.66 \\ 1 & 2424 & 37.97 \\ 1 & 659 & 40.10 \end{bmatrix} \quad \text{and the vectors,} \quad \mathbf{y} = \begin{bmatrix} 47.5 \\ 52.3 \\ 56.8 \\ 48.4 \\ 54.2 \\ 55.1 \\ 54.4 \\ 48.8 \\ 50.5 \\ 52.7 \end{bmatrix} \quad \text{and} \quad \hat{\boldsymbol{\beta}} = \begin{bmatrix} \hat{\beta}_0 \\ \hat{\beta}_1 \\ \hat{\beta}_2 \end{bmatrix}$$

from which we obtain, through matrix operations, Eqs. 8.24a through 8.26:

$$\mathbf{X}^T\mathbf{X} = \begin{bmatrix} 10 & 15{,}350 & 388 \\ 15{,}350 & 3.36 \times 10^7 & 5.92 \times 10^5 \\ 388 & 5.92 \times 10^5 & 15{,}061 \end{bmatrix}$$

$$(\mathbf{X}^T\mathbf{X})^{-1} = \begin{bmatrix} 357.12 & -0.0038 & -9.051 \\ -0.0038 & 1.372 \times 10^{-7} & 9.237 \times 10^5 \\ -9.051 & 9.237 \times 10^5 & 0.2296 \end{bmatrix}$$

and

$$\mathbf{X}^T\mathbf{y} = \begin{bmatrix} 520.7 \\ 772{,}104 \\ 20{,}208 \end{bmatrix} \quad \text{Hence,} \quad \hat{\boldsymbol{\beta}} = (\mathbf{X}^T\mathbf{X})^{-1}(\mathbf{X}^T\mathbf{y}) = \begin{bmatrix} 121.05 \\ -0.0034 \\ -1.644 \end{bmatrix}$$

and the regression coefficients are $\hat{\beta}_o = 121.05$, $\hat{\beta}_1 = -0.0034$, and $\hat{\beta}_2 = -1.644$. Thus, we obtain the multiple linear regression of Y on X_1 and X_2 as

$$y' = E(Y|x_1, x_2) = 121.05 - 0.0034x_1 - 1.644x_2$$

With the above regression equation, we summarize the results of the calculations in Table E8.5b. From these results, we also obtain the following. The respective sample means

$$\bar{x}_1 = \frac{15{,}350}{10} = 1535; \qquad \bar{x}_2 = \frac{388.01}{10} = 38.80; \qquad \bar{y} = \frac{520.7}{10} = 52.07$$

and respective sample standard deviations:

$$s_{X_1} = \sqrt{\frac{1}{9}(33{,}552{,}622 - 10 \times 1535^2)} = 1054 \text{ ft};$$

$$s_{X_2} = \sqrt{\frac{1}{9}(15{,}061 - 10 \times 38.80^2)} = 0.87°\text{N}, \quad \text{and} \quad s_Y = \sqrt{\frac{1}{9}(27{,}202 - 10 \times 52.07^2)} = 3.15°\text{F}$$

Also, according to Eq. 8.27, the conditional standard deviation of Y is

$$s_{Y|x_1,x_2} = \sqrt{\frac{3.80}{10 - 2 - 1}} = 0.74°\text{F}$$

Table E8.5b   Calculations for Example 8.5

Elevation x_{i1}, (ft)	Latitude x_{i2}, °N	Mean Temp. y_i °F	y_i'	$(y_i - y_i')^2$
2375	39.27	47.5	48.43	0.86
1459	39.00	52.3	51.99	0.10
604	38.35	56.8	55.97	0.69
3242	37.58	48.4	48.27	0.02
550	39.38	54.2	54.45	0.06
675	38.05	55.1	56.22	1.25
635	39.65	54.4	53.72	0.46
2727	38.66	48.8	48.24	0.31
2424	37.97	50.5	50.41	0.01
659	40.10	52.7	52.89	0.04
$\sum$ 15,350	388.01	520.7		$\Delta^2 = 3.80$

Applied to Gary, West Virginia, which is located at an elevation of 1426 ft and at a latitude of 37.37°N, the expected mean annual temperature would be

$$E(Y|1426, 37.37) = 121.3 - 0.0034 \times 1426 - 4.65 \times 37.37 = 54.80°\text{F}$$

with a conditional standard deviation of 0.74°F.

If the mean annual temperature in Gary can be assumed to be Gaussian, the 10-percentile value of the temperature, $y_{0.10}$, would be determined as follows:

$$P(Y < y_{0.10}) = \Phi\left(\frac{y_{0.10} - 54.80}{0.74}\right) = 0.10$$

from which, we obtain $y_{0.10} = 54.80 + 0.74\Phi^{-1}(0.10) = 54.80 - 0.74 \times 1.28 = 53.9°\text{F}$.   ◄

Multiple Correlations

In multiple regression, the dependent variable Y would be a function of two or more independent or control variables. Therefore, Y may be correlated with each of the independent variables, e.g., between Y and X_j, with corresponding correlation coefficient ρ_{Y,X_j}. Moreover, each pair of the independent variables may also be mutually correlated with a correlation coefficient ρ_{X_i,X_j} between X_i and X_j. From the set of observed data of sample size n, the estimates of the respective correlation coefficients between Y and X_j, following Eq. 8.9, would be

$$\hat{\rho}_{Y,X_j} = \frac{1}{n-1}\frac{\sum_{i=1}^{n} x_{ij}y_i - n\bar{x}_j \bar{y}}{s_{X_j} \cdot s_Y}$$

▶ **8.6   NONLINEAR REGRESSION**

Functional relationships between, or among, engineering variables are not always linear, or may not always be adequately described by linear models. Observed data, from field measurements or experiments, for the variables may show nonlinear trends in the scattergrams of the data points. In such cases, a nonlinear relationship between the variables would be more appropriate. The determination of the appropriate nonlinear relationships on the basis of observational data involves *nonlinear regression analysis.*

Nonlinear regression is usually based on an assumed nonlinear function of the mean-value of the dependent variable, Y, as a function of the independent (or control) variable X, with certain undetermined coefficients that must be evaluated on the basis of the observed data. The simplest type of nonlinear functions for the regression of Y on X is

$$E(Y|x) = \alpha + \beta g(x) \tag{8.28}$$

where $g(x)$ is a predetermined nonlinear function of x. For example, $g(x)$ may be a polynomial function, $x + x^2$; an exponential function, e^x; or a logarithmic function, $\ln x$, or any other nonlinear function of x. This is usually coupled with a constant conditional variance, $\text{Var}(Y|x) = \text{constant}$, or a conditional variance that is a function of x, i.e., $g(x)$.

By defining a new variable $x' = g(x)$, Eq. 8.28 becomes

$$E(Y|x') = \alpha + \beta x' \tag{8.29}$$

We can then see that Eq. 8.29 is of the same mathematical form as the linear regression equation of Eq. 8.1. Furthermore, if the observed data points (x_i, y_i) are also transformed to (x_i', y_i), the original nonlinear regression of Y on X is, therefore, converted to a corresponding linear regression of Y on X', and the regression coefficients α and β can be estimated with Eqs. 8.3 and 8.4, whereas the conditional variance $s_{Y|x'}^2$ can also be estimated with Eq. 8.6.

▶ **EXAMPLE 8.6**

Data of the average all-day parking cost (in the 1960s) in the central business districts of 15 U.S. cities were collected as listed in Table E8.6, and the data are plotted against the city population. The data points in the scattergram of Fig. E8.6a clearly show a nonlinear trend. However, if we use $x' = \ln x$ in place of x, the scattergram would be as shown in Fig. E8.6b, which would then show a reasonably linear trend. On the latter basis, we may model the average all-day parking cost as a linear function of $\ln x$ or with the following nonlinear regression equation:

$$E(Y|x) = \alpha + \beta \ln x$$

where

$Y = $ average all-day parking cost, in 1960s $

$x = $ urban population, in 1000s

TABLE E8.6   Summary of Data and Computations for Example 8.6

City No.	Population x_i (in 1000)	Parking Cost y_i (in $)	$x_i' = \ln x_i$	$x_i' y_i$	$x_i'^2$	y_i^2	y_i'	$(y_i - y_i')^2$
1	190	0.50	5.25	2.62	27.5	0.25	0.51	0.000
2	310	0.48	5.74	2.75	32.9	0.23	0.63	0.023
3	270	0.53	5.60	2.97	31.3	0.28	0.60	0.005
4	320	0.58	5.77	3.35	33.3	0.34	0.63	0.003
5	460	0.60	6.13	3.68	37.6	0.36	0.72	0.014
6	340	0.67	5.83	3.91	34.0	0.45	0.65	0.000
7	380	0.69	5.94	4.10	35.3	0.48	0.68	0.000
8	520	0.75	6.25	4.69	39.1	0.56	0.75	0.000
9	310	0.80	5.74	4.59	32.9	0.64	0.63	0.029
10	400	0.80	5.99	4.79	35.9	0.64	0.69	0.012
11	470	0.81	6.15	4.98	37.9	0.66	0.73	0.006
12	840	0.92	6.73	6.19	45.3	0.85	0.87	0.003
13	1910	0.92	7.56	6.95	57.1	0.85	1.07	0.023
14	3290	1.40	8.10	11.34	65.6	1.96	1.21	0.036
15	3600	1.12	8.19	9.17	67.1	1.25	1.23	0.012
$\sum$		11.57	94.97	76.08	612.8	9.80		$\Delta^2 = 0.166$

From the calculations summarized in Table E8.6, we also obtain the following. The sample means

$$\bar{x}' = \frac{94.97}{15} = 6.33 \quad \text{and} \quad \bar{y} = \frac{11.57}{15} = 0.771$$

and sample standard deviations

$$s_{x'} = \sqrt{\frac{1}{14}\left(612.8 - 15 \times 6.33^2\right)} = 0.92 \quad \text{and} \quad s_Y = \sqrt{\frac{1}{14}\left(9.80 - 15 \times 0.771^2\right)} = 0.25$$

Then, with Eqs. 8.3 and 8.4, we obtain the regression coefficients,

$$\hat{\beta} = \frac{76.08 - 15 \times 6.33 \times 0.771}{612.8 - 15 \times 6.33^2} = 0.244 \quad \text{and} \quad \hat{\alpha} = 0.771 - 0.244 \times 6.33 = -0.773$$

Hence, the regression of Y on $\ln X$ is

$$E(Y|\ln x) = -0.773 + 0.244 \ln x$$

which is displayed in Fig. E8.6b, whereas the corresponding nonlinear regression of Y on X is shown in Fig. E8.6a.

The conditional standard deviation of Y given $x' = \ln x$ is

$$s_{Y|x'} = \sqrt{\frac{0.166}{15 - 2}} = \sqrt{0.0128} = 0.113$$

and the corresponding correlation coefficient, for the linear regression of Y on $\ln X$, according to Eq. 8.9, is

$$\hat{\rho} = \frac{1}{14}\left(\frac{76.08 - 15 \times 6.33 \times 0.771}{0.92 \times 0.25}\right) = 0.89$$

which shows that there is a reasonable linear relation between Y and $\ln X$.

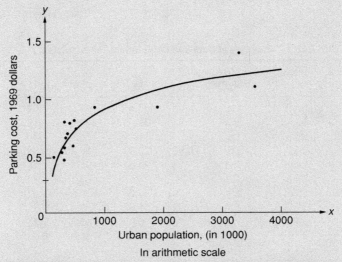

Figure 8.6a Data points of parking cost vs. population (in arithmetic scale) and regression equation of Y on X. (Data from Wynn, 1969.)

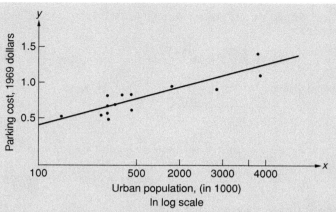

Figure 8.6b Data points of parking cost vs. population (in semilog scale) and regression of Y on ln X. (Data from Wynn, 1969.)

◀

▶ **EXAMPLE 8.7**

In order to predict the average dissolved oxygen (DO) concentration in a pool based on the average pool temperature, T, measured data for DO (in mg/l) and corresponding pool temperature T (in °C) were obtained as summarized in Table E8.7. In this case, an exponential model may be an appropriate functional relation between the DO and the temperature T; that is,

$$DO = \alpha e^{\beta T}$$

By taking the natural logarithm of both sides of the above equation, we have

$$\ln DO = \ln \alpha + \beta T$$

Therefore, using $X = T$ and $Y = \ln DO$, we obtain the linear regression of Y on X,

$$E(Y|X) = \ln \alpha + \beta X \qquad \text{or} \qquad E(\ln DO|T) = \ln \alpha + \beta T$$

TABLE E8.7 Summary of Data and Calculations for Example 8.7

Ave. Temp., x_i (°C)	Ave. DO, y_i (mg/l)	ln y_i	x_i ln y_i	x_i^2	$(\ln y_i)^2$	ln $y_i' = 3.86$ $-0.11 x_i$	$(\ln y_i - \ln y_i')^2$
23.2	4.2	1.44	33.41	538.2	2.07	1.31	0.0169
23.4	3.7	1.31	30.65	547.6	1.72	1.29	0.0004
23.8	3.8	1.34	31.89	566.4	1.80	1.24	0.0100
24.1	2.5	0.92	22.17	580.8	0.85	1.21	0.0841
24.4	3.1	1.13	27.57	595.4	1.28	1.18	0.0025
24.6	3.2	1.16	28.53	605.2	1.34	1.15	0.0001
25.5	2.9	1.06	27.03	650.3	1.12	1.06	0.0000
27.2	2.5	0.92	25.02	739.8	0.85	0.87	0.0025
27.3	3.0	1.10	30.03	745.3	1.21	0.86	0.0576
27.5	2.3	0.83	22.83	756.3	0.69	0.84	0.0001
28.7	1.6	0.47	13.49	823.7	0.22	0.70	0.0529
$\sum$ 279.7	32.8	11.68	292.62	7,149.0	13.15		$\Delta^2 = 0.2271$

From the calculations summarized in Table E8.7, we obtain the following. The respective sample means:

$$\overline{T} = \frac{279.7}{11} = 25.43 \quad \text{and} \quad \overline{\ln DO} = \frac{11.68}{11} = 1.06$$

and the respective sample standard deviations

$$s_T = \sqrt{\frac{1}{10}(7149.0 - 11 \times 25.43^2)} = 1.883; \quad \text{and} \quad s_{\ln DO} = \sqrt{\frac{1}{10}(13.15 - 11 \times 1.06^2)} = 0.281$$

We also obtain the estimates for the regression coefficients

$$\hat{\beta} = \frac{292.62 - 11 \times 25.43 \times 1.06}{7149.0 - 11 \times 25.43^2} = -0.11 \quad \text{and} \quad \ln \hat{\alpha} = 1.06 + 0.11 \times 25.43 = 3.86$$

Hence, the linear regression of $\ln DO$ on T is

$$E(\ln DO|T) = 3.86 - 0.11T$$

which is shown in the semilogarithmic graph of Fig. E8.7, superimposed on the data points. The equivalent nonlinear regression of DO on T would be

$$E(DO|T) = \exp(3.86 - 0.11T) = 47.5e^{-0.11T}$$

The estimated correlation coefficient associated with the above linear regression equation of $\ln DO$ on T is, according to Eq. 8.9,

$$\hat{\rho} = \frac{1}{10}\left(\frac{292.62 - 11 \times 25.43 \times 1.06}{1.883 \times 0.281}\right) = -0.74$$

or by Eq. 8.10,

$$\hat{\rho} = (-0.11)\frac{1.883}{0.281} = -0.74$$

and the corresponding conditional standard deviation of $\ln DO$ for given T is

$$s_{\ln DO|T} = \sqrt{\frac{0.2271}{11 - 2}} = 0.159$$

We may also construct the confidence interval of the linear regression of $\ln DO$ on T using the following discrete values of the temperature T: $T = 23.5°$, $25.0°$, $26.5°$, and $28.0°$C.

At $T = 23.5°$C: $\quad \langle \mu_{\ln DO}\rangle_{0.95} = 1.275 \pm 2.262 \times 0.159\sqrt{\dfrac{1}{11} + \dfrac{(23.5 - 25.43)^2}{(7149 - 11 \times 25.43^2)}}$

$$= (1.12 \to 1.43)$$

At $T = 25.0°$C: $\quad \langle \mu_{\ln DO}\rangle_{0.95} = 1.110 \pm 2.262 \times 0.159\sqrt{\dfrac{1}{11} + \dfrac{(25.0 - 25.43)^2}{(7149 - 11 \times 25.43^2)}}$

$$= (1.00 \to 1.22)$$

At $T = 26.5°$C: $\quad \langle \mu_{\ln DO}\rangle_{0.95} = 0.945 \pm 2.262 \times 0.159\sqrt{\dfrac{1}{11} + \dfrac{(26.5 - 25.43)^2}{(7149 - 11 \times 25.43^2)}}$

$$= (0.82 \to 1.07)$$

At $T = 28.0°$C: $\quad \langle \mu_{\ln DO}\rangle_{0.95} = 0.780 \pm 2.262 \times 0.159\sqrt{\dfrac{1}{11} + \dfrac{(28.0 - 25.43)^2}{(7149 - 11 \times 25.43^2)}}$

$$= (0.59 \to 0.97)$$

With the above four intervals for the corresponding discrete values of T, the 95% confidence interval for the linear regression line may be constructed as shown in Fig. E8.7.

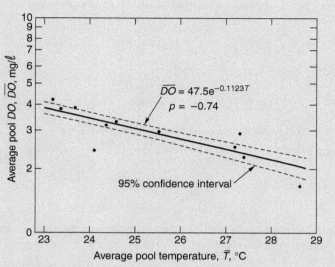

Figure E8.7 Dissolved oxygen (DO) and temperature (T) relationship. (Data from Butts, Schnepper, and Evans, 1970.)

◀

▶ **EXAMPLE 8.8**

In mountainous regions, the hazard of glacier lake outburst is of major concern. The maximum discharge and runout distance of an outburst is a function of the volume of a glacier lake. Through remote sensing data, the area of a lake, A, can be determined and from empirical data on mean depths, D, of glacier lakes, the required volume of a lake may be estimated. Available data for the Swiss Alps are tabulated in Table E8.8 (after Huggel et al., 2002).

TABLE E8.8   Summary of Data and Calculations for Example 8.8

Name of Lake	Lake Area A, m$^2 \times 10^6$	Depth D, m	$x_i = \text{Log}A_i$	$y_i = \text{Log }D_i$	x_iy_i	x_i^2	y_i^2	y_i'	$(y_i - y_i')^2$
Ice Cave Lake	0.0035	2.9	3.5441	0.4624	1.6388	12.56	0.2138	0.5428	0.0065
Gruben Lake 5	0.01	5	4.0000	0.699	2.796	16.00	0.4886	0.732	0.001
Crusoe-Baby Lake	0.017	4.7	4.2304	0.6721	2.8432	17.8962	0.4517	0.8276	0.0242
Gruben Lake 3	0.021	7.1	4.3222	0.8512	3.6791	18.6814	0.7245	0.8657	0.0002
Gruben Lake 1	0.023	10.4	4.3617	1.017	4.4358	19.0244	1.0343	0.8821	0.0182
MT' Lake	0.0416	12	4.6191	1.0792	4.9849	21.336	1.1646	0.9889	0.0082
Lac d'Arsine	0.059	13.6	4.7708	1.1335	5.4077	22.7605	1.2848	1.0519	0.0066
Nostetuko Lake	0.2622	28.6	5.4186	1.4564	7.8916	29.3612	2.1211	1.3207	0.0184
Between Lake	0.4	18.8	5.6021	1.2742	7.1382	31.3835	1.6236	1.3968	0.015
Abmachimai Lake	0.565	34.3	5.752	1.5353	8.831	33.0855	2.3571	1.4591	0.0058
Gjanupsvatn	0.6	33.3	5.7782	1.5224	9.0000	33.3876	2.3177	1.4699	0.0028
Quongzonk Co	0.753	27.9	5.8768	1.4456	9.4955	34.5368	2.0898	1.5109	0.0043
Laguna Paron	1.6	46.9	6.2041	1.6712	10.3683	38.4908	2.7929	1.6467	0.0006
Summit Lake	5	50	6.699	1.699	11.3816	44.8766	2.8866	1.8521	0.0234
Phontom Lake	6	83.3	6.7781	1.9206	13.018	45.9426	3.6887	1.8849	0.0013
$\sum$	15.3553	378.8	77.9572	18.4391	101.706	419.323	25.2398	6.8946	$\Delta^2 = 0.1365$

In this case, regression may be performed for log D on log A; i.e.,

$$\log D = \alpha + \beta \log A$$

which we can observe is linear in log–log space. The details of the calculations are summarized in Table E8.8.

From Table E8.8, we obtain the sample means of lake areas and depths as

$$\overline{A} = \frac{15.355}{15} = 1.024 \times 10^6 \, \text{m}^2 \quad \text{and} \quad \overline{D} = \frac{378.8}{15} = 25.25 \, \text{m}$$

and the respective sample standard deviations of log A and log D are

$$s_{\log A} = \sqrt{\frac{1}{14} \left[(419.323) - 15 \left(\frac{77.957}{15} \right)^2 \right]} = 1.006$$

$$s_{\log D} = \sqrt{\frac{1}{14} \left[(25.240) - 15 \left(\frac{18.439}{15} \right)^2 \right]} = 0.429$$

Estimates for the regression coefficients are

$$\hat{\beta} = \frac{101 - 15 \times \dfrac{77.957}{15} \times \dfrac{18.439}{15}}{419.323 - 15 \times \left(\dfrac{77.957}{15} \right)^2} = 0.415 \qquad \hat{\alpha} = \frac{18.439}{15} - 0.415 \times \frac{77.957}{15} = -0.928$$

Therefore, the regression equation for log D on log A is

$$E(\log D \mid \log A) = -0.928 + 0.415 \log A$$

and in terms of the original variables, the mean depth, D, and area, A, the relevant nonlinear relation is

$$D = 0.118 A^{0.415}$$

The graph of the data points and the above linear regression equation of log D on log A are shown in Fig. E8.8.

The correlation coefficient associated with the regression of log D on log A is, according to Eq. 8.9,

$$\hat{\rho} = \frac{101 - 15 \left(\dfrac{77.957}{15} \right) \left(\dfrac{18.439}{15} \right)}{1.006 \times 0.429} = 0.86$$

and the corresponding conditional standard deviation of log D for given log A is

$$s_{\log D \mid \log A} = \sqrt{\frac{0.1365}{15 - 2}} = 0.102$$

We may also construct the 95% confidence bounds for the regression of log D on log A at the following selected values of A:

$A = 5000 \, \text{m}^2$ (or log $A = 3.699$):

$$\langle \mu_{\log D} \rangle_{.95} = 0.607 \pm 2.160 \times 0.102 \sqrt{\frac{1}{15} + \frac{(3.699 - 5.197)^2}{(419.323 - 15 \times 5.197^2)}} = (0.383 \rightarrow 0.831)$$

or

$$\langle \mu_D \rangle_{.95} = (2.41 \rightarrow 6.78) \, \text{m}$$

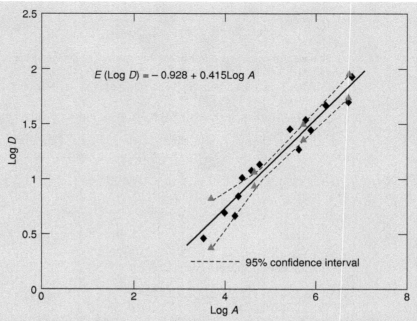

Figure E8.8 Relationship between log D and log A for Glacier Lakes. (After Huggel et al., 2002.)

$A = 50,000 \text{ m}^2$ (or log $A = 4.699$)

$$\langle \mu_{\log D} \rangle_{.95} = 1.022 \pm 2.160 \times 0.102 \sqrt{\frac{1}{15} + \frac{(4.699 - 5.197)^2}{(419.323 - 15 \times 5.197^2)}} = (0.958 \rightarrow 1.086)$$

or

$$\langle \mu_D \rangle_{.95} = (9.08 \rightarrow 12.19)\,\text{m}$$

$A = 500,000 \text{ m}^2$ (or log $A = 5.699$)

$$\langle \mu_{\log D} \rangle_{.95} = 1.437 \pm 2.160 \times 0.102 \sqrt{\frac{1}{15} + \frac{(5.699 - 5.197)^2}{(419.323 - 15 \times 5.197^2)}} = (1.373 \rightarrow 1.501)$$

or

$$\langle \mu_D \rangle_{.95} = (23.60 \rightarrow 31.70)\,\text{m}$$

$A = 5,000,000 \text{ m}^2$ (or log $A = 6.699$)

$$\langle \mu_{\log D} \rangle_{.95} = 1.852 \pm 2.160 \times 0.102 \sqrt{\frac{1}{15} + \frac{(6.699 - 5.197)^2}{(419.323 - 15 \times 5.197^2)}} = (1.747 \rightarrow 1.957)$$

or

$$\langle \mu_D \rangle_{.95} = (55.85 \rightarrow 90.57)\,\text{m}$$

which can be used to establish the 95% confidence intervals of the regression line as shown in dash lines in Fig. E8.8. ◀

The form of the nonlinear functions assumed in Eq. 8.28 can be generalized as follows:

$$E(Y|x) = \alpha + \beta_1 g_1(x) + \beta_2 g_2(x) + \cdots + \beta_k g_k(x) \qquad (8.30)$$

where $g_j(x)$, $j = 1, 2, \ldots, k$ are predetermined functions of the independent variable x. An example of Eq. 8.30 is the following polynomial relation,

$$E(Y|x) = \alpha + \beta_1 x + \beta_2 x^2 + \cdots + \beta_k x^k \qquad (8.31)$$

We now observe that by converting each of the polynomial terms in Eq. 8.30 into the respective transformed variables, $z_j = g_j(x)$, Eq. 8.30 becomes

$$E(Y|x) = \alpha + \beta_1 z_1 + \beta_2 z_2 + \cdots + \beta_k z_k \qquad (8.32)$$

Hence, by evaluating each of the transformed variables $z_j = g_j(x)$ on the basis of the original data set, the nonlinear problem of Eq. 8.30 is transformed to that of a multiple linear regression problem of Eq. 8.32, and the procedure presented earlier in Sect. 8.5 should apply.

▶ 8.7 APPLICATIONS OF REGRESSION ANALYSIS IN ENGINEERING

Regression analyses have been used widely in practically all branches of engineering, either to develop empirical relations between two (or more) variables or to confirm a theoretical relationship and evaluate the needed constants on the basis of observed data. Oftentimes the necessary theoretical relationship between engineering variables cannot be derived on the basis of theoretical considerations; in these cases, the required relationship may be established empirically on the basis of experimental or field observations. For example, by plotting the logarithm of the observed fatigue life N (in cycles to failure) of a material versus the logarithm of the applied stress range (maximum minus minimum stress) S, a linear trend can be observed as shown in Fig. 8.4.

This linear trend may be represented by the linear regression equation,

$$\log N = a - b \log S$$

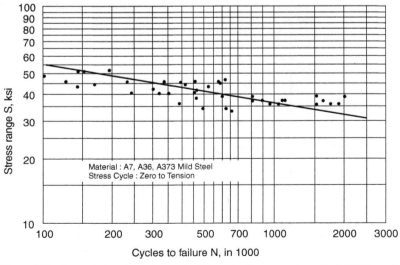

Figure 8.4 *S–N* relation for fatigue of mild steel. (Data courtesy of W. H. Munse.)

in which the constants a and b can be evaluated as estimates of the regression coefficients based on the set of measured data. The above regression equation, therefore, yields the so-called S–N relation of the form

$$NS^b = a$$

There are also situations in which the mathematical form of a required relationship among the principal variables may be postulated from heuristic considerations; regression analysis may then be applied to assess the validity (statistically) of the mathematical equation or to evaluate the values of the parameters on the basis of the observed data. For example. Smeed (1968) postulated that the peak flow of traffic [in *passenger car units* (pcu)] into the center of a city is

$$Q = \alpha f A^{1/2}$$

where f = the fraction of the city center that is occupied by roadways; A = the area of the city center in ft²; and α = a constant that depends on the speed of the traffic and efficiency of the road system. Basically, this equation is based on the hypothesis that the volume of traffic (in pcu) that can enter the central area is proportional to the circumference of the central area. Data from 35 cities, including 20 from the UK, are shown plotted on a log–log graph in Fig. 8.5.

The least-squares linear regression of log (Q/f) on log A yields a slope of 0.53 for the regression line. Also, from the regression line of Fig. 8.5, the constant α can be evaluated as the value of (Q/f) at $A = 1$.

There are also situations in engineering in which it is difficult to measure a quantity (or variable) of interest directly, but may be obtained indirectly through its relationship with another variable. For example, in determining the maximum stress at the extreme fiber of a steel beam, it is difficult to measure the stress directly; however, through the stress-strain relationship of the beam material, the maximum stress can be determined by measuring the corresponding strain at the extreme fiber of the beam. Also, some engineering variables

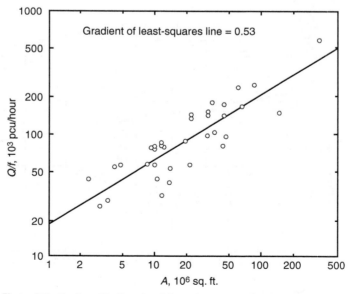

Figure 8.5 Peak traffic flow (pcu) into city center of area A. (After Miller, 1970.)

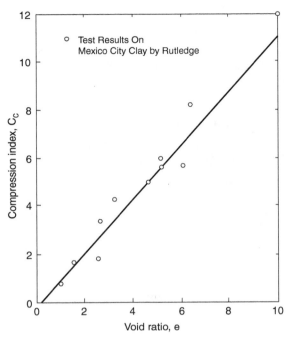

Figure 8.6 Compression index vs. void ratio of soil. (After Nishida, 1956.)

can be measured more readily and economically than others; for example, the initial void ratios of clay samples can be measured inexpensively in a laboratory, whereas the direct measurement of the compression index of the same soil would be much more costly and would require considerable effort and time. In such a case, if an empirical relation is established between the void ratio and the compression index of soils, such as the relation illustrated in Fig. 8.6, we can simply measure the void ratio and predict the compression index of a soil by applying the appropriate regression equation.

Another example in this regard is the determination of the fully cured 28-day concrete strength; obviously, this will require testing the concrete specimens after 28 days. At the rate that construction progresses nowadays, 28 days would be too long; methods of early determination of concrete strength are highly desirable for quality assurance and have been suggested, such as using an accelerated strength obtained through an accelerated curing process. A linear regression has been developed, e.g., by Malhotra and Zoldners (1969), as shown in Fig. 8.7, between the 28-day strength and the accelerated strength of concrete, based on data from nine construction jobs across Canada.

In traffic engineering, Heathington and Tutt (1971) have developed linear relationships between short-interval observed traffic volume and long-term traffic volume in six cities in Texas; some of these results are shown in Fig. 8.8. Obviously, there is considerable benefit in using short-term observations for predicting long-term traffic conditions.

Multiple linear regression also finds many applications in engineering; for example, Martin et al. (1963) used multiple linear regression to determine the expected number of trips generated, Y, per dwelling unit in a community as a function of automobile ownership X_1, population density X_2, distance from the central business district X_3, and family income X_4 as follows:

$$E(Y|x_1, x_2, x_3, x_4) = 4.33 + 3.89x_1 - 0.005x_2 - 0.128x_3 - 0.012x_4$$

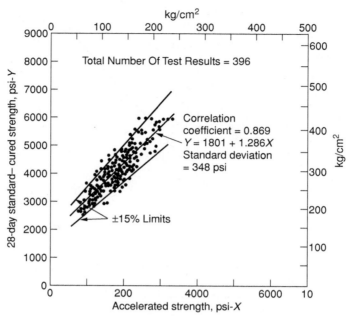

Figure 8.7 Relation between 28-day strength and accelerated strength of concrete. (After Malhotra and Zoldners, 1969.)

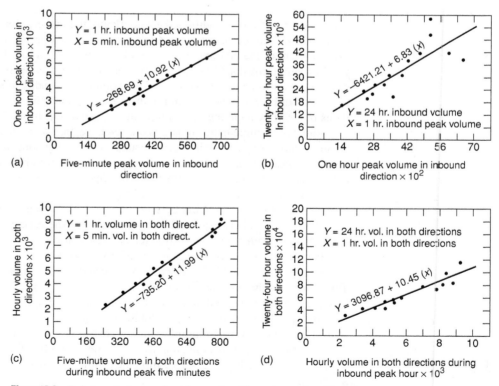

Figure 8.8 Relations between short-interval and long-interval traffic volumes. (After Heathington and Tutt, 1971.)

TABLE 8.1 Multiple Regression of Trip Generation

Independent Variables	Regression Equation	$s_{Y\|x_1,...,x_k}$	r (Eq. 8.7)
X_1, X_2, X_3, X_4	$y' = 4.33 + 3.89x_1 - 0.005x_2 - 0.128x_3 - 0.012x_4$	0.87	0.837
X_1, X_2	$y' = 3.80 + 3.79x_1 - 0.003x_2$	0.87	0.835
X_2, X_4	$y' = 5.49 - 0.0089x_2 + 0.227x_4$	1.02	0.764
X_1	$y' = 2.88 + 4.60x_1$	0.89	0.827
X_2	$y' = 7.22 - 0.013x_2$	1.10	0.718
X_4	$y' = 3.07 + 0.44x_4$	1.20	0.655
X_3	$y' = 3.55 + 0.74x_3$	1.30	0.575

$y' = E(Y|x_1, \ldots, x_k) = $ Expected number of trips per dwelling unit.
$X_1 = $ Automobile ownership (no. per dwelling unit).
$X_2 = $ Population density (no. per net residential acre).
$X_3 = $ Distance from central business district (in miles).
$X_4 = $ Family income (in 1960s US$1000).

Alternative multiple linear regressions, using fewer independent variables, were also performed for this problem. The respective results of these different regression analyses are summarized in Table 8.1. We can observe from the values of r (see Eq. 8.7) in Table 8.1 that the reduction in the unconditional variance of Y generally increases with the number of independent variables included in the regression analysis.

Nonlinear regression is also widely used in engineering. Besides those illustrated earlier in Examples 8.6 and 8.8, Fig. 8.9 shows an application involving the logarithmic transformation, in which data points of the average stress per cycle of repeated loading are plotted against the logarithm of the number of cycles to failure of concrete beams. Superimposed on the data points is the linear regression for determining the S–N relation for predicting the expected fatigue life of concrete beams. In this case, because of the large variability of concrete strengths, a wide scatter in the data points can be observed.

The application of a double logarithmic transformation is illustrated in Fig. 8.10, where data points of the logarithm of river flows and the logarithm of distance downstream are plotted, yielding a linear regression equation for the transformed variables.

Another application of the double logarithmic transformation is shown in Fig. 8.11 in which data points of the maximum sustained wind speed and the radial distance from the center of a hurricane are plotted, with the corresponding linear regression line superimposed.

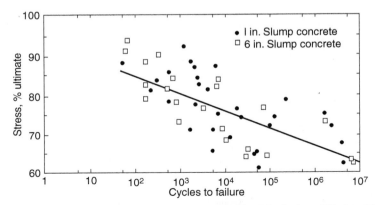

Figure 8.9 S–N diagram of concrete beams. (After Murdock and Kesler, 1958.)

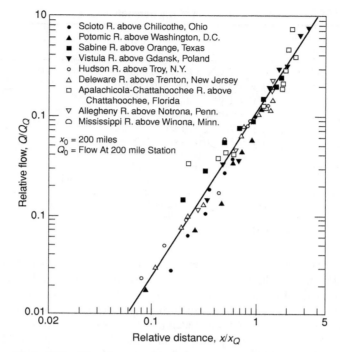

Figure 8.10 River flow vs. distance downstream. (After Shull and Gloyna, 1969.)

Polynomial functions are also often used in nonlinear regression. For example, a third-degree polynomial function is fitted, through regression analysis, to the data points of vehicle speeds and corresponding traffic densities in Fig. 8.12.

Numerous other examples of applications of regression analysis can be observed also in Chapter 1.

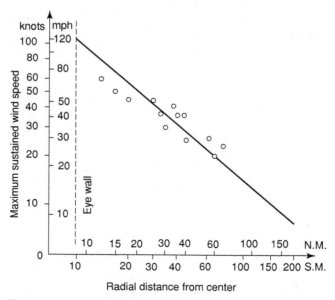

Figure 8.11 Surface hurricane wind profile. (After Goldman and Ushijima, 1974.)

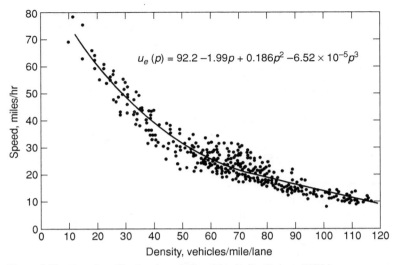

$$u_e(p) = 92.2 - 1.99p + 0.186p^2 - 6.52 \times 10^{-5}p^3$$

Figure 8.12 Speed-traffic density relationship. (After Payne, 1973.)

▶ 8.8 CONCLUDING SUMMARY

The statistical relationship (particularly the mean and variance) between a dependent variable as a function of one or more independent (or control) variables may be determined empirically through *regression analysis*. This is generally limited to a linear equation of the mean value of the dependent variable as a function of the independent variable(s) with unknown constants known as *regression coefficients*. Regardless of the form of the functional relationship, the required regression analysis can be limited to the linear case, as any nonlinear relationship can be converted into a linear function through proper transformations of the variables for the purpose of regression analyses. On the basis of least squares, regression analysis provides a systematic approach for empirically estimating the regression coefficients from a set of observed data. The strength of linearity underlying a regression equation is measured by the corresponding *correlation coefficient*; a high correlation coefficient (i.e., close to ±1.0) implies the existence of a strong linear relationship between the variables, whereas a low correlation (close to zero) would mean the lack of a linear relationship (although there may be a nonlinear relationship). Confidence intervals of a linear regression equation may also be determined with a specified confidence level.

Regression and correlation analyses have applications in many areas of engineering and are especially significant in situations where the necessary relationships must be developed empirically. Moreover, even in situations in which the functional form may be known from physical considerations, regression analyses can be useful also for evaluating the unknown constants based on the observed data. We should emphasize, however, that the relationship developed from a regression analysis may not necessarily imply any causal effect between the variables; it simply establishes the statistical relationship on the basis of the observed data.

▶ PROBLEMS

8.1 A tensile load test was performed on an aluminum specimen. The applied tensile force and the corresponding elongation of the specimen at various stages of the test are recorded as shown in the table on p. 340.

(a) We may assume that the force–elongation relation over the range of the applied loads is linear. On this basis, determine the least-squares estimate of the Young's modulus of elasticity of the aluminum specimen, which is the slope of the stress-strain curve. The cross-sectional area of the specimen is 0.10 in^2, and the length of the specimen is 10 in.

Tensile Force Elongation

X (kip)	Y (10^{-3} in.)
1	9
2	20
3	28
4	41
5	52
6	63

(b) Under zero load, the elongation of the specimen must also be zero. Therefore, the intercept, α, of the regression equation should be $\alpha = 0$. In this case, what would be the best estimate of the Young's modulus?

8.2 The distance Y necessary for stopping a vehicle is a function of the speed of travel of the vehicle X. Suppose the following set of data were observed for 12 vehicles traveling at different speeds as shown in the table below.

Vehicle No.	Speed, kph	Stopping Distance, m
1	40	15
2	9	2
3	100	40
4	50	15
5	15	4
6	65	25
7	25	5
8	60	25
9	95	30
10	65	24
11	30	8
12	125	45

(a) Plot the stopping distance versus the speed of travel.
(b) Assume that the stopping distance is a linear function of the speed, i.e., $E(Y|x) = \alpha + \beta x$.

Estimate the regression coefficients, α and β, and the conditional standard deviation $s_{Y|x}$. Also, determine the correlation coefficient between Y and X.
(c) Determine the 90% confidence interval of the above regression equation.

8.3 The peak-hour traffic volume and the daily traffic volume on a given highway bridge have been observed for 7 days. The observed data are as shown below:
(a) Plot a graph showing the peak-hour traffic volume vs. the daily traffic volume.
(b) Estimate the correlation coefficient between the peak-hour traffic volume and the daily traffic volume.

(c) Suppose that the total traffic count on a given day has been observed to be 55,000 vehicles. What is the probability that the peak-hour traffic volume on that day exceeded 7000 vehicles per hour? (*Hint:* Use an appropriate regression analysis.)

X	Y
Peak-hour Traffic Volume (in 1000)	Daily traffic volume (in 10,000)
1.5	0.6
4.6	3.4
3.0	2.5
5.5	2.8
7.8	4.8
6.8	6.4
6.3	5.0

8.4 Data for the per capita energy consumption and per capita GNP output for eight different countries have been complied by Meadows et al, (1972) as tabulated below:

Country #	Per Capita GNP, X	Per Capita Energy Consumption, Y
1	600	1000
2	2700	700
3	2900	1400
4	4200	2000
5	3100	2500
6	5400	2700
7	8600	2500
8	10300	4000

In the above table,

X = the per capita GNP of a country in 1972 U.S. dollar equivalent

Y = the per capita energy consumption in kilos of coal equivalent.

(a) Plot the data on a two-dimensional graph.
(b) Determine the regression equation for predicting the per capita energy consumption on the basis of a country's per capita GNP output, and draw the regression line on the graph of Part (a).
(c) Determine the correlation coefficient between X and Y.
(d) Evaluate the conditional standard deviation $s_{Y|X=x}$.
(e) Determine the 95% confidence interval of the regression line, and sketch the upper and lower interval lines on the graph.
(f) Similarly, determine the regression equation for predicting the per capita GNP on the basis of the per capita energy consumption, and evaluate the corresponding conditional standard deviation $s_{X|Y=y}$.

8.5 A survey of passenger car weight (kip) and corresponding gasoline mileage (miles per gallon) gives the following:

Car	Gasoline Mileage (mpg)	Weight (kip)
1	25	2.5
2	17	4.2
3	20	3.6
4	21	3.0

These four cars represent a random sample of the entire passenger car population.

(a) Suppose passenger car weight (in kips) is normally distributed and (by analysis using probability paper) its mean and standard deviation are estimated to be 3.33 and 1.04, respectively. Find the probability that another car picked at random from the population will weigh more than 4.5 kips.

(b) Using linear regression analysis with constant variance, answer the following:

If you buy a car that weighs 2.3 kips, what is the probability that it will have a gasoline mileage of more than 28 mpg?

(c) Develop the 95% confidence interval of the resulting regression equation, and sketch the interval on the graph.

8.6 The error incurred in a given type of measurement by a surveyor appears to be affected by the surveyor's years of experience. The following is the data observed for five surveyors.

Surveyor	Years of Experience, Y	Measurement Error M in.
1	3	1.5
2	5	0.8
3	10	1.0
4	20	0.8
5	25	0.5

On the basis of the above information, answer the following and state your assumptions:

(a) For a surveyor with 15 years of experience, what is the probability that his measurement error will be less than 1 in.? (*Ans. 0.713*)

(b) For a 65-year-old surveyor who has 30 years of experience, can you estimate the probability that his measurement error will be less than 1 in.? Please elaborate.

8.7 Dissolved oxygen (*DO*) concentration in ppm (parts per million) in a stream is found to decrease with the time of travel, *T*, downstream (Thayer and Krutchkoff, 1966). The data shown in the next table below represent a set of measurements of the *DO* versus *T* for a particular stream:

Dissolved Oxygen DO (ppm)	Time of Travel T (days)
0.28	0.5
0.29	1
0.29	1.6
0.18	1.8
0.17	2.6
0.18	3.2
0.1	3.8
0.12	4.7

(a) Determine the least-squares regression of *DO* on *T*, i.e., the regression equation for predicting the dissolved oxygen concentration based on the travel time of the water downstream.
(b) Evaluate the correlation coefficient between *DO* and *T*, and also the conditional standard deviation of *DO* for given *T*.
(c) Determine the 95% confidence interval of the regression line.

8.8 The actual concrete strength, *Y*, in a structure is generally higher than that measured on a specimen, *X*, from the same batch of concrete. Data show that a regression equation for predicting the actual concrete strength is

$$E(Y|x) = 1.12x + 0.05 \text{ (ksi)}; \quad 0.1 < x < 0.5$$

and

$$\text{Var}(Y|x) = 0.0025 \text{ (ksi)}^2$$

Assume that *Y* follows a normal distribution for a given value of x.
(a) For a given job, in which the measured strength is 0.35 ksi, what is the probability that the actual strength will exceed the requirement of 0.3 ksi?
(b) Suppose the engineer has lost the data on the measured strength on the concrete specimen. However, he recalls that it is either 0.35 or 0.40 with the relative likelihood of 1 to 4. What is the probability that the actual strength will exceed the requirement of 0.3 ksi?
(c) Suppose the measured values of concrete strength at two sites A and B are 0.35 and 0.4 ksi, respectively. What is the probability that the actual strength for the concrete structure at site A will be higher than that at site B? You may assume that the predicted actual concrete strength between the sites are statistically independent.

8.9 Data on the construction costs for three houses in a given residential area are as follows:

Floor Area (1000 sq. ft.)	Cost ($1,000)
1.05	63
1.83	92
3.14	204

Plot the above data on a piece of *x-y* paper.
(a) Determine by linear regression the relation between the construction cost of houses as a function of the floor area. Sketch this on the graph.
(b) Estimate the standard deviation of construction cost for the given floor area.
(c) How good is the linear relation between cost and floor area? The answer depends on the correlation coefficient.
(d) If you wish to build a house with 2500 sq. ft. of floor area, what is the probability that the construction cost will not exceed $180,000 (based on the information given above)? Assume that the cost for a given floor area is normally distributed.

8.10 In order to determine the reliability of the rated mileage (in mpg) of cars, 6 different makes of cars were driven for the same distance over combined city and highway roads, with the following results:

Make of Car	Rated Mileage, mpg	Actual Mileage, mpg
A	20	16
B	25	19
C	30	25
D	30	22
E	25	18
F	15	12

(a) Find the linear regression equation for determining the mean actual mileage for the given rated mileage of a car; i.e., develop the equation for

$$E(Y|X=x)$$

in which

$$Y = \text{the actual mileage, in mpg; and}$$
$$X = \text{the rated mileage, in mpg.}$$

(b) Evaluate the conditional standard deviation of Y for a given $X = x$, $s_{Y|x}$, and the correlation coefficient between X and Y.
(c) Suppose two cars, models Q and R, have rated mileage of 22 mpg and 24 mpg, respectively. What is the probability that the model Q car will have better actual mileage than that of model R car?
(d) Plot the data points on an *x-y* graph, and sketch in the regression line. Develop and sketch also the 90% confidence interval of the regression line.

8.11 In Example 8.3, we show the data of observed settlements of pile groups and the corresponding calculated settlements obtained with a nonlinear model (proposed by Viggiani, 2001) as shown earlier in Table E8.3 for varying levels of loading (Columns 2, 3, and 4).
(a) Perform the linear regression of the corresponding calculated settlements on the observed settlements, i.e., $E(X|y)$, and evaluate the conditional standard deviation $s_{X|y}$.
(b) Evaluate the correlation coefficient between X and Y.

(c) Sketch the regression line and determine its 95% confidence interval; sketch also this interval.

8.12 Suppose a survey of the effect of a fare increase on the loss of ridership for mass transit systems in the United States reveals the data tabulated below.

% Fare Increase, X	% Loss in Ridership, Y
5	1.5
35	12
20	7.5
15	6.3
4	1.2
6	1.7
18	7.2
23	8
38	11.1
8	3.6
12	3.7
17	6.6
17	4.4
13	4.5
7	2.8
23	8

(a) Plot the above data for the percentage loss in ridership versus the percentage fare increase in an *x-y* graph.
(b) Perform a linear regression analysis for predicting the expected percentage loss in ridership as a function of the percentage fare increase for a mass transit system in the United States.
(c) Evaluate the correlation coefficient between X and Y, and estimate the constant conditional standard deviation of the loss in ridership for a given fare increase.
(d) Determine the 90% confidence interval of the regression equation of Part (b) above.

8.13 Seismic damage to an urban area will depend on the intensity of a given earthquake. Based on a regression analysis of past earthquake damage data, the expected damage in a given area during an earthquake of intensity I was determined to be

$$E(D|I) = 10.5 + 15I$$

in which $D =$ damage loss in $ million. The corresponding conditional standard deviation was found to be $s_{D|I} = 30$, which is constant for all I.
(a) On the assumption that the seismic damage loss in the area, D, is a Gaussian variate, what is the probability that the damage loss will exceed $150 million if an earthquake of intensity $I = 6$ should occur?
(b) If damage-causing earthquakes in the area are limited to intensities of $I = 6, 7$, and 8 with relative likelihoods of occurrence of 0.6, 0.3, and 0.1, respectively, what would be the expected seismic damage loss of the urban area in the next earthquake?

8.14 Several simply supported timber beams were tested experimentally under varying loads P to determine its deflections D.

The measured deflections at the midspan of the test beams were measured as follows:

P, tons	D, cm
8.4	4.8
6.7	2.9
4.0	2.0
10.2	5.5

(a) On the basis of the above test results, determine the linear regression of the deflection on the load, and the associated conditional standard deviation (assume this is constant under all loads).

(b) Develop also the corresponding 90% confidence interval of the above regression equation.

(c) What would be the mean deflection of the beam under a load $P = 8$ tons? Also, assuming normal distribution, what would be the 75-percentile deflection under this load?

8.15 Test data on the deformation and Brinell hardness of a certain type of steel have been obtained as follows:

D (deformation, mm):	6	11	13	22	28	35
H (Brinell hardness, kg/mm²):	68	65	53	44	37	32

(a) Estimate the correlation coefficient between deformation and Brinell hardness of the particular steel.

(b) Assuming that the Brinell hardness varies linearly with the deformation, determine the regression for $E(H|D=d)$ and the associated constant conditional standard deviation $s_{H|d}$.

(c) Suppose that the Brinell hardness corresponding to a given deformation may be modeled as a normal variate. What is the probability that the Brinell hardness for a deformation of 20 mm will be between 40 and 50 kg/mm²?

8.16 As expected, the stopping distance, D, for a car depends on the speed of travel, V, and on the condition of the road surface. However, there is also variability in the stopping distance at a given speed even under the same road condition. From several road tests on a dry pavement, the following were observed:

Stopping Distance Travel Speed

Car #	D(ft)	V(mph)
1	46	25
2	6	5
3	110	60
4	46	30
5	16	10
6	75	45
7	16	15
8	76	40
9	90	45
10	32	20

(a) Evaluate the correlation coefficient between V and D. On this basis, can we say that there is a reasonable linear relationship between the stopping distance and the speed of travel?

(b) Plot the above stopping distance versus travel speed on an x-y graph.

(c) From the above graph, a nonlinear function may be suggested to model the stopping distance–speed relationship as follows:

$$E(D|V = v) = a + bv + cv^2$$

Estimate the regression coefficients a, b, and c, and evaluate also the conditional standard deviation $s_{D|V=v}$.

(d) Determine the expected stopping distance for a car traveling at a speed of 50 mph. However, if the driver wants a 90% probability of stopping the car traveling at 50 mph, what distance should he allow?

8.17 Data on the daily consumption of water in the Midwest have been collected for seven towns with varying population as follows:

Town #	Population, X	Total Consumption, (10⁶ gal/day)
1	12,000	1.2
2	40,000	5.2
3	60,000	7.8
4	90,000	12.8
5	120,000	18.5
6	135,000	22.3
7	180,000	31.5

(a) Based on the above observed data, develop the regression of the per capita water consumption on the total population; i.e., estimate the regression coefficients α and β in the linear equation

$$E(Y|x) = \alpha + \beta x$$

in which Y is the per capita water consumption.

(b) Estimate the constant conditional standard deviation $s_{Y|X=x}$.

(c) Determine the 98% confidence interval of the above regression equation.

(d) An engineer is interested in studying the daily consumption of water in a town A, with a population of 100,000. If he assumes that the per capita consumption of water is a normal variate for a given population, what is the probability (on the basis of the above regression equation) that the demand for water in the town will exceed 17,000,000 gal/day?

8.18 The rate of oxygenation from the atmospheric reaeration process for a stream depends on the mean velocity of the stream flow and the average depth of the stream bed. Data for 12 streams have been recorded as shown in the next table (after Thayer and Krutchkoff, 1966).

Suppose the following nonlinear relationship has been suggested for the mean oxygenation rate:

$$E(X|V, H) = \alpha V^{\beta_1} H^{\beta_2}$$

Mean Oxygenation Rate, X (ppm/day)	Mean Velocity, V (fps)	Mean Depth, H (ft)
2.272	3.07	3.27
1.44	3.69	5.09
0.981	2.1	4.42
0.496	2.68	6.14
0.743	2.78	5.66
1.129	2.64	7.17
0.281	2.92	11.41
3.361	2.47	2.12
2.794	3.44	2.93
1.568	4.65	4.54
0.455	2.94	9.5
0.389	2.51	6.29

With the data given in this table, determine the regression coefficients α, β_1 and β_2, and evaluate the corresponding conditional standard deviation.

8.19 The peak-hour traffic volume and the 24-hour daily traffic volume on a toll bridge have been recorded for 14 days as follows:

Peak Hour Traffic Vol., X (in 10^3)	24-hour Traffic Vol., Y (in 10^4)
1.4	1.6
2.2	2.3
2.4	2
2.7	2.2
2.9	2.6
3.1	2.6
3.6	2.1
4.1	3
3.4	3
4.3	3.8
5.1	5.1
5.9	4.2
6.4	3.8
4.6	4.2

Assume that the conditional standard deviation $s_{Y|X=x}$ varies in a quadratic form with x from the origin.

(a) Determine the regression of Y on X.

(b) Estimate the prediction error about the regression line; i.e., $s_{Y|X=x}$.

(c) Determine the 98% confidence interval for the regression equation of Part (a).

(d) If the peak-hour traffic volume on a certain morning was observed to be 3500 vehicles, what is the probability that more than 30,000 vehicles will be crossing the toll bridge on that day?

▶ **REFERENCES**

Butts, T. A., Schnepper, D. H., and Evans, R. L., "Statistical Assessment of DO in Navigation Pool," *Jour. of Sanitary Engineering,* ASCE, Vol. 96, April 1970.

Galligan, W. L., and Snodgrass, D. V., "Machine Stress Rated Lumber: Challenge to Design," *Jour. of the Structural Div.,* ASCE, Vol. 96, December 1970.

Goldman, J. L., and Ushijima, T., "Decrease in Hurricane Winds after Landfall," *Jour. of the Structural Div.,* ASCE, Vol. 100, January 1974.

Hald, A., *Statistical Theory with Engineering Applications,* J. Wiley & Sons, New York, 1952.

Heathington, K. W., and Tutt, P. R., "Traffic Volume Characteristics on Urban Freeway," *Transportation Engineering Jour.,* ASCE, Vol. 97, February 1971.

Huggel, C., Kaab, A., Haeberli, W., Teysseire, P., and Paul, F., "Remote Sensing Based Assessment of Hazards from Glacier Lake Outbursts: A Case Study in the Swiss Alps," *Canadian Geotechnical Jour.,* Vol. 39, March 2002.

Lambe, T. W., and Whitman, R. V., *Soil Mechanics,* John Wiley & Sons, Inc., New York, 1969, p. 375.

Malhotra, V. M., and Zoldners, N. G., "Some Field Experience in the Use of an Accelerated Method of Estimating 28-Day Strength of Concrete," *Jour. of American Concrete Institute,* November 1969.

Martin, B. V., Memmott, F. W., and Bone, A. J., "Principles and Techniques of Predicting Future Demand for Urban Area Transportation," Research Rept. No, 38, Dept. of Civil Engineering, MIT, Cambridge, MA, January 1963.

Meadows, D. H., Meadows, D. L., Randers, J., and Behrens, W. W., *The Limits of Growth,* Universe Books, New York, 1972.

Miller, A. J., "The Amount of Traffic Which Can Enter a City Center During Peak Periods," *Transportation Science,* ORSA, Vol. 4, 1970, pp. 409–411.

Moulton, L. K., and Schaub, J. H., "Estimation of Climatic Parameters for Frost Depth Predictions," *Jour. of Transportation Engineering,* ASCE, Vol. 85, November 1969.

Murdock, J. W., and Kesler, C. E., "Effects of Range of Stress on Fatigue Strength of Plain Concrete Beams," *Jour. of American Concrete Inst.,* Vol. 30, August 1958.

Nishida, Y. K., "A Brief Note on Compression Index of Soil," *Jour. of the Soil Mechanics and Foundation Div.,* ASCE, Vol. SM3, July 1956.

Payne, H. J., "Freeway Traffic Control and Surveillance Model," *Transportation Engineering Jour.,* ASCE, Vol. 99, November 1973.

Shull, R. D., and Gloyna, E. F., "Transport of Dissolved Water in Rivers," *Jour. of the Sanitary Engineering Div.,* ASCE, Vol. 95, December 1969.

Smeed, R. J., "Traffic Studies and Urban Congestion," *Jour. Transportation Economics and Policy*, Vol. 2, No. 1, 1968, pp. 33–70.

Thayer, R. P., and Krutchkoff, R. G., "A Stochastic Model for Pollution and Dissolved Oxygen in Streams," Water Resources Research Center, Virginia Polytechnic Inst., Blacksburg, VA, 1966.

Viggiani, C., "Analysis and Design of Piled Foundations," *Rivista Italiana di Geotecnica*, Vol. 35, 2001.

Wynn, F. H., "Shortcut Modal Split Formula," *Highway Research Record*, No. 283, Highway Research Board, National Research Council, 1969.

CHAPTER 9

The Bayesian Approach

▷ 9.1 INTRODUCTION

In engineering, we often need to use whatever information is available in formulating a sound basis for making decisions. This may include observed data (field as well as experimental), information derived from theoretical models, and expert judgments based on experience. The various sources and types of information must often be combined regardless of their respective qualities. Moreover, available information may need to be updated as new information or data are acquired. When the available information is in statistical form, or contains variability, which is invariably the case with engineering information, as we saw in Chapter 1, the proper tools for combining and updating the available information is embodied in the *Bayesian approach*. In this chapter, we shall present the fundamentals of the Bayesian approach to typical engineering problems involving probability and statistics.

In Chapter 1, we identified two broad types of uncertainties—the *aleatory* uncertainty that is associated with the inherent variability of information, and the *epistemic* uncertainty that is associated with the imperfections in our knowledge or ability to make predictions. The aleatory uncertainty gives rise to a calculated probability, whereas the epistemic uncertainty leads to a lack of confidence in the calculated probability. In this regard, the Bayesian approach can be relevant in two ways: (1) to systematically update the existing aleatory and epistemic uncertainties as additional information or data for each type of uncertainty becomes available; and (2) to provide an alternative basis for combining the two types of uncertainties for the purpose of decision making or formulating bases for design (see Ang and Tang, 1984).

An essential topic concerns the estimation of the parameters of an underlying probability model; we shall see that the Bayesian approach provides another logical basis for estimating the parameters, which is different from the *classical approach* that we presented in Chapter 6. There is also a role for the Bayesian approach in the regression and correlation analyses between two (or more) random variables.

9.1.1 Estimation of Parameters

In Chapter 6, we presented the methods of point and interval estimations of the parameters of a given probability distribution, based on the classical statistical approach. Such an approach assumes that the parameters are constants (but unknown) and that sample statistics are used as estimators of these parameters. Because the estimators are invariably imperfect, errors of estimation are unavoidable; in the classical approach, confidence intervals may be used to express the degree of these errors.

As implied earlier, accurate estimates of the parameters require large amounts of data. When the observed data are limited, as is often the case in engineering, the statistical estimates have to be supplemented (or may even be superseded) by judgmental informa-tion. With the classical statistical approach there is no provision for combining judgmental information with observational data in the estimation of the parameters.

For illustration, consider a case in which a traffic engineer wishes to determine the effectiveness of the road improvement at an intersection. Based on his experience with similar sites and traffic conditions, and on a traffic-accident model, he estimated that the average occurrences of accidents at the improved intersection would be about once a year. However, during the first month after the improved intersection is opened to traffic, an accident occurs at the intersection. A dichotomy, therefore, may arise: The engineer may hold strongly to his judgmental belief, in which case he would insist that the accident is only a chance occurrence and the average accident rate remains *once a year*, in spite of the most recent accident. On the other hand, if he considers only the observed data, he would estimate the average accident rate to be *once a month*. Intuitively, it would seem that both types of information are relevant and ought to be used in determining the average accident rate. Within the classical method of statistical estimation, however, there is no formal basis for such analysis. Problems of this type may be resolved effectively by combining the observed data and the judgmental information; formally, this is the subject of Bayesian estimation.

The *Bayesian method* approaches the estimation problem from another point of view. In this case, the unknown parameters of a distribution are assumed (or modeled) also as random variables. In this way, all sources of uncertainty associated with the estimation of the parameters can be combined formally (through the total probability theorem). With this approach, subjective judgments based on intuition, experience, or indirect information are incorporated systematically with observed data (through the Bayes Theorem) to obtain a balanced estimation. The Bayesian method is particularly helpful in cases where there is a strong basis for such judgments. We introduce the basic concepts of the Bayesian approach in the following sections.

▶ 9.2 BASIC CONCEPTS—THE DISCRETE CASE

The *Bayesian approach* has special significance to engineering design where available information is invariably limited and subjective judgment is often necessary. In the case of parameter estimation, we may often have some knowledge (perhaps inferred intuitively from experience) of the possible values, or range of values, of a parameter; moreover, we may also have some intuitive judgment on the values that are more likely to occur than others. For simplicity, suppose that the possible values of a parameter θ were assumed to be a set of discrete values θ_i, $i = 1, 2, \ldots, k$, with *prior* relative likelihoods $p_i = P(\Theta = \theta_i)$ as illustrated in Fig. 9.1 (Θ is the random variable whose values represent possible values of the parameter θ).

As additional information becomes available (such as the results of a series of tests or experiments), the prior assumptions on the parameter θ may be modified formally through the Bayes' theorem as follows.

Let ε denote the observed outcome of the experiment. Then applying Bayes' theorem of Eq. 2.20, we obtain the updated PMF for θ as

$$P(\Theta = \theta_i | \varepsilon) = \frac{P(\varepsilon | \Theta = \theta_i) P(\Theta = \theta_i)}{\displaystyle\sum_{i=1}^{k} P(\varepsilon | \Theta = \theta_i) P(\Theta = \theta_i)} \qquad i = 1, 2, \ldots, k \qquad (9.1)$$

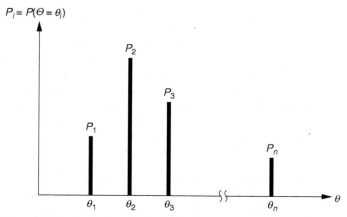

Figure 9.1 Prior PMF of parameter θ.

The various terms in Eq. 9.1 can be interpreted as follows:

$P(\varepsilon|\Theta = \theta_i)$ is the likelihood of the experimental outcome ε if $\Theta = \theta_i$—that is, the conditional probability of obtaining a particular experimental outcome assuming that the parameter is θ_i.

$P(\Theta = \theta_i)$ is the *prior* probability of $\Theta = \theta_i$—that is, prior to the availability of the experimental information ε.

$P(\Theta = \theta_i|\varepsilon)$ is the *posterior* probability of $\Theta = \theta_i$—that is, the probability that has been revised in light of the experimental outcome ε.

Denoting the prior and posterior probabilities as $P'(\Theta = \theta_i)$ and $P''(\Theta = \theta_i)$, respectively, Eq. 9.1 becomes

$$P''(\Theta = \theta_i) = \frac{P(\varepsilon|\Theta = \theta_i)P'(\Theta = \theta_i)}{\displaystyle\sum_{i=1}^{k} P(\varepsilon|\Theta = \theta_i)P'(\Theta = \theta_i)} \tag{9.1a}$$

Equation 9.1a, therefore, gives the posterior probability mass function of Θ. (In general, we shall use $'$ and $''$ to denote the prior and posterior information.)

The *expected value* of Θ is then commonly used as the Bayesian estimator of the parameter; that is,

$$\hat{\theta}'' = E(\Theta|\varepsilon) = \sum_{i=1}^{k} \theta_i P''(\Theta = \theta_i) \tag{9.2}$$

We may point out that in Eqs. 9.1 and 9.2, the observational data ε and any judgmental information are both used and combined in a systematic way to estimate the underlying parameter θ.

In the Bayesian framework, the significance of judgmental information is reflected also in the calculation of the relevant probabilities. In the preceding case, where subjective judgments were used in the estimation of the parameter θ, such judgments would be reflected in the calculation of the probability associated with the basic random variable X; for example, $P(X \le a)$ is obtained through the theorem of total probability using the posterior PMF of Eq. 9.1a. That is,

$$P(X \le a) = \sum_{i=1}^{k} P(X \le a|\Theta = \theta_i)P''(\Theta = \theta_i) \tag{9.3}$$

This represents the updated probability of the event $(X \leq a)$ based on all the available information. It may be emphasized that in Eq. 9.3 the uncertainty associated with the error of estimating the parameter [as reflected in $P''(\Theta = \theta_i)$] is combined formally and systematically with the inherent variability of the random variable X. In contrast, the combination will be difficult in the classical statistical approach (presented in Chapter 6), in which the parameter uncertainty may be described in terms of confidence intervals.

To further clarify the general concepts introduced above, consider the following examples.

▶ **EXAMPLE 9.1**

Reinforced concrete piles for a building foundation could be subject to defects resulting from poor construction quality. Some of the common defects would include insufficient bonding, inadequate length, cracks, and voids in the concrete. An engineer would like to estimate the proportion of piles that are defective in a given project that may consist of hundreds of piles. Suppose that from the engineer's experience with a range of construction quality for various pile foundation contractors in the region of the project site, he estimated (judgmentally) that the proportion of defective piles, p, for the site would range from 0.2 to 1.0 with 0.4 as the most likely value; more specifically, p is described by the prior PMF as shown in Fig. E9.1a. The values of p are discretized at 0.2 intervals to simplify the illustration.

On the basis of this prior PMF, which is based entirely on the engineer's judgment, the estimated probability of a defective pile would be (by virtue of the total probability theorem)

$$p' = (0.2)(0.3) + (0.4)(0.4) + (0.6)(0.15) + (0.8)(0.10) + (1.0)(0.05) = 0.44$$

In order to supplement his judgment, the engineer ordered a pile to be selected for inspection. The outcome of the inspection shows that the pile is *defective*. Based on the result of this single inspection, the PMF of p would be revised according to Eq. 9.1a, obtaining the posterior PMF as follows:

$$P''(p = 0.2) = \frac{(0.2)(0.3)}{(0.2)(0.3) + (0.4)(0.4) + (0.6)(0.15) + (0.8)(0.1) + (1.0)(0.05)} = 0.136$$

similarly, we obtain the posterior probabilities for the other values of p as follows:

$$P''(p = 0.4) = 0.364$$
$$P''(p = 0.6) = 0.204$$
$$P''(p = 0.8) = 0.182$$
$$P''(p = 1.0) = 0.114$$

These are shown graphically in Fig. E9.1b.

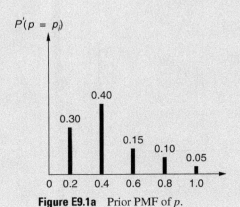

Figure E9.1a Prior PMF of p.

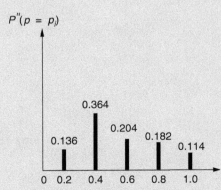

Figure E9.1b PMF of p after one inspection.

The Bayesian estimate for p, Eq. 9.2, therefore, is

$$\hat{p}'' = E(p|\varepsilon) = 0.2(0.136) + 0.4(0.364) + 0.6(0.204) + 0.8(0.182) + 1.0(0.114) = 0.55$$

In Fig. E9.1b, we see that because a defective pile was discovered from the single inspection, the probabilities for the higher values of p_i are increased from those of the prior distribution, resulting in a higher estimate for p, namely, $\hat{p}'' = E(p|\varepsilon) = 0.55$, whereas the prior estimate was 0.44. Observe that the inspection of a single pile, indicating a defective pile, does not imply that all piles will be defective; instead, the inspection result merely serves to increase the estimated proportion of defective piles by 0.11 (from 0.44 to 0.55). Figure E9.1c illustrates how the PMF of p changes with increasing number of consecutive defective piles observed; the distribution shifts toward $p = 1.0$ as $n \to \infty$.

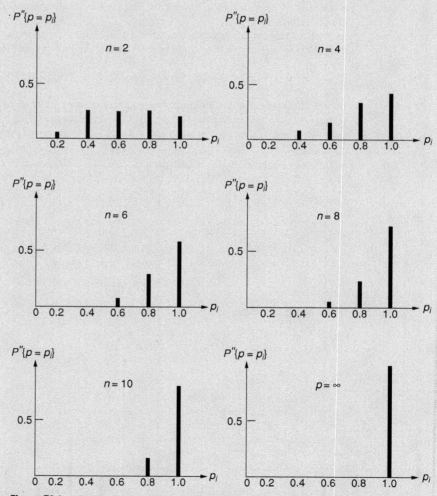

Figure E9.1c PMF of p for increasing number of defective piles inspected.

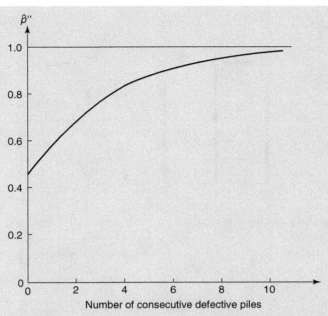

Figure E9.1d $\hat{p}''$ versus number of consecutive defective piles inspected.

Figure E9.1d shows the corresponding Bayesian estimate for p; we observe that after a sequence of six consecutive defective piles, the estimate for p is 0.90. If a long sequence of defective piles is observed, the Bayesian estimate of p approaches 1.0—a result that tends to the classical estimate; in such a case, there is an overwhelming amount of observed data to supersede any prior judgment. Ordinarily, however, where observational data are limited, judgment would be important and is reflected properly in the Bayesian estimation process. We may point out that if a person has a strong feeling about certain values of the parameter, it would imply a narrow prior distribution. In this case, it would require a large amount of observed information to override his or her prior judgment.

Now suppose that each of the main columns in the building is supported on a group of three piles. The column will suffer settlement problems if all the piles in the group are defective. Consider the case right after a test pile was found to be defective. Based on the posterior PMF of Fig. E9.1b, and using Eq. 9.3 with X denoting the number of defective piles, the probability that a main column will have a settlement problem is

$$\begin{aligned}
P(X = 3) &= P(X = 3 | p = 0.2)P''(p = 0.2) + P(X = 3 | p = 0.4)P''(p = 0.4) \\
&\quad + \cdots + P(X = 3 | p = 1.0)P''(p = 1.0) \\
&= (0.2)^3(0.136) + (0.4)^3(0.364) + (0.6)^3(0.204) + (0.8)^3(0.182) + (1)(0.114) \\
&= 0.255
\end{aligned}$$ ◀

▶ **EXAMPLE 9.2**

A traffic engineer is interested in the average rate of accidents v at an improved road intersection. Suppose that from his previous experience with similar road design configurations and traffic conditions, he deduced that the expected accident rate would be between 1 and 3 per year, with an average of 2, and postulated the prior PMF shown in Fig. E9.2. The occurrence of accidents is assumed to be a Poisson process.

During the first month after completion of the intersection, one accident occurred.

(a) In the light of this observation, i.e., *one accident in the first month*, revise the estimate for v.

(b) Using the result of part (a), determine the probability of no accident in the next 6 months.

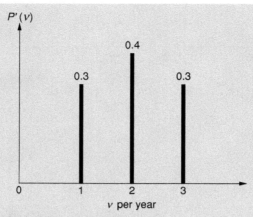

Figure E9.2 Prior distribution of v.

SOLUTIONS (a) Let ε be the event that an accident occurred in one month. The posterior probabilities according to Eq. 9.1a are therefore,

$$P''(v = 1) = \frac{P(\varepsilon|v = 1)P'(v = 1)}{P(\varepsilon|v = 1)P'(v = 1) + P(\varepsilon|v = 2)P'(v = 2) + P(\varepsilon|v = 3)P'(v = 3)}$$

$$= \frac{e^{-1/12}(1/12)(0.3)}{e^{-1/12}(1/12)(0.3) + e^{-1/6}(1/6)(0.4) + e^{-1/4}(1/4)(0.3)} = 0.166$$

Note that the probability of the observed event for a given value of v is proportional to the exponential PDF evaluated at an occurrence time of 1 month. Similarly,

$$P''(v = 2) = 0.411$$
$$P''(v = 3) = 0.423$$

Hence, the updated value of v is $\hat{v}'' = E(v|\varepsilon) = (0.166)(1) + (0.411)(2) + (0.423)(3) = 2.26$ accidents per year.

(b) Let A be the event of no accidents in the next 6 months. Then

$$P(A) = P(A|v = 1)P''(v = 1) + P(A|v = 2)P''(v = 2) + P(A|v = 3)P''(v = 3)$$
$$= e^{-1/2}(0.166) + e^{-1}(0.411) + e^{-3/2}(0.423)$$
$$= 0.346$$ ◀

▶ 9.3   THE CONTINUOUS CASE

9.3.1   General Formulation

In Sect. 9.2, the possible values of the parameter θ (such as p in Example 9.1 and v in Example 9.2) were limited to a discrete set of values; this was purposely assumed to simplify the presentation of the concepts underlying the Bayesian method of estimation.

In many situations, however, the value of a parameter could be a continuum of possible values. Thence, it would be appropriate to assume the parameter to be a continuous random variable in the Bayesian estimation. In this case, we develop the corresponding results, analogous to Eqs. 9.1 through 9.3, as follows.

Let Θ be the random variable for the parameter of a distribution, with a prior PDF $f'(\theta)$ as shown in Fig. 9.2. The prior probability that Θ will be between θ_i and $\theta_i + \Delta\theta$ is $f'(\theta_i)\Delta\theta$.

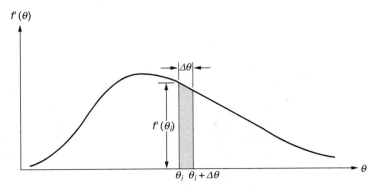

Figure 9.2 Continuous prior distribution of parameter Θ.

Then, if ε is an observed experimental outcome, the prior distribution $f'(\theta)$ can be revised in the light of ε using the Bayes' theorem, obtaining the posterior probability that θ will be in $(\theta_i, \theta_i + \Delta\theta)$ as follows:

$$f''(\theta_i)\Delta\theta = \frac{P(\varepsilon|\theta_i)f'(\theta_i)\Delta\theta}{\displaystyle\sum_{i=1}^{k} P(\varepsilon|\theta_i)f'(\theta_i)\Delta\theta}$$

where $P(\varepsilon|\theta_i) = P(\varepsilon|\theta_i < \theta \le \theta_i + \Delta\theta)$. In the limit, the above becomes

$$f''(\theta_i) = \frac{P(\varepsilon|\theta)f'(\theta)}{\int_{-\infty}^{\infty} P(\varepsilon|\theta)f'(\theta)d\theta} \tag{9.4}$$

The term $P(\varepsilon|\theta)$ is the conditional probability, or likelihood, of observing the experimental outcome ε assuming that the value of the parameter is θ. Hence, $P(\varepsilon|\theta)$ is a function of θ and is commonly referred to as the *likelihood function* of θ and denoted $L(\theta)$. The denominator is independent of θ; this is simply a normalizing constant required to insure that $f''(\theta)$ is a proper PDF. Equation 9.4, therefore, may be expressed also as

$$f''(\theta) = kL(\theta)f'(\theta) \tag{9.5}$$

where

$k = \left[\int_{-\infty}^{\infty} L(\theta)f'(\theta)d\theta\right]^{-1}$ is the normalizing constant

$L(\theta) = $ the likelihood of observing the experimental outcome ε assuming a given θ.

We observe from Eq. 9.5 that both the prior distribution and the likelihood function contribute to the posterior distribution of θ. In this way, as in the discrete case, the significance of judgment and of any observational data is combined properly and systematically; the former through $f'(\theta)$ and the latter embedded in $L(\theta)$.

Analogous to the discrete case, Eq. 9.2, the expected value of θ is commonly used as the point estimator of the parameter. Hence, the updated estimate of the parameter θ, in the light of the observational data ε, is given by

$$\hat{\theta}'' = E(\Theta|\varepsilon) = \int_{-\infty}^{\infty} \theta f''(\theta)d\theta \tag{9.6}$$

The uncertainty in the estimation of the parameter can be included in the calculation of the probability associated with a value of the underlying random variable. For example, if X is a random variable whose probability distribution has a parameter θ, its probability is

$$P(X \leq a) = \int_{-\infty}^{\infty} P(X \leq a|\theta) f''(\theta) d\theta \qquad (9.7)$$

Physically, Eq. 9.7 is the average probability of $(X \leq a)$ weighted by the posterior probabilities of the parameter θ.

▶ **EXAMPLE 9.3**

Let us consider again the problem of Example 9.1, in which the proportion of defective piles at the site is of concern; this time, however, we will assume that the probability p is a continuous random variable. If there is no (prior) factual information on p, a uniform prior distribution may be assumed (known as the *diffuse prior*), namely,

$$f'(p) = 1.0 \qquad 0 \leq p \leq 1$$

On the basis of a single inspection of one pile, revealing that the pile is defective, the likelihood function is simply that the probability of the event $\varepsilon = 1$ pile selected for inspection is defective, which is simply p. Hence, the posterior distribution of p, according to Eq. 9.5, is

$$f''(p) = kp(1.0) \qquad 0 \leq p \leq 1$$

in which the normalizing constant is

$$k = \left[\int_0^1 p\,dp\right]^{-1} = 2$$

Thus, the posterior PDF of p is

$$f''(p) = 2p \qquad 0 \leq p \leq 1$$

and the Bayesian estimate of p is then

$$\hat{p}'' = E(p|\varepsilon) = \int_0^1 p \cdot 2p\,dp = 0.667$$

If a sequence of n piles were inspected, out of which r piles were found to be defective, then the likelihood function is the probability of observing r failures among the n piles inspected. If the proportion of defective piles, or equivalently the probability of each pile being defective is p, and statistical independence is assumed between the piles, the likelihood function would be defined by the binomial PMF; i.e.,

$$L(p) = \binom{n}{r} p^r (1-p)^{n-r}$$

Then, with the diffuse prior, the posterior distribution of p becomes

$$f''(p) = k \binom{n}{r} p^r (1-p)^{n-r} \qquad 0 \leq p \leq 1$$

where

$$k = \left[\int_0^1 \binom{n}{r} p^r (1-p)^{n-r}\,dp\right]^{-1}$$

Thus, the Bayesian estimator is

$$\hat{P}'' = E(p|\varepsilon) = \frac{\int_0^1 p \binom{n}{r} p^r(1-p)^{n-r}\,dp}{\int_0^1 \binom{n}{r} p^r(1-p)^{n-r}\,dp}$$

$$= \frac{\int_0^1 p^{r+1}(1-p)^{n-r}\,dp}{\int_0^1 p^r(1-p)^{n-r}\,dp}$$

Repeated integration-by-parts of the above integrals yields

$$\hat{P}'' = \frac{r+1}{n}\frac{\int_0^1 (p^n - p^{n+1})\,dp}{\int_0^1 (p^{n-1} - p^n)\,dp} = \frac{r+1}{n+2}$$

From the above result, we may observe that as the number of inspections n increases (with the ratio r/n remaining constant), the Bayesian estimate for p approaches that of the classical estimate; that is,

$$\frac{r+1}{n+2} \to \frac{r}{n} \qquad \text{for large } n \qquad\qquad ◀$$

▶ **EXAMPLE 9.4**

An engineer is designing a temporary structure subjected to wind load on a newly developed island in the Pacific. Of interest is the probability p that the annual maximum wind speed will not exceed 120 km/hr. Records for the annual maximum wind speed in the island are available only for the last 5 years, and among these, the 120 km/hr wind was exceeded only once. An island in this region has a longer record of wind speeds; however, it is at some distance away. After a comparative study of the geographical condition in the two islands, the engineer inferred from this longer record that the average value of p for the newly developed island is 2/3 with a c.o.v. of 27%. Since p is bounded between 0 and 1.0, the following beta distribution (consistent with the above statistics) is also assumed for the prior distribution:

$$f'(p) = 20p^3(1-p) \qquad 0 \le p \le 1$$

In this case, the likelihood that the annual maximum wind speed will exceed 120 kph in 1 out of 5 years is

$$L(p) = \binom{5}{4} p^4(1-p)$$

Hence, the posterior PDF of p is

$$f''(p) = kL(p)f'(p)$$

$$= k\left[\binom{5}{4} p^4(1-p)\right][20p^3(1-p)] = 100kp^7(1-p)^2$$

where

$$k = \left[\int_0^1 100 p^7 (1 - p)^2 dp \right]^{-1} = 3.6$$

Thus,

$$f''(p) = 360 p^7 (1 - p)^2 \qquad 0 \le p \le 1$$

In this case, the prior PDF is equivalent to the assumption of one exceedance in 4 years, whereas the resulting posterior distribution is tantamount to two exceedances in 9 years. In fact, the above posterior distribution is the same as that obtained for a case in which two exceedances were observed in nine years and assuming a diffused (i.e., uniform) prior distribution. This example should serve also to illustrate a property of the Bayesian approach—namely, that information from sources other than the observed data can be useful in the estimation process.

The relation between the likelihood function and the prior and posterior distributions of the parameter p is illustrated in Fig. E9.4. Observe that the posterior distribution is "sharper" than either the prior distribution or the likelihood function. This implies that more information is "contained" in the posterior distribution than in either the prior or the likelihood function.

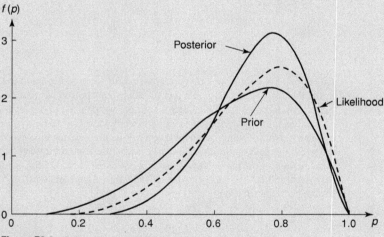

Figure E9.4 Prior, likelihood, and posterior functions.

▶ **EXAMPLE 9.5**

The occurrences of earthquakes may be modeled as a Poisson process with a mean occurrence rate v. Suppose that the historical record for a region A shows that n_0 earthquakes have occurred in the past t_0 years. The corresponding likelihood function is then given by

$$L(v) = P(n_0 \text{ quakes in } t_0 \text{ years} | v)$$
$$= \frac{(v t_0)^{n_0}}{n_0!} e^{-v t_0} \qquad v \ge 0$$

If there is no other information for estimating v, a diffuse prior may be assumed; this implies that $f'(v)$ is independent of the values of v and thus can be absorbed into the normalizing constant k. Then the posterior distribution of v becomes

$$f''(v) = kL(v)$$
$$= k\frac{(vt_0)^{n_0}}{n_0!}e^{-vt_0} \qquad v \geq 0$$

Upon normalization, $k = t_0$. The resulting $f''(v)$ may be compared with the gamma density function of Eq. 3.44 (for the random variable v), thus indicating that the posterior distribution of v follows the gamma distribution. The probability of the event ($E = n$ earthquakes in the next t years in region A) is then given by Eq. 9.7 as follows:

$$P(E) = \int_0^\infty P(E|v)f''(v)dv$$
$$= \int_0^\infty \frac{(vt)^n}{n!}e^{-vt} \cdot \frac{t_0(vt_0)^{n_0}}{n_0!}e^{-vt_0}dv$$
$$= \left(\int_0^\infty \frac{(t+t_0)[v(t+t_0)]^{n+n_0}}{(n+n_0)!}e^{-v(t+t_0)}dv\right)\frac{(n+n_0)!}{n!n_0!}\frac{t^n t_0^{(n_0+1)}}{(t+t_0)^{n+n_0+1}}$$

Since the integrand inside the parentheses is a gamma density function, the integral is equal to 1.0. Hence,

$$P(E) = \frac{(n+n_0)!}{n!n_0!}\frac{t^n t_0^{(n_0+1)}}{(t+t_0)^{n+n_0+1}} = \frac{(n+n_0)!}{n!n_0!}\frac{(t/t_0)^n}{(1+t/t_0)^{n+n_0+1}}$$

a result that was first derived by Benjamin (1968).

For a numerical illustration, suppose that historical records in the region A show that two earthquakes with intensity exceeding VI (MM scale) had occurred in the last 60 years. The probability that there will be no earthquakes with this intensity in the next 20 years, therefore, is

$$P(E) = \frac{(0+2)!}{0!2!}\frac{(20/60)^0}{(1+20/60)^3} = 0.42 \qquad ◀$$

9.3.2 A Special Application of the Bayesian Updating Process

An interesting application of the Bayesian updating process is in the inspection and detection of material defects (Tang, 1973). Fatigue and fracture failures in metal structures are frequently the result of unchecked propagation of flaws or cracks in the joints (welds) or base metals. Periodic inspection and repair can be used to minimize the risk of fracture failure by limiting the existing flaw sizes. Methods of detecting flaws, such as nondestructive testing (NDT) are often used. However, such detection methods are invariably imperfect; consequently, not all flaws may be detected during an inspection.

The probability of defecting a flaw generally increases with the flaw size and the detection power of the device. An example of a detectability curve for ultrasonics method is shown in Fig. 9.3. Hence, even when a structure is inspected and all detected flaws are repaired, it is difficult to ensure that there are no flaws larger than some specified size.

Suppose that an NDT device is used to inspect a set of welds in a structure and all detected flaws are fully repaired. On the basis of this assumption, the flaws that remain in

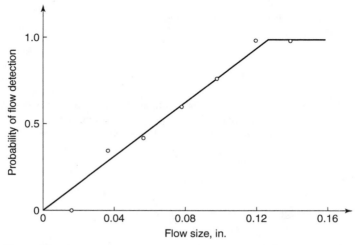

Figure 9.3 Detectability versus actual flaw depth. (Data from Packman et al., 1968.)

the weld would be those that were not detected. Let X be the flaw size and D the event that a flaw is detected. The probability that a flaw size (for example, depth) will be between x and $x + dx$ given that the flaw was not detected is, therefore,

$$P(x < X \leq x + dx | \overline{D}) = \frac{P(\overline{D} | x < X \leq x + dx) P(x < X \leq x + dx)}{P(\overline{D})}$$

This can be expressed also in terms of the PDF as

$$f_X(x | \overline{D}) = k P(\overline{D} | x) f_X(x) \tag{9.8}$$

in which $f_X(x)$ is the distribution of the flaw size prior to inspection and repair, whereas $f_X(x | \overline{D})$ is the distribution of the flaw size after inspection and repair have been performed. Also, $P(\overline{D} | x) = 1 - P(D | x)$, where $P(D | x)$ is simply the probability of detecting a flaw with depth x, which is the function defined by the detectability curve, such as that shown in Fig. 9.3. Comparing Eq. 9.8 with Eq. 9.5, we observe that Eq. 9.8 is of the same form as Eq. 9.5, with the following equivalences:

$$f_X(x | \overline{D}) \quad \sim \text{ the posterior distribution}$$
$$P(\overline{D} | x) \quad \sim \text{ the likelihood function}$$
$$f_X(x) \quad \sim \text{ the prior distribution}$$

▶ **EXAMPLE 9.6**

As an illustration, suppose that the initial (prior) distribution of the flaw depths X in a series of welds has a triangular shape described as follows (see Fig. E9.6):

$$f_X(x) = \begin{cases} 208.3x; & 0 < x \leq 0.06 \\ 20 - 125x; & 0.06 < x \leq 0.16 \\ 0; & x > 0.16 \end{cases}$$

Assume also that the NDT device used in the inspection has the detectability curve shown in Fig. 9.3; mathematically, this curve is given by

$$P(D|x) = \begin{cases} 0; & x \leq 0 \\ 8x; & 0 < x \leq 0.125 \\ 1.0; & x > 0.125 \end{cases}$$

Substituting the appropriate expressions for each interval of X into Eq. 9.8, we obtain the updated PDF of flaw depths as follows:

$$f_X(x|\overline{D}) = \begin{cases} 0 & x \leq 0 \\ k(1 - 8x)(208.3x) & 0 < x \leq 0.06 \\ k(1 - 8x)(20 - 125x) & 0.06 < x \leq 0.125 \\ 0 & x > 0.125 \end{cases}$$

which, after normalization, becomes

$$f_X(x|\overline{D}) = \begin{cases} 0 & x \leq 0 \\ 495x - 3964x^2 & 0 < x \leq 0.06 \\ 47.6 - 678x + 2379x^2 & 0.06 < x \leq 0.125 \\ 0 & x > 0.125 \end{cases}$$

The above prior, likelihood, and posterior functions are plotted in Fig. E9.6. It can be observed that the likelihood function, which is the "complementary function" of Fig. 9.3, behaves as a filter, it cuts off flaws larger than 0.125 in., and it also eliminates many of the remaining larger flaws. Thus, after the inspection and repair process, the distribution of flaw depths is shifted toward smaller flaw sizes.

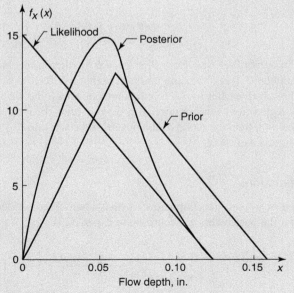

Figure E9.6 Distribution of flaw depth.

▶ 9.4   BAYESIAN CONCEPT IN SAMPLING THEORY

9.4.1   General Formulation

If the experimental outcome ε in Eq. 9.4 is a set of observed values $(x_1, x_2, \ldots, x_n)$, representing a random sample (see Sect. 6.2.1) from a population X with an underlying PDF $f_X(x)$, the probability of observing this particular set of values, assuming that the parameter of the distribution is θ, is

$$P(\varepsilon|\theta) = \prod_{i=1}^{n} f_X(x_i|\theta)dx$$

Then, if the prior PDF of θ is $f'(\theta)$, the corresponding posterior density function becomes, according to Eq. 9.4,

$$f''(\theta) = \frac{\left[\prod_{i=1}^{n} f_X(x_i|\theta)dx\right] f'(\theta)}{\int_{-\infty}^{\infty} \left[\prod_{i=1}^{n} f_X(x_i|\theta)dx\right] f'(\theta)d\theta}$$
$$= kL(\theta)f'(\theta) \tag{9.9}$$

in which the normalizing constant is

$$k = \left[\int_{-\infty}^{\infty} \left(\prod_{i=1}^{n} f_X(x_i|\theta)\right) f'(\theta)d\theta\right]^{-1}$$

whereas the likelihood function $L(\theta)$ is the product of the PDF of X evaluated at $(x_1, x_2, \ldots, x_n)$, or

$$L(\theta) = \prod_{i=1}^{n} f_X(x_i|\theta) \tag{9.10}$$

Using the posterior PDF of θ, Eq. 9.9, in Eq. 9.6 we, therefore, obtain the Bayesian estimator of the parameter θ. It is interesting to observe that the likelihood function of Eq. 9.10 is the same as that given earlier in Eq. 6.7 in connection with the classical method of maximum likelihood estimation. Furthermore, if a diffuse prior distribution is assumed (e.g., as in Eq. 9.13 below), then the mode of the posterior distribution, Eq. 9.9, would give the maximum likelihood estimator.

9.4.2   Sampling from Normal Populations

In the case of a Gaussian population with known standard deviation, σ, the likelihood function for the parameter, μ, according to Eq. 9.10, is

$$L(\mu) = \prod_{i=1}^{n} \frac{1}{\sqrt{2\pi}\sigma} \exp\left[-\frac{1}{2}\left(\frac{x_i - \mu}{\sigma}\right)^2\right] = \prod_{i=1}^{n} N_\mu(x_i, \sigma)$$

where $N_\mu(x_i, \sigma)$ denotes the PDF of μ with mean value x_i and standard deviation σ. It can be shown (e.g., Tang, 1971) that the product of m normal PDFs with respective means μ_i

and standard deviations σ_i is also a normal PDF with mean and variance

$$\mu^* = \frac{\sum_{i=1}^{m} \left(\mu_i/\sigma_i^2\right)}{\sum_{i=1}^{m} 1/\sigma_i^2} \quad \text{and} \quad \left(\sigma^*\right)^2 = \frac{1}{\sum_{i=1}^{m} 1/\sigma_i^2} \tag{9.11}$$

Therefore, the likelihood function $L(\mu)$ becomes

$$L(\mu) = N_\mu \left(\frac{\sum_{i=1}^{n} \left(x_i/\sigma^2\right)}{\sum_{i=1}^{n} \left(1/\sigma^2\right)}, \frac{1}{\sqrt{\sum_{i=1}^{n} \left(1/\sigma^2\right)}} \right) = N_\mu \left(\frac{\left(1/\sigma^2\right) \sum_{i=1}^{n} x_i}{n/\sigma^2}, \frac{1}{\sqrt{n/\sigma^2}} \right)$$

$$= N_\mu \left(\overline{x}, \frac{\sigma}{\sqrt{n}} \right) \tag{9.12}$$

where $\overline{x}$ is the sample mean of Eq. 6.1.

Without Prior Information

In the absence of prior information on μ, a diffuse prior distribution may be assumed. In such a case, we obtain the posterior distribution of μ, as

$$f''(\mu) = kL(\mu)$$

$$= kN_\mu \left(\overline{x}, \frac{\sigma}{\sqrt{n}} \right)$$

$$= N_\mu \left(\overline{x}, \frac{\sigma}{\sqrt{n}} \right) \tag{9.13}$$

where k is necessarily equal to 1.0 upon normalization. Therefore, without prior information, the posterior distribution of μ is Gaussian with a mean value equal to the sample mean $\overline{x}$ and standard deviation $\sigma/\sqrt{n}$.

Using the expected value of μ as the Bayesian estimator, we obtain, in accordance with Eq. 9.6,

$$\hat{\mu}'' = E(\mu|\varepsilon) = \overline{x}$$

That is, the sample mean $\overline{x}$ is the point estimate of the population mean. We recognize that this is the same as the classical estimate of Eq. 6.1. Therefore, in the absence of prior information, the Bayesian and classical methods give the same estimates for the population mean. Conceptually, however, the Bayesian basis for this estimate differs from that of the classical approach. Whereas Eq. 9.13 says that the posterior distribution of μ is Gaussian with mean $\overline{x}$ and standard deviation $\sigma/\sqrt{n}$, the classical approach (of Sect. 6.2) says that the sample mean $\overline{X}$ is a Gaussian random variable with mean μ and standard deviation $\sigma/\sqrt{n}$.

Significance of Prior Information

In contrast to the classical approach, however, prior judgmental information can be included in the estimation of the parameter μ. This is accomplished explicitly through the prior distribution $f'(\mu)$; we demonstrate this for the case of a Gaussian population as follows.

When X is Gaussian with known variance, it is mathematically convenient to assume also a Gaussian prior (see Sect. 9.4.4). Suppose that $f'(\mu)$ is $N(\mu', \sigma')$. Then, with the likelihood function of Eq. 9.12, the posterior distribution of μ becomes

$$f''(\mu) = kL(\mu)f'(\mu)$$
$$= kN_\mu\left(\bar{x}, \frac{\sigma}{\sqrt{n}}\right)N_\mu(\mu', \sigma')$$

which is a product of two normal PDFs. Again, it can be shown from Eq. 9.11 that $f''(\mu)$ is also Gaussian with mean

$$\mu'' = \frac{[\bar{x}/(\sigma/\sqrt{n})^2] + [\mu'/(\sigma')^2]}{[1/(\sigma/\sqrt{n})^2] + [1/(\sigma')^2]} = \frac{\bar{x}(\sigma')^2 + \mu'(\sigma^2/n)}{(\sigma')^2 + (\sigma^2/n)} \tag{9.14}$$

and standard deviation

$$\sigma'' = \sqrt{\frac{(\sigma')^2(\sigma^2/n)}{(\sigma')^2 + (\sigma^2/n)}} \tag{9.15}$$

In this case, the Bayesian estimator of μ, Eq. 9.6, yields

$$\hat{\mu}'' = \mu''$$

That is, the Bayesian estimate of the mean value is an average of the prior mean μ' and the sample mean $\bar{x}$, weighted inversely by the respective variances, as indicated in Eq. 9.14. As expected, the posterior mean approaches the sample mean for large sample size n. However, it will also approach the sample mean if the basis for the judgment is relatively weak (i.e., large σ') or if the population shows small scatter (i.e., small σ). In these cases, a given sampling plan will become more effective.

Equation 9.14 is an example of how prior information is combined systematically with observed data, in the present case, for estimating the mean value μ.

It is important to observe that the posterior variance of μ, as given by Eq. 9.15, is always less than* $(\sigma')^2$ or (σ^2/n); that is, the variance of the posterior distribution is always less than that of the prior distribution or of the likelihood function.

On the basis of the posterior distribution of μ, i.e., $N_\mu(\mu'', \sigma'')$, and with Eqs. 9.14 and 9.15, we may also determine the probability that μ is between a and b by

$$P(a < \mu \le b) = \int_a^b f''(\mu)d\mu$$

9.4.3 Error in Estimation

Any error in the estimation of a parameter θ can be combined with the inherent variability of the underlying random variable, for example, X, to obtain the total uncertainty associated with X. Accounting for the error in the estimation of θ, the PDF of X becomes (by virtue

---

*Since $(\sigma')^2 \ge 0$ and $\sigma^2/n \ge 0$,

$$(\sigma')^2\left(\sigma'^2 + \frac{\sigma^2}{n}\right) \ge (\sigma'^2)\left(\frac{\sigma^2}{n}\right)$$

or

$$(\sigma')^2 \ge \frac{(\sigma'^2)(\sigma^2/n)}{(\sigma'^2) + \sigma^2/n} = \sigma''^2$$

Similarly, it can be shown that $(\sigma'')^2 \le \sigma^2/n$.

of the total probability theorem)

$$f_X(x) = \int_{-\infty}^{\infty} f_X(x|\theta) f''(\theta) d\theta \tag{9.16}$$

In the case of a Gaussian variate X, with known σ, and μ estimated from a sample data,

$$f_X(x) = \int_{-\infty}^{\infty} f_X(x|\mu) f''(\mu) d\mu$$

where $f_X(x|\mu) = N_X(\mu, \sigma)$, and $f''(\mu)$ is given by Eq. 9.13. Again it can be shown (e.g., Tang, 1971) that the last integral above yields the normal PDF

$$f_X(x) = N(\overline{x}, \sqrt{\sigma^2 + \sigma^2/n}) \tag{9.17}$$

▶ **EXAMPLE 9.7**

A toll bridge was recently opened to traffic. For the past 2 weeks, records on rush-hour traffic during the last 10 work days showed a sample mean of 1535 vehicles per hour (vph). Suppose that rush-hour traffic has a normal distribution with a standard deviation of 164 vph. Based on this observed information, the posterior distribution of the mean rush-hour traffic μ is, according to Eq. 9.13, $N(1535, 164/\sqrt{10})$ or N(1535, 51.9) vph. The point estimate of μ, therefore, is 1535 vph.

The probability that μ will be between 1500 and 1600 vph is given by

$$P(1500 < \mu \leq 1600) = \Phi\left(\frac{1600 - 1535}{51.9}\right) - \Phi\left(\frac{1500 - 1535}{51.9}\right)$$
$$= \Phi(1.253) - \Phi(-0.674)$$
$$= 0.6445$$

Of greater interest are probabilities associated with the rush-hour traffic (rather than its mean) on a given work day. Suppose that for the present toll collection procedure, problems could arise if the rush-hour traffic exceeds 1700 vph on a given day. Then the probability that this will occur on any given day, based on Eq. 9.17, is given by

$$P(X > 1700) = 1 - \Phi\left(\frac{1700 - 1535}{\sqrt{(164)^2 + (51.9)^2}}\right)$$
$$= 1 - \Phi(0.958)$$
$$= 0.169$$

In other words, in about 17% of the work days, the present toll collection system will be inadequate during rush hours. Observe that the error in the estimation of μ has been included in computing this probability.

Now suppose that before the toll bridge was opened for traffic, simulation modeling was performed to predict the rush-hour traffic on the bridge. Based on the simulation results alone, it was estimated that the mean rush-hour traffic on a work day would be 1500 ± 100 with 90% confidence. How can this information be used with the observed traffic flow in the estimation of μ?

Assuming a Gaussian prior and with the foregoing simulation results, we obtain the prior distribution of the mean rush-hour traffic μ to be $N(1500, 60.8)$ vph. Then, applying Eqs. 9.14 and 9.15, the posterior distribution of μ is also Gaussian with

$$\mu'' = \frac{1535(60.8)^2 + 1500(51.9)^2}{(60.8)^2 + (51.9)^2} = 1520 \text{ vph}$$

and

$$\sigma'' = \sqrt{\frac{(60.8)^2 (51.9)^2}{(60.8)^2 + (51.9)^2}} = 39.5 \, \text{vph}$$

Therefore, by incorporating the result of simulation, the estimated mean rush-hour traffic is 1520 vph and the corresponding standard deviation is 39.5 vph. ◀

▶ **EXAMPLE 9.8**

Five repeated measurements of the elevation (relative to a fixed datum) on the top of a bridge pier under construction were performed as follows:

20.45 m

20.38 m

20.51 m

20.42 m

20.46 m

Assume that the measurement error is Gaussian with zero mean and a standard deviation of 0.08 m.

(a) Estimate the actual elevation of the pier based on the given measurements.

(b) Suppose that the elevation of the pier was previously measured by another surveying crew; the elevation was estimated to be 20.42 ± 0.02 m (that is, the mean measurement was 20.42 m with a standard error of 0.02 m). Estimate the elevation of the pier, taking advantage of this additional prior information.

SOLUTION The estimation of an actual dimension δ in surveying and photogrammetry is equivalent to the estimation of the mean value of a random variable (see Sect. 6.2.3). Measurement error is invariably assumed to be Gaussian with zero mean; this means tacitly that a set of measurements constitute a sample from a normal population. Therefore, the results derived in Sect. 9.4.2 are applicable to the estimation of geometric quantities in surveying and photogrammetry.

(a) The sample mean of the five measurements is

$$\bar{d} = \frac{1}{5}(20.45 + 20.38 + 20.51 + 20.42 + 20.46)$$

$$= 20.444 \, \text{m}$$

and a sample standard deviation of 0.08 m.

Hence, on the basis of the five observations, the actual elevation of the pier has a Gaussian distribution $N(20.444, 0.08/\sqrt{5})$ or $N(20.444, 0.036)$ m. In the convention of surveying and photogrammetry, the elevation of the pier would be given as 20.444 ± 0.036 m.

(b) In the case where prior information is available, such information can be incorporated through a prior distribution of δ. In the present case, using the pier elevation estimated earlier by the other crew, the prior distribution of δ may be modeled as $N(20.42, 0.020)$ m. Then applying Eqs. 9.14 and 9.15, the Bayesian estimate of the elevation is

$$d'' = \frac{(20.420)(0.036)^2 + (20.444)(0.020)^2}{(0.036)^2 + (0.020)^2} = 20.426 \, \text{m}$$

and the corresponding standard error is

$$\sigma_d'' = \sqrt{\frac{(0.036)^2 \, (0.020)^2}{(0.036)^2 + (0.020)^2}} = 0.017 \, \text{m}$$

◄

► **EXAMPLE 9.9**

The annual maximum flow of a stream has been recorded for the last 5 years as follows:

$$21.5, \; 19.2, \; 23.4, \; 20.1, \; 18.1 \; \left(\text{in } 100 \, \text{m}^3/\text{sec} \right)$$

Based on extensive data from nearby streams, it was determined that the annual maximum stream flow may be modeled by a lognormal distribution. Assume that the parameter ζ in the lognormal distribution may be estimated from the above five sample values as follows. The natural logarithm of the above data values are, respectively, 3.07, 2.96, 3.15, 3.00, and 2.90 from which we obtain the sample mean $\bar{x} = 3.016$, and the sample standard deviation $\zeta = 0.097$.

Without any prior information, the posterior distribution of λ, according to Eq. 9.13, is $N(\bar{x}, \zeta/\sqrt{5})$ or $N(3.016, 0.097/\sqrt{5}) = N(3.016, 0.043)$. However, if prior information is available, it can be incorporated through the prior distribution of λ. For example, suppose that $f'(\lambda)$ is assumed to be $N(2.9, 0.06)$; then from Eqs. 9.14 and 9.15 the posterior distribution, $f''(\lambda)$, will be normal with

$$\mu_\lambda'' = \frac{3.016(0.06)^2 + 2.9(0.043)^2}{(0.06)^2 + (0.043)^2} = 2.98$$

and

$$\sigma_\lambda'' = \sqrt{\frac{(0.06)^2(0.043)^2}{(0.06)^2 + (0.043)^2}} = 0.035$$

That is, in this latter case, the posterior distribution of λ is $N(2.98, 0.035)$. ◄

9.4.4 The Utility of Conjugate Distributions

In deriving the posterior distribution of a parameter by Eq. 9.5 or 9.9, considerable mathematical simplification can be achieved if the distribution of the parameter is appropriately chosen with respect to that of the underlying random variable. We saw this in Sect. 9.4.2 in the case of the Gaussian random variable X with known σ; by assuming the prior distribution of μ to be also Gaussian, the posterior distribution of μ remains Gaussian. This was similarly demonstrated for the discrete case in Example 9.4, in which the random variable has a binomial distribution and the prior distribution for its parameter p was assumed to be a standard beta distribution (with parameters $q' = 4$ and $r' = 2$). The resulting posterior distribution for p is also a standard beta distribution, with updated parameters $q'' = 8$ and $r'' = 3$.

Such pairs of distributions are known in the Bayesian terminology as *conjugate pairs* or *conjugate distributions*. By choosing a prior distribution that is a conjugate of the distribution of the underlying random variable, a convenient posterior distribution, which is usually of the same mathematical form as the prior, is obtained. Therefore, by selecting appropriate pairs of conjugate distributions, considerable mathematical simplifications can be achieved. Table 9.1 summarizes some of the conjugate pairs of common distributions.

It should be emphasized that conjugate distributions are selected solely for mathematical convenience and simplicity. For a random variable with a specified distribution, its conjugate prior distribution may be adopted if there is no other basis for the choice of the prior distribution. However, if there is evidence to support a particular prior distribution, then such a distribution ought to be used, mathematical complications notwithstanding.

TABLE 9.1 Conjugate Distributions

Basic Random Variables	Parameter	Prior and Posterior Distributions of Parameter
Binomial $$p_X(x) = \binom{n}{x}\theta^x(1-\theta)^{n-x}$$	θ	Beta $$f_\Theta(\theta) = \frac{\Gamma(q+r)}{\Gamma(q)\Gamma(r)}\theta^{q-1}(1-\theta)^{r-1}$$
Exponential $$f_X(x) = \lambda e^{-\lambda x}$$	λ	Gamma $$f_\Lambda(\lambda) = \frac{\nu(\nu\lambda)^{k-1}e^{-\nu\lambda}}{\Gamma(k)}$$
Normal $$f_X(x) = \frac{1}{\sqrt{2\pi}\sigma}\exp\left[-\frac{1}{2}\left(\frac{x-\mu}{\sigma}\right)^2\right]$$ (with known σ)	μ	Normal $$f_M(\mu) = \frac{1}{\sqrt{2\pi}\sigma_\mu}\exp\left[-\frac{1}{2}\left(\frac{\mu-\mu_\mu}{\sigma_\mu}\right)^2\right]$$
Normal $$f_X(x) = \frac{1}{\sqrt{2\pi}\sigma}\exp\left[-\frac{1}{2}\left(\frac{x-\mu}{\sigma}\right)^2\right]$$	μ, σ	Gamma-Normal $$f(\mu,\sigma) = \left\{\frac{1}{\sqrt{2\pi}\sigma/n}\exp\left[-\frac{1}{2}\left(\frac{\mu-m}{\sigma/\sqrt{n}}\right)^2\right]\right\}$$ $$\cdot\left\{\frac{[(n-1)/2]^{(n+1)/2}}{\Gamma[(n+1)/2]}\left(\frac{u}{\sigma^2}\right)^{(n-1)/2}\exp\left(-\frac{n-1}{2}\frac{u}{\sigma^2}\right)\right\}$$
Poisson $$p_X(x) = \frac{(\mu t)^x}{x!}e^{-\mu t}$$	μ	Gamma $$f_M(\mu) = \frac{\nu(\nu\mu)^{k-1}e^{-\nu\mu}}{\Gamma(k)}$$
Lognormal $$f_X(x) = \frac{1}{\sqrt{2\pi}\zeta x}\exp\left[-\frac{1}{2}\left(\frac{\ln x-\lambda}{\zeta}\right)^2\right]$$ (with known ζ)	λ	Normal $$f_\Lambda(\lambda) = \frac{1}{\sqrt{2\pi}\sigma}\exp\left[-\frac{1}{2}\left(\frac{\lambda-\mu}{\sigma}\right)^2\right]$$

▶ **EXAMPLE 9.10**

The occurrence of flaws in a welded joint may be modeled by a Poisson process with a mean occurrence rate of μ flaws per meter of weld. Actual observations performed with a powerful device (assume it would not miss detecting any significant flaw) detected five flaws in a weld of 9.2 m. However, from previous experience with a similar type of welds, the mean flaw rate would vary with the quality of workmanship on a specific job, and based on this information, the engineer believes that the mean flaw rate in the present job has a mean of 0.5 flaw/m with a c.o.v. of 40%. Determine the mean and c.o.v. of μ for this type of weld, using the observed data as well as the information from prior experience.

Since the number of flaws in a given weld length is described by the Poisson distribution, it is convenient, according to Table 9.1, to prescribe its conjugate gamma distribution as the prior distribution for the parameter μ. From the information given above, and observing from Sect. 3.2.8 (see also Column 4 of Table 9.1) for the mean and variance of the gamma distribution, we have the mean of μ

$$E'(\mu) = \frac{k'}{\nu'} = 0.5$$

and c.o.v. of μ

$$\delta'(\mu) = \frac{\sqrt{k'/\nu'^2}}{k'/\nu'} = \frac{1}{\sqrt{k'}} = 0.4$$

Thus, the prior parameters of the gamma distribution are $k' = 6.25$ and $\nu' = 12.5$.

It follows then that the posterior distribution of μ is also gamma distributed. From the relationship given in Table 9.1 (column 5) between the prior and posterior statistics, and the sample data, we evaluate the parameters k'' and ν'' of the posterior gamma distribution as follows:

Mean and Variance of Parameter	Posterior Statistics
$E(\Theta) = \dfrac{q}{q+r}$	$q'' = q' + x$
$\mathrm{Var}(\Theta) = \dfrac{qr}{(q+r)^2(q+r+1)}$	$r'' = r' + n - x$
$E(\lambda) = \dfrac{k}{v}$	$v'' = v' + \sum_i x_i$
$\mathrm{Var}(\lambda) = \dfrac{k}{v^2}$	$k'' = k' + n$
$E(\mu) = \mu_\mu$	$\mu''_\mu = \dfrac{\mu'_\mu(\sigma^2/n) + \bar{x}\sigma'^2_\mu}{\sigma^2/n + (\sigma'_\mu)^2}$
$\mathrm{Var}(\mu) = \sigma^2_\mu$	$\sigma''_\mu = \sqrt{\dfrac{(\sigma'_\mu)^2(\sigma^2/n)}{(\sigma'_\mu)^2 + \sigma^2/n}}$
$E(\mu) = m$	$n'' = n' + n$
$\mathrm{Var}(\mu) = \dfrac{uv}{(v-2)n}\quad v>2$	$m'' = (n'm' + nm)/n''$
$E(\sigma) = \sqrt{\dfrac{uv}{2}}\,\dfrac{\Gamma[(v-1)/2]}{\Gamma[v/2]}\quad v>1$	$v'' = v' + v + 1$
$\mathrm{Var}(\sigma) = \dfrac{uv}{v-2} - E^2(\sigma)\quad v>2$	$u'' = [(v'u' + n'm'^2) + (vu + nm^2) - n''m''^2]/v''$
$E(\mu) = \dfrac{k}{v}$	$v'' = v' + t$
$\mathrm{Var}(\mu) = \dfrac{k}{v^2}$	$k'' = k' + x$
$E(\lambda) = \mu$	$\mu'' = \dfrac{\mu'(\zeta^2/n) + \sigma^2\overline{\ln x}}{\zeta^2/n + \sigma^2}$
$\mathrm{Var}(\lambda) = \sigma^2$	$\sigma'' = \sqrt{\dfrac{\sigma^2(\zeta^2/n)}{\sigma^2 + \zeta^2/n}}$

With $x = 5$ flaws, and $t = 9.2$ m,

$$k'' = k' + x = 6.25 + 5 = 11.25$$
$$v'' = v' + t = 12.5 + 9.2 = 21.7$$

Hence, the updated mean and c.o.v. of the average flaw rate μ are

$$E''(\mu) = \frac{k''}{v''} = \frac{11.25}{21.7} = 0.52 \text{ flaw/m}$$

$$\delta''(\mu) = \frac{1}{\sqrt{k''}} = \frac{1}{\sqrt{11.25}} = 0.30 \qquad \blacktriangleleft$$

▶ **EXAMPLE 9.11**

Suppose the maximum water elevation of a river, H, during a flood can be described by an exponential distribution as follows:

$$f_H(h) = \lambda e^{-\lambda h} \qquad h \geq 0$$

No official record on flood occurrences exists at a given site. However, the local inhabitants recall that only two floods have occurred in recent years, whose elevations are at least 5 ft. Assume that the flood elevations between floods are statistically independent, and there is no other information for estimating the parameter λ.

(a) Determine the distribution of λ, its mean value, and c.o.v. of H based on the given information.

(b) What is the probability that the next flood will exceed 5 ft?

(c) After a period of time, three floods occurred at the given site and the maximum water elevations were recorded as 3, 4, and 5 feet, respectively. With this new information, what would be the updated distribution of λ, its mean value, and c.o.v.?

SOLUTIONS (a) In this part, since the observed information is in terms of exceedance (instead of exactly equal to) of a given value, the use of conjugate concepts for Bayesian updating would not be applicable. We need to start with the basic updating equation of Eq. 9.5. The likelihood function in this case is the probability of water elevation exceeding 5 ft in two floods in terms of λ. For one flood, the probability of water elevation exceeding 5 ft is

$$p = P(H > 5) = 1 - F_H(5) = e^{-5\lambda}$$

Hence, the likelihood function becomes, with ε = two floods exceeding 5 ft,

$$L(\lambda) = P(\varepsilon|\lambda) = p^2 = (e^{-5\lambda})^2 = e^{-10\lambda}$$

By assuming a diffused prior distribution of Eq. 9.5, the updated distribution for λ is

$$f''(\lambda) = ke^{-10\lambda}$$

Upon normalization, we obtain $k = 10$, and the posterior distribution of λ becomes

$$f''(\lambda) = 10e^{-10\lambda}; \qquad \lambda \geq 0$$

Hence, f'' is another exponential distribution with a parameter of 10. The corresponding mean and c.o.v. are, respectively, 0.1 and 1. Note that this is also a gamma distribution with parameters $v = 10$ and $k = 1$ (see Table 9.1).

(b) The probability that the next flood will exceed 5 feet is determined by integrating over all possible values of λ, yielding the updated probability as

$$P''(H > 5) = \int_0^\infty P''(H > 5|\lambda)f''(\lambda)d\lambda = \int_0^\infty e^{-5\lambda}(10e^{-10\lambda})d\lambda = \frac{10}{15}\int_0^\infty 15e^{-15\lambda}d\lambda = 0.667$$

(c) Since the basic random variable X is the flood level, which is exponentially distributed, the gamma distribution (with $v' = 10$ and $k' = 1$) obtained in Part (a) above would serve as a suitable prior distribution for the parameter λ for this part of the updating. Applying the formula in Table 9.1, the posterior distribution of λ would also follow the gamma distribution with

$$v'' = v' + \sum x_i = 10 + 3 + 4 + 5 = 22 \qquad \text{and} \qquad k'' = k' + n = 1 + 3 = 4$$

Hence, we obtain the updated posterior PDF of λ as

$$f''(\lambda) = (22^4/4!)\lambda^3 e^{-22\lambda} = 9,761\lambda^3 e^{-22\lambda}; \qquad \lambda \geq 0$$

whose mean and c.o.v. are, respectively, $4/22 = 0.182$ and $1/2 = 0.5$. ◀

▶ 9.5 ESTIMATION OF TWO PARAMETERS

The Bayesian approach described in the earlier sections can be extended to include the case of more than one parameter. Consider the case of two parameters in which θ_1 and θ_2 denote the two parameters to be estimated. In this case, the general Bayesian updating equation of Eq. 9.5 becomes

$$f''(\theta_1, \theta_2) = kL(\theta_1, \theta_2)f'(\theta_1, \theta_2) \tag{9.18}$$

where $f'(\theta_1, \theta_2)$ and $f''(\theta_1, \theta_2)$ are, respectively, the prior and posterior joint PDFs of θ_1 and θ_2; $L(\theta_1, \theta_2)$ is the likelihood of the observed information for given values of θ_1 and

θ_2. If θ_1 and θ_2 are statistically independent, $f'(\theta_1, \theta_2)$ can be expressed as $f'(\theta_1) \cdot f'(\theta_2)$ to facilitate the assessment of the prior joint distribution. However, as will be shown later in an example, θ_1 and θ_2 may not necessarily remain statistically independent after incorporation of the observed information through the likelihood function. Again, k is the normalizing constant to ensure that the posterior joint distribution $f''(\theta_1, \theta_2)$ is a proper PDF. It would be determined from

$$\int_{-\infty}^{\infty} \int_{-\infty}^{\infty} f''(\theta_1, \theta_2) d\theta_1 d\theta_2 = 1$$

or

$$\int_{-\infty}^{\infty} \int_{-\infty}^{\infty} k L(\theta_1, \theta_2) f'(\theta_1, \theta_2) d\theta_1 d\theta_2 = 1 \tag{9.19}$$

With the joint PDF of θ_1 and θ_2 defined, the respective marginal distributions of θ_1 and θ_2 can be determined according to Eqs. 3.67 and 3.68 as

$$f''(\theta_1) = \int_{-\infty}^{\infty} f''(\theta_1, \theta_2) d\theta_2 \tag{9.20a}$$

$$f''(\theta_2) = \int_{-\infty}^{\infty} f''(\theta_1, \theta_2) d\theta_1 \tag{9.20b}$$

Moreover, the corresponding Bayesian estimators can be obtained accordingly from

$$\hat{\theta}_1'' = E''(\theta_1) = \int_{-\infty}^{\infty} \theta_1 f''(\theta_1) d\theta_1 \tag{9.21a}$$

$$\hat{\theta}_2'' = E''(\theta_2) = \int_{-\infty}^{\infty} \theta_2 f''(\theta_2) d\theta_2 \tag{9.21b}$$

▶ **EXAMPLE 9.12**

Boulders are frequently found embedded in a certain type of soil stratum. The occurrence of these boulders may be modeled according to a Poisson process in a spatial domain with a mean rate μ per unit volume of the soil deposit. To gain information on the density of boulders in a given deposit, an engineer has penetrated a rod (of negligible cross-sectional area) to a depth of 40 ft into the stratum. If the rod has not encountered any boulder, what can be said about the mean rate μ of boulders in the soil deposit?

It may be obvious that the size of the boulders would also play a role in interpreting the observed result. To simplify the problem, we may assume that the boulders are spherical and all have the same radius R at this site. However, R is highly variable between sites. Prior to the penetration test, the radius R of the boulder is believed to be uniformly distributed between 1 and 4 ft, whereas no other information is available on the parameter μ. We would be interested in determining the updated distribution of μ and R in light of the penetration test.

The ordinate of the prior PDF of the radius R is a constant 1/3 over the range from 1 to 4 ft. By assuming a diffused prior distribution for μ, $f'(\mu)$ may be simply absorbed into the normalizing constant. The event of not encountering any boulder along the rod is equivalent to not finding any boulders whose centers are within a distance, $R = r$, from the rod. Hence, the likelihood function is equivalent to the probability of zero occurrence in a cylindrical volume of $\pi r^2 (40)$. Since the centers of the boulders occur according to a Poisson process with a mean rate of μ per unit volume of soil, the likelihood function is

$$L(\varepsilon | \mu, r) = P(X = 0 \text{ in a 40-ft cylinder with radius } r | \mu)$$
$$= \exp[-(40\pi r^2 \mu)] \qquad 1 \le r \le 4; \qquad 0 < \mu < \infty$$

The updated joint distribution of μ and r becomes

$$f''(\mu, r) = \frac{k}{4 - 1} \exp[-(40\pi r^2 \mu)]$$

in which k may be determined from

$$k = \left[\frac{1}{3} \int_1^4 \int_0^\infty e^{-\mu\pi r^2(40)} \, d\mu dr \right]^{-1}$$

$$= \left[\frac{1}{3} \int_1^4 \frac{1}{40\pi r^2} dr \right]^{-1} = 40\pi(4) = 160\pi$$

Hence,

$$f''(\mu, r) = \frac{160\pi}{3} e^{-40\pi\mu r^2}; \qquad 1 \le r \le 4; \qquad 0 \le \mu \le \infty$$

from which we obtain the marginal PDF of the boulder radius as

$$f''(r) = \int_0^\infty \frac{160\pi}{3} \cdot e^{-40\pi\mu r^2} d\mu = \frac{4}{3r^2}$$

whose mean value is

$$E''(R) = \int_1^4 r \cdot \frac{4}{3r^2} dr = \frac{4}{3}\ln 4 = 1.85 \text{ ft}$$

The prior and the updated posterior marginal distributions of the radius R are as shown in Fig. E9.12a.

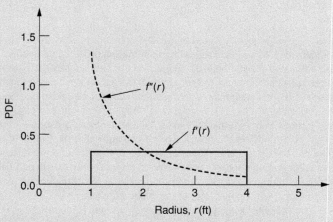

Figure E9.12a　Prior and updated distribution of boulder radius.

Therefore, based on the observed event of "no encounter by a 40-ft rod," the expected boulder radius reduces from 2.5 ft to 1.85 ft.

Similarly, the marginal distribution of the mean rate μ can be determined, through a transformation of variable, as follows,

$$f''(\mu) = \sqrt{\frac{1}{40}} \frac{160\pi}{3\sqrt{\mu}} [\Phi(4\sqrt{80\pi\mu}) - \Phi(\sqrt{80\pi\mu})]$$

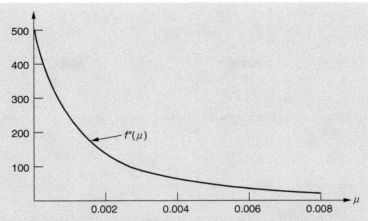

Figure E9.12b Updated distribution of mean boulder rate (per cu ft).

which is a decreasing function as shown in Fig. E9.12b, from which we obtain the corresponding updated expected mean boulder rate of 0.0035 per cubic foot. In order to ensure a lower frequency and size of boulders in this soil stratum, the event of no encounter by one or more rods with a deeper penetration would be needed. ◄

Geologic anomalies in a soil deposit are often causes of geotechnical failures. Hence, the detection of these anomalies through a site exploration program is important. As demonstrated in Example 9.12, Bayesian methods can systematically incorporate observed site exploration results with available judgmental information to infer the characteristics, such as the occurrence probability or frequency and size distributions, of relevant anomalies. Further literature on this subject can be found in Tang et al. (1983, 1986, 1990, 1991, and 1993).

► **EXAMPLE 9.13**

A set of 10 tests for the bending strength of Douglas fir lumber shows a sample mean of 6 ksi and a sample standard deviation of 1.2 ksi. Prior to these tests, the engineer believed that the mean bending strength of this species of wood has a mean of 5 ksi and a c.o.v. of 16.3%; and the standard deviation of the bending strength to have a mean of 1.253 ksi and a c.o.v. of 52.3% based on previous experiences with this kind of material. Assume that the bending strength follows a normal distribution $N(\mu, \sigma)$.

The engineer now wishes to update these prior estimates of the parameters μ and σ by combining the observed test results with his previous estimates. For simplicity, we will apply the conjugate approach for the updating process. Since both μ and σ need to be estimated, the joint Gamma-Normal distribution may be used as a convenient prior for this purpose according to Table 9.1. The Gamma-Normal distribution is defined by four parameters, namely, m, u, n, and v where $v = n - 1$. The first step is to evaluate the prior parameters consistent with the engineer's prior experience.

In summary, the prior information is as follows: $E'(\mu) = 5$; $\delta'(\mu) = 0.163$; $E'(\sigma) = 1.253$; and $\delta'(\sigma) = 0.523$. By using the formulas in Column 4 of Table 9.1, we evaluate the values of m', u', n', and v' to be 5, 1, 3, and 4, respectively.

In the next step, we would use the formula in Column 5 of Table 9.1 to combine the prior and the observed information. Recall that the observed statistics are $n = 10$; $\bar{x} = 6$; $s = 1.2$; $v = n - 1 = 9$. Hence, we have

$$n'' = n' + n = 3 + 10 = 13$$

and

$$m'' = \left(n'm' + n\bar{x}\right)\frac{1}{n''} = (3 \times 5 + 10 \times 6)\frac{1}{13} = 5.769$$

Also,
$$v'' = v' + v + 1 = 4 + 9 + 1 = 14$$

and $u'' = \dfrac{1}{v''} \left[(v'u' + n'm'^2) + (vs^2 + n\bar{x}^2) - n''m''^2 \right]$

$$= \frac{1}{14} \left[(4 \times 1 + 3 \times 5^2) + (9 \times 1.2^2 + 10 \times 6^2) - 13 \times 5.769^2 \right] = 1.379$$

Finally, the posterior statistics may be evaluated by using the formulas in Column 4 of Table 9.1, yielding

$$E''(\mu) = m'' = 5.769$$

$$\text{Var}''(\mu) = \frac{u''v''}{(v'' - 2)n''} = \frac{1.379 \times 14}{(14 - 2)\,13} = 0.124$$

The c.o.v., therefore, is

$$\delta''(\mu) = \frac{\sqrt{0.124}}{5.769} = 0.061$$

The updated mean of 5.769 ksi is between the prior estimate and the sample mean. It may be observed that the uncertainty level represented by the c.o.v. of μ is reduced from 16.3% to 6.1%. For the estimated standard deviation, s, further mathematical manipulation with the formulas in Column 4 of Table 9.1 would yield an updated mean and standard deviation of 1.242 and 0.257 ksi, respectively. Hence, the uncertainty level, c.o.v., of s is also reduced from 52.3% to 20.7%. ◀

▶ 9.6 BAYESIAN REGRESSION AND CORRELATION ANALYSES

9.6.1 Linear Regression

In Chapter 8, we presented the classical approach to regression analysis in which the statistical relationship between two (or more) variables is described. For the case of linear regression between two variables X and Y with constant conditional variance, the required relationships are described in Eqs. 8.1 through 8.6, which are summarized here for convenient reference as follows. The linear regression equation is

$$E(Y|x) = \alpha + \beta x \tag{9.22}$$

and the constant conditional variance is

$$\text{Var}(Y|x) = \sigma^2 \tag{9.23}$$

in which the parameters α, β, and σ^2 are estimated from a set of sampled data. Based on least squares errors, the formulas for estimating these parameters are

$$\hat{\beta} = \frac{\sum x_i y_i - n\bar{x}\,\bar{y}}{\sum x_i^2 - n\bar{x}^2} \tag{9.24}$$

and

$$\hat{\alpha} = \bar{y} - \beta\bar{x} \tag{9.25}$$

with which the mean-value function of Y for given x becomes

$$E(Y|x) = \hat{\alpha} + \hat{\beta}x \tag{9.26}$$

and

$$\hat{\sigma}^2 = \frac{\sum (y_i - \overline{y})^2 - \beta^2 \sum (x_i - \overline{x})^2}{n - 2} \tag{9.27}$$

in which $\sum = \sum_{i=1}^{n}$, $n =$ the sample size, and $\overline{x}$ and $\overline{y}$ are the sample means of X and Y, respectively. When the sampled data are limited, there may be significant errors (or epistemic uncertainty) in the estimated values of these parameters α, β, and σ^2. Consequently, the estimated mean-value function of Eq. 9.26 would also include these errors. Moreover, the overall variability of the predicted value of Y for given x should reflect such epistemic uncertainty in addition to the basic scatter σ^2 about the regression line. Following the Bayesian approach presented in this chapter, the parameters α, β, σ^2, and $E(Y|x)$ can be considered as random variables themselves. In this way, these epistemic uncertainties due to sampling error can be directly incorporated into the probabilistic analysis.

The likelihood of observing a set of y_i values at given values of x_i given the values of α, β, σ^2 and the random variable Y at a given value of x is normally distributed with the mean and variance given in Eqs. 9.26 and 9.27. Hence,

$$L(\alpha, \beta, \sigma^2) = \prod_{i=1}^{n} \frac{1}{\sqrt{2\pi}\sigma} e^{-\frac{1}{2}\left(\frac{y_i - \alpha - \beta x_i}{\sigma}\right)^2} \tag{9.28}$$

By combining the likelihood function with an assumed prior distribution of the parameters according to Eq. 9.18, the updated distribution of the parameters is

$$f''(\alpha, \beta, \sigma^2) = kL(\alpha, \beta, \sigma^2)f'(\alpha, \beta, \sigma^2) \tag{9.29}$$

which can be subsequently used to determine the updated distribution of the mean-value function, $\theta_x = E(Y|x)$, as follows:

$$f''(\theta_x) = \int\int\int f''(\theta_x | \alpha, \beta, \sigma^2) f''(\alpha, \beta, \sigma^2) d\alpha \, d\beta \, d\sigma \tag{9.30}$$

and the updated distribution of Y at given x is

$$f''(Y_x) = \int\int\int N_Y(\alpha + \beta x, \sigma) f''(\alpha, \beta, \sigma^2) d\alpha \, d\beta \, d\sigma \tag{9.31}$$

in which $N_Y(\alpha + \beta x, \sigma)$ denotes a normal PDF for Y with mean $(\alpha + \beta x)$ and standard deviation σ. The above formulation is generally difficult to carry out analytically, although the use of compatible prior distributions could be helpful. Numerical procedures would generally be required to obtain the pertinent results. A set of useful results is presented in Tang (1980) for the case of databased prediction (i.e., no prior information on the regression coefficients including σ) as follows:

$$E(\theta_x) = \alpha + \beta x \tag{9.32}$$

$$\text{Var}(\theta_x) = \frac{n-1}{n-3}\left[\frac{1}{n}\left\{1 + \frac{n}{n-1}\frac{(x-\overline{x})^2}{s_x^2}\right\}\right]\sigma^2 \tag{9.33}$$

in which α, β, and σ^2 are estimated from the observed statistics according to Eqs. 9.24 through 9.27, and $s_x^2 =$ the sample variance of each x_i. Observe that the (epistemic) uncertainty in the regression line depends on the degree of extrapolation as measured by $(x - \overline{x})^2/s_x^2$ and the number of data points n in evaluating the regression line. A larger extrapolation or a smaller sample size will result in a larger variance $\text{Var}(\theta_x)$. The factor $(n-1)/(n-3)$ accounts for the contribution of the uncertainty in the estimate of the basic scatter σ^2; its effect diminishes as the sample size increases. Finally, these uncertainties on the parameters can be combined with the basic scatter in evaluating the overall uncertainty on the predicted value of Y for a given value of x. Indeed, by using a formula presented in Raiffa and Schlaifer (1961) for a general normal regression process, Tang (1980) has shown that

$$\text{Var}(Y|x) = \frac{n-1}{n-3}(1+\gamma)\sigma^2 \tag{9.34}$$

in which γ stands for the quantity within the square brackets in Eq. 9.33. It represents the uncertainty contributed by the variance in the mean-value function or regression equation. In the extreme case, when n goes to infinity, $\text{Var}(Y|x)$ approaches σ^2 as expected, since the epistemic uncertainty will not be significant. The expected value of Y remains that given by the regression equation of Eq. 9.32.

▶ **EXAMPLE 9.14**

Consider the precipitation-runoff data presented in Example 8.2, where the values of α, β, and σ^2 here have been determined to be $\alpha = -0.14$, $\beta = 0.435$, and $\sigma^2 = 0.075$. Hence, for a storm with precipitation x (in.), the runoff will have a mean

$$E(Y|x) = -0.14 + 0.435x$$

According to Eq. 9.26, its variance can be determined from Eq. 9.34 in which γ refers to the quantity within the square brackets in Eq. 9.33. Hence, the variance is

$$\text{Var}(Y|x) = \frac{24}{22}\left(1+\left[\frac{1}{25}\left\{1+\frac{25}{24}\frac{(x-2.16)^2}{36.8}\right\}\right]\right)0.075$$

As an example, for a storm with 4 in. of precipitation, the runoff would have a mean of

$$E(Y|x=4) = -0.14 + (0.435)(4) = 1.60\,\text{in.}$$

and a variance of

$$\text{Var}(Y|x=4) = \frac{24}{22}\left(1+\left[\frac{1}{25}\left\{1+\frac{25}{24}\frac{(4-2.16)^2}{36.8}\right\}\right]\right)0.075 = 0.0854\,\text{in}^2$$

If the value of the basic scatter σ^2 can be assumed to be equal to 0.075, there would not be any contribution from the uncertain σ^2. Hence, $\text{Var}(Y|x)$ in Eq. 9.34 becomes $(1+\gamma)0.075$, yielding 0.078 for the variance of Y at $x = 4$. Moreover, the runoff Y will be normally distributed. The probability that the runoff will exceed 2 in. in this storm with 4 in. of precipitation may be determined as

$$P(Y>2|x=4) = 1 - \Phi\left(\frac{2-1.60}{\sqrt{0.078}}\right) = 1 - \Phi(0.435) = 0.332$$

◀

9.6.2 Updating the Regression Parameters

It is generally not common practice to update the statistics of the parameters α, β, and σ^2 based on different sources of information. For example, a prior set of data that is used to estimate these parameters may simply be combined with a new set of data into a single set for a Bayesian regression analysis. The mean and variance of the regression equation can then be determined accordingly. However, if the information on the prior data set is not available, judgment would be needed to develop an equivalent set of key statistical parameters describing the information contained in the prior.

One may refer to Lee (1989) for a pragmatic procedure of carrying out this work. A more common situation is perhaps to estimate the mean value of a random variable Y in which limited data on Y are available. To supplement these data or to obtain a preliminary estimate of the mean value of Y, one may explore if a regression relation exists between Y and another variable. As an example, consider the runoff Y in Example 9.13 for a storm with a precipitation of 4 in. Suppose only two runoff values have been observed associated with storms having precipitation of 4 in. In this case, Bayesian regression analysis can be used to provide the prior mean and variance of $E(Y|x=4)$, which can be updated with the directly measured runoffs for storms with a precipitation of 4 in. The updating here would reduce to the case of updating the mean value of the random variable Y at $x=4$.

▶ **EXAMPLE 9.15**

If the runoff values observed for two storms with precipitation of 4 in. each at a given location are 1.5 in. and 2.5 in., respectively, what are the updated estimates of the mean and variance of the mean runoff from storms with 4 in. of precipitation?

Let θ_4 denote the mean runoff for a storm with a precipitation of 4 in. From the Bayesian regression analysis on the runoff–precipitation data in Example 8.2 or 9.14, it was shown that

$$E(\theta_4) = -0.14 + (0.435)(4) = 1.60 \text{ in.}$$

and variance

$$\text{Var}(\theta_4) = \frac{24}{22}\left[\frac{1}{25}\left\{1 + \frac{25}{24}\frac{(4-2.16)^2}{36.8}\right\}\right]0.075 = 0.0036 \text{ in}^2$$

by using Eqs. 9.32 and 9.33. These two values will be the prior statistics of θ_4. For the directly measured values, we have $\bar{y} = 2$. Suppose the value of runoff Y at $x=4$ in. is normally distributed with variance assumed to be known, equal to 0.075, then the updated statistics of θ_4 can be determined from Eqs. 9.14 and 9.15 as

$$E''(\theta_4) = \frac{(2)(0.0036) + (1.60)(0.075/2)}{(0.0036) + (0.075/2)} = 1.64$$

$$\text{Var}''(\theta_4) = \frac{(0.0036)(0.075/2)}{(0.0036) + (0.075/2)} = 0.0033$$

We may observe that the updated mean is closer to 1.60 than 2, indicating that in this case the directly measured runoff values have less weight than that given by the regression equation. ◀

9.6.3 Correlation Analysis

The dependence between two normal random variables X and Y is measured by the correlation coefficient ρ. Consider a simple case in which the mean and variance of both X and Y are assumed known; the likelihood function of ρ is simply the probability of observing the set of n pairs of values (x_i, y_i), $i=1$ to n, for given $\mu_X, \mu_Y, \sigma_X, \sigma_Y$. Hence

$$L(\rho) = \prod_{i=1}^{n} f_{Y|x}(y_i|x_i) f_X(x_i)$$
$$= \prod_{i=1}^{n} \frac{1}{\sqrt{2\pi}\sigma_{Y|x}} \exp\left\{-\frac{1}{2}\left(\frac{y_i - \mu_{Y|x_i}}{\sigma_{Y|x}}\right)^2\right\} \frac{1}{\sqrt{2\pi}\sigma_X} \exp\left\{-\frac{1}{2}\left(\frac{x_i - \mu_X}{\sigma_X}\right)^2\right\} \tag{9.35}$$

in which

$$\mu_{Y|x_i} = \mu_Y + \rho\frac{\sigma_Y}{\sigma_X}(x_i - \mu_X) \tag{9.36}$$

$$\sigma_{Y|x}^2 = \sigma_Y^2(1 - \rho^2) \tag{9.37}$$

Invoking the Bayesian updating equation, the updated distribution of the correlation coefficient is given by

$$f''(\rho) = kL(\rho)f'(\rho) \tag{9.38}$$

which can be determined for an assumed prior distribution $f'(\rho)$.

Even for this simplified case, the mathematical procedure for carrying out the updating process is analytically cumbersome, as a convenient mathematical form of the prior distribution is not obvious. Often, one has to resort to numerical procedures or approximate methods in obtaining results of interest. For example, Lee (1989) has suggested an approximate method based on the hyperbolic tangent substitution of a variable.

▶ **EXAMPLE 9.16**

The data in Example 9.13 show that there is significant correlation between the precipitation and runoff at the Monocacy River. Suppose prior information on the correlation coefficient is also available. Determine the corresponding updated distribution of the correlation coefficient. Numerical calculations based on Eqs. 9.36 through 9.38 are used for this task. Three separate prior distributions were considered to describe the information on the correlation coefficient before collecting the precipitation-runoff data. They are:

Case (a): Upper triangular distribution between 0 and 1

Case (b): Uniform distribution between 0 and 1

Case (c): Uniform distribution between 0.5 and 1

Results for the updated (posterior) distributions corresponding to each of these three cases are shown in Fig. E9.16. The statistics of the correlation coefficient before and after incorporating the observed data are summarized in Table E9.16. Since the observed data consist of 25 pairs of measured values and the data follow closely a linear trend, the posterior distribution of ρ is dominated by the

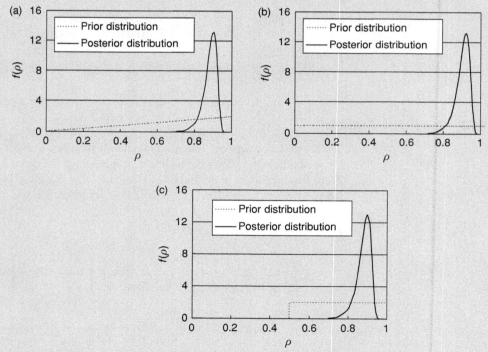

Figure E9.16 Prior and posterior distributions of correlation coefficient.

observed data; in this case, therefore, the different prior assumptions have minimal influence on the updated distribution and statistics.

TABLE E9.16   Prior and Posterior Statistics of Correlation Coefficient

Case	Prior		Posterior	
	μ'_ρ	σ'_ρ	μ''_ρ	σ''_ρ
(a)	0.667	0.056	0.88289	0.001296
(b)	0.50	0.083	0.881315	0.001386
(c)	0.75	0.021	0.881321	0.001383

◀

▶ 9.7   CONCLUDING SUMMARY

In the process of engineering planning and design, judgmental assumptions and inferential information are often useful and necessary. The significance of such prior information and its role (in combination with observational data) in the process of estimation are formally the subject of Bayesian statistics. The basic concepts of the Bayesian approach have been introduced here with special reference to sampling and estimation. These concepts can be extended to numerous applications in engineering, including applications in Bayesian statistical decisions; the latter can be found in Ang and Tang (1984).

Philosophically, there are fundamental differences between the Bayesian and the classical approaches to probability and statistics. Within the Bayesian context, a probability or a probability statement is an expression of the *degree-of-belief*, whereas in the classical sense, probability is (strictly speaking) a measure of relative frequency. Furthermore, in estimation, the Bayesian approach assumes that a parameter is a random variable, whereas in the classical approach it is an unknown constant.

Relative to engineering planning and design, the Bayesian approach may offer the following advantages:

1. It provides the formal framework for incorporating engineering judgment (expressed in probability terms) with observational data.

2. It systematically combines uncertainties associated with randomness (aleatory) and those arising from errors of estimation and prediction (epistemic), as discussed in Chapter 1.

3. It provides a formal procedure for systematically updating information.

▶ PROBLEMS

9.1 A new structure is subjected to proof testing. Assume that the maximum proof load is specified at a reasonably high level so that the calculated probability of the structure surviving the maximum proof load is 0.90. However, it is felt that this calculation is only 70% reliable, and there is a 25% chance that the true probability may be 0.50; moreover, there is even a 5% chance that it may be only 0.10.

(a) What is the expected probability of survival before the proof test?

(b) If only one structure is proof tested, and it survives the maximum proof load, determine the updated distribution of the survival probability.

(c) What is the expected probability of survival after the proof test?

(d) If three structures were proof tested, and two of the structures survived whereas one failed under the maximum proof load, determine the updated expected probability of survival.

9.2 A new waste-treatment process has just been developed. In order to evaluate its effectiveness, the treatment process is installed for a trial period. Each day the output from the treatment process is inspected to see if it satisfies the specified standard. Suppose that the outputs between days are statistically independent, and there is a probability p that the daily output will be acceptable. If the prior PMF is as shown in the figure below, determine the posterior distribution of p with each of the following observations.

(a) The output on the first day of the trial period is of unacceptable quality.

(b) For a 3-day trial period, the quality is unacceptable in only one day.

(c) For a 3-day trial period, the first two days are satisfactory, whereas the quality is unacceptable on the third day.

In each case, determine also the Bayesian estimate for p. (*Ans. 0.536, 0.617, 0.617.*)

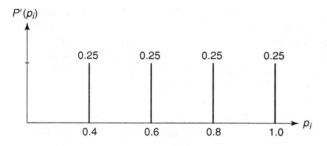

9.3 A hazardous street intersection has been improved by changing the geometric design to reduce the accident and fatality rates. For simplicity, assume that accident and fatality rates can be classified as high H or low L, leading to the following possible conditions: $H_A H_F$ (high accident rate, high fatality rate), $H_A L_F$, $L_A H_F$, and $L_A L_F$. Preliminary evaluation revealed that the relative likelihoods for these four conditions are 3:3:2:2.

An accident rate prediction model, for example, the Tharp model, was used to obtain a better evaluation of the accident potential at this (improved) intersection. Because of possible inaccuracies in the prediction model, a predicted condition may not be actually realized. Furthermore, the probability of a correct prediction depends on the underlying actual condition, as indicated in the following table of conditional probabilities.

Predicted	Actual $H_A H_F$	$H_A L_F$	$L_A H_F$	$L_A L_F$
$H'_A H'_F$	0.30	0.40	0.20	0.25
$H'_A L'_F$	0.30	0.30	0.20	0.25
$L'_A H'_F$	0.20	0.20	0.50	0.25
$L'_A L'_P$	0.20	0.10	0.10	0.25

(a) What is the probability that the model will indicate $H'_A H'_F$?

(b) Suppose that the model predicted $H'_A H'_F$, what is the probability that the condition of the improved intersection will actually be $H_A H_F$?

(c) If the model predicted $L'_A L'_F$, what is the updated relative likelihoods of the four possible conditions?

9.4 An instrument is used to check the accuracy of a set of measurements. However, it can only record three readings, namely, $x = 1, 2,$ or 3. The reading $x = 2$ implies that the previously measured value is within a tolerable error, whereas $x = 1$ and $x = 3$ denote that the measurement is on the low and high side, respectively. Suppose the distribution of the underlying random variable X is as follows:

$$p_X(x_i) = \begin{cases} \dfrac{1-m}{2} & \text{for } x_i = 1 \\ m & \text{for } x_i = 2 \\ \dfrac{1-m}{2} & \text{for } x_i = 3 \end{cases}$$

where m is the parameter. For a particular set of measurements, the engineer estimated that the value of m would be 0.4 or 0.8 with equal likelihood. However, on checking a set of measurements, the first one indicates $x = 2$.

(a) What should be the engineer's revised distribution of m?

(b) Estimate the probability that a least two out of the next three measurements will be accurate.

9.5 An engineer plans to build a log cabin in the middle of a forest where logs of similar size are available. He assumes that the bending capacity M of each log follows a Rayleigh distribution

$$f_M(m) = \frac{m}{\lambda^2} e^{-(1/2)(m/\lambda)^2} \quad m \geq 0$$

where the parameter λ is the modal value of the distribution.

From previous experience with similar logs, he feels that λ would be 4 (kip-ft) with probability 0.4 or 5 (kip-ft) with probability 0.6. Not entirely satisfied with these subjective probabilities, he decides to get a better measure of the parameter λ. Being pressed for time and with limited supply of logs, he can only afford to test the bending capacity of two logs by a simple load test on the site. The test results yielded 4.5 kip-ft and 5.2 kip-ft for the two tests.

(a) Determine the posterior distribution (discrete) for the parameter λ.

(b) Derive the distribution of the bending capacity of the logs M, using the posterior distribution of λ.

(c) What is the probability that M will be less than 2 kip-ft?

9.6 The absolute error E (in cm) of each measurement from a surveying instrument is governed by the triangular distribution shown in the figure below, where α denotes the upper limit of the error.

Two measurements were made, and the errors are 1 and 2 cm, respectively.

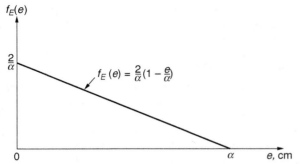

$f_E(e)$

$\frac{2}{\alpha}$

$f_E(e) = \frac{2}{\alpha}\left(1 - \frac{e}{\alpha}\right)$

α e, cm

0

(a) Suppose that α is assumed to be 2 or 3 cm with equal likelihood prior to the two measurements; determine the updated distribution of α. Estimate the value of α based on this updated distribution.

(b) Now suppose that the prior PDF of α is uniform between 2 and 3; determine and plot the updated distribution of α, and evaluate the corresponding Bayesian estimate for α.

(c) Repeat Part (b) if the PDF of α is a triangular distribution between 2 and 3 with the mode at 2.

9.7 Suppose that the prior PDF of the mean accident rate v in Example 9.2 is

$$f'(v) = \frac{0.271}{v} \qquad 0.5 \le v \le 20$$
$$= 0 \qquad \text{elsewhere}$$

(a) Determine the posterior density function of v based on the observation that an accident was recorded during the first month of operation.

(b) Compare the probability of an "accident free" month before and after the observation in Part (a).

9.8 In Problem 9.1, suppose that the survival probability has a prior density function as follows.

  (i) Uniform between $p = 0$ to 0.9.

  (ii) Uniform between $p = 0.9$ to 1.0.

  (iii) It is more likely that p will exceed 0.9 than be less than 0.9; the relative likelihood between these two possibilities is 7 to 3.

(a) Determine the prior density function of p.

(b) If three structures were proof tested, and all three survived the maximum proof load, determine and plot the posterior PDF of p.

(c) What is the estimated value of p in light of the results in Part (b)?

9.9 The occurrence of fires in a city may be modeled by a Poisson process. Suppose the average occurrences of fires v is assumed to be 15 or 20 times a year; the likelihood of 20 fires a year (on the average) is twice that of 15 fires a year.

(a) Determine the probability that there will be 20 occurrences of fire in the next year.

(b) If there are exactly 20 fires in the next year, what will be the updated PMF of v?

(c) What is the probability that there will be 20 occurrences of fire in the year after next, in the light of Part (b)?

9.10 For a certain type of construction project in a hazardous environment, the time T (in days) between two successive accidents follows an exponential distribution,

$$f_T(t) = \lambda e^{-t\lambda} \qquad t \ge 0$$

where the parameter λ depends on the quality of supervision at the site. It is believed that λ could be either 1/5 or 1/10 with relative likelihood of 1:2. For a particular site, accidents were reported on days 2 and 5, respectively, from the commencement date of the project.

(a) What is the updated probability that λ is 1/10?

(b) What is the probability that no accidents will occur for at least 10 days after the second accident?

(c) Suppose, after the second accident, the construction project has entered its 15th day from commencement without having a third accident. What is the probability that λ is 1/10 now?

(d) For the next 10 days of construction (i.e., between day 15 and day 25), if the number of accidents exceeds a certain number (n_C), then one would be more inclined towards the belief that λ is 1/5 than 1/10. Determine this critical number n_C.

(e) Rather than the discrete values 1/5 and 1/10, suppose λ can actually take on continuous values. Use only the observed information given prior to Part (a) (i.e., accidents on days 2 and 5) to determine the PDF for λ.

9.11 Holes can occur in geotextiles, and their occurrences may be modeled by a Poisson process with a mean rate of M per acre. Suppose M has a mean of 10 per acre and a c.o.v. of 50% based on previous experience with this kind of fabric. Inspection of 0.6 acre of the material at a given site has revealed only one hole.

(a) In light of the inspection result, determine the updated mean and c.o.v. of M for the fabric at the site.

(b) What is the probability that there will not be any holes in the remaining 0.4 acre of this piece of synthetic fabric?

9.12 Consider a case where the mean compression index of a soil sample is $N(\mu, \sigma)$ and σ is assumed to be equal to 0.16. Laboratory tests on four specimens show the following compression index values: 0.75, 0.89, 0.91, and 0.81.

(a) What is the posterior distribution of μ if there is no other information except the observed data?

(b) Suppose there is prior information to indicate that μ is Gaussian with mean of 0.8 and a c.o.v. of 25%. What will be the posterior distribution of μ if this prior information is taken into account?

(c) What is the probability that μ will be less than 0.95, using the data from Part (b)? (*Ans. 0.938*)

9.13 An air passenger commutes between San Francisco and Los Angeles regularly. Lately, he started recording the time of each flight. He computed the average flight time from his five previous trips to be 65 min. Suppose the flight time T is a Gaussian random variable with known standard deviation of 10 min.

(a) Based purely on the data, what is the posterior distribution of μ_T?

(b) The passenger is now on a plane from San Francisco to Los Angeles. By coincidence, the passenger sitting next to him also has been keeping track of the flight time. From his record of 10 previous trips, he obtained an average of 60 min. Assume that these two passengers have never before taken a plane together. With this additional information, what would be the updated distribution of μ_T?

(c) What is the probability that their current flight will take more than 80 min?

9.14 Six measurements were made of an angle as follows:

$$
\begin{array}{cc}
32°04' & 32°05' \\
31°59' & 31°57' \\
32°01' & 32°00' \\
\end{array}
$$

Assume that the measurement error is Gaussian with zero mean; and the standard deviation of each measurement can be represented by the sample standard deviation of the six measurements above.

(a) Estimate the angle.

(b) Subsequently, the engineer discovered that the angle has been measured before and recorded as $32°00' \pm 2'$. Estimate the actual angle using both sets of measurements.

9.15 A distance L is measured independently by three surveyors with three sets of instruments. The respective measurements are 2.15, 2.20, and 2.18 km. Suppose the ratio of the standard error of the three measurements is 1:2:3. Estimate the actual distance L on the basis of the three sets of measurements. Assume that measurement error is Gaussian with zero mean.

9.16 In designing reinforced concrete structural members to resist ultimate loads, a capacity reduction factor ϕ is often used. Suppose that the structural member is a beam element and is designed for pure flexure. The conventional value of ϕ is 0.9. However, a committee is investigating the effect on the probability of failure of beams against ultimate load if ϕ is increased to 0.95. Twelve beams are designed using $\phi = 0.95$, and each of them is subjected to the designed ultimate load in the laboratory. It is desired to estimate the probability of failure, p.

Suppose that prior experiences show that p has a mean of 0.1 and a standard deviation of 0.06. The laboratory test results show that one out of 12 beams tested failed the ultimate load. Suggest a suitable prior distribution and determine the mean and variance of p from these data.

9.17 A newly developed construction material was tested for its fire resistance. Results from testing two samples show that one had serious fire damage after being exposed to only 2 hr of fire, whereas the other lasted for 3 hr. Suppose the duration of fire, T, for damage of a material follows a normal distribution $N(\mu, 0.5)$. Before the tests, the mean duration until damage, μ, for this material was expected to be 3 hr with a c.o.v. of 20% based on a theoretical study.

(a) Determine the updated mean and c.o.v. of the mean duration until damage for this material.

(b) With the above observed information, what is the probability that a wall constructed using this material will be damaged in less than 1 hr of fire?

9.18 A contractor is trying out a new markup policy in preparing his bids. With his new strategy, he expects that the probability of winning a given job, p, is 0.5 with a variance of 0.05.

(a) Determine a suitable PDF for the probability p.

(b) Suppose the contractor has submitted bids for four jobs using this new markup strategy. The outcome of the first two jobs revealed that he has won one but lost the other. What would be the updated distribution of p?

(c) What is the probability that the contractor will win the remaining two jobs?

9.19 The time between breakdowns of a certain type of construction equipment follows an exponential distribution with mean $1/\lambda$ where the mean rate of failure λ was rated by the manufacturer to have a mean of 0.5 per year and a c.o.v. of 20%. A contractor owns two pieces of this construction equipment. The operational times until breakdown of the equipments were subsequently observed to be 12 and 18 months, respectively. Determine the updated mean and c.o.v. of the parameter by using the conjugates approach.

9.20 Cracks are observed on a concrete bridge deck after an earthquake. In order to assess the condition of the bridge, the sizes of cracks are measured. Suppose six cracks were measured and their lengths are as follows:

Crack Number	Crack Length (cm)
1	3
2	5
3	at most 6
4	4
5	at least 4
6	8

Suppose the length of a crack follows the exponential distribution with parameter λ as follows:

$$ f_X(x) = \lambda e^{-\lambda x} $$

Assume that there is no other information on the crack length. Determine the distribution (PDF) of the parameter λ.

9.21 An engineer has moved to a new town. He discovered that the town lies within a tornado zone. The information he has gathered for tornadoes in this town is as follows:

(i) Two tornadoes were observed within the last 20 yr.

(ii) The maximum wind speeds of these two tornadoes were 190 and 160 kph, respectively.

(iii) From an independent source, it is believed that there is a 95% probability that the mean maximum wind speed for tornadoes in the region will be between 145 and 175 kph.

(iv) The standard deviation of maximum wind speed may be assumed to be 15 kph.

(v) The occurrence of tornadoes may be assumed to follow a Poisson process.

(vi) The maximum wind speed in a tornado may be assumed to follow a normal distribution.

Assume that no other information is available. The engineer intends to live in this town for 5 years, and he estimates that his house should be strong enough to withstand a 120 kph wind without damage.

Answer the following questions:

(a) What is the probability that the town will be hit by a tornado during the next 5 years?

(b) What is the probability that the maximum wind speed of the next tornado will exceed 220 kph?

(c) What is the probability that his house will be undamaged during the next 5-yr period?

9.22 A given type of accident during the operation of a nuclear power plant may involve the leakage of radioactive material, which could be hazardous to the immediate neighborhood. Let X be the distance (in km) from the power plant beyond which health hazards will be below an acceptable level in the event of an accident. The PDF of X is uniformly distributed as

$$f_X(x) = \frac{1}{h} \qquad 0 < x < h$$

in which h is the parameter, denoting the maximum possible dispersion distance of the radioactive material. Based on theoretical studies, a prior distribution of h has been established as

$$f'_H(h) = 0.003\,h^2 \qquad 0 < h < 10$$

(a) Determine the expected distance of the hazardous zone based on the given information. (*Ans. 3.75*)

(b) Suppose an accident has occurred and the hazardous zone was observed to be extending to a distance of 4 km. What is the range of h now?

(c) Determine the updated distribution of H. (*Ans.* $f''_H = 0.023h$; $4 < h < 10$)

(d) What is the probability that the hazardous zone will not exceed 4 km in the next accident? (*Ans: 0.571*)

9.23 Grouted soil nails are often used to strengthen soil slopes against stability failure. However, full grouting along the entire length of the nail is generally not achieved in certain cases. On a given project, an engineer estimates the average fraction of length grouted to be within 0.75 to 0.95 with 95% confidence, based on his previous experiences with the particular geologic formation and construction crew. To improve his estimate, the engineer tested a sample of four soil nails for the given job. The average grouted length measured is 0.89, while the standard deviation of measured grouted length is 0.06.

(a) Determine the updated mean and c.o.v. of the average grouted length of soil nails for the site. State any assumptions used.

(b) Final approval of the construction job requires inspection and testing by the regulatory agency. Suppose a sample of two nails will be selected and tested by the agency. Acceptance requires that each of the following two conditions be satisfied.

1. The sample average grouted length must be at least 0.85.

2. The minimum grouted lengths of nails tested should not be lower than 0.83.

(c) Which condition does the engineer think is more stringent (less likely) to be met for this site?

9.24 For Problem 9.13, suppose that the standard deviation of T is not known, and the sample standard deviation from the 5 trips is 10 minutes. Based solely on the observed data, determine the posterior statistics of μ_T and σ_T.

9.25 Consider the data given in Problem 10 of Chapter 8 for the rated and actual mileage of 6 cars. By using the Bayesian regression approach, what is the probability that a car with rated mileage of 24 mpg will have an actual mileage less than 18 mpg? How would this probability be changed if the value of the basic scatter σ^2 can be assumed equal to 1.73 mpg?

▶ REFERENCES

Ang, A. H-S., and Tang, W., *Probability Concepts in Engineering Planning and Design—Decision Risk and Reliability*, Vol II. John Wiley & Sons, New York, 1984.

Benjamin, J. R. "Probabilistic Model for Seismic Force Design." *Jour. of Structural Division*, ASCE, Vol. 94, May 1968, pp. 1175–1196.

Halim, I. S., and Tang, W. H. "Bayesian Method for Characterization of Geological Anomaly," *Proceedings, ISUMA 1990, The First International Symposium on Uncertainty Modeling and Analysis*, Maryland, pp. 585–594, 1990.

Halim, I. S., and Tang, W. H. "Reliability of Undrained Clay Slope Considering Geologic Anomaly," *Proceedings, ICASP*, 1991, Mexico City, pp. 776–783.

Halim, I. S., and Tang, W. H. "Site Exploration Strategy for Geologic Anomaly Characterization," *Jour. of Geotechnical Engineering,*

ASCE, Vol. 119, No. GT2, pp. 195–213, February 1993.

Lee, Peter M., *Bayesian Statistics: An Introduction*, Edward Arnold Publisher, London, 1989.

Packman, P. F., Pearson, H. S., Owens, J. S., and Marchese, G. B., "The Applicability of a Fracture Mechanics—Nondestructive Testing Design Criteria." *Technical Report, AFML-TR-68-32*, Air Force Materials Laboratory, Wright-Patterson Air Force Base, Ohio, May 1968.

Raiffa, H., and Schlaifer, R. *Applied Statistical Decision Theory*. Division of Research, Harvard Business School, Boston, MA., 1961.

Tang, W. H., "A Bayesian Evaluation of Information for Foundation Engineering Design," *Proc., 1st International Conference on Applications of Statistics and Probability*, Hong Kong Univ. Press, Sept. 1971, pp. 173–185.

Tang, W. H. "Bayesian Frequency Analysis," *Jour. of the Hydraulics Division, ASCE*, Vol. 106, No. HY7, 1980, pp. 1203–1218.

Tang, W. H., and Saadeghvaziri, A. "Updating Distribution of Anomaly Size and Fraction," in *Recent Advances in Engineering Mechanics and Their Impact on Civil Engineering Practice*, Vol. II, edited by W. F. Chen and A. D. M. Lewis, 1983, pp. 895–898.

Tang, W. H., and Quek, S. T. "Statistical Model of Boulder Size and Fraction," *Jour. of Geotechnical Engineering, ASCE*, Vol. 1, 1986.

Probability Tables

TABLE A.1 Standard Normal Probabilities (pg. 1 of 4) $\Phi(x) = \dfrac{1}{\sqrt{2\pi}} \displaystyle\int_{-\infty}^{x} e^{-y^2/2} dy$

x	$\Phi(x)$	x	$\Phi(x)$	x	$\Phi(x)$
0.00	0.500000000	0.40	0.655421742	0.80	0.788144601
0.01	0.503989356	0.41	0.659097026	0.81	0.791029912
0.02	0.507978314	0.42	0.662757273	0.82	0.793891946
0.03	0.511966473	0.43	0.666402179	0.83	0.796730608
0.04	0.515953437	0.44	0.670031446	0.84	0.799545807
0.05	0.519938806	0.45	0.673644780	0.85	0.802337457
0.06	0.523922183	0.46	0.677241890	0.86	0.805105479
0.07	0.527903170	0.47	0.680822491	0.87	0.807849798
0.08	0.531881372	0.48	0.684386303	0.88	0.810570345
0.09	0.535856393	0.49	0.687933051	0.89	0.813267057
0.10	0.539827837	0.50	0.691462461	0.90	0.815939875
0.11	0.543795313	0.51	0.694974269	0.91	0.818588745
0.12	0.547758426	0.52	0.698468212	0.92	0.821213602
0.13	0.551716787	0.53	0.701944035	0.93	0.823814458
0.14	0.555670005	0.54	0.705401484	0.94	0.826391220
0.15	0.559617692	0.55	0.708840313	0.95	0.828943874
0.16	0.563559463	0.56	0.712260281	0.96	0.831472393
0.17	0.567494932	0.57	0.715661151	0.97	0.833976754
0.18	0.571423716	0.58	0.719042691	0.98	0.836456941
0.19	0.575345435	0.59	0.722404675	0.99	0.838912940
0.20	0.579259709	0.60	0.725746882	1.00	0.841344746
0.21	0.583166163	0.61	0.729069096	1.01	0.843752355
0.22	0.587064423	0.62	0.732371107	1.02	0.846135770
0.23	0.590954115	0.63	0.735652708	1.03	0.848494997
0.24	0.594834872	0.64	0.738913700	1.04	0.850830050
0.25	0.598706326	0.65	0.742153889	1.05	0.853140944
0.26	0.602568113	0.66	0.745373085	1.06	0.855427700
0.27	0.606419873	0.67	0.748571105	1.07	0.857690346
0.28	0.610261248	0.68	0.751747770	1.08	0.859928910
0.29	0.614091881	0.69	0.754902906	1.09	0.862143428
0.30	0.617911422	0.70	0.758036348	1.10	0.864333939
0.31	0.621719522	0.71	0.761147932	1.11	0.866500487
0.32	0.625515835	0.72	0.764237502	1.12	0.868643119
0.33	0.629300019	0.73	0.767304908	1.13	0.870761888
0.34	0.633071736	0.74	0.770350003	1.14	0.872856849
0.35	0.636830651	0.75	0.773372648	1.15	0.874928064
0.36	0.640576433	0.76	0.776372708	1.16	0.876975597
0.37	0.644308755	0.77	0.779350054	1.17	0.878999516
0.38	0.648027292	0.78	0.782304562	1.18	0.880999893
0.39	0.651731727	0.79	0.785236116	1.19	0.882976804

TABLE A.1 **Standard Normal Probabilities (pg. 2 of 4)** $\Phi(x) = \dfrac{1}{\sqrt{2\pi}} \displaystyle\int_{-\infty}^{x} e^{-y^2/2}dy$

x	$\Phi(x)$	x	$\Phi(x)$	x	$\Phi(x)$
1.20	0.884930330	1.70	0.955434537	2.20	0.986096552
1.21	0.886860554	1.71	0.956367063	2.21	0.986447419
1.22	0.888767563	1.72	0.957283779	2.22	0.986790616
1.23	0.890651448	1.73	0.958184862	2.23	0.987126279
1.24	0.892512303	1.74	0.959070491	2.24	0.987454539
1.25	0.894350226	1.75	0.959940843	2.25	0.987775527
1.26	0.896165319	1.76	0.960796097	2.26	0.988089375
1.27	0.897957685	1.77	0.961636430	2.27	0.988396208
1.28	0.899727432	1.78	0.962462020	2.28	0.988696156
1.29	0.901474671	1.79	0.963273044	2.29	0.988989342
1.30	0.903199515	1.80	0.964069681	2.30	0.989275890
1.31	0.904902082	1.81	0.964852106	2.31	0.989555923
1.32	0.906582491	1.82	0.965620498	2.32	0.989829561
1.33	0.908240864	1.83	0.966375031	2.33	0.990096924
1.34	0.909877328	1.84	0.967115881	2.34	0.990358130
1.35	0.911492009	1.85	0.967843225	2.35	0.990613294
1.36	0.913085038	1.86	0.968557237	2.36	0.990862532
1.37	0.914656549	1.87	0.969258091	2.37	0.991105957
1.38	0.916206678	1.88	0.969945961	2.38	0.991343681
1.39	0.917735561	1.89	0.970621002	2.39	0.991575814
1.40	0.919243341	1.90	0.971283440	2.40	0.991802464
1.41	0.920730159	1.91	0.971933393	2.41	0.992023740
1.42	0.922196159	1.92	0.972571050	2.42	0.992239746
1.43	0.923641490	1.93	0.973196581	2.43	0.992450589
1.44	0.925066300	1.94	0.973810155	2.44	0.992656369
1.45	0.926470740	1.95	0.974411940	2.45	0.992857189
1.46	0.927854963	1.96	0.975002105	2.46	0.993053149
1.47	0.929219123	1.97	0.975580815	2.47	0.993244347
1.48	0.930563377	1.98	0.976148236	2.48	0.993430881
1.49	0.931887882	1.99	0.976704532	2.49	0.993612845
1.50	0.933192799	2.00	0.977249868	2.50	0.993790335
1.51	0.934478288	2.01	0.977784406	2.51	0.993963442
1.52	0.935744512	2.02	0.978308306	2.52	0.994132258
1.53	0.936991636	2.03	0.978821730	2.53	0.994296874
1.54	0.938219823	2.04	0.979324837	2.54	0.994457377
1.55	0.939429242	2.05	0.979817785	2.55	0.994613854
1.56	0.940620059	2.06	0.980300730	2.56	0.994766392
1.57	0.941792444	2.07	0.980773828	2.57	0.994915074
1.58	0.942946567	2.08	0.981237234	2.58	0.995059984
1.59	0.944082597	2.09	0.981691100	2.59	0.995201203
1.60	0.945200708	2.10	0.982135579	2.60	0.995338812
1.61	0.946301072	2.11	0.982570822	2.61	0.995472889
1.62	0.947383862	2.12	0.982996977	2.62	0.995603512
1.63	0.948449252	2.13	0.983414193	2.63	0.995730757
1.64	0.949497417	2.14	0.983822617	2.64	0.995854699
1.65	0.950528532	2.15	0.984222393	2.65	0.995975411
1.66	0.951542774	2.16	0.984613665	2.66	0.996092967
1.67	0.952540318	2.17	0.984996577	2.67	0.996207438
1.68	0.953521342	2.18	0.985371269	2.68	0.996318892
1.69	0.954486023	2.19	0.985737882	2.69	0.996427399

TABLE A.1 Standard Normal Probabilities (pg. 3 of 4) $\Phi(x) = \dfrac{1}{\sqrt{2\pi}} \displaystyle\int_{-\infty}^{x} e^{-y^2/2}\,dy$

x	$\Phi(x)$	x	$\Phi(x)$	x	$\Phi(x)$
2.70	0.996533026	3.20	0.999312862	3.70	0.999892200
2.71	0.996635840	3.21	0.999336325	3.71	0.999896370
2.72	0.996735904	3.22	0.999359047	3.72	0.999900389
2.73	0.996833284	3.23	0.999381049	3.73	0.999904260
2.74	0.996928041	3.24	0.999402352	3.74	0.999907990
2.75	0.997020237	3.25	0.999422975	3.75	0.999911583
2.76	0.997109932	3.26	0.999442939	3.76	0.999915043
2.77	0.997197185	3.27	0.999462263	3.77	0.999918376
2.78	0.997282055	3.28	0.999480965	3.78	0.999921586
2.79	0.997364598	3.29	0.999499063	3.79	0.999924676
2.80	0.997444807	3.30	0.999516576	3.80	0.999927652
2.81	0.997522925	3.31	0.999533520	3.81	0.999930517
2.82	0.997598818	3.32	0.999549913	3.82	0.999933274
2.83	0.997672600	3.33	0.999565770	3.83	0.999935928
2.84	0.997744323	3.34	0.999581108	3.84	0.999938483
2.85	0.997814039	3.35	0.999595942	3.85	0.999940941
2.86	0.997881795	3.36	0.999610288	3.86	0.999943306
2.87	0.997947641	3.37	0.999624159	3.87	0.999945582
2.88	0.998011624	3.38	0.999637571	3.88	0.999947772
2.89	0.998073791	3.39	0.999650537	3.89	0.999949878
2.90	0.998134187	3.40	0.999663071	3.90	0.999951904
2.91	0.998192856	3.41	0.999675186	3.91	0.999953852
2.92	0.998249843	3.42	0.999686894	3.92	0.999955726
2.93	0.998305190	3.43	0.999698209	3.93	0.999957527
2.94	0.998358939	3.44	0.999709143	3.94	0.999959259
2.95	0.998411130	3.45	0.999719707	3.95	0.999960924
2.96	0.998461805	3.46	0.999729912	3.96	0.999962525
2.97	0.998511001	3.47	0.999739771	3.97	0.999964064
2.98	0.998558758	3.48	0.999749293	3.98	0.999965542
2.99	0.998605113	3.49	0.999758490	3.99	0.999966963
3.00	0.998650102	3.50	0.999767371		
3.01	0.998693762	3.51	0.999775947		
3.02	0.998736127	3.52	0.999784227		
3.03	0.998777231	3.53	0.999792220		
3.04	0.998817109	3.54	0.999799936		
3.05	0.998855793	3.55	0.999807384		
3.06	0.998893315	3.56	0.999814573		
3.07	0.998929706	3.57	0.999821509		
3.08	0.998964997	3.58	0.999828203		
3.09	0.998999218	3.59	0.999834661		
3.10	0.999032397	3.60	0.999840891		
3.11	0.999064563	3.61	0.999846901		
3.12	0.999095745	3.62	0.999852698		
3.13	0.999125968	3.63	0.999858289		
3.14	0.999155261	3.64	0.999863681		
3.15	0.999183648	3.65	0.999868880		
3.16	0.999211154	3.66	0.999873892		
3.17	0.999237805	3.67	0.999878725		
3.18	0.999263625	3.68	0.999883383		
3.19	0.999288636	3.69	0.999887873		

TABLE A.1   **Standard Normal Probabilities (pg. 4 of 4)** $\Phi(x) = \dfrac{1}{\sqrt{2\pi}} \displaystyle\int_{-\infty}^{x} e^{-y^2/2} dy$

x	$1 - \Phi(x)$	x	$1 - \Phi(x)$	x	$1 - \Phi(x)$
4.0	3.17E-05	6.0	9.87E-10	8.0	6.66E-16
4.1	2.07E-05	6.1	5.30E-10	8.1	2.22E-16
4.2	1.33E-05	6.2	2.82E-10	8.2	1.11E-16
4.3	8.54E-06	6.3	1.49E-10	8.3	0
4.4	5.41E-06	6.4	7.77E-11	8.4	0
4.5	3.40E-06	6.5	4.02E-11	8.5	0
4.6	2.11E-06	6.6	2.06E-11	8.6	0
4.7	1.30E-06	6.7	1.04E-11	8.7	0
4.8	7.93E-07	6.8	5.23E-12	8.8	0
4.9	4.79E-07	6.9	2.60E-12	8.9	0
5.0	2.87E-07	7.0	1.28E-12		
5.1	1.70E-07	7.1	6.24E-13		
5.2	9.96E-08	7.2	3.01E-13		
5.3	5.79E-08	7.3	1.44E-13		
5.4	3.33E-08	7.4	6.81E-14		
5.5	1.90E-08	7.5	3.19E-14		
5.6	1.07E-08	7.6	1.48E-14		
5.7	5.99E-09	7.7	6.77E-15		
5.8	3.32E-09	7.8	3.11E-15		
5.9	1.82E-09	7.9	1.44E-15		

TABLE A.2 CDF of the Binomial Distribution $\displaystyle\sum_{k=0}^{x}\binom{n}{k}p^k(1-p)^{n-k}$

p	n=2, x=0	2, 1	3, 0	3, 1	3, 2	4, 0	4, 1	4, 2	4, 3	5, 0	5, 1	5, 2	5, 3	5, 4
0.01	0.9801	0.9999	0.9703	0.9997	1	0.9606	0.9994	1	1	0.9510	0.9990	1	1	1
0.05	0.9025	0.9975	0.8574	0.9927	0.9999	0.8145	0.9860	0.9995	1	0.7738	0.9774	0.9988	1	1
0.10	0.8100	0.9900	0.7290	0.972	0.999	0.6561	0.9477	0.9963	0.9999	0.5905	0.9185	0.9914	0.9995	1
0.15	0.7225	0.9775	0.6141	0.9393	0.9966	0.522	0.8905	0.9880	0.9995	0.4437	0.8352	0.9734	0.9978	0.9999
0.20	0.6400	0.9600	0.5120	0.8960	0.9920	0.4096	0.8192	0.9728	0.9984	0.3277	0.7373	0.9421	0.9933	0.9997
0.25	0.5625	0.9375	0.4219	0.8438	0.9844	0.3164	0.7383	0.9492	0.9961	0.2373	0.6328	0.8965	0.9844	0.9990
0.30	0.4900	0.9100	0.3430	0.7840	0.9730	0.2401	0.6517	0.9163	0.9919	0.1681	0.5282	0.8369	0.9692	0.9976
0.35	0.4225	0.8775	0.2746	0.7182	0.9571	0.1785	0.5630	0.8735	0.9850	0.1160	0.4284	0.7648	0.9460	0.9947
0.40	0.3600	0.8400	0.2160	0.6480	0.9360	0.1296	0.4752	0.8208	0.9744	0.0778	0.3370	0.6826	0.9130	0.9898
0.45	0.3025	0.7975	0.1664	0.5748	0.9089	0.0915	0.3910	0.7585	0.9590	0.0503	0.2562	0.5931	0.8688	0.9815
0.50	0.2500	0.7500	0.1250	0.5000	0.8750	0.0625	0.3125	0.6875	0.9375	0.0313	0.1875	0.5000	0.8125	0.9687
0.55	0.2025	0.6975	0.0911	0.4253	0.8336	0.0410	0.2415	0.6090	0.9085	0.0185	0.1312	0.4069	0.7438	0.9497
0.60	0.1600	0.6400	0.0640	0.3520	0.7804	0.0256	0.1792	0.5248	0.8704	0.0102	0.0870	0.3174	0.6630	0.9222
0.65	0.1225	0.5775	0.0429	0.2818	0.7254	0.0150	0.1265	0.4370	0.8215	0.0053	0.0540	0.2352	0.5716	0.8840
0.70	0.0900	0.5100	0.0270	0.2160	0.6570	0.0081	0.0837	0.3483	0.7599	0.0024	0.0308	0.1631	0.4718	0.8319
0.75	0.0625	0.4375	0.0156	0.1563	0.5781	0.0039	0.0508	0.2617	0.6836	0.0010	0.0156	0.1035	0.3672	0.7627
0.80	0.0400	0.3600	0.0080	0.1040	0.4880	0.0016	0.0272	0.1808	0.5904	0.0003	0.0067	0.0579	0.2627	0.6723
0.85	0.0225	0.2775	0.0034	0.0608	0.3859	0.0005	0.0120	0.1095	0.4780	0.0001	0.0022	0.0266	0.1648	0.5563
0.90	0.0100	0.1900	0.0010	0.0280	0.2710	0.0001	0.0037	0.0523	0.3439	0	0.0005	0.0086	0.0815	0.4095
0.95	0.0025	0.0975	0.0001	0.0073	0.1426	0	0.0005	0.0140	0.1855	0	0	0.0012	0.0226	0.2262
0.99	0	0.0199	0	0	0.0297	0	0	0.0006	0.0394	0	0	0	0.0010	0.0490

(Continued)

TABLE A.2 (*Continued*)

p	n = 6, x = 0	6, 1	6, 2	6, 3	6, 4	6, 5	8, 0	8, 1	8, 2	8, 3	8, 4	8, 5	8, 6	8, 7
0.01	0.9415	0.9985	1	1	1	1	0.9227	0.9973	0.9999	1	1	1	1	1
0.05	0.7351	0.9672	0.9978	0.9999	1	1	0.6634	0.9428	0.9942	0.9996	1	1	1	1
0.10	0.5314	0.8857	0.9842	0.9987	0.9999	1	0.4305	0.8131	0.9619	0.995	0.9996	1	1	1
0.15	0.3771	0.7765	0.9527	0.9941	0.9996	1	0.2725	0.6572	0.8948	0.9786	0.9971	0.9998	1	1
0.20	0.2621	0.6554	0.9011	0.983	0.9984	0.9999	0.1678	0.5033	0.7969	0.9437	0.9896	0.9988	0.9999	1
0.25	0.178	0.5339	0.8306	0.9624	0.9954	0.9998	0.1001	0.3671	0.6785	0.8862	0.9727	0.9958	0.9996	1
0.30	0.1176	0.4202	0.7443	0.9295	0.9891	0.9993	0.0576	0.2553	0.5518	0.8059	0.942	0.9887	0.9987	0.9999
0.35	0.0754	0.3191	0.6471	0.8826	0.9777	0.9982	0.0319	0.1691	0.4278	0.7064	0.8939	0.9747	0.9964	0.9998
0.40	0.0467	0.2333	0.5443	0.8208	0.959	0.9959	0.0168	0.1064	0.3154	0.5941	0.8263	0.9502	0.9915	0.9993
0.45	0.0277	0.1636	0.4415	0.7447	0.9308	0.9917	0.0084	0.0632	0.2201	0.477	0.7396	0.9115	0.9819	0.9983
0.50	0.0156	0.1094	0.3438	0.6563	0.8906	0.9844	0.0039	0.0352	0.1445	0.3633	0.6367	0.8555	0.9648	0.9961
0.55	0.0083	0.0692	0.2553	0.5585	0.8364	0.9723	0.0017	0.0181	0.0885	0.2604	0.523	0.7799	0.9368	0.9916
0.60	0.0041	0.041	0.1792	0.4557	0.7667	0.9533	0.0007	0.0085	0.0498	0.1737	0.4059	0.6846	0.8936	0.9832
0.65	0.0018	0.0223	0.1174	0.3529	0.6809	0.9246	0.0002	0.0036	0.0253	0.1061	0.2936	0.5722	0.8309	0.9681
0.70	0.0007	0.0109	0.0705	0.2557	0.5798	0.8824	0.0001	0.0013	0.0113	0.058	0.1941	0.4482	0.7447	0.9424
0.75	0.0002	0.0046	0.0376	0.1694	0.4661	0.822		0.0004	0.0042	0.0273	0.1138	0.3215	0.6329	0.8999
0.80	0.0001	0.0016	0.017	0.0989	0.3446	0.7379		0.0001	0.0012	0.0104	0.0563	0.2031	0.4967	0.8322
0.85	0	0.0004	0.0059	0.0473	0.2235	0.6229	0	0	0.0002	0.0029	0.0214	0.1052	0.3428	0.7275
0.90	0	0.0001	0.0013	0.0159	0.1143	0.4686	0	0	0	0.0004	0.005	0.0381	0.1869	0.5695
0.95	0	0	0.0001	0.0022	0.0328	0.2649	0	0	0	0	0.0004	0.0058	0.0572	0.3366
0.99	0	0	0	0	0.0015	0.0585	0	0	0	0	0	0	0.0027	0.0773

(*Continued*)

TABLE A.2 *(Continued for n = 10)*

p	x = 0	1	2	3	4	5	6	7	8	9
0.01	0.9044	0.9957	0.9999	0.9999	1	1	1	1	1	1
0.05	0.5987	0.9139	0.9885	0.999	0.9999	1	1	1	1	1
0.10	0.3487	0.7361	0.9298	0.9872	0.9984	0.9999	1	1	1	1
0.15	0.1969	0.5443	0.8202	0.95	0.9901	0.9986	0.9999	1	1	1
0.20	0.1074	0.3758	0.6778	0.8791	0.9672	0.9936	0.9991	0.9999	1	1
0.25	0.0563	0.244	0.5256	0.7759	0.9219	0.9803	0.9965	0.9996	1	1
0.30	0.0282	0.1493	0.3828	0.6496	0.8497	0.9527	0.9894	0.9984	0.9999	1
0.35	0.0135	0.086	0.2616	0.5138	0.7515	0.9051	0.974	0.9952	0.9995	1
0.40	0.006	0.0464	0.1673	0.3823	0.6331	0.8338	0.9452	0.9877	0.9983	0.9999
0.45	0.0025	0.0233	0.0996	0.266	0.5044	0.7384	0.898	0.9726	0.9955	0.9997
0.50	0.001	0.0107	0.0547	0.1719	0.377	0.623	0.8281	0.9453	0.9893	0.999
0.55	0.0003	0.0045	0.0274	0.102	0.2616	0.4956	0.734	0.9004	0.9767	0.9975
0.60	0.0001	0.0017	0.0123	0.0548	0.1662	0.3669	0.6177	0.8327	0.9536	0.994
0.65	0	0.0005	0.0048	0.026	0.0949	0.2485	0.4862	0.7384	0.914	0.9865
0.70	0	0.0001	0.0016	0.0106	0.0473	0.1503	0.3504	0.6172	0.8507	0.9718
0.75	0	0	0.0004	0.0035	0.0197	0.0781	0.2241	0.4744	0.756	0.9437
0.80	0	0	0.0001	0.0009	0.0064	0.0328	0.1209	0.3222	0.6242	0.8926
0.85	0	0	0	0.0001	0.0014	0.0099	0.05	0.1798	0.4557	0.8031
0.90	0	0	0	0	0.0001	0.0016	0.0128	0.0702	0.2639	0.6513
0.95	0	0	0	0	0	0.0001	0.001	0.0115	0.0861	0.4013
0.99	0	0	0	0	0	0	0	0.0001	0.0043	0.0956

(Continued)

TABLE A.2 (*Continued for n = 15*)

p	x = 0	1	2	3	4	5	6	7	8	9	10	11	12	13	14
0.01	0.8601	0.9904	0.9996	1	1	1	1	1	1	1	1	1	1	1	1
0.05	0.4633	0.829	0.9638	0.9945	0.9994	0.9999	1	1	1	1	1	1	1	1	1
0.10	0.2059	0.549	0.8159	0.9444	0.9873	0.9978	0.9997	1	1	1	1	1	1	1	1
0.15	0.0874	0.3186	0.6042	0.8227	0.9383	0.9832	0.9964	0.9994	0.9999	1	1	1	1	1	1
0.20	0.0352	0.1671	0.398	0.6482	0.8358	0.9389	0.9819	0.9958	0.9992	0.9999	1	1	1	1	1
0.25	0.0134	0.0802	0.2361	0.4613	0.6865	0.8516	0.9434	0.9827	0.9958	0.9992	0.9999	1	1	1	1
0.30	0.0047	0.0353	0.1268	0.2969	0.5155	0.7216	0.8689	0.95	0.9848	0.9963	0.9993	0.9999	1	1	1
0.35	0.0016	0.0142	0.0617	0.1727	0.3519	0.5643	0.7548	0.8868	0.9578	0.9876	0.9972	0.9995	0.9999	1	1
0.40	0.0005	0.0052	0.0271	0.0905	0.2173	0.4032	0.6098	0.7869	0.905	0.9662	0.9907	0.9981	0.9997	1	1
0.45	0.0001	0.0017	0.0107	0.0424	0.1204	0.2608	0.4522	0.6535	0.8182	0.9231	0.9745	0.9937	0.9989	0.9999	1
0.50	0	0.0005	0.0037	0.0176	0.0592	0.1509	0.3036	0.5	0.6964	0.8491	0.9408	0.9824	0.9963	0.9995	1
0.55	0	0.0001	0.0011	0.0063	0.0255	0.0769	0.1818	0.3465	0.5478	0.7392	0.8796	0.9576	0.9893	0.9983	0.9999
0.60	0	0	0.0003	0.0019	0.0093	0.0338	0.095	0.2131	0.3902	0.5968	0.7827	0.9095	0.9729	0.9948	0.9995
0.65	0	0	0.0001	0.0005	0.0028	0.0124	0.0422	0.1132	0.2452	0.4357	0.6481	0.8273	0.9383	0.9858	0.9984
0.70	0	0	0	0.0001	0.0007	0.0037	0.0152	0.05	0.1311	0.2784	0.4845	0.7031	0.8732	0.9647	0.9953
0.75	0	0	0	0	0.0001	0.0008	0.0042	0.0173	0.0566	0.1484	0.3135	0.5387	0.7639	0.9198	0.9866
0.80	0	0	0	0	0	0.0001	0.0008	0.0042	0.0181	0.0611	0.1642	0.3518	0.602	0.8329	0.9648
0.85	0	0	0	0	0	0	0.0001	0.0006	0.0036	0.0168	0.0617	0.1773	0.3958	0.6814	0.9126
0.90	0	0	0	0	0	0	0	0	0.0003	0.0022	0.0127	0.0556	0.1841	0.451	0.7941
0.95	0	0	0	0	0	0	0	0	0	0.0001	0.0006	0.0055	0.0362	0.171	0.5367
0.99	0	0	0	0	0	0	0	0	0	0	0	0	0.0004	0.0096	0.1399

(*Continued*)

TABLE A.2 (*Continued for n = 20*)

p	x=0	1	2	3	4	5	6	7	8	9	10	11	12	13	14	15	16	17	18	19
0.01	0.818	0.983	0.999	1	1	1	1	1	1	1	1	1	1	1	1	1	1	1	1	1
0.05	0.358	0.736	0.925	0.984	0.997	1	1	1	1	1	1	1	1	1	1	1	1	1	1	1
0.10	0.122	0.392	0.677	0.867	0.957	0.989	0.998	1	1	1	1	1	1	1	1	1	1	1	1	1
0.15	0.039	0.176	0.405	0.648	0.83	0.933	0.978	0.994	0.999	1	1	1	1	1	1	1	1	1	1	1
0.20	0.012	0.069	0.206	0.411	0.63	0.804	0.913	0.968	0.99	0.997	0.999	1	1	1	1	1	1	1	1	1
0.25	0.003	0.024	0.091	0.225	0.415	0.617	0.786	0.898	0.959	0.986	0.996	0.999	1	1	1	1	1	1	1	1
0.30	8E-04	0.008	0.036	0.107	0.238	0.416	0.608	0.772	0.887	0.952	0.983	0.995	0.999	1	1	1	1	1	1	1
0.35	2E-04	0.002	0.012	0.044	0.118	0.245	0.417	0.601	0.762	0.878	0.947	0.98	0.994	0.999	1	1	1	1	1	1
0.40	0	5E-04	0.004	0.016	0.051	0.126	0.25	0.416	0.596	0.755	0.873	0.944	0.979	0.994	0.998	1	1	1	1	1
0.45	0	1E-04	9E-04	0.005	0.019	0.055	0.13	0.252	0.414	0.591	0.751	0.869	0.942	0.979	0.994	0.999	1	1	1	1
0.50	0	0	2E-04	0.001	0.006	0.021	0.058	0.132	0.252	0.412	0.588	0.748	0.868	0.942	0.979	0.994	0.999	1	1	1
0.55	0	0	0	3E-04	0.002	0.006	0.021	0.058	0.131	0.249	0.409	0.586	0.748	0.87	0.945	0.981	0.995	0.999	1	1
0.60	0	0	0	0	3E-04	0.002	0.007	0.021	0.057	0.128	0.245	0.404	0.584	0.75	0.874	0.949	0.984	0.996	1	1
0.65	0	0	0	0	0	3E-04	0.002	0.006	0.02	0.053	0.122	0.238	0.399	0.583	0.755	0.882	0.956	0.988	0.998	1
0.70	0	0	0	0	0	0	3E-04	0.001	0.005	0.017	0.048	0.113	0.228	0.392	0.584	0.763	0.893	0.965	0.992	0.999
0.75	0	0	0	0	0	0	0	2E-04	9E-04	0.004	0.014	0.041	0.102	0.214	0.383	0.585	0.775	0.909	0.976	0.997
0.80	0	0	0	0	0	0	0	0	1E-04	6E-04	0.003	0.01	0.032	0.087	0.196	0.37	0.589	0.794	0.931	0.989
0.85	0	0	0	0	0	0	0	0	0	0	2E-04	0.001	0.006	0.022	0.067	0.17	0.352	0.595	0.824	0.961
0.09	0	0	0	0	0	0	0	0	0	0	0	1E-04	4E-04	0.002	0.011	0.043	0.133	0.323	0.608	0.878
0.95	0	0	0	0	0	0	0	0	0	0	0	0	0	0	3E-04	0.003	0.016	0.076	0.264	0.642
0.99	0	0	0	0	0	0	0	0	0	0	0	0	0	0	0	0	0	0.001	0.017	0.182

TABLE A.3   Critical Values of *t*-Distribution at Confidence Level $(1-\alpha) = p$

d.o.f.	$p = 0.900$	$p = 0.950$	$p = 0.975$	$p = 0.990$	$p = 0.995$	$p = 0.999$
1	3.0777	6.3138	12.7062	31.8205	63.6567	318.3088
2	1.8856	2.9200	4.3027	6.9646	9.9248	22.3271
3	1.6377	2.3534	3.1824	4.5407	5.8409	10.2145
4	1.5332	2.1318	2.7764	3.7469	4.6041	7.1732
5	1.4759	2.0150	2.5706	3.3649	4.0321	5.8934
6	1.4398	1.9432	2.4469	3.1427	3.7074	5.2076
7	1.4149	1.8946	2.3646	2.9980	3.4995	4.7853
8	1.3968	1.8595	2.3060	2.8965	3.3554	4.5008
9	1.3803	1.8331	2.2622	2.8214	3.2498	4.2968
10	1.3722	1.8125	2.2281	2.7638	3.1693	4.1437
11	1.3634	1.7959	2.2001	2.7181	3.1058	4.0247
12	1.3562	1.7823	2.1788	2.6810	3.0545	3.9296
13	1.3502	1.7709	2.1604	2.6503	3.0123	3.8520
14	1.3450	1.7613	2.1448	2.6245	2.9768	3.7874
15	1.3406	1.7531	2.1314	2.6025	2.9467	3.7328
16	1.3368	1.7459	2.1199	2.5835	2.9208	3.6862
17	1.3334	1.7396	2.1098	2.5669	2.8982	3.6458
18	1.3304	1.7341	2.1009	2.5524	2.8784	3.6105
19	1.3277	1.7291	2.0930	2.5395	2.8609	3.5794
20	1.3253	1.7247	2.0860	2.5280	2.8453	3.5518
21	1.3232	1.7207	2.0796	2.5176	2.8314	3.5272
22	1.3212	1.7171	2.0739	2.5083	2.8188	3.5050
23	1.3195	1.7139	2.0687	2.4999	2.8073	3.4850
24	1.3178	1.7109	2.0639	2.4922	2.7969	3.4668
25	1.3163	1.7081	2.0595	2.4851	2.7874	3.4502
26	1.3150	1.7056	2.0555	2.4786	2.7787	3.4350
27	1.3137	1.7033	2.0518	2.4727	2.7707	3.4210
28	1.3125	1.7011	2.0484	2.4671	2.7633	3.4082
29	1.3114	1.6991	2.0452	2.4620	2.7564	3.3962
30	1.3104	1.6973	2.0423	2.4573	2.7500	3.3852
31	1.3095	1.6955	2.0395	2.4528	2.7440	3.3749
32	1.3086	1.6939	2.0369	2.4487	2.7385	3.3653
33	1.3077	1.6924	2.0345	2.4448	2.7333	3.3563
34	1.3070	1.6909	2.0322	2.4411	2.7284	3.3479
35	1.3062	1.6896	2.0301	2.4377	2.7238	3.3400
36	1.3055	1.6883	2.0281	2.4345	2.7195	3.3326
37	1.3049	1.6871	2.0262	2.4314	2.7154	3.3256
38	1.3042	1.6806	2.0244	2.4286	2.7116	3.3190
39	1.3036	1.6849	2.0227	2.4258	2.7079	3.3128
40	1.3031	1.6839	2.0211	2.4233	2.7045	3.3069
45	1.3006	1.6794	2.0141	2.4121	2.6896	3.2815
50	1.2987	1.6759	2.0086	2.4033	2.6778	3.2614
55	1.2971	1.6703	2.0040	2.3961	2.6682	3.2451
60	1.2958	1.6706	2.0003	2.3901	2.6603	3.2317
70	1.2938	1.6794	1.9944	2.3808	2.6479	3.2108
80	1.2922	1.6759	1.9901	2.3739	2.6387	3.1953
90	1.2910	1.6750	1.9867	2.3685	2.6316	3.1833
∞	1.2824	1.6449	1.9600	2.3264	2.5759	3.0903

TABLE A.4 Critical Values of the χ^2 Distribution at Probability Level α

d.o.f.	$\alpha = 0.001$	$\alpha = 0.005$	$\alpha = 0.01$	$\alpha = 0.025$	$\alpha = 0.05$	$\alpha = 0.10$	$\alpha = 0.20$
1	0	0	0.0002	0.0010	0.0039	0.0158	0.0642
2	0.0020	0.0100	0.0201	0.0506	0.1026	0.2107	0.4463
3	0.0243	0.0717	0.1148	0.2158	0.3518	0.5844	1.0052
4	0.0908	0.2070	0.2971	0.4844	0.7107	1.0636	1.6488
5	0.2102	0.4117	0.5543	0.8312	1.1455	1.6103	2.3425
6	0.3811	0.6757	0.8721	1.2373	1.6354	2.2041	3.0701
7	0.5985	0.9893	1.2390	1.6899	2.1673	2.8331	3.8223
8	0.8571	1.3444	1.6465	2.1797	2.7326	3.4895	4.5936
9	1.1519	1.7349	2.0879	2.7004	3.3251	4.1682	5.3801
10	1.4787	2.1559	2.5582	3.2470	3.9403	4.8652	6.1791
11	1.8339	2.6032	3.0535	3.8157	4.5748	5.5778	6.9887
12	2.2142	3.0738	3.5706	4.4038	5.2260	6.3038	7.8073
13	2.6172	3.5650	4.1069	5.0088	5.8919	7.0415	8.6339
14	3.0407	4.0747	4.6604	5.6287	6.5706	7.7895	9.4673
15	3.4827	4.6009	5.2293	6.2621	7.2609	8.5468	10.3070
16	3.9416	5.1422	5.8122	6.9077	7.9616	9.3122	11.1521
17	4.4161	5.6972	6.4078	7.5642	8.6718	10.0852	12.0023
18	4.9048	6.2648	7.0149	8.2307	9.3905	10.8649	12.8570
19	5.4068	6.8440	7.6327	8.9065	10.1170	11.6509	13.7158
20	5.9210	7.4338	8.2604	9.5908	10.8508	12.4426	14.5784
21	6.4467	8.0337	8.8972	10.2829	11.5913	13.2396	15.4446
22	6.9830	8.6427	9.5425	10.9823	12.3380	14.0415	16.3140
23	7.5292	9.2604	10.1957	11.6886	13.0905	14.8480	17.1865
24	8.0849	9.8862	10.8564	12.4012	13.8484	15.6587	18.0618
25	8.6493	10.5197	11.5240	13.1197	14.6114	16.4734	18.9398
26	9.2221	11.1602	12.1981	13.8439	15.3792	17.2919	19.8202
27	9.8028	11.8076	12.8785	14.5734	16.1514	18.1139	20.7030
28	10.3909	12.4613	13.5647	15.3079	16.9279	18.9392	21.5880
29	10.9861	13.1211	14.2565	16.0471	17.7084	19.7677	22.4751
30	11.5880	13.7867	14.9535	16.7908	18.4927	20.5992	23.3641
31	12.1963	14.4578	15.6555	17.5387	19.2806	21.4336	24.2551
32	12.8107	15.1340	16.3622	18.2908	20.0719	22.2706	25.1478
33	13.4309	15.8153	17.0735	19.0467	20.8665	23.1102	26.0422
34	14.0567	16.5013	17.7891	19.8063	21.6643	23.9523	26.9383
35	14.6878	17.1918	18.5089	20.5694	22.4650	24.7967	27.8359
36	15.3241	17.8867	19.2327	21.3359	23.2686	25.6433	28.7350
37	15.9653	18.5858	19.9602	22.1056	24.0749	26.4921	29.6355
38	16.6112	19.2889	20.6914	22.8785	24.8839	27.3430	30.5373
39	17.2616	19.9959	21.4262	23.6543	25.6954	28.1958	31.4405
40	17.9164	20.7065	22.1643	24.4330	26.5093	29.0505	32.3450
45	21.2507	24.3110	25.9013	28.3662	30.6123	33.3504	36.8844
50	24.6739	27.9907	29.7067	32.3574	34.7643	37.6886	41.4492
55	28.1731	31.7348	33.5705	36.3981	38.9580	42.0596	46.0356
60	31.7383	35.5345	37.4849	40.4817	43.1880	46.4589	50.6406
65	35.3616	39.3831	41.4436	44.6030	47.4496	50.8829	55.2620
70	35.3616	43.2752	45.4417	48.7576	51.7393	55.3289	59.8978
75	42.7573	47.2060	49.4750	52.9419	56.0541	59.7946	64.5466
80	46.5199	51.1719	53.5401	57.1532	60.3915	64.2778	69.2069
90	54.1552	59.1963	61.7541	65.6466	69.1260	73.2911	78.5584
100	61.9179	67.3276	70.0649	74.2219	77.9295	82.3581	87.9453

(Continued)

TABLE A.4 (*Continued*) Critical Values of the χ^2 Distribution at Probability Level $(1-\alpha) = p$

d.o.f.	p = 0.800	p = 0.900	p = 0.950	p = 0.975	p = 0.990	p = 0.995	p = 0.999
1	1.6424	2.7055	3.8415	5.0239	6.6349	7.8794	10.8276
2	3.2189	4.6052	5.9915	7.3778	9.2103	10.5966	13.8155
3	4.6416	6.2514	7.8147	9.3484	11.3449	12.8382	16.2662
4	5.9886	7.7794	9.4877	11.1433	13.2767	14.8603	18.4668
5	7.2893	9.2364	11.0705	12.8325	15.0863	16.7496	20.5150
6	8.5581	10.6446	12.5916	14.4494	16.8119	18.5476	22.4577
7	9.8032	12.0170	14.0671	16.0128	18.4753	20.2777	24.3219
8	11.0301	13.3616	15.5073	17.5345	20.0902	21.9550	26.1245
9	12.2421	14.6837	16.9190	19.0228	21.6660	23.5894	27.8772
10	13.4420	15.9872	18.3070	20.4832	23.2093	25.1882	29.5883
11	14.6314	17.2750	19.6751	21.9200	24.7250	26.7568	31.2641
12	15.8120	18.5493	21.0261	23.3367	26.2170	28.2995	32.9095
13	16.9848	19.8119	22.3620	24.7356	27.6882	29.8195	34.5282
14	18.1508	21.0641	23.6848	26.1189	29.1412	31.3193	36.1233
15	19.3107	22.3071	24.9958	27.4884	30.5779	32.8013	37.6973
16	20.4651	23.5418	26.2962	28.8454	31.9999	34.2672	39.2524
17	21.6146	24.7690	27.5871	30.1910	33.4087	35.7185	40.7902
18	22.7595	25.9894	28.8693	31.5264	34.8053	37.1565	42.3124
19	23.9004	27.2036	30.1435	32.8523	36.1909	38.5823	43.8202
20	25.0375	28.4120	31.4104	34.1696	37.5662	39.9968	45.3147
21	26.1711	29.6151	32.6706	35.4789	38.9322	41.4011	46.7970
22	27.3015	30.8133	33.9244	36.7807	40.2894	42.7957	48.2679
23	28.4288	32.0069	35.1725	38.0756	41.6384	44.1813	49.7282
24	29.5533	33.1962	36.4150	39.3641	42.9798	45.5585	51.1786
25	30.6752	34.3816	37.6525	40.6465	44.3141	46.9279	52.6197
26	31.7946	35.5632	38.8851	41.9232	45.6417	48.2899	54.0520
27	32.9117	36.7412	40.1133	43.1945	46.9629	49.6449	55.4760
28	34.0266	37.9159	41.3371	44.4608	48.2782	50.9934	56.8923
29	35.1394	39.0875	42.5570	45.7223	49.5879	52.3356	58.3012
30	36.2502	40.2560	43.7730	46.9792	50.8922	53.6720	59.7031
31	37.3591	41.4217	44.9853	48.2319	52.1914	55.0027	61.0983
32	38.4663	42.5847	46.1943	49.4804	53.4858	56.3281	62.4872
33	39.5718	43.7452	47.3999	50.7251	54.7755	57.6484	63.8701
34	40.6756	44.9032	48.6024	51.9660	56.0609	58.9639	65.2472
35	41.7780	46.0588	49.8018	53.2033	57.3421	60.2748	66.6188
36	42.8788	47.2122	50.9985	54.4373	58.6192	61.5812	67.9852
37	43.9782	48.3634	52.1923	55.6680	59.8925	62.8833	69.3465
38	45.0763	49.5126	53.3835	56.8955	61.1621	64.1814	70.7029
39	46.1730	50.6598	54.5722	58.1201	62.4281	65.4756	72.0547
40	47.2685	51.8051	55.7585	59.3417	63.6907	66.7660	73.4020
45	52.7288	57.5053	61.6562	65.4102	69.9568	73.1661	80.0767
50	58.1638	63.1671	67.5048	71.4202	76.1539	79.4900	86.6608
55	63.5772	68.7962	73.3115	77.3805	82.2921	85.7490	93.1675
60	68.9721	74.3970	79.0819	83.2977	88.3794	91.9517	99.6072
65	74.3506	79.9730	84.8206	89.1771	94.4221	98.1051	105.9881
70	79.7146	85.5270	90.5312	95.0232	100.4252	104.2149	112.3169
75	85.0658	91.0615	96.2167	100.8393	106.3929	110.2856	118.5991
80	90.4053	96.5782	101.8795	106.6286	112.3288	116.3211	124.8392
90	101.0537	107.565	113.1453	118.1359	124.1163	128.2989	137.2084
100	111.6667	118.498	124.3421	129.5612	135.8067	140.1695	149.4493

TABLE A.5   Critical Values of D_n^α at Significance Level α in the K-S Test

d.o.f. $= n$	$\alpha = 0.20$	$\alpha = 0.10$	$\alpha = 0.05$	$\alpha = 0.01$
5	0.45	0.51	0.56	0.67
10	0.32	0.37	0.41	0.49
15	0.27	0.30	0.34	0.40
20	0.23	0.26	0.29	0.36
25	0.21	0.24	0.27	0.32
30	0.19	0.22	0.24	0.29
35	0.18	0.20	0.23	0.27
40	0.17	0.19	0.21	0.25
45	0.16	0.18	0.20	0.24
50	0.15	0.17	0.19	0.23
>50	$1.07/\sqrt{n}$	$1.22/\sqrt{n}$	$1.36/\sqrt{n}$	$1.63/\sqrt{n}$

TABLE A.6   Critical Values of the Anderson-Darling Goodness-of-Fit Test

Table A.6a Critical Values of c_α at Significance Level α of the A-D Test
  for the Normal Distribution (μ and σ Estimated from
    Sample Size n)

Significance Level α	a_α	b_0	b_1
0.2	0.5091	-0.756	-0.39
0.1	0.6305	-0.75	-0.8
0.05	0.7514	-0.795	-0.89
0.025	0.8728	-0.881	-0.94
0.01	1.0348	-1.013	-0.93
0.005	1.1578	-1.063	-1.34

Excerpted from D'Agostino and Stephens, 1986 (see References in Chapter 7).

Table A.6b Critical Values of c_α at Significance Level α
  of the A-D Test for the Exponential
    Distribution (Parameter λ Estimated from)
      Sample Size n)

Significance Level α	c_α
0.25	0.736
0.20	0.816
0.15	0.916
0.10	0.162
0.05	1.321
0.025	1.591
0.01	1.959
0.005	2.244
0.0025	2.534

Excerpted from Petitt, 1977, Pearson and Hartley, 1972, and
D'Agostino and Stephens, 1986 (see References in Chapter 7).

Table A.6c Critical Values of c_α at Significance Level α of the A-D Test for the Gamma Distribution (Parameters Estimated from Sample Data)

k	Significance Level α					
	0.25	0.10	0.05	0.025	0.01	0.005
1	0.486	0.657	0.786	0.917	1.092	1.227
2	0.477	0.643	0.768	0.894	1.062	1.190
3	0.475	0.639	0.762	0.886	1.052	1.178
4	0.473	0.637	0.759	0.883	1.048	1.173
5	0.472	0.635	0.758	0.881	1.045	1.170
6	0.472	0.635	0.757	0.880	1.043	1.168
8	0.471	0.634	0.755	0.878	1.041	1.165
10	0.471	0.633	0.754	0.877	1.040	1.164
12	0.471	0.633	0.754	0.876	1.038	1.162
15	0.470	0.632	0.754	0.876	1.038	1.162
20	0.470	0.632	0.753	0.875	1.037	1.161
∞	0.470	0.631	0.752	0.873	1.035	1.159

Excerpted from Lockhart and Stephens, 1985 (see References in Chapter 7).

Table A.6d Critical Values of c_α at Significance Level α of the A-D Test for the Gumbel and Weibull Distribution (Parameters Estimated from Sample Size n)

Significance Level α	c_α
0.25	0.474
0.10	0.637
0.05	0.757
0.025	0.877
0.01	1.038

Excerpted from Stephens, 1977 (see References in Chapter 7).

Combinatorial Formulas

In probability problems involving discrete and finite sample spaces, the definition of events and of the underlying sample spaces entails the enumeration of sets or subsets of sample points. For this purpose, the techniques of *combinatorial analysis* are often useful. We summarize here some of the basic elements of combinatorial analysis.

▶ B.1 THE BASIC RELATION

If there are k positions in a sequence, and n_1 distinguishable elements can occupy position 1, n_2 elements can occupy position 2, ..., and n_k elements can occupy position k, the number of distinct sequences of k elements each is

$$N(k|n_1, n_2, \ldots, n_k) = n_1 n_2 \cdots n_k \tag{B.1}$$

▶ **EXAMPLES**

(1) If an engineering design involves three parameters ϕ_1, ϕ_2, ϕ_3 and there are, respectively, 2, 3, 4 values of these parameters, the number of feasible designs is

$$N(3|2, 3, 4) = 2 \times 3 \times 4 = 24$$

(2) In a 3-dimensional Cartesian coordinate system with x, y, z axes, if 10 discrete values are specified for each of the three axes, e.g., $x = 0, 1, 2, \ldots, 9$, then the total number of coordinate positions is

$$N(3|10, 10, 10) = 10^3 = 1,000 \qquad ◀$$

▶ B.2 ORDERED SEQUENCES

In a set of n distinct elements, the number of ordered sequences or *arrangements*, each with k elements, is

$$(n)_k = n(n-1)(n-2) \cdots (n-k+1) = \frac{n!}{(n-k)!} \tag{B.2}$$

We can observe that for the first position in the sequence of k elements, there are n elements available to occupy it; but there are only $(n-1)$ elements available for the second position since one of the n elements has been used for the first position and only $(n-2)$ elements are available for the third position, and so on. Thence, by virtue of the basic relation, Eq. B.1,

$$(n)_k = N(k|n, n-1, n-2, \ldots, n-k+1)$$

thus, we obtain Eq. B.2.

▶ **EXAMPLES**

(1) If no digits are repeated, the number of four-digit figures is

$$(10)_4 = 10 \times 9 \times 8 \times 7 = 5040$$

whereas if the digits can be repeated, the number of four-digit figures would be $(10)^4 = 10,000$. This latter case would include 0000 as one of the figures.

(2) In taking samples sequentially from a discrete sample space (population), the sampling may be performed *with replacements* or *without replacements*; i.e., when an element is drawn from the population it is either returned or not returned to the population before the next element is drawn. In a population of n elements, the number of ordered samples of size r is n^r for sampling with replacements, whereas in sampling without replacements, the corresponding number of ordered samples would be $(n)_r$. ◀

▷ B.3 THE BINOMIAL COEFFICIENT

In a set of n distinguishable elements, the number of possible subsets of k different elements each (regardless of order) is given by the *binomial coefficient*,

$$\binom{n}{k} = \frac{(n)_k}{k!} \tag{B.3}$$

It may be emphasized that in Eq. B.2 the ordering of the k elements is of significance (i.e., different orderings of the same elements constitute different sequences or arrangements), whereas in Eq. B.3 the order is irrelevant. In a set of k elements, the positions of the elements can be permuted $k!$ times; hence, by virtue of Eq. B.2, we obtain Eq. B.3 as the number of different k-element subsets (disregarding order).

Equation B.3 is defined only for $k \leq n$. Using Eq. B.2 in Eq. B.3 we also have

$$\binom{n}{k} = \frac{n!}{k!(n-k)!} \tag{B.3a}$$

From Eq. B.3a, we can observe that

$$\binom{n}{k} = \binom{n}{n-k} \tag{B.4}$$

Equation B.3 or B.3a is known as the *binomial coefficient* because this is precisely the coefficient in the binomial expansion of $(x+y)^n$, namely,

$$(x+y)^n = \binom{n}{0}x^n y^0 + \binom{n}{1}x^{n-1}y + \binom{n}{2}x^{n-2}y^2 + \cdots + \binom{n}{n}x^0 y^n$$

▶ **EXAMPLES**

(1) From a population of n elements, the number of different samples of size r is $\binom{n}{r}$. This is the same as sampling without replacement, except that the order is disregarded; therefore, the number of samples is

$$\frac{(n)_r}{r!} = \binom{n}{r}$$

(1) Among 25 concrete cylinders marked 1, 2,..., 25, the total number of possible samples of five cylinders each is

$$\binom{25}{5} = \frac{25!}{5!20!} = 53,120$$

◀

▶ B.4   THE MULTINOMIAL COEFFICIENT

If n distinguishable elements are divided into r different groups of $k_1, k_2, \ldots, k_r$ elements, respectively, so that $k_1 + k_2 + \cdots + k_r = n$, the number of ways that such r groups can be formed is given by the *multinomial coefficient*

$$\binom{n}{k_1, k_2, \ldots, k_r} = \frac{n!}{k_1!k_2! \cdots k_r!} \tag{B.5}$$

Among the n elements, the first group of k_1 elements can be selected in $\binom{n}{k_1}$ ways. The second group of k_2 elements can be selected from the remaining $(n - k_1)$ elements in $\binom{n - k_1}{k_2}$ ways, and so on. The total number of ways of dividing the n elements into r groups of $k_1, k_2, \ldots, k_r$, therefore, is

$$\binom{n}{k_1}\binom{n - k_1}{k_2} \cdots \binom{n - k_1 - k_2 - \cdots - k_{r-2}}{k_{r-1}}\binom{k_r}{k_r} = \frac{n!}{k_1!k_2! \cdots k_r!}$$

▶ **EXAMPLES**

(1) In a given seismic region, suppose six earthquakes of intensities (in MM scale) V, VI, and VII may occur in the next 10 years. The number of different sequences in which three quakes of intensity V, two of intensity VI, and one of intensity VII can occur is

$$\frac{6!}{3! \times 2! \times 1!} = 60$$

(2) Suppose a construction company wishes to acquire three types of construction equipment consisting of two bulldozers, four road graders, and six backhoes for a total fleet of 12 pieces of equipment for a project. The number of different sequences that the company may be able to acquire the fleet of equipment is

$$\frac{12!}{2! \times 4! \times 6!} = 13,860$$

◀

▶ B.5   STIRLING'S FORMULA

An important and useful formula for approximately computing the factorial of a large number is the *Stirling's formula*:

$$n! = \sqrt{2\pi}(n)^{n+\frac{1}{2}}e^{-n} \tag{B.6}$$

A derivation of Eq. B.6 can be found in Feller (1957)[*]. The approximation is good, even for n as small as 10 (error is less than 1%).

---

[*]Feller, W., *An Introduction to Probability and Its Applications*, Vol. 1, 2nd Ed., J. Wiley and Sons, New York, 1957.

Derivation of the Poisson Distribution

In Chapter 3, Sect. 3.2.6, we saw that the Poisson distribution describes the probability mass function (PMF) of the number of occurrences of an event within a specified interval of time or space. It is the result of an underlying counting process $X(t)$, known as a *Poisson process*, which is a model of the random occurrences of an event in time (or space) t.

The Poisson process model is based on the following assumptions:

1. At any instant of time (or point in space), there can be at most one occurrence of an event; in other words, the probability of n occurrences of the event over a small interval Δt is of order $\mathbf{o}(\Delta t)^n$.

2. The occurrences of an event in nonoverlapping time (or space) intervals are statistically independent; this is the assumption of *independent increments*.

3. The probability of the occurrence of an event in the interval $(t, t + \Delta t)$ is proportional to Δt; that is,

$$P[X(\Delta t) = 1] = v\,\Delta t$$

in which v is a proportionality constant.

On the basis of the second assumption above, we obtain with the theorem of total probability,

$$P[X(t + \Delta t) = x] = P[X(t) = x]P[X(\Delta t) = 0] + P[X(t) = x - 1]P[X(\Delta t) = 1]$$
$$+ P[X(t) = x - 2]P[X(\Delta t) = 2] + \cdots$$

Then, using the notation $p_x(t) = P[X(t) = x]$, and observing that $P[X(t) = 0] = \{1 - P[X(t) = 1] - P[X(t) = 2] - \cdots\}$, we obtain on the basis of the first and third assumptions above,

$$p_x(t + \Delta t) = p_x(t)[1 - v\,\Delta t - \mathbf{o}(\Delta t)^2 - \cdots]$$
$$+ (v\,\Delta t)p_{x-1}(t) + \mathbf{o}(\Delta t)^2 p_{x-2}(t) + \cdots$$

Neglecting the higher order terms, the above equation becomes

$$\frac{p_x(t + \Delta t) - p_x(t)}{\Delta t} = -v p_x(t) + v p_{x-1}(t)$$

In the limit, as $\Delta t \to 0$, we obtain the following differential equation for $p_x(t)$:

$$\frac{dp_x(t)}{dt} = -v p_x(t) + v p_{x-1}(t) \tag{C.1}$$

We should recognize that Eq. C.1 applies for $x \geq 1$. For $x = 0$, Eq. C.1 becomes

$$\frac{dp_o(t)}{dt} = -v p_o(t) \tag{C.2}$$

If the counting process starts at $t = 0$, the initial conditions associated with Eqs. C.1 and C.2 are

$$p_o(0) = 1.0 \qquad \text{and} \qquad p_x(0) = 0$$

The solution of Eq. C.2 with the first of the above initial conditions yields, for $x = 0$,

$$p_o(t) = e^{-vt}$$

Whereas for $x \geq 1$ the solutions to Eq. C.1 are

$$p_1(t) = vt e^{-vt}; \qquad p_2(t) = \frac{(vt)^2}{2!} e^{-vt}$$

and for a general x,

$$p_x(t) = \frac{(vt)^x}{x!} e^{-vt} \tag{C.3}$$

which is the PMF of the Poisson distribution, where v is the *mean occurrence rate* of the event.

Index